INTRODUCTION TO MATHEMATICAL LOGIC

Volume I

PRINCETON MATHEMATICAL SERIES

Edited by PHILLIP A. GRIFFITHS, MARSTON MORSE, and ELIAS M. STEIN

1. The Classical Groups, by HERMANN WEYL
3. An Introduction to Differential Geometry, by LUTHER PFAHLER EISENHART
4. Dimension Theory, by W. HUREWICZ and H. WALLMAN
6. The Laplace Transform, by D. V. WIDDER
7. Integration, by EDWARD J. MCSHANE
8. Theory of Lie Groups: 1, by C. CHEVALLEY
9. Mathematical Methods of Statistics, by HARALD CRAMÉR
10. Several Complex Variables, by S. BOCHNER and W. T. MARTIN
11. Introduction to Topology, by S. LEFSCHETZ
12. Algebraic Geometry and Topology, edited by R. H. FOX, D. C. SPENCER, and A. W. TUCKER
14. The Topology of Fibre Bundles, by NORMAN STEENROD
15. Foundations of Algebraic Topology, by SAMUEL EILENBERG and NORMAN STEENROD
16. Functionals of Finite Riemann Surfaces, by MENAHEM SCHIFFER and DONALD C. SPENCER
17. Introduction to Mathematical Logic, Vol. 1, by ALONZO CHURCH
19. Homological Algebra, by H. CARTAN and S. EILENBERG
20. The Convolution Transform, by I. I. HIRSCHMAN and D. V. WIDDER
21. Geometric Integration Theory, by H. WHITNEY
22. Qualitative Theory of Differential Equations, by V. V. NEMYTSKII and V. V. STEPANOV
23. Topological Analysis, by GORDON T. WHYBURN (revised 1964)
24. Analytic Functions, by AHLFORS, BEHNKE and GRAUERT, BERS, et al.
25. Continuous Geometry, by JOHN VON NEUMANN
26. Riemann Surfaces, by L. AHLFORS and L. SARIO
27. Differential and Combinatorial Topology, edited by S. S. CAIRNS
28. Convex Analysis, by R. T. ROCKAFELLAR
29. Global Analysis, edited by D. C. SPENCER and S. IYANAGA
30. Singular Integrals and Differentiability Properties of Functions, by E. M. STEIN
31. Problems in Analysis, edited by R. C. GUNNING

VOLUME I

INTRODUCTION TO MATHEMATICAL LOGIC

BY ALONZO CHURCH

PRINCETON, NEW JERSEY

PRINCETON UNIVERSITY PRESS

1956

PUBLISHED 1956, BY PRINCETON UNIVERSITY PRESS
ISBN 0-691-07984-6

Sixth Printing, 1970

PRINTED IN THE UNITED STATES OF AMERICA

Preface

This is a revised and much enlarged edition of *Introduction to Mathematical Logic, Part I*, which was published in 1944 as one of the Annals of Mathematics Studies. In spite of extensive additions, it remains an introduction rather than a comprehensive treatise. It is intended to be used as a textbook by students of mathematics, and also within limitations as a reference work. As a textbook it offers a beginning course in mathematical logic, but presupposes some substantial mathematical background.

An added feature in the new edition is the inclusion of many exercises for the student. Some of these are of elementary character, straightforward illustrations serving the purpose of practice; others are in effect brief sketches of difficult developments to which whole sections of the main text might have been devoted; and still others occupy various intermediate positions between these extremes. No attempt has been made to classify exercises systematically according to difficulty. But for routine use by beginning students the following list is tentatively suggested as a basis for selection: 12.3–12.9, 14.0–14.8, 15.0–15.3, 15.9, 15.10, 18.0–18.3, 19.0–19.7, 19.9, 19.10, 23.1–23.6, 24.0–24.5, 30.0–30.4 (with assistance if necessary), 34.0, 34.3–34.6, 35.1, 35.2, 38.0–38.5, 39.0, 41.0, 43.0, 43.1, 43.4, 45.0, 45.1, 48.0–48.11, 52.0, 52.1, 54.2–54.6, 55.1, 55.2, 55.22, 56.0–56.2, 57.0–57.2.

The book has been cut off rather abruptly in the middle, in order that Volume I may be published, and at many places there are references forward to passages in the still unwritten Volume II. In order to make clear at least the general intent of such references, a tentative table of contents of Volume II has been added at the end of the table of contents of the present volume, and references to Volume II should be understood in the light of this.

Volume I has been written over a period of years, beginning in 1947, and as portions of the work were completed they were made available in manuscript form in the Fine Hall Library of Princeton University. The work was carried on during regular leave of absence from Princeton University from September, 1947, to February 1, 1948, and then under a contract of Princeton University with the United States Office of Naval Research from February 1 to June 30, 1948. To this period should be credited the Introduction and Chapters I and II—although some minor changes have been made in this material since then, including the addition of exercises 15.4, 18.3, 19.12, 24.10, 26.3(2), 26.3(3), 26.8, 29.2, 29.3, 29.4, 29.5, as well as changes designed to correct errors or to take into account newly published papers. The remainder of the work was done during 1948–1951 with the aid of grants from the Scientific Research Fund of Princeton University and the Eugene Higgins Trust Fund, and credit is due to these Funds for making possible the writing of the latter half of the volume.

For individual assistance, I am indebted still to the persons named in the Preface of the edition of 1944, especially to C. A. Truesdell — whose notes on the lectures of 1943 have continued to be of great value, both in the writing of Volume I and in the preliminary work which has been done towards the writing of Volume II, and notwithstanding the extensive changes which have been made from the content and plan of the original lectures. I am also indebted to many who have read the new manuscript or parts of it and have supplied valuable suggestions and corrections, including especially E. Adler, A. F. Bausch, W. W. Boone, Leon Henkin, J. G. Kemeny, Maurice L'Abbé, E. A. Maier, Paul Meier, I. L. Novak, and Rulon Wells.

ALONZO CHURCH

Princeton,, New Jersey
August 31, 1951

(*Added November 28, 1955.*) For suggestions which could be taken into account only in the proof I am indebted further to A. N. Prior, T. T. Robinson, Hartley Rogers, Jr., J. C. Shepherdson, F. O. Wyse, and G. Zubieta Russi; for assistance in the reading of the proof itself, to Michael Rabin and to Zubieta; and especially for their important contribution in preparing the indexes, to Robinson and Zubieta.

(*Added January 17, 1958.*) In the second printing, additional corrections which were necessary have been made in the text as far as possible, and those which could not be fitted into the text have been included in a list of *Errata* at the end of the book. For some of these corrections I am indebted to Max Black, S. C. Kleene, E. J. Lemmon, Walter Stuermann, John van Heijenoort; for the observation that exercise 55.3(3) would be better placed as 55.2(3), to D. S. Geiger; and for important corrections to 38.8(10) and footnote 550, to E. W. Beth. For assistance in connection with Wajsberg's paper (see the correction to page 142) I am further indebted to T. T. Robinson.

ALONZO CHURCH

Contents

TENTATIVE TABLE OF CONTENTS OF VOLUME TWO

Introduction

This introduction contains a number of preliminary explanations, which it seems are most suitably placed at the beginning, though many will become clearer in the light of the formal development which follows. The reader to whom the subject is new is advised to read the introduction through once, then return to it later after a study of the first few chapters of the book. Footnotes may in general be omitted on a first reading.

00. Logic. Our subject is *logic*—or, as we may say more fully, in order to distinguish from certain topics and doctrines which have (unfortunately) been called by the same name, it is *formal logic*.

Traditionally, (formal) logic is concerned with the analysis of sentences or of propositions[1] and of proof[2] with attention to the *form* in abstraction from the *matter*. This distinction between form and matter is not easy to make precise immediately, but it may be illustrated by examples.

To take a relatively simple argument for illustrative purposes, consider the following:

I Brothers have the same surname; Richard and Stanley are brothers; Stanley has surname Thompson; therefore Richard has surname Thompson.

Everyday statement of this argument would no doubt leave the first of the three premisses[3] tacit, at least unless the reasoning were challenged; but

---

[1]See §04.

[2]In the light both of recent work and of some aspects of traditional logic we must add here, besides proof, such other relationships among sentences or propositions as can be treated in the same manner, i.e., with regard to form in abstraction from the matter. These include (e.g.) disproof, compatibility; also partial confirmation, which is important in connection with inductive reasoning (cf. C. G. Hempel in *The Journal of Symbolic Logic*, vol. 8 (1943), pp. 122—143).

But no doubt these relationships both can and should be reduced to that of proof, by making suitable additions to the object language (§07) if necessary. E.g., in reference to an appropriate formalized language as object language, disproof of a proposition or sentence may be identified with proof of its negation. The corresponding reduction of the notions of compatibility and confirmation to that of proof apparently requires modal logic—a subject which, though it belongs to formal logic, is beyond the scope of this book.

[3]Following C. S. Peirce (and others) we adopt the spelling *premiss* for the logical term to distinguish it from *premise* in other senses, in particular to distinguish the plural from the legal term *premises*.

for purposes of logical analysis all premisses must be set down explicitly. The argument, it may be held, is valid from its form alone, independently of the matter, and independently in particular of the question whether the premisses and the conclusion are in themselves right or wrong. The reasoning may be right though the facts be wrong, and it is just in maintaining this distinction that we separate the form from the matter.

For comparison with the foregoing example consider also:

II Complex numbers with real positive ratio have the same amplitude; $i - \sqrt{3}/3$ and ω are complex numbers with real positive ratio; ω has amplitude $2\pi/3$; therefore $i - \sqrt{3}/3$ has amplitude $2\pi/3$.

This may be held to have the same form as I, though the matter is different, and therefore to be, like I, valid from the form alone.

Verbal similarity in the statements of I and II, arranged at some slight cost of naturalness in phraseology, serves to highlight the sameness of form. But, at least in the natural languages, such linguistic parallelism is not in general a safe guide to sameness of logical form. Indeed, the natural languages, including English, have been evolved over a long period of history to serve practical purposes of facility of communication, and these are not always compatible with soundness and precision of logical analysis.

To illustrate this last point, let us take two further examples:

III I have seen a portrait of John Wilkes Booth; John Wilkes Booth assassinated Abraham Lincoln; thus I have seen a portrait of an assassin of Abraham Lincoln.

IV I have seen a portrait of somebody; somebody invented the wheeled vehicle; thus I have seen a portrait of an inventor of the wheeled vehicle.

The argument III will be recognized as valid, and presumably from the logical form alone, but IV as invalid. The superficial linguistic analogy of the two arguments as stated is deceptive. In this case the deception is quickly dispelled upon going beyond the appearance of the language to consider the meaning, but other instances are more subtle, and more likely to generate real misunderstanding. Because of this, it is desirable or practically necessary for purposes of logic to employ a specially devised language, a *formalized language* as we shall call it, which shall reverse the tendency of the natural languages and shall follow or reproduce the logical form—at the expense, where necessary, of brevity and facility of communication. To particular formalized language thus involves adopting a particul or system of logical analysis. (This must be regarded as the essenti of a formalized language, not the more conspicuous but theoreti important feature that it is found convenient to replace the spelled words of most (written) natural languages by single letters and various special symbols.)

01. Names. One kind of expression which is familiar in the natural languages, and which we shall carry over also to formalized languages, is the *proper name*. Under this head we include not only proper names which are arbitrarily assigned to denote in a certain way—such names, e.g., as "Rembrandt," "Caracas," "Sirius," "the Mississippi," "The Odyssey," "eight"—but also names having a structure that expresses some analysis of the way in which they denote.[4] As examples of the latter we may cite: "five hundred nine," which denotes a certain prime number, and in the way expressed by the linguistic structure, namely as being five times a hundred plus nine; "the author of *Waverley*," which denotes a certain Scottish novelist, namely Sir Walter Scott, and in the particular way expressed by the linguistic structure, namely as having written *Waverley*; "Rembrandt's birthplace"; "the capital of Venezuela"; "the cube of 2."

The distinction is not always clear in the natural languages between the two kinds of proper names, those which are arbitrarily assigned to have a certain meaning (primitive proper names, as we shall say in the case of a formalized language), and those which have a linguistic structure of meaningful parts. E.g., "The Odyssey" has in the Greek a derivation from "Odysseus," and it may be debated whether this etymology is a mere matter of past history or whether it is still to be considered in modern English that the name "The Odyssey" has a structure involving the name "Odysseus." This uncertainty is removed in the case of a formalized language by fixing and making explicit the formation rules of the language (§07).

There is not yet a theory of the meaning of proper names upon which

---

[4]We extend the usual meaning of *proper name* in this manner because such alternative terms as *singular name* or *singular term* have traditional associations which we wish to avoid. The single word *name* would serve the purpose except for the necessity of distinguishing from the *common names* (or *general names*) which occur in the natural languages, and hereafter we shall often say simply *name*.

We do use the word *term*, but in its everyday meaning of an item of terminology, and not with any reference to the traditional doctrine of "categorical propositions" or the like.

general agreement has been reached as the best. Full discussion of the question would take us far beyond the intended scope of this book. But it is necessary to outline briefly the theory which will be adopted here, due in its essentials to Gottlob Frege.[5]

The most conspicuous aspect of its meaning is that a proper name always is, or at least is put forward as, a *name of* something. We shall say that a proper name *denotes*[6] or *names*[7] that of which it is a name. The relation between a proper name and what it denotes will be called the *name relation*,[8]

---

[5]See his paper, "Ueber Sinn und Bedeutung," in *Zeitschrift für Philosophie und philosophische Kritik*, vol. 100 (1892), pp. 25–50. (There are an Italian translation of this by L. Geymonat in *Gottlob Frege, Aritmetica e Logica* (1948), pp. 215–252, and English translations by Max Black in *The Philosophical Review*, vol. 57 (1948), pp. 207–230, and by Herbert Feigl in *Readings in Philosophical Analysis* (1949), pp. 85–102. See reviews of these in *The Journal of Symbolic Logic*, vol. 13 (1948), pp. 152–153, and vol. 14 (1949), pp. 184–185.)

A similar theory, but with some essential differences, is proposed by Rudolf Carnap in his recent book *Meaning and Necessity* (1947).

A radically different theory is that of Bertrand Russell, developed in a paper in *Mind*, vol. 14 (1905), pp. 479–493; in the Introduction to the first volume of *Principia Mathematica* (by A. N. Whitehead and Bertrand Russell, 1910); and in a number of more recent publications, among them Russell's book, *An Inquiry into Meaning & Truth* (1940). The doctrine of Russell amounts very nearly to a rejection of proper names as irregularities of the natural languages which are to be eliminated in constructing a formalized language. It falls short of this by allowing a narrow category of proper names which must be names of sense qualities that are known by acquaintance, and which, in Fregean terms, have *Bedeutung* but not *Sinn*.

[6]In the usage of J. S. Mill, and of others following him, not only a singular name (proper name in our terminology) but also a common or general name is said to denote, with the difference that the former denotes only one thing, the latter, many things. E.g., the common name "man" is said to denote Rembrandt; also to denote Scott; also to denote Frege; etc.

In the formalized languages which we shall study, the nearest analogues of the common name will be the *variable* and the *form* (see §02). And we prefer to use a different terminology for variables and forms than that of denoting—in particular because we wish to preserve the distinction of a proper name, or constant, from a form which is concurrent to a constant (in the sense of § 02), and from a variable which has one thing only in its range. In what follows, therefore, we shall speak of *proper names only* as denoting.

From another point of view common names may be thought of as represented in the formalized languages, not by variables or forms, but by proper names of classes (class constants). Hence the usage has also arisen according to which a proper name of a class is said to denote the various members of the class. We shall not follow this, but shall speak of a proper name of a class as denoting the class itself. (Here we agree with Mill, who distinguishes a singular collective name, or proper name of a class, from a common or general name, calling the latter a "name of a class" only in the distributive sense of being a name of each individual.)

[7]We thus translate Frege's *bedeuten* by *denote* or *name*. The verb to *mean* we reserve for general use, in reference to possible different kinds of meaning.

[8]The name relation is properly a ternary relation, among a language, a word or phrase of the language, and a denotation. But it may be treated as binary by fixing the language in a particular context. Similarly one should speak of the denotation of a name *with respect to a language*, omitting the latter qualification only when the language has been fixed or when otherwise no misunderstanding can result.

and the thing[9] denoted will be called the *denotation.* For instance, the proper name "Rembrandt" will thus be said to denote or name the Dutch artist Rembrandt, and he will be said to be the denotation of the name "Rembrandt." Similarly, "the author of *Waverley*" denotes or names the Scottish author, and he is the denotation both of this name and of the name "Sir Walter Scott."

That the meaning of a proper name does not consist solely in its denotation may be seen by examples of names which have the same denotation though their meanings are in some sense different. Thus "Sir Walter Scott" and "the author of *Waverley*" have the same denotation; it is contained in the meaning of the first name, but not of the second, that the person named is a knight or baronet and has the given name "Walter" and surname "Scott";[10] and it is contained in the meaning of the second name, but not of the first, that the person named wrote *Waverley* (and indeed as sole author, in view of the definite article and of the fact that the phrase is put forward as a proper name). To bring out more sharply the difference in meaning of the two names let us notice that, if two names are *synonymous* (have the same meaning in all respects), then one may always be substituted for the other without change of meaning. The sentence, "Sir Walter Scott is the author of *Waverley*," has, however, a very different meaning from the sentence, "Sir Walter Scott is Sir Walter Scott": for the former sentence conveys an important fact of literary history of which the latter gives no hint. This difference in meaning may lead to a difference in truth when the substitution of one name for the other occurs within certain contexts.[11] E.g., it is true that "George IV once demanded to know whether Scott was the author of *Waverley*"; but false that "George IV once demanded to know whether Scott was Scott."[12]

---

[9]The word *thing* is here used in its widest sense, in short for anything namable.

[10]The term *proper name* is often restricted to names of this kind, i.e., which have as part of their meaning that the denotation is so called or is or was entitled to be so called. As already explained, we are not making such a restriction.

Though it is, properly speaking, irrelevant to the discussion here, it is of interest to recall that Scott did make use of "the author of *Waverley*" as a pseudonym during the time that his authorship of the Waverley Novels was kept secret.

[11]Contexts, namely, which render the occurrences of the names *oblique* in the sense explained below.

[12]The particular example is due to Bertrand Russell; the point which it illustrates, to Frege.

This now famous question, put to Scott himself in the indirect form of a toast "to the author of *Waverley*" at a dinner at which Scott was present, was met by him with a flat denial, "Sire, I am not the author of *Waverley*." We may therefore enlarge on the example by remarking that Scott, despite a pardonable departure from the truth, did not mean to go so far as to deny his self-identity (as if he had said "I am not I"). And his hearers surely did not so understand him, though some must have shrewdly guessed the deception as to his authorship of *Waverley*.

Therefore, besides the denotation, we ascribe to every proper name another kind of meaning, the *sense*,[13] saying, e.g., that "Sir Walter Scott" and "the author of *Waverley*" have the same denotation but different senses.[14] Roughly, the sense is what is grasped when one understands a name,[15] and it may be possible thus to grasp the sense of a name without having knowledge of its denotation except as being determined by this sense. If, in particular, the question "Is Sir Walter Scott the author of *Waverley*?" is used in an intelligent demand for new information, it must be that the questioner knows the senses of the names "Sir Walter Scott" and "the author of *Waverley*" without knowing of their denotations enough to identify them certainly with each other.

We shall say that a name *denotes* or *names* its denotation and *expresses*[16] its sense. Or less explicitly we may speak of a name just as *having* a certain denotation and *having* a certain sense. Of the sense we say that it *determines* the denotation, or *is a concept*[17] of the denotation.

Concepts[17] we think of as non-linguistic in character—since synonymous names, in the same or different languages, express the same sense or concept —and since the same name may also express different senses, either in different languages or, by equivocation, in the same language. We are even

---

[13]We adopt this as the most appropriate translation of Frege's *Sinn*, especially since the technical meaning given to the word *sense* thus comes to be very close indeed to the ordinary acceptation of the sense of an expression. (Russell and some others following him have used "meaning" as a translation of Frege's *Sinn*.)

[14]A similar distinction is made by J. S. Mill between the denotation and the connotation of a name. And in fact we are prepared to accept *connotation* as an alternative translation of *Sinn*, although it seems probable that Frege did not have Mill's distinction in mind in making his own. We do not follow Mill in admitting names which have denotation without connotation, but rather hold that a name must always point to its denotation *in some way*, i.e., through some sense or connotation, though the sense may reduce in special cases just to the denotation's being called so and so (e.g., in the case of personal names), or to its being what appears here and now (as sometimes in the case of the demonstrative "this"). Because of this and other differences, and because of the more substantial content of Frege's treatment, we attribute the distinction between sense and denotation to Frege rather than to Mill. Nevertheless the discussion of names in Mill's *A System of Logic* (1843) may profitably be read in this connection.

[15]It is not meant by this to imply any psychological element in the notion of sense. Rather, a sense (or a concept) is a postulated abstract object, with certain postulated properties. These latter are only briefly indicated in the present informal discussion; and in particular we do not discuss the assumptions to be made about equality of senses, since this is unnecessary for our immediate purpose.

[16]This is our translation of Frege's *drückt aus*. Mill's term *connotes* is also acceptable here, provided that care is taken not to confuse Mill's meaning of this term with other meanings which it has since acquired in common English usage.

[17]This use of *concept* is a departure from Frege's terminology. Though not identical with Carnap's use of *concept* in recent publications, it is closely related to it, and was suggested to the writer by correspondence with Carnap in 1943. It also agrees well with Russell's use of *class-concept* in *The Principles of Mathematics* (1903)—cf. §69 thereof.

prepared to suppose the existence of concepts of things which have no name in any language in actual use. But every concept of a thing is a sense of some name of it in some (conceivable) language.

The possibility must be allowed of concepts which are not concepts of any actual thing, and of names which express a sense but have no denotation. Indeed such names, at least on one very plausible interpretation, do occur in the natural languages such as English: e.g., "Pegasus,"[18] "the king of France in A.D. 1905." But, as Frege has observed, it is possible to avoid such names in the construction of formalized languages.[19] And it is in fact often convenient to do this.

To understand a language fully, we shall hold, requires knowing the senses of all names in the language, but not necessarily knowing which senses determine the same denotation, or even which senses determine denotations at all.

In a well constructed language of course every name should have just one sense, and it is intended in the formalized languages to secure such univ-

[18]While the exact sense of the name "Pegasus" is variable or uncertain, it is, we take it, roughly that of the winged horse who took such and such a part in such and such supposed events—where only such minimum essentials of the story are to be included as it would be necessary to verify in order to justify saying, despite the common opinion, that "Pegasus did after all exist."

We are thus maintaining that, in the present actual state of the English language, "Pegasus" is not just a personal name, having the sense of who or what was called so and so, but has the more complex sense described. However, such questions regarding the natural languages must not be supposed always to have one final answer. On the contrary, the present actual state (at any time) tends to be indeterminate in a way to leave much debatable.

[19]For example, in the case of a formalized language obtained from one of the logistic systems of Chapter X (or of a paper by the writer in *The Journal of Symbolic Logic*, vol. 5 (1940), pp. 56–68) by an interpretation retaining the principal interpretation of the variables and of the notations λ (abstraction) and ( ) (application of function to argument), it is sufficient to take the following precautions in assigning senses to the primitive constants. For a primitive constant of type o or ι the sense must be such as—on the basis of accepted presuppositions—to assure the existence of a denotation in the appropriate domain, $\mathfrak{D}$ (of truth-values) or $\mathfrak{I}$ (of individuals). For a primitive constant of type $\alpha\beta$ the sense must be such as—on the same basis—to assure the existence of a denotation which is in the domain $\mathfrak{A}\mathfrak{B}$, i.e., which is a function from the (entire) domain $\mathfrak{B}$ which is taken as the range of variables of type β, to the domain $\mathfrak{A}$ which is taken as the range of variables of type α.

Then every well-formed formula without free variables will have a denotation, as indeed it must if such interpretation of the logistic system is to accord with formal properties of the system.

As in the case, e.g., of $\iota_{\alpha(o\alpha)}$, it may happen that the most immediate or naturally suggested interpretation of a primitive constant of type $\alpha\beta$ makes it denote a function from a proper part of the domain $\mathfrak{B}$ to the domain $\mathfrak{A}$. In such a case the definition of the function must be extended, by artificial means if necessary, over the remainder of the domain $\mathfrak{B}$, so as to obtain a function having the entire domain $\mathfrak{B}$ as its range. The sense assigned to the primitive constant must then be such as to determine this latter function as denotation, rather than the function which had only a proper part of $\mathfrak{B}$ as its range.

ocacy. But this is far from being the case in the natural languages. In particular, as Frege has pointed out, the natural languages customarily allow, besides the *ordinary* (*gewöhnlich*) use of a name, also an *oblique* (*ungerade*) use of the name, the sense which the name would express in its ordinary use becoming the denotation when the name is used obliquely.[20]

Supposing univocacy in the use of names to have been attained (this ultimately requires eliminating the oblique use of names by introducing special names to denote the senses which other names express[21]), we make, with Frege, the following assumptions, about names which have a linguistic structure and contain other names as constituent parts: (1) when a con-

---

[20]For example, in "Scott is the author of *Waverley*" the names "Scott," "*Waverley*," "the author of *Waverley*" have ordinary occurrences. But in "George IV wished to know whether Scott was the author of *Waverley*" the same three names have oblique occurrences (while "George IV" has an ordinary occurrence). Again, in "Schliemann sought the site of Troy" the names "Troy" and "the site of Troy" occur obliquely. For to seek the site of some other city, determined by a different concept, is not the same as to seek the site of Troy, not even if the two cities should happen as a matter of fact (perhaps unknown to the seeker) to have had the same site.

According to the Fregean theory of meaning which we are advocating, "Schliemann sought the site of Troy" asserts a certain relation as holding, not between Schliemann and the site of Troy (for Schliemann might have sought the site of Troy though Troy had been a purely fabulous city and its site had not existed), but between Schliemann and a certain concept, namely that of the site of Troy. This is, however, not to say that "Schliemann sought the site of Troy" means the same as "Schliemann sought the concept of the site of Troy." On the contrary, the first sentence asserts the holding of a certain relation between Schliemann and the concept of the site of Troy, and is true; but the second sentence asserts the holding of a like relation between Schliemann and the concept of the concept of the site of Troy, and is very likely false. The relation holding between Schliemann and the concept of the site of Troy is not quite that of having sought, or at least it is misleading to call it that—in view of the way in which the verb *to seek* is commonly used in English.

(W. V. Quine—in *The Journal of Philosophy*, vol. 40 (1943), pp. 113–127, and elsewhere—introduces a distinction between the "meaning" of a name and what the name "designates" which parallels Frege's distinction between sense and denotation, also a distinction between "purely designative" occurrences of names and other occurrences which coincides in many cases with Frege's distinction between ordinary and oblique occurrences. For a discussion of Quine's theory and its differences from Frege's see a review by the present writer, in *The Journal of Symbolic Logic*, vol. 8 (1943), pp. 45–47; also a note by Morton G. White in *Philosophy and Phenomenological Research*, vol. 9, no. 2 (1948), pp. 305–308.)

[21]As an indication of the distinction in question we shall sometimes (as we did in the second paragraph of footnote 20) use such phrases as "the concept of Sir Walter Scott," "the concept of the author of *Waverley*," "the concept of the site of Troy" to *denote* the same concepts which are *expressed* by the respective names "Sir Walter Scott," "the author of *Waverley*," "the site of Troy." The definite article "the" sufficiently distinguishes the phrase (e.g.) "the concept of the site of Troy" from the similar phrase "a concept of the site of Troy," the latter phrase being used as a common name to refer to any one of the many different concepts of this same spot.

This device is only a rough expedient to serve the purpose of informal discussion. It does not do away with the oblique use of names because, when the phrase "the concept of the site of Troy" is used in the way described, it contains an oblique occurrence of "the site of Troy."

stituent name is replaced by another having the same sense, the sense of the entire name is not changed; (2) when a constituent name is replaced by another having the same denotation, the denotation of the entire name is not changed (though the sense may be).[22]

We make explicit also the following assumption (of Frege), which, like (1) and (2), has been implicit in the foregoing discussion: (3) The denotation of a name (if there is one) *is a function of* the sense of the name, in the sense of §03 below; i.e., given the sense, the existence and identity of the denotation are thereby fixed, though they may not necessarily therefore be known to every one who knows the sense.

02. Constants and variables. We adopt the mathematical usage according to which a proper name of a number is called a *constant*, and in connection with formalized languages we extend this usage by removing the restriction to numbers, so that the term *constant* becomes synonymous with *proper name having a denotation.*

However, the term *constant* will often be applied also in the construction of uninterpreted calculi—logistic systems in the sense of §07—some of the symbols or expressions being distinguished as constants just in order to treat them differently from others in giving the rules of the calculus. Ordinarily the symbols or expressions thus distinguished as constants will in fact become proper names (with denotation) in at least one of the possible interpretations of the calculus.

As already familiar from ordinary mathematical usage, a *variable* is a symbol whose meaning is like that of a proper name or constant except that the single denotation of the constant is replaced by the possibility of various *values* of the variable.

Because it is commonly necessary to restrict the values which a variable may take, we think of a variable as having associated with it a certain non-empty range of possible values, the *range of* the variable as we shall call it. Involved in the meaning of a variable, therefore, are the kinds of meaning which belong to a proper name of the range.[23] But a variable must not be

---

[22]To avoid serious difficulties, we must also assume when a constituent name has no denotation that the entire name is then likewise without denotation. In the natural languages such apparent examples to the contrary as "the myth of *Pegasus*," "the search by Ponce de Leon for *the fountain of youth*" are to be explained as exhibiting oblique occurrences of the italicized constituent name.

[23]Thus the distinction of sense and denotation comes to have an analogue for variables Two variables with ranges determined by different concepts have to be considered as variables of different kinds, even if the ranges themselves should be identical. However, because of the restricted variety of ranges of variables admitted, this question does not arise in connection with any of the formalized languages which are actually considered below.

identified with a proper name of its range, since there are also differences of meaning between the two.[24]

The meaning which a variable does possess is best explained by returning to the consideration of complex names, containing other names as constituent parts. In such a complex name, having a denotation, let one of the constituent names be replaced at one or more (not necessarily all) of its occurrences by a variable, say x. To avoid complications, we suppose that x is a variable which does not otherwise occur,[25] and that the denotation of the constituent name which x replaces is in the range of x. The resulting expression (obtained from the complex name by thus replacing one of the constituent names by a variable) we shall call a *form*.[26] Such a form, for each value of x within the range of x, or at least for certain such values of x, has a *value*. Namely, the value of the form, for a given value of x, is the same as the denotation of the expression obtained from the form by substituting everywhere for x a name of the given value of x (or, if the expression so obtained

---

[24]That such an identification is impossible may be quickly seen from the point of view of the ordinary mathematical use of variables. For two proper names of the range are fully interchangeable if only they have the same sense; but two distinct variables must be kept distinct even if they have the same range determined by the same concept. E.g., if each of the letters x and y is a variable whose range is the real numbers, we are obliged to distinguish the two inequalities $x(x + y) \geqq 0$ and $x(x + x) \geqq 0$ as different —indeed the second inequality is universally true, the first one is not.

[25]This is for momentary convenience of explanation. We shall apply the name *form* also to expressions which are similarly obtained but in which the variable x may otherwise occur, provided the expression has at least one occurrence of x as a free variable (see footnote 28 and the explanation in §06 which is there referred to).

[26]This is a different use of the word *form* from that which appeared in §00 in the discussion of form and matter. We shall distinguish the latter use, when necessary, by speaking more explicitly of *logical form*.

Our present use of the word *form* is similar to that which is familiar in algebra, and in fact may be thought of as obtained from it by removing the restriction to a special kind of expressions (polynomials, or homogeneous polynomials). For the special case of propositional forms (see §04), the word is already usual in logic in this sense, independently of its use by algebraists—see, e.g., J. N. Keynes, *Formal Logic*, 4th edn., 1906, p. 53; Hugh MacColl in *Mind*, vol. 19 (1910), p. 193; Susanne K. Langer, *Introduction to Symbolic Logic*, 1937, p. 91; also Heinrich Scholz, *Vorlesungen über Grundzüge der Mathematischen Logik*, 1949 (for the use of *Aussageform* in German).

Instead of the word *form*, we might plausibly have used the word *variable* here, by analogy with the way in which we use *constant*. I.e., just as we apply the term *constant* to a complex name containing other names (constants) as constituent parts, so we might apply the term *variable* to an appropriate complex expression containing variables as constituent parts. This usage may indeed be defended as having some sanction in mathematical writing. But we prefer to preserve the better established usage according to which a variable is always a single symbol (usually a letter or letter with subscripts).

The use, by some recent authors, of the word *function* (with or without a qualifying adjective) for what we here call a form is, in our opinion, unfortunate, because it tends to conflict with and obscure the abstract notion of a function which will be explained in §03.

has no denotation, then the form has no value for that value of x).[27]

A variable such as x, occurring in the manner just described, is called a *free variable*[28] of the expression (form) in which it occurs.

Likewise suppose a complex name, having a denotation, to contain two constituent names neither of which is a part of the other, and let these two constituent names be replaced by two variables, say x and y respectively, each at one or more (not necessarily all) of its occurrences. For simplicity suppose that x and y are variables which do not occur in the original complex name, and that the denotations of the constituent names which x and y replace are in the ranges of x and y respectively. The resulting expression (obtained by the substitution described) is a *form*, with two *free variables* x and y. For certain pairs of values of x and y, within the ranges of x and y respectively, the form has a *value*. Namely, the value of the form, for given values of x and y, is the same as the denotation of the expression obtained from the form by substituting everywhere for x and y names of their re-

---

[27]It follows from assumption (2), at the end of §01, that the value thus obtained for the form is independent of the choice of a particular name of the given value of x.

The distinction of sense and denotation is, however, relevant here. For in addition to a *value* of the form in the sense explained in the text (we may call it more explicitly a *denotation value*), a complete account must mention also what we may call a *sense value* of the form. Namely, a sense value of the form is determined by a concept of some value of x, and is the same as the sense of the expression obtained from the form by substituting everywhere for x a name having this concept as its sense.

It should also be noted that a form, in a particular language, may have a value even for a value of x which is without a name in that language: it is sufficient that the given value of x shall have a name in some suitable extension of the language—say, that obtained by adding to the vocabulary of the language a name of the given value of x, and allowing it to be substitutable for x wherever x occurs as a free variable. Likewise a form may have a sense value for a given concept of a value of x if some suitable extension of the language contains a name having that concept as its sense.

It is indeed possible, as we shall see later by particular examples, to construct languages of so restricted a vocabulary as to contain no constants, but only variables and forms. But it would seem that the most natural way to arrive at the meaning of forms which occur in these languages is by contemplating languages which are extensions of them and which do contain constants—or else, what is nearly the same thing, by allowing a temporary change in the meaning of the variables ("fixing the values of the variables") so that they become constants.

[28]We adopt this term from Hilbert (1922), Wilhelm Ackermann (1924), J. v. Neumann (1927), Hilbert and Ackermann (1928), Hilbert and Bernays (1934). For what we here call a free variable the term *real variable* is also familiar, having been introduced by Giuseppe Peano in 1897 and afterward adopted by Russell (1908), but is less satisfactory because it conflicts with the common use of "real variable" to mean a variable whose range is the real numbers.

As we shall see later (§06), a free variable must be distinguished from a *bound variable* (in the terminology of the Hilbert school) or *apparent variable* (Peano's terminology). The difference is that an expression containing x as a free variable has values for various values of x, but an expression, containing x as a bound or apparent variable only, has a meaning which is independent of x—not in the sense of having the same value for every value of x, but in the sense that the assignment of particular values to x is not a relevant procedure.

spective values (or, if the expression so obtained has no denotation, then the form has no value for these particular values of x and y).

In the same way forms with three, four, and more free variables may be obtained. If a form contains a single free variable, we shall call it a *singulary*[29] form, if just two free variables, *binary*, if three, *ternary*, and so on. A form with exactly n different free variables is an *n-ary* form.

Two forms will be called *concurrent* if they agree in value—i.e., either have the same value or both have no value—for each assignment of values to their free variables. (Since the two forms may or may not have the same free variables, all the variables are to be considered together which have free occurrences in either form, and the forms are concurrent if they agree in value for every assignment of values to these variables.) A form will be called *concurrent* to a constant if, for every assignment of values to its free variables, its value is the same as the denotation of the constant. And two constants will be called *concurrent* if they have the same denotation.

Using the notion of concurrence, we may now add a fourth assumption, or principle of meaning, to the assumptions (1)–(3) of the last two paragraphs of §01. This is an extension of (2) to the case of forms, as follows: (4) In any constant or form, when a constituent constant or form is replaced by another concurrent to it, the entire resulting constant or form is concurrent to the original one.[30] The significance of this principle will become clearer in connection with the use of operators and bound variables, explained in §06 below. It is to be taken, like (2), as a part of our explanation of the name relation, and thus a part of our theory of meaning.

As in the case of *constant*, we shall apply the terms *variable* and *form* also in the construction of uninterpreted calculi, introducing them by special definition for each such calculus in connection with which they are to be used. Ordinarily the symbols and expressions so designated will be ones which become variables and forms in our foregoing sense under one of the principal interpretations of the calculus as a language (see §07).

It should be emphasized that a variable, in our usage, is a symbol of a

---

[29]We follow W. V. Quine in adopting this etymologically more correct term, rather than the presently commoner "unary."

[30]For completeness—using the notion of sense value explained in footnote 27 and extending it in obvious fashion to n-ary forms—we must also extend the assumption (1) to the case of forms, as follows. Let two forms be called *sense-concurrent* if they agree in sense value for each system of concepts of values of their free variables; let a form be called *sense-concurrent* to a constant if, for every system of concepts of values of its free variables, its sense value is the same as the sense of the constant; and let two constants be called *sense-concurrent* if they express the same sense. Then: (5) In any constant or form, when a constituent constant or form is replaced by another which is sense-concurrent to it, the entire resulting constant or form is sense-concurrent to the original one.

certain kind[31] rather than something (e.g., a number) which is denoted or otherwise meant by such symbol. Mathematical writers do speak of "variable real numbers," or oftener "variable quantities," but it seems best not to interpret these phrases literally. Objections to the idea that real numbers are to be divided into two sorts or classes, "constant real numbers" and "variable real numbers," have been clearly stated by Frege[32] and need not be repeated here at length.[33] The fact is that a satisfactory theory has never been developed on this basis, and it is not easy to see how it might be done.

The mathematical theory of real numbers provides a convenient source of examples in a system of notation[34] whose general features are well established. Turning to this theory to illustrate the foregoing discussion, we cite as particular examples of constants the ten expressions:

$$0,\ -\frac{1}{2},\ e,\ -\frac{1}{2\pi},\ \frac{1-4+1}{4\pi},\ 4e^4,\ e^e,\ e-e,\ -\frac{\pi}{2\pi},\ \frac{\sin \pi/7}{\pi/7}.$$

Let us say that x and y are variables whose range is the real numbers, and m, n, r are variables whose range is the positive integers.[35] The following are examples of forms:

---

[31]Therefore, a variable (or more precisely, particular instances or occurrences of a variable) can be written on paper—just as the figure 7 can be written on paper, though the number 7 cannot be so written except in the indirect sense of writing something which denotes it.

And similarly constants and forms are symbols or expressions of certain kinds. It is indeed usual to speak also of numbers and physical quantities as "constants"—but this usage is not the same as that in which a constant can be contrasted with a variable, and we shall avoid it in this book.

[32]See his contribution to *Festschrift Ludwig Boltzmann Gewidmet*, 1904. (Frege's theory of functions as "ungesättigt," mentioned at the end of his paper, is another matter, not necessarily connected with his important point about variables. It will not be adopted in this book, but rather we shall take a function—see §03—to be more nearly what Frege would call "Werthverlauf einer Function.")

[33]However, we mention the following parallel to one of Frege's examples. Shall we say that the usual list of seventeen names is a complete list of the Saxon kings of England, or only that it is a complete list of the constant Saxon kings of England, and that account must be taken in addition of an indefinite number of variable Saxon kings? One of these variable Saxon kings would appear to be a human being of a very striking sort, having been, say, a grown man named Alfred in A.D. 876, and a boy named Edward in A.D. 976.

According to the doctrine we would advocate (following Frege), there are just seventeen Saxon kings of England, from Egbert to Harold, and neither a variable Saxon king nor an indeterminate Saxon king is to be admitted to swell the number. And the like holds for the positive integers, for the real numbers, and for all other domains abstract and concrete. Variability or indeterminacy, where such exists, is a matter of language and attaches to symbols or expressions.

[34]We say "system of notation" rather than "language" because only the specifically numerical notations can be regarded as well established in ordinary mathematical writing. They are usually supplemented (for the statement of theorems and proofs) by one or another of the natural languages, according to the choice of the particular writer.

[35]Every positive integer is also a real number. I.e., the terms must be so understood for purposes of these illustrations.

$$y,\ -\frac{1}{y},\ -\frac{1}{x},\ -\frac{1}{2x},\ \frac{1-4+1}{4x},\ 4e^x,\ xe^x,\ x^x,$$

$$x-x,\ n-n,\ -\frac{x}{2x},\ -\frac{r}{2r},\ \frac{\sin x}{x},\ \frac{\sin r}{r},$$

$$ye^x,\ -\frac{y}{xy},\ -\frac{r}{xr},\ \frac{x-m+1}{m\pi}.$$

The forms on the first two lines are singulary, each having one free variable, y, x, n, or r as the case may be. The forms on the third line are binary, the first two having x and y as free variables, the third one x and r, the fourth one x and m.[36]

The constants

$$-\frac{1}{2\pi} \text{ and } \frac{1-4+1}{4\pi}$$

are not identical. But they are concurrent, since each denotes the same number.[37] Similarly the constants $e-e$ and 0, though not identical, are concurrent because the numbers $e-e$ and 0 are identical. Similarly $-\pi/2\pi$ and $-1/2$.

The form xe^x, for the value 0 of x, has the value 0. (Of course it is the number 0 that is here in question, not the constant 0, so that it is equally correct to say that the form xe^x, for the value 0 of x, has the value $e-e$; or that, for the value $e-e$ of x, it has the value 0; etc.) For the value 1 of x the form xe^x has the value e. For the value 4 of x its value is $4e^4$, a real number for which (as it happens) no simpler name is in standard use.

The form ye^x, for the values 0 and 4 of x and y respectively, has the value 4.

---

[36]To illustrate the remark of footnote 28, following are some examples of expressions containing bound variables:

$$\int_0^2 x^x dx, \qquad \lim_{x\to 0}\frac{\sin x}{x}, \qquad \sum_{n=1}^{\infty}\prod_{m=1}^{m=n}\frac{x-m+1}{m\pi}.$$

The first two of these are constants, containing x as a bound variable. The third is a singulary form, with x as a free variable and m and n as bound variables.

A variable may have both free and bound occurrences in the same expression. An example is $\int_0^x x^x dx$, the double use of the letter x constituting no ambiguity. Other examples are the variable Δx in $(D_x \sin x)\Delta x$ and the variable x in $xE(k)$, if the notations $D_x \sin x$ and $E(k)$ are replaced by their equivalents

$$\lim_{\Delta x\to 0}\frac{\sin(x+\Delta x)-\sin x}{\Delta x}$$

and

$$\int_0^1 \frac{\sqrt{1-k^2x^2}}{\sqrt{1-x^2}}dx$$

respectively.

[37]Whether these two constants have the same sense (as well as the same denotation) is a question which depends for its answer on a general theory of equality of senses, such as we have not undertaken to discuss here—cf. footnote 15. It is clear that Frege, though he formulates no complete theory of equality of senses, would regard these two constants as having different senses. But a plausible case might be made out for supposing that the two constants have the same sense, on some such ground as that the equation between them expresses a necessary proposition or is true on logical grounds alone or the like. No doubt there is more than one meaning of "sense," according to the criterion adopted for equality of senses, and the decision among them is a matter of convention and expediency.

For the values 1 and 1 of x and y it has the value $1e^1$; or, what is the same thing, it has the value e.

The form $-y/xy$, for the values e and 2 of x and y respectively, has the value $-1/e$. For the values e and e of x and y, it has again the value $-1/e$. For the values e and 0 of x and y it has no value, because of the non-existence of a quotient of 0 by 0.

The form $-r/xr$, for the values e and 2 of x and r respectively, has the value $-1/e$. But there is no value for the values e and e of x and r, because e is not in the range of r (e is not one of the possible values of r).

The forms

$$-\frac{1}{2x} \text{ and } \frac{1-4+1}{4x}$$

are concurrent, since they are both without a value for the value 0 of x, and they have the same value for all other values of x. The forms $-1/x$ and $-y/xy$ fail to be concurrent, since they disagree for the value 0 of y (if the value of x is not 0). But the forms $-1/x$ and $-r/xr$ are concurrent.

The forms $-1/y$ and $-1/x$ are not concurrent, as they disagree, e.g., for the values 1 and 2 of x and y respectively.

The forms $x - x$ and $n - n$ are concurrent to the same constant, namely 0,[38] and are therefore also concurrent to each other.

The forms $-x/2x$ and $-r/2r$ are non-concurrent because of disagreement for the value 0 of x. The latter form, but not the former, is concurrent to a constant, namely to $-1/2$.

03. Functions. By a *function*—or, more explicitly, a *one-valued singulary* function—we shall understand an operation[39] which, when applied to something as *argument*, yields a certain thing as the *value* of the function *for* that argument. It is not required that the function be applicable to every possible thing as argument, but rather it lies in the nature of any given function to be applicable to certain things and, when applied to one of them as argument, to yield a certain value. The things to which the function is applicable constitute the *range of* the function (or the *range of arguments of* the function) and the values constitute the *range of values of* the function. The function itself consists in the yielding or determination[39] of a value from each argument in the range of the function.

As regards equality or identity of functions we make the decision which is

[38]Or also to any other constant which is concurrent to 0.

[39]Of course the words "operation," "yielding," "determination" as here used are near-synonyms of "function" and therefore our statement, if taken as a definition, would be open to the suspicion of circularity. Throughout this Introduction, however, we are engaged in informal explanation rather than definition, and, for this purpose, elaboration by means of synonyms may be a useful procedure. Ultimately, it seems, we must take the notion of function as primitive or undefined, or else some related notion, such as that of a class. (We shall see later how it is possible to think of a class as a special case of a function, and also how classes may be used, in certain connections or for certain purposes, to replace and do the work of functions in general.)

usual in mathematics. Namely, functions are identical if they have the same range and have, for each argument in the range, the same value. In other words, we take the word "function" to mean what may otherwise be called a *function in extension*. If the way in which a function yields or produces its value from its argument is altered without causing any change either in the range of the function or in the value of the function for any argument, then the function remains the same; but the associated *function concept*, or concept determining the function (in the sense of §01), is thereby changed.

We shall speak of a function *from* a certain class *to* a certain class to mean a function which has the first class as its range and has all its values in the second class (though the second class may possibly be more extensive than the range of values of the function).

To denote the value of a function for a given argument, it is usual to write a name of the function, followed by a name of the argument between parentheses. And of course the same notation applies (*mutatis mutandis*) with a variable or a form in place of either one or both of the names. Thus if f is a function and x belongs to the range of f, then $f(x)$ is the value of the function f for the argument x.[40]

This is the usual notation for application of a function to an argument, and we shall often employ it. In some contexts (see Chapter X) we find it convenient to alter the notation by changing the position of the parentheses, so that we may write in the altered notation: if f is a function and x belongs to the range of f, then (fx) is the value of the function f for the argument x.

So far we have discussed only *one-valued singular functions* (and have used the word "function" in this sense). Indeed no use will be made in this book of many-valued functions,[41] and the reader must always understand

[40]This sentence exemplifies the use of variables to make general statements, which we assume is understood from familiar mathematical usage, though it has not yet been explained in this Introduction. (See the end of §06.)

[41]It is the idea of a many-valued (singulary) function that, for a fixed argument, there may be more than one value of the function. If a name of the function is written, followed by a name of an argument between parentheses, the resulting expression is a common name (see footnote 6) denoting the values of the function for that argument.

Though many-valued functions seem to arise naturally in the mathematical theories of real and complex numbers, objections immediately suggest themselves to the idea as just explained and are not easily overcome. Therefore it is usual to replace such many-valued functions in one way or another by one-valued functions. One method is to replace a many-valued singulary function by a corresponding one-valued binary propositional function or relation (§04). Another method is to replace the many-valued function by a one-valued function whose values are classes, namely, the value of the one-valued function for a given argument is the class of the values of the many-valued function for that argument. Still another method is to change the range of the function, an argument for which the function has n values giving way to n different arguments for each of which the function has a different one of those n values (this is the standard role of the Riemann surface in the theory of complex numbers).

"function" to mean a one-valued function. But we go on to explain functions of more than one argument.

A *binary function*, or function *of two arguments*,[42] is characterized by being applicable to two arguments in a certain order and yielding, when so applied, a certain value, the *value of* the function *for* those two arguments in that order. It is not required that the function be applicable to every two things as arguments; but rather, the function is applicable in certain cases to an ordered pair of things as arguments, and all such ordered pairs constitute the *range of* the function. The values constitute the *range of values of* the function.

Binary functions are identical (i.e., are the same function) if they have the same range and have, for each ordered pair of arguments which lies in that range, the same value.

To denote the value of a binary function for given arguments, it is usual to write a name of the function and then, between parentheses and separated by a comma, names of the arguments in order. Thus if f is a binary function and the ordered pair of x and y belongs to the range of f, then $f(x, y)$ is the value of the function f for the arguments x and y in that order.

In the same way may be explained the notion of a ternary function, of a quaternary function, and so on. In general, an n-ary function is applied to n arguments in an order, and when so applied yields a value, provided the ordered system of n arguments is in the range of the function. The value of an n-ary function for given arguments is denoted by a name of the function followed, between parentheses and separated by commas, by names of the arguments in order.

Two binary functions ϕ and ψ are called *converses*, each of the other, in case the two following conditions are satisfied: (1) the ordered pair of x and y belongs to the range of ϕ if and only if the ordered pair of y and x belongs to the range of ψ; (2) for all x, y such that the ordered pair of x and y belongs to the range of ϕ,[43]

$$\phi(x, y) = \psi(y, x).$$

A binary function is called *symmetric* if it is identical with its converse. The notions of converse and of symmetry may also be extended to n-ary functions, several different converses and several different kinds of symme-

[42]Though it is in common use we shall avoid the phrase "function of two variables" (and "function of three variables" etc.) because it tends to make confusion between *arguments* to which a function is applied and *variables* taking such arguments as values.

[43]The use of the sign $=$ to express that things are identical is assumed familiar to the reader. We do not restrict this notation to the special case of numbers, but use it for identity generally.

try appearing when the number of arguments is three or more (we need not stop over details of this).

We shall speak of a function *of* things of a certain kind to mean a function such that all the arguments to which it is applicable are of that kind. Thus a singulary function of real numbers, for instance, is a function from some class of real numbers to some (arbitrary) class. A binary function of real numbers is a binary function whose range consists of ordered pairs of real numbers (not necessarily all ordered pairs of real numbers).

We shall use the phrase "____ is a function of ____," filling the blanks with forms,[44] to mean what is more fully expressed as follows: "There exists a function f such that

$$\_\_\_\_ = f(\_\_\_\_)$$

for all ____," where the first two blanks are filled, in order, with the same forms as before, and the third blank is filled with a complete list of the free variables of those forms. Similarly we shall use "____ is a function of ____ and ____," filling the three blanks with forms, to stand for: "There exists a binary function f such that

$$\_\_\_\_ = f(\_\_\_\_, \_\_\_\_)$$

for all ____," where the first three blanks are filled, in order, with the same forms as before, and the last blank is filled with a complete list of the free variables of those forms.[45] And similar phraseology will also be used where the reference is to a function f of more than two arguments.

The phraseology just explained will also be used with the added statement of a condition or restriction. For example, "____ is a function of ____ and ____ if ____," where the first three blanks are filled with forms, and the fourth is filled with the statement of a condition involving some or all of the free variables of those forms,[46] stands for: "There exists a binary function f such that

$$\_\_\_\_ = f(\_\_\_\_, \_\_\_\_)$$

for all ____ for which ____," where the first three blanks are filled, in order,

---

[44]Our explanation assumes that neither of these forms has the particular letter f as one of its free variables. In the contrary case, the explanation is to be altered by using in place of the letter f as it appears in the text some variable (with appropriate range) which is not a free variable of either form.

[45]The theory of real numbers again serving as a source of examples, it is thus true that $x^3 + y^3$ is a function of $x + y$ and xy. But it is false that $x^3 + x^2y - xy^2 + y^3$ is a function of $x + y$ and xy (as is easily seen on the ground that the form $x^3 + x^2y - xy^2 + y^3$ is not symmetric). Again, $x^4 + y^4 + z^4 + 4x^3y + 4xy^3 + 4x^3z + 4xz^3 + 4y^3z + 4yz^3$ is a function of $x + y + z$ and $xy + xz + yz$. But $x^4 + y^4 + z^4$ is not a function of $x + y + z$ and $xy + xz + yz$.

[46]Thus with a *propositional form* in the sense of §04 below.

with the same forms as before, the fourth blank is filled with a complete list of the free variables of those forms, and the fifth blank is filled in the same way as the fourth blank was before.[47]

Also the same phraseology, explained in the two preceding paragraphs, will be used with common names[48] in place of forms. In this case the forms which the common names represent have to be supplied from the context. For example, the statement that "*The density of helium gas* is a function of *the temperature* and *the pressure*" is to be understood as meaning the same as "*The density of h* is a function of *the temperature of h* and *the pressure of h*," where the three italicized forms replace the three original italicized common names, and where h is a variable whose values are instantaneous bits of helium gas (and whose range consists of all such). Or to avoid introducing the variable h with so special a range, we may understand instead: "The density of b is a function of the temperature of b and the pressure of b if b is an instantaneous bit of helium gas." Similarly the statement at the end of §01 that the denotation of a name is a function of the sense means more explicitly (the reference being to a fixed language) that there exists a function f such that

$$\text{denotation of } N = f(\text{sense of } N)$$

for all names N for which there is a denotation.

It remains now to discuss the relationship between *functions*, in the abstract sense that we have been explaining, and *forms*, in the sense of the preceding section (§02).

If we suppose the language fixed, every singulary form has corresponding to it a function f (which we shall call the *associated function* of the form) by the rule that the value of f for an argument x is the same as the value of the form for the value x of the free variable of the form, the range of f consisting of all x's such that the form has a value for the value x of its free variable.[49]

---

[47]Accordingly it is true, for example, that: $x^3 + x^2y - xy^2 + y^3$ is a function of $x + y$ and xy if $x \geqq y$. For the special case that the variables have a range consisting of real or complex numbers, a geometric terminology is often used, thus: $x^3 + x^2y - xy^2 + y^3$ is a function of $x + y$ and xy in the half-plane $x \geqq y$.

[48]See footnotes 4, 6.

[49]For example, in the theory of real numbers, the form $\frac{1}{2}(e^x - e^{-x})$ determines the function sinh as its associated function, by the rule that the value of sinh for an argument x is $\frac{1}{2}(e^x - e^{-x})$. The range of sinh then consists of all x's (i.e., all real numbers x) for which $\frac{1}{2}(e^x - e^{-x})$ has a value. In other words, as it happens in this particular case, the range consists of all real numbers.

Of course the free variable of the form need not be the particular letter x, and indeed it may be clearer to take an example in which the free variable is some other letter.

Thus the form $\frac{1}{2}(e^y - e^{-y})$ determines the function sinh as its associated function, by the rule that the value of sinh for an argument x is the same as the value of the form $\frac{1}{2}(e^y - e^{-y})$ for the value x of the variable y. (I.e., in particular, the value of sinh for

But, still with reference to a fixed language, not every function is necessarily the associated function of some form.[50]

It follows that two concurrent singulary forms with the same free variable have the same associated function. Also two singulary forms have the same associated function if they differ only by alphabetic change of the free variable,[51] i.e., if one is obtained from the other by substituting everywhere for its free variable some other variable with the same range—with, however, the proviso (the need of which will become clearer later) that the substituted variable must remain a free variable at every one of its occurrences resulting from the substitution.

As a notation for (i.e., to denote) the associated function of a singulary form having, say, x as its free variable, we write the form itself with the letters λx prefixed. And of course likewise with any other variable in place of x.[52] Parentheses are to be supplied as necessary.[53]

---

the argument 2 is the same as the value of the form $\frac{1}{2}(e^y - e^{-y})$ for the value 2 of the variable y; and so on for each different argument x that may be assigned.)

Ordinarily, just the equation

$$\sinh(x) = \tfrac{1}{2}(e^x - e^{-x})$$

is written as sufficient indication of the foregoing. And this equation may even be called a *definition* of sinh, in the sense of footnote 168, (1) or (3).

[50]According to classical real-number theory, the singulary functions from real numbers to real numbers (or even just the analytic singulary functions) are non-enumerable. Since the forms in a particular language are always enumerable, it follows that there is no language or system of notation in which every singulary function from real numbers to real numbers is the associated function of some form.

Because of the non-enumerability of the real numbers themselves, it is even impossible in any language to provide proper names of all the real numbers. (Such a thing as, e.g., an infinite decimal expansion must not be considered a *name* of the corresponding real number, as of course an infinite expansion cannot ever be written out in full, or included as a part of any actually written or spoken sentence.)

[51]E.g., as appears in footnote 49, the forms $\frac{1}{2}(e^x - e^{-x})$ and $\frac{1}{2}(e^y - e^{-y})$ have the same associated function.

[52]Thus the expressions $\lambda x(\frac{1}{2}(e^x - e^{-x}))$, $\lambda y(\frac{1}{2}(e^y - e^{-y}))$, sinh are all three synonymous, having not only the same denotation (namely the function sinh), but also the same sense, even under the severest criterion of sameness of sense.

(In saying this we are supposing a language or system of notation in which the two different expressions sinh and $\lambda x(\frac{1}{2}(e^x - e^{-x}))$ both occur. However, the very fact of synonymy shows that the expression sinh is dispensable in principle: except for considerations of convenience, it could always be replaced by the longer expression $\lambda x(\frac{1}{2}(e^x - e^{-x}))$. In constructing a formalized language, we prefer to avoid such duplications of notation so far as readily possible. See §11.)

The expressions $\lambda x(\frac{1}{2}(e^x - e^{-x}))$ and $\lambda y(\frac{1}{2}(e^y - e^{-y}))$ contain the variables x and y respectively, as *bound* variables in the sense of footnotes 28, 36 (and of §06 below). For, according to the meaning just explained for them, these expressions are constants, not singulary forms. But of course the expression $\frac{1}{2}(e^x - e^{-x})$ is a singulary form, with x as a free variable.

The meaning of such an expression as $\lambda x(ye^x)$, formed from the binary form ye^x by prefixing λx, now follows as a consequence of the explanation about variables and forms in §02. In this expression, x is a bound variable and y is a free variable, and the

As an obvious extension of this notation, we shall also prefix the letters λx (λy, etc.) to any constant as a notation for the function whose value is the same for all arguments and is the denotation of the constant, the range of the function being the same as the range of the variable x.[54] This function will be called an *associated singulary function* of the constant, by analogy with the terminology "associated function of a form," though there is the difference that the same constant may have various associated functions with different ranges. Any function whose value is the same for all arguments will be called a *constant function* (without regard to any question whether it is an associated function of a constant, in some particular language under consideration).[55]

Analogous to the associated function of a singulary form, a binary form has two associated binary functions, one for each of the two orders in which the two free variables may be considered—or better, one for each of the two ways in which a pair of arguments of the function may be assigned as values to the two free variables of the form.

The two associated functions of a binary form are identical, and thus reduce to one function, if and only if they are symmetric. In this case the binary form itself is also called *symmetric*.[56]

Likewise an n-ary form has $n!$ associated n-ary functions, one for each of the permutations of its free variables. Some of these associated functions are identical in certain cases of symmetry.

Likewise a constant has associated m-ary functions, for $m = 1, 2, 3, \ldots$, by an obvious extension of the explanation already made for the special case $m = 1$. And by a still further extension of this we may speak of the associated m-ary functions of an n-ary form, when $m > n$. In particular a

---

expression is a singulary form whose values are singulary functions. From it, by prefixing λy, we obtain a constant, denoting a singulary function, and the range of values of this singulary function consists of singulary functions.

[53]In constructing a formalized language, the manner in which parentheses are to be put in has to be specified with more care. As a matter of fact this will be done, as we shall see, not by associating parentheses with the notation λx, but by suitable provision for parentheses (or brackets) in connection with various other notations which may occur in the form to which λx is prefixed.

[54]Thus in connection with real-number theory we use $\lambda x2$ as a notation for the function whose range consists of all real numbers and whose value is 2 for every argument.

[55]Note should also be taken of expressions in which the variable after λ is not the same as the free variable of the form which follows; thus, for example, $\lambda y(\frac{1}{2}(e^x - e^{-x}))$. As is seen from the explanation in §02, this expression is a singulary form with x as its free variable, the values of the form being constant functions. For the value 0 of x, e.g., the form $\lambda y(\frac{1}{2}(e^x - e^{-x}))$ has as its value the constant function $\lambda y0$.

In both expressions, $\lambda y(\frac{1}{2}(e^x - e^{-x}))$ and $\lambda y0$, y is a bound or apparent variable.

[56]We have already used this term, as applied to forms, in footnote 45, assuming the reader's understanding of it as familiar mathematical terminology.

singulary form has not only an associated singulary function but also associated binary functions, associated ternary functions, and so on. (When, however, we speak simply of *the* associated function of a singulary form, we shall mean the associated singulary function.)

The notation by means of λ for the associated functions of a form, as introduced above for singulary functions, is readily extended to the case of m-ary functions,[57] but we shall not have occasion to use such extension in this book. The passage from a form to an associated function (for which the λ-notation provides a symbolism) we shall speak of as *abstraction* or, more explicitly, *m-ary functional abstraction* (if the associated function is m-ary).

Historically the notion of a function was of gradual growth in mathematics, and its beginning is difficult to trace. The particular word "function" was first introduced by G. W. v. Leibniz and was adopted from him by Jean Bernoulli. The notation $f(x)$, or fx, with a letter such as f in the role of a function variable, was introduced by A. C. Clairaut and by Leonhard Euler. But early accounts of the notion of *function* do not sufficiently separate it from that of an expression containing free variables (or a *form*). Thus Euler explains a *function of a variable quantity* by identifying it with an analytic expression,[58] i.e., a form in some standard system of mathematical notation. The abstract notion of a function is usually attributed by historians of mathematics to G. Lejeune Dirichlet, who in 1837 was led by his study of Fourier series to a major generalization in freeing the idea of a function from its former dependence on a mathematical expression or law of circumscribed kind.[59] Dirichlet's notion of a function was adopted by Bernhard Riemann (1851),[60] by Hermann Hankel (1870),[61] and indeed by mathematicians generally. But two important steps remained to be taken by

[57]This has been done by Carnap in *Notes for Symbolic Logic* (1937) and elsewhere.

[58]"*Functio quantitatis variabilis est expressio analytica quomodocunque composita ex illa quantitate variabili et numeris seu quantitatibus constantibus.* Omnis ergo expressio analytica, in qua praeter quantitatem variabilem z omnes quantitates illam expressionem componentes sunt constantes, erit functio ipsius z . . . *Functio ergo quantitatis variabilis ipsa erit quantitas variabilis.*" *Introductio in Analysin Infinitorum* (1748), p. 4; *Opera*, ser. 1, vol. 8, p. 18. See further footnote 62.

[59]See his *Werke*, vol. 1, p. 135. It is not important that Dirichlet restricts his statement at this particular place to continuous functions, since it is clear from other passages in his writings that the same generality is allowed to discontinuous functions. On page 132 of the same volume is his well-known example of a function from real numbers to real numbers which has exactly two values, one for rational arguments and one for irrational arguments.

Dirichlet's generalization had been partially anticipated by Euler in 1749 (see an account by H. Burkhardt in *Jahresbericht der Deutschen Mathematiker-Vereinigung*, vol. 10 part 2 (1908), pp. 13–14) and later by J. B. J. Fourier (see his *Oeuvres*, vol. 1, pp. 207, 209, 230–232).

[60]*Werke*, pp. 3–4.

[61]In a paper reprinted in the *Mathematische Annalen*, vol. 20 (1882), pp. 63–112.

Frege (in his *Begriffsschrift* of 1879 and later publications): (i) the elimination of the dubious notion of a variable quantity in favor of the variable as a kind of symbol;[62] (ii) the admission of functions of arbitrary range by removing the restriction that the arguments and values of a function be numbers. Closely associated with (ii) is Frege's introduction of the *propositional function* (in 1879), a notion which we go on to explain in the next section.

04. Propositions and propositional functions. According to grammarians, the unit of expression in the natural languages is the *sentence*, an aggregation of words which makes complete sense or expresses a complete thought. When the complete thought expressed is that of an assertion, the sentence is called a *declarative sentence*. In what follows we shall have occasion to refer only to declarative sentences, and the simple word "sentence" is to be understood always as meaning a declarative sentence.[63]

We shall carry over the term *sentence* from the natural languages also to the formalized languages. For logistic systems in the sense of §07—uninterpreted calculi—the term *sentence* will be introduced by special definition in each case, but always with the intention that the expressions defined to be sentences are those which will become sentences in our foregoing sense under interpretations of the calculus as a formalized language.[64]

In order to give an account of the meaning of sentences, we shall adopt a theory due to Frege according to which sentences are names of a certain kind. This seems unnatural at first sight, because the most conspicuous use of sentences (and indeed the one by which we have just identified or

---

[62]The passage quoted from Euler in footnote 58 reads as if his *variable quantity* were a kind of symbol or expression. But this is not consistent with statements made elsewhere in the same work which are essential to Euler's use of the notion of function —e.g., "*Si fuerit y functio quaecunque ipsius z, tum vicissim z erit functio ipsius y*" (*Opera*, p. 24), "Sed omnis transformatio consistit in alio modo eandem functionem exprimendi, quemadmodum ex Algebra constat eandem quantitatem per plures diversas formas exprimi posse" (*Opera*, p. 32).

[63]The question may be raised whether, say, an interrogative or an imperative logic is possible, in which interrogative or imperative sentences and what they express (questions or commands) have roles analogous to those of declarative sentences and propositions in logic of ordinary kind. And some tentative proposals have in fact been made towards an imperative logic, and also towards an optative logic or logic of wishes. But these matters are beyond the scope of this book.

[64]Cf. the explanation in §02 regarding the use in connection with logistic systems of the terms *constant, variable, form*. An analogous explanation applies to a number of terms of like kind to be introduced below—in particular, *propositional variable, propositional form, operator, quantifier, bound variable, connective*.

described them) is not barely to name something but to make an assertion. Nevertheless it is possible to regard sentences as names by distinguishing between the assertive use of a sentence on the one hand, and its non-assertive use, on the other hand, as a name and a constituent of a longer sentence (just as other names are used). Even when a sentence is simply asserted, we shall hold that it is still a name, though used in a way not possible for other names.[65]

An important advantage of regarding sentences as names is that all the ideas and explanations of §§01–03 can then be taken over at once and applied to sentences, and related matters, as a special case. Else we should have to develop independently a theory of the meaning of sentences; and in the course of this, it seems, the developments of these three sections would be so closely paralleled that in the end the identification of sentences as a kind of names (though not demonstrated) would be very forcefully suggested as a means of simplifying and unifying the theory. In particular we shall require variables for which sentences may be substituted, forms which become sentences upon replacing their free variables by appropriate constants, and associated functions of such forms—things which, on the theory of sentences as names, fit naturally into their proper place in the scheme set forth in §§02–03.

Granted that sentences are names, we go on, in the light of the discussion in §01, to consider the denotation and the sense of sentences.

As a consequence of the principle (2), stated in the next to last paragraph of §01, examples readily present themselves of sentences which, though in some sense of different meaning, must apparently have the same denotation. Thus the denotation (in English) of "Sir Walter Scott is the author of *Waverley*" must be the same as that of "Sir Walter Scott is Sir Walter Scott,"

---

[65]To distinguish the non-assertive use of a sentence and the assertive use, especially in a formalized language, Frege wrote a horizontal line, —, before the sentence in the former case, and the character ⊢ before it in the latter case, the addition of the vertical line thus serving as a sign of assertion. Russell, and Whitehead and Russell in *Principia Mathematica*, did not follow Frege's use of the horizontal line before non-asserted sentences, but did take over the character ⊢ in the role of an assertion sign.

(Frege also used the horizontal line before names other than sentences, the expression so formed being a false sentence. But this is a feature of his notation which need not concern us here.)

In this book we shall not make use of a special assertion sign, but (in a formalized language) shall employ the mere writing of a sentence displayed on a separate line or lines as sufficient indication of its assertion. This is possible because sentences used non-assertively are always constituent parts of asserted sentences, and because of the availability of a two-dimensional arrangement on the printed page. (In a one-dimensional arrangement the assertion sign would indeed be necessary, if only as punctuation.)

The sign ⊢ which is employed below, in Chapter I and later chapters, is not the Frege-Russell assertion sign, but has a wholly different use.

the name "the author of *Waverley*" being replaced by another which has the same denotation. Again the sentence "Sir Walter Scott is the author of *Waverley*" must have the same denotation as the sentence "Sir Walter Scott is the man who wrote twenty-nine Waverley Novels altogether," since the name "the author of *Waverley*" is replaced by another name of the same person; the latter sentence, it is plausible to suppose, if it is not synonymous with "The number, such that Sir Walter Scott is the man who wrote that many Waverley Novels altogether, is twenty-nine," is at least so nearly so as to ensure its having the same denotation; and from this last sentence in turn, replacing the complete subject by another name of the same number, we obtain, as still having the same denotation, the sentence "The number of counties in Utah is twenty-nine."

Now the two sentences, "Sir Walter Scott is the author of *Waverley*" and "The number of counties in Utah is twenty-nine," though they have the same denotation according to the preceding line of reasoning, seem actually to have very little in common. The most striking thing that they do have in common is that both are true. Elaboration of examples of this kind leads us quickly to the conclusion, as at least plausible, that all true sentences have the same denotation. And parallel examples may be used in the same way to suggest that all false sentences have the same denotation (e.g., "Sir Walter Scott is not the author of *Waverley*" must have the same denotation as "Sir Walter Scott is not Sir Walter Scott").

Therefore, with Frege, we postulate[66] two abstract objects called *truth-values*, one of them being *truth* and the other one *falsehood*. And we declare all true sentences to denote the truth-value truth, and all false sentences to denote the truth-value falsehood. In alternative phraseology, we shall also speak of a sentence as *having* the truth-value truth (if it is true) or *having* the truth-value falsehood (if it is false).[67]

The sense of a sentence may be described as that which is grasped when one understands the sentence, or as that which two sentences in different languages must have in common in order to be correct translations each of the other. As in the case of names generally, it is possible to grasp the sense

---

[66]To Frege, as a thoroughgoing Platonic realist, our use of the word "postulate" here would not be acceptable. It would represent his position better to say that the situation indicates that *there are* two such things as truth and falsehood (*das Wahre* and *das Falsche*).

[67]The explicit use of two truth-values appears for the first time in a paper by C. S. Peirce in the *American Journal of Mathematics*, vol. 7 (1885), pp. 180–202 (or see his *Collected Papers*, vol. 3, pp. 210–238). Frege's first use of truth-values is in his *Funktion und Begriff* of 1891 and in his paper of 1892 which is cited in footnote 5; it is in these that the account of sentences as names of truth-values is first put forward.

of a sentence without therefore necessarily having knowledge of its denotation (truth-value) otherwise than as determined by this sense. In particular, though the sense is grasped, it may sometimes remain unknown whether the denotation is truth.

Any concept of a truth-value, provided that *being a truth-value* is contained in the concept, and whether or not it is the sense of some actually available sentence in a particular language under consideration, we shall call a *proposition*, translating thus Frege's *Gedanke.*

Therefore a proposition, as we use the term, is an abstract object of the same general category as a class, a number, or a function. It has not the psychological character of William of Ockham's *propositio mentalis* or of the traditional *judgment*: in the words of Frege, explaining his term *Gedanke*, it is "nicht das subjective Thun des Denkens, sondern dessen objectiven Inhalt, der fähig ist, gemeinsames Eigenthum von Vielen zu sein."

Traditional (post-Scholastic) logicians were wont to define a proposition as a judgment expressed in words, thus as a linguistic entity, either a sentence or a sentence taken in association with its meaning.[68] But in non-technical English the word has long been used rather for the meaning (in our view the sense) of a sentence,[69] and logicians have latterly come to accept this as the technical meaning of "proposition." This is the happy result of a process which, historically, must have been due in part to sheer confusion between the sentence in itself and the meaning of the sentence. It provides in English a distinction not easily expressed in some other languages, and makes possible a translation of Frege's *Gedanke* which is less misleading than the word "thought."[70]

According to our usage, every proposition determines or is a concept of

---

[68]E.g., in Isaac Watts's *Logick*, 1725: "A *Proposition* is a Sentence wherein two or more *Ideas* or *Terms* are joined or disjoined by one Affirmation or Negation. . . . In describing a *Proposition* I use the Word *Terms* as well as *Ideas*, because when mere Ideas are join'd in the Mind without Words, it is rather called a *Judgment*; but when clothed with Words, it is called a *Proposition*, even tho' it be in the Mind only, as well as when it is expressed by speaking or Writing." Again in Richard Whately's *Elements of Logic*, 1826: "The second part of Logic treats of the *proposition*; which is, '*Judgment expressed in words*.' A Proposition is defined logically 'a sentence indicative,' i.e. affirming or denying; (this excludes *commands* and *questions*.)" Here Whately is following in part the Latin of Henry Aldrich (1691). In fact these passages show no important advance over Petrus Hispanus, who wrote a half millennium earlier, but they are quoted here apropos of the history of the word "proposition" in English.

[69]Consider, for example, the incongruous result obtained by substituting the words "declarative sentence" for the word "proposition" in Lincoln's Gettysburg Address.

[70]For a further account of the history of the matter, we refer to Carnap's *Introduction to Semantics*, 1942, pp. 235–236; and see also R. M. Eaton, *General Logic*, 1931.

(or, as we shall also say, has) some truth-value. It is, however, a somewhat arbitrary decision that we deny the name *proposition* to senses of such sentences (of the natural languages) as express a sense but have no truth-value.[71] To this extent our use of *proposition* deviates from Frege's use of *Gedanke*. But the question will not arise in connection with the formalized languages which we shall study, as these languages will be so constructed that every name—and in particular every sentence—has a denotation.

A proposition is then *true* if it determines or has the truth-value truth, *false* if it has the truth-value falsehood. When a sentence expressing a proposition is asserted we shall say that the proposition itself is thereby *asserted*.[72]

A variable whose range is the two truth-values—thus a variable for which sentences (expressing propositions) may appropriately be substituted—is called a *propositional variable*. We shall not have occasion to use variables

---

[71]By the remark of footnote 22, such are sentences which contain non-obliquely one or more names that express a sense but lack a denotation—or so, following Frege, we shall take them. Examples are: "The present king of France is bald"; "The present king of France is not bald"; "The author of *Principia Mathematica* was born in 1861." (As to the last example, it is true that the phrase "the author of *Principia Mathematica*" in some appropriate supporting context may be an ellipsis for something like "the author of *Principia Mathematica* who was just mentioned" and therefore have a denotation; but we here suppose that there is no such supporting context, so that the phrase can only mean "the one and only author of *Principia Mathematica*" and therefore have no denotation.)

To sentences as a special case of names, of course the second remark of footnote 22 also applies. Thus we understand as true (and containing oblique occurrences of names) each of the sentences: "Lady Hamilton was like Aphrodite in beauty"; "The fountain of youth is not located in Florida"; "The present king of France does not exist." Cases of doubt whether a sentence has a truth-value or not are also not difficult to find in this connection, the exact meaning of various phraseologies in the natural languages being often insufficiently determinate for a decision.

[72]Notice the following distinction. The statement that a certain proposition was asserted (say on such and such an occasion) need not reveal what language was used nor make any reference to a particular language. But the statement that a certain sentence was asserted does not convey the meaning of the transaction unless it is added what language was used. For not only may the same proposition be expressed by different sentences in different languages, but also the same sentence may be used to assert different propositions according to what language the user intends. It is beside the point that the latter situation is comparatively rare in the principal known natural languages; it is not rare when all possible languages are taken into account.

Thus, if the language is English, the statement, "Seneca said that man is a rational animal," conveys the proposition that Seneca asserted but not the information what language he used. On the other hand the statement, "Seneca wrote, 'Rationale enim animal est homo,' " gives only the information what succession of letters he set down, not what proposition he asserted. (The reader may guess or know from other sources that Seneca used Latin, but this is neither said nor implied in the given statement—for there are many languages besides Latin in which this succession of letters spells a declarative sentence and, for all that thou and I know, one of them may once have been in actual use.)

whose values are propositions, but we would suggest the term *intensional propositional variable* for these.

A form whose values are truth-values (and which therefore becomes a sentence when its free variables are replaced by appropriate constants) is a *propositional form*. Usage sanctions this term[73] rather than "truth-value form," thus naming the form rather by what is expressed, when constants replace the variables, than by what is denoted.

A propositional form is said to be *satisfied by* a value of its free variable, or a system of values of its free variables, if its value for those values of its free variables is truth. (More explicitly, we should speak of a system of values of variables as satisfying a given propositional form *in a given language*, but the reference to the particular language may often be omitted as clear from the context.) A propositional form may also be said to be *true* or *false* for a given value of its free variable, or system of values of its free variables, according as its value for those values of its free variables is truth or falsehood.

A function whose range of values consists exclusively of truth-values, and thus in particular any associated function of a propositional form, is a *propositional function*. Here again, established usage sanctions "propositional function"[74] rather than "truth-value function," though the latter term would be the one analogous to, e.g., the term "numerical function" for a function whose values are numbers.

A propositional function is said to be *satisfied by* an argument (or ordered system of arguments) if its value for that argument (or ordered system of arguments) is truth. Or synonymously we may say that a propositional function *holds for* a particular argument or ordered system of arguments.

From its use in mathematics, we assume that the notion of a *class* is already at least informally familiar to the reader. (The words *set* and *aggregate* are ordinarily used as synonymous with *class*, but we shall not follow this usage, because in connection with the Zermelo axiomatic set

---

[73]Cf. footnote 26.

[74]This statement seems to be on the whole just, though the issue is much obscured by divergencies among different writers as to the theory of meaning adopted and in the accounts given of the notions of function and proposition. The idea of the propositional function as an analogue of the numerical function of mathematical analysis originated with Frege, but the term "propositional" function is originally Russell's. Russell's early use of this term is not wholly clear. In his introduction to the second edition of *Principia Mathematica* (1925) he decides in favor of the meaning which we are adopting here, or very nearly that.

theory[75] we shall wish later to give the word *set* a special meaning, somewhat different from that of *class*.) We recall that a class is something which has or may have *members*, and that classes are considered identical if and only if they have exactly the same members. Moreover it is usual mathematical practice to take any given singulary propositional form as having associated with it a class, namely the class whose members are those values of the free variable for which the form is true.

In connection with the functional calculi of Chapters III–VI, or rather, with the formalized languages obtained from them by adopting one of the indicated principal interpretations (§07), it turns out that we may secure everything necessary about classes by just identifying a class with a singulary propositional function, and membership in the class with satisfaction of the singulary propositional function. We shall consequently make this identification, on the ground that no purpose is served by maintaining a distinction between classes and singulary propositional functions.

We must add at once that the notion of a class obtained by thus identifying classes with singulary propositional functions does not quite coincide with the informal notion of a class which we first described, because it does not fully preserve the principle that classes are identical if they have the same members. Rather, it is necessary to take into account also the *range-members* of a class (constituting, i.e., the range of the singulary propositional function). And only when the range-members are given to be the same is the principle preserved that classes are identical if they have the same members. This or some other departure from the informal notion of a class is in fact necessary, because, as we shall see later,[76] the informal notion—in the presence of some other assumptions difficult to avoid—is self-inconsistent and leads to antinomies. (The *sets* of Zermelo set theory preserve the principle that sets having the same members are identical, but at the sacrifice of the principle that an arbitrary singulary propositional form has an associated set.)

Since, then, a class is a singulary propositional function, we speak of the *range of* the class just as we do of the propositional function (i.e., it is the same thing). We think of the range as being itself a class, having as members the range-members of the given class, and having the same range-members.

(In any particular discussion hereafter in which classes are introduced,

---

[75]Chapter XI.
[76]In Chapter VI.

and in the absence of any indication to the contrary, it is to be understood that there is a fixed range determined in advance and that all classes have this same range.)

Relations may be similarly accounted for by identifying them with binary propositional functions, the relation being said to *hold between* an ordered pair of things (or the things being said to *stand in* that relation, or to *bear* that relation one to the other) if the binary propositional function is satisfied by the ordered pair. Given that the ranges are the same, this makes two relations identical if and only if they hold between the same ordered pairs, and to indicate this we may speak more explicitly of a *relation in extension*—using this term as synonymous with *relation*.

A *property*, as ordinarily understood, differs from a class only or chiefly in that two properties may be different though the classes determined by them are the same (where the class determined by a property is the class whose members are the things that have that property). Therefore we identify a property with a *class concept*, or concept of a class in the sense of §01. And two properties are said to *coincide in extension* if they determine the same class.

Similarly, a *relation in intension* is a *relation concept*, or concept of a relation in extension.

To turn once more for illustrative purposes to the theory of real numbers and its notations, the following are examples of propositional forms:

$$\sin x = 0, \qquad \sin x = 2,$$

$$e^x > 0, \qquad e^x > 1, \qquad x > 0,$$

$$\varepsilon > 0, \qquad \varepsilon < 0,$$

$$x^3 + y^3 = 3xy, \qquad x \neq y,$$

$$|x - y| < t, \qquad |x - y| < \varepsilon,$$

$$\text{If } |x - y| < \delta \text{ then } |\sin x - \sin y| < \varepsilon.$$

Here we are using x, y, t as variables whose range is the real numbers, and ε and δ as variables whose range is the positive real numbers. The seven forms on the first three lines are examples of singulary propositional forms. Those on the fourth line are binary, on the fifth line ternary, while on the last line is an example of a quaternary propositional form.

Each of the singulary propositional forms has an associated class. Thus with the form $\sin x = 0$ is associated the class of those real numbers whose sine is 0, i.e., the class whose range is the real numbers and whose members are 0, π, $-\pi$, 2π, -2π, 3π, and so on. As explained, we identify this class with the propositional function $\lambda x(\sin x = 0)$, or in other words the function from real numbers to truth-values which has for any argument x the value $\sin x = 0$.

The two propositional forms $e^x > 1$ and $x > 0$ have the same associated class, namely, the class whose range is the real numbers and whose members are the positive real numbers. This class is identified with either $\lambda x(e^x > 1)$ or $\lambda x(x > 0)$, these two propositional functions being identical with each other by the convention about identity of functions adopted in §03.

Since the propositional form $\sin x = 2$ has the value falsehood for every value of x, the associated class $\lambda x(\sin x = 2)$ has no members.

A class which has no members is called a *null class* or an *empty class*. From our conventions about identity of propositional functions and of classes, if the range is given, it follows that there is only one null class. But, e.g., the range of the null class associated with the form $\sin x = 2$ and the range of the null class associated with the form $\varepsilon < 0$ are not the same: the former range is the real numbers, and the latter range is the positive real numbers.[77] We shall speak respectively of the "null class of real numbers" and of the "null class of positive real numbers."

A class which coincides with its range is called a *universal class*. For example, the class associated with the form $e^x > 0$ is the universal class of real numbers; and the class associated with the form $\varepsilon > 0$ is the universal class of positive real numbers.

The binary propositional forms $x^3 + y^3 = 3xy$ and $x \neq y$ are both symmetric and therefore each have one associated binary propositional function or relation. In particular, the associated relation of the form $x \neq y$ is the relation of diversity between real numbers; or in other words the relation which has the pairs of real numbers as its range, which any two different real numbers bear to each other, and which no real number bears to itself.

The ternary propositional forms $|x - y| < t$ and $|x - y| < \varepsilon$ have each three associated ternary propositional functions[78] (being symmetric in x and y). All six of these propositional functions are different; but an appropriately chosen pair of them, one associated with each form, will be found to agree in value for all ordered triples of arguments which are in the range of both, differing only in that the first one has the value falsehood for certain ordered triples of arguments which are not in the range of the other.

05. Improper symbols, connectives.

When the expressions, especially the sentences, of a language are analyzed into the single symbols of which they consist, symbols which may be regarded as indivisible in the sense that

---

[77]According to the informal notion that classes with the same members are identical, it would be true absolutely that there is only one null class. The distinction of null classes with different ranges was introduced by Russell in 1908 as a part of his theory of types (see Chapter VI). The same thing had previously been done by Ernst Schröder in the first volume of his *Algebra der Logik* (1890), though with a very different motivation.

[78]We may also occasionally use the term *ternary relation* (and *quaternary relation* etc). But the simple term *relation* will be reserved for the special case of a binary relation, or binary propositional function.

no division of them into parts has relevance to the meaning,[79] we have seen that there are two sorts of symbols which may in particular appear, namely primitive proper names and variables. These we call *proper symbols*, and we regard them as having meaning in isolation, the primitive names as denoting (or at least purporting to denote) something, the variables as having (or at least purporting to have) a non-empty range. But in addition to proper symbols there must also occur symbols which are *improper*—or in traditional (Scholastic and pre-Scholastic) terminology, *syncategorematic*—i.e., which have no meaning in isolation but which combine with proper symbols (one or more) to form longer expressions that do have meaning in isolation.[80]

Conspicuous among improper symbols are parentheses and brackets of various kinds, employed (as familiar in mathematical notation) to show the way in which parts of an expression are associated. These parentheses and brackets occur as constituents in certain combinations of improper symbols such as we now go on to consider—either exclusively to show association and in connection with other improper symbols which carry the burden of showing the particular character of the notation,[81] or else sometimes in a way that combines the showing of association with some special meaning-producing character.[82]

Connectives are combinations of improper symbols which may be used together with one or more constants to form or produce a new constant.

---

[79]The formalized languages are to be so constructed as to make such analysis into single symbols precisely possible. In general it is possible in the natural languages only partially and approximately—or better, our thinking of it as possible involves a certain idealization.

In written English (say), the single symbols obtained are not just the letters with which words are spelled, since the division of a word into letters has or may have no relevance to the meaning. Frequently the single symbols are words. In other cases they are parts of words, since the division, e.g., of "books" into "book" and "s" or of "colder" into "cold" and "er" does have relevance to the meaning. In still other cases the linguistic structure of meaningful parts is an idealization, as when "worse" is taken to have an analysis parallel to that of "colder," or "I went" an analysis parallel to that of "I shall go," or "had I known" parallel to that of "if I should hear." (Less obvious and more complex examples may be expected to appear if analysis is pressed more in detail.)

[80]Apparently the case may be excluded that several improper symbols combine without any proper symbols to form an expression that has meaning in isolation. For the division of that expression into the improper symbols as parts could then hardly be said to have relevance to the meaning.

[81]Thus in the expression $(t - (x - y))$ we may say that the inner parentheses serve exclusively to show the association together of the part $x - y$ of the expression, and that they are used in connection with the sign $-$, which serves to show subtraction.

[82]In real number theory, the usual notation $|\quad|$ for the absolute value is an obvious example of this latter. Again it may be held that the parentheses have such a double use in either of the two notations introduced in §03 for application of a singulary function to its argument.

Then, as follows from the discussion in §02, if we replace one or more of the constants each by a form which has the denotation of that constant among its values, the resulting expression becomes a form (instead of a constant); and the free variables of this resulting form are the free variables of all the forms (one or more) which were united by means of the connective (with each other and possibly also with some constants) to produce the resulting form. In order to give completely the meaning-producing character of a particular connective in a particular language, not only is it necessary to give the denotation[83] of the new constant in every permissible case that the connective is used together with one or more constants to form such a new constant, but also, for every case that the connective may be used with forms or forms and constants to produce a resulting form, it is necessary to give the complete scheme of values of this resulting form for values of its free variables. And this must all be done in a way to conform to the assumptions about sense and denotation at the end of §01, and to the conventions about meaning and values of variables and forms as these were described in §02. Connectives may then be used not only in languages which contain constants but also in languages whose only proper symbols are variables.[84]

The constants or forms, united by means of a connective to produce a new constant or form, are called the *operands*. A connective is called *singulary, binary, ternary,* etc., according to the number of its operands.

A singulary connective may be used with a variable of appropriate range as the operand (this falls under our foregoing explanation since, of course, a variable is a special case of a form). The form so produced is called an *associated form* of the connective if the range of the variable includes the denotations of all constants which may be used as operands of the connective and all the relevant values of all the forms which may be used as operands of the connective (where by a *relevant* value of a form used as operand is meant a value corresponding to which the entire form, consisting of connective and operand, has a value). And the *associated function* of a singulary connective is the associated function of any associated form. The associated function as thus defined is clearly unique.

---

[83]It is not necessary (or possible) to give the sense of the new constant separately, since the way in which the denotation is given carries with it a sense—the same phrase which is used to name the denotation must also express a sense.

Further questions arise if, besides constants, names having a sense but no denotation are allowed. Such names seem to be used with connectives in the natural languages and in usual systems of mathematical notation, and indeed some illustrations which we have employed depend on this. However, as already explained, we avoid this in the formalized languages which we shall consider.

[84]Cf. footnote 27.

The notion of the associated function of a singulary connective is possible also in the case of a language containing no variable with a range of the kind required to produce an associated form, namely we may consider an extension of the language obtained by adding such a variable.

In the same way an n-ary connective may be used together with n different variables as operands to produce a form; and this is called an *associated form* of the connective if, for each variable, the range includes both the denotations of all constants and all relevant values of all forms which may be used as operands at that place. The *associated function* of the connective is that one of the associated n-ary functions of an associated form which is obtained by assigning the arguments of the function, in their order, as values to the free variables of the form in their left-to-right order of occurrence in the form.

In general the meaning-producing character of a connective is most readily given by just giving the associated function, this being sufficient to fix the use of the connective completely.[85]

Indeed there is a close relationship between connectives and *functional constants* or proper names of functions. Differences are that (a) a functional constant *denotes* a function whereas a connective *is associated with* a function, (b) a connective is never replaced by a variable, and (c) the notation for application of a function to its arguments may be paralleled by a different notation when a corresponding connective takes the place of a functional constant. But these differences are from some points of view largely nonessential because (a) notations of course have such meaning as we choose to give them (within limitations imposed by requirements of consistency and adequacy), (b) languages are possible which do not contain variables with functions as values and in which functional constants are never replaced by variables, and (c) the notation for application of a function to its arguments may, like any other, be changed—or even duplicated

---

[85]For example, the familiar notation (—) for subtraction of real numbers may be held to be a connective. That is, the combination of symbols which consists of a left parenthesis, a minus sign, and a right parenthesis, in that order, may be considered as a connective—where the understanding is that an appropriate constant or form is to be filled in at each of two places, namely immediately before and immediately after the minus sign. To give completely the meaning-producing character of this connective, it is necessary to give the denotation of the resulting constant when constants are filled in at the two places, and also to give the complete scheme of values of the resulting form when forms are filled in at the two places, or a form at one place and a constant at the other. In order to do this in a way to conform to §§01, 02, it may often be most expeditious first to introduce (by whatever means may be available in the particular context) the binary function of real numbers that is called *subtraction*, and then to declare this to be the associated function of the connective.

by introducing several synonymous notations into the same language.[86]

In the case of a language having notations for application of a function to its arguments, it is clear that a connective may often be eliminated or dispensed with altogether by employing instead a name of the associated function—by modifying the language, if necessary, to the extent of adding such a name to its vocabulary. However, the complete elimination of all connectives from a language can never be accomplished in this way. For the notations for application of a singulary function to its argument, for application of a binary function to its arguments, and so on (e.g., the notations for these which were introduced in §03) are themselves connectives. And though these connectives, like any other, no doubt have their associated functions,[87] nevertheless not all of them can ever be eliminated by the device in question.[88]

---

[86]Thus, to use once more the example of the preceding footnote, we may hold that the notation (—) is a connective and that the minus sign has no meaning in isolation. Or alternatively we may hold that the minus sign denotes (is a name of) the binary function, *subtraction*, and that in such expressions as, e.g., $(x - y)$ or $(5 - 2)$ we have a special notation for application of a binary function to its arguments, different from the notation for this which was introduced in §03. The choice would seem to be arbitrary between these two accounts of the meaning of the minus sign. But from one standpoint it may be argued that, if we are willing to invent some name for the binary function, then this name might just as well, and would most simply, be the minus sign.

[87]As explained below, we are for expository purposes temporarily ignoring difficulties or complications which may be caused by the theory of types or by such alternative to the theory of types as may be adopted. On this basis, for the connective which is the notation for application of a singulary function to its argument, we explain the associated function by saying that it is the binary function whose value for an ordered pair of arguments f, x is $f(x)$. But if a name of this associated function is to be used for the purpose of eliminating the connective, then another connective is found to be necessary, the notation, namely, for application of a binary function to its arguments. If the latter connective is to be eliminated by using a name of its associated function, then the notation for application of a ternary function to its arguments becomes necessary. And so on. Obviously no genuine progress is being made in these attempts.

(After studying the theory of types the reader will see that the foregoing statement, and others we have made, remain in some sense essentially true on the basis of that theory. It is only that the connective, e.g., which is the notation for application of a singulary function to its argument must be thought of as replaced by many different connectives, corresponding to different types, and each of these has its own associated function. Or alternatively, if we choose to retain this connective as always the same connective, regardless of considerations of type, then there may well be no variable in the language with a range of the kind required to produce an associated form: an extension of the language by adding such a variable can be made to provide an associated form, but not so easily a name of the associated function. See Carnap, *The Logical Syntax of Language* (cited in footnote 131), examples at the end of §53, and references there given; also Bernard Notcutt's proposal of "intertypical variables" in *Mind*, n.s. vol. 43 (1934), pp. 63–77; and remarks by Tarski in the appendix to his *Wahrheitsbegriff* (cited in footnote 140).)

[88]There is, however, a device which may be used in appropriate context (cf. Chapter X) to eliminate all the connectives except the notation for application of a singulary function to its argument. This is done by reconstruing a binary function as a singulary function whose values are singulary functions; a ternary function as a singulary function whose values are binary functions in the foregoing sense; and so on. For it turns out

Connectives other than notations for application of a function to its arguments are apparently always eliminable in the way described by a sufficient extension of the language in which they occur (including if necessary the addition to the language of notations for application of a function to its arguments). Nevertheless such other connectives are often used—especially in formalized languages of limited vocabulary, where it may be preferred to preserve this limitation of vocabulary, so as to use the language as a means of singling out for separate consideration some special branch of logic (or other subject).

In particular we shall meet with *sentence connectives* in Chapter I. Namely, these are connectives which are used together with one or more sentences to produce a new sentence; or when propositional forms replace some or all of the sentences as operands, then a propositional form is produced rather than a sentence.

The chief *singulary* sentence connective we shall need is one for negation. In this role we shall use, in formalized languages, the single symbol $\sim$, which, when prefixed to a sentence, forms a new sentence that is the negation of the first one. The associated function of this connective is the function from truth-values to truth-values whose value for the argument *falsehood* is *truth*, and whose value for the argument *truth* is *falsehood*. For convenience in reading orally expressions of a formalized language, the symbol $\sim$ may be rendered by the word "not" or by the phrase "it is false that."

The principal *binary* sentence connectives are indicated in the table which follows. The notation which we shall use in formalized languages is shown in the first column of the table, with the understanding that each of the two blanks is to be filled by a sentence of the language in question. In the second column of the table a convenient oral reading of the connective is suggested, or sometimes two alternative readings; here the understanding is that the two blanks are to be filled by oral readings of the same two sentences (in the same order) which filled the two corresponding blanks in the first column; and words which appear between parentheses are words which

---

that n-ary functions in the sense thus obtained can be made to serve all the ordinary purposes of n-ary functions (in any sense).

The alternative device of reducing (e.g.) a binary function to a singulary function by reconstruing it as a singulary function whose arguments are ordered pairs is also useful in certain contexts (e.g., in axiomatic set theory). This device does not (at least *prima facie*) serve to reduce the number of connectives to one, as besides the notation for application of a singulary function to its argument there will be required also a connective which unites the names of two things to form a name of their ordered pair (or at least some notation for this latter purpose). Nevertheless it is a device which may sometimes be used to accomplish a reduction, especially where other connectives—or operators (§06)—are available.

may ordinarily be omitted for brevity, but which are to be supplied whenever necessary to avoid a misunderstanding or to emphasize a distinction. In the third column the associated function of the connective is indicated by means of a code sequence of four letters: in doing this, t is used for truth and f for falsehood, and the first letter of the four gives the value of the function for the arguments t, t, the second letter gives the value for the arguments t, f, the third letter for the arguments f, t, the fourth letter for the arguments f, f. In many cases there is an English name in standard use, which may denote either the connective or its associated function. This is indicated in a fourth column of the table; where alternative names are in use, both are given, and in some cases where none is in use a suggested name is supplied.

[___ ∨ ___]	___ or ___ (or both).	tttf	(Inclusive) disjunction, alternation.
[___ ⊂ ___]	___ if ___.[89]	ttft	Converse implication.
[___ ⊃ ___]	If ___ then ___,[89]	tftt	The (truth-functional) conditional,[90]
	___ (materially) implies ___.[89]		(material) implication.
[___ ≡ ___]	___ if and only if ___,[89]	tfft	The (truth-functional) biconditional,[90]
	___ is (materially) equivalent to ___.[89]		(material) equivalence.
[___ ___]	___ and ___.	tfff	Conjunction.
[___ | ___]	Not both ___ and ___.	fttt	Non-conjunction, Sheffer's stroke.
[___ ≢ ___]	___ or ___ but not both, ___ is not (materially) equivalent to ___.[89]	fttf	Exclusive disjunction, (material) non-equivalence.
[___ ⊅ ___]	___ but not ___.	ftff	(Material) non-implication.
[___ ⊄ ___]	Not ___ but ___.	fftf	Converse non-implication.
[___ $\bar{\lor}$ ___]	Neither ___ nor ___.	ffft	Non-disjunction.

[89] The use of the English words "if," "implies," "equivalent" in these oral readings must not be taken as indicating that the meanings of these English words are faithfully

The notations which we use as sentence connectives—and those which we use as quantifiers (see below)—are adaptations of those in Whitehead and Russell's *Principia Mathematica* (some of which in turn were taken from Peano). Various other notations are in use,[91] and the student who would

rendered by the corresponding connectives in all, or even in most, cases. On the contrary, the meaning-producing character of the connectives is to be learned with accuracy from the third column of the table, where the associated functions are given, and the oral readings supply at best a rough approximation.

As a matter of fact, the words "if . . . then" and "implies" as used in ordinary non-technical English often seem to denote a relation between propositions rather than between truth-values. Their possible meanings when employed in this way are difficult to fix precisely and we shall make no attempt to do so. But we select the one use of the words "if . . . then" (or "implies")—their material use, we shall call it—in which they may be construed as denoting a relation between truth-values, and we assign this relation as the associated function for the connective [$\supset$].

As examples of the material use of "if . . . then," consider the four following English sentences:

(i) If Joan of Arc was a patriot then Nathan Hale was a patriot.
(ii) If Joan of Arc was a patriot then Vidkun Quisling was a patriot.
(iii) If Vidkun Quisling was a patriot then attar of roses is a perfume.
(iv) If Vidkun Quisling was a patriot then Limburger cheese is a perfume.

For the sake of the illustration let us suppose examination of the historical facts to reveal that Joan of Arc and Nathan Hale were indeed patriots and that Vidkun Quisling was not a patriot. Then (i), (iii), and (iv) are true, and (ii) is false; and to reach these conclusions no examination is necessary of the characteristics of either attar of roses or Limburger cheese. (If the reader is inclined to question the truth of, e.g., (iii) on the ground of complete lack of connection between Vidkun Quisling and attar of roses, then this means that he has in mind some other use of "if . . . then" than the material use.)

[90]These terms were introduced by Quine, who uses them for "the mode of composition described in" the list of truth-values as given in the third column of the table—i.e., in effect, and in our terminology, for the associated function of the connective rather than for the connective itself. See his *Mathematical Logic*, 1940, pp. 15, 20.

We prefer the better established terms *material implication* and *material equivalence*, from which the adjective *material* may be omitted whenever there is no danger of confusion with other kinds of implication or equivalence—as, for example, with formal implication and formal equivalence (§06), or with kinds of implication and equivalence (belonging to modal logic) which are relations between propositions rather than between truth-values.

[91]Worthy of special remark is the parenthesis-free notation of Jan Łukasiewicz. In this, the letters N, A, C, E, K are used in the roles of negation, disjunction, implication, equivalence, conjunction respectively. Further letters may be introduced if desired (R has been employed as non-equivalence, D as non-conjunction). In use as a sentence connective, the letter is written first and then in order the sentences or propositional forms together with which it is used. No parentheses or brackets or other notations specially to show association are necessary. E.g., the propositional form

$$[[p \supset [q \vee r]] \supset \sim p]$$

(where p, q, r are propositional variables) becomes, in the Łukasiewicz notation,

$$CCpAqrNp.$$

It is of course possible to apply the same idea to other connectives, in particular to the notation for application of a singulary function to its argument. Hence (see footnote 88) parentheses and brackets may be avoided altogether in a formalized language. The possibility of this is interesting. But the notation so obtained is unfamiliar, and less perspicuous than the usual one.

compare the treatments of different authors must learn a certain facility in shifting from one system of notation to another.

The brackets which we indicate as constituents in these notations may in actual use be found unnecessary at certain places, and we may then just omit them at such places (though only as a practically convenient abbreviation).

We shall use the term *truth-function*[92] for a propositional function of truth-values which has as range, if it is n-ary, all ordered systems of n truth-values. Thus every associated function of a sentence connective is a truth-function. And likewise every associated function of a form built up from propositional variables solely by iterated use of sentence connectives.[93]

06. Operators, quantifiers. An *operator* is a combination of improper symbols which may be used together with one or more variables—the *operator variables* (which must be fixed in number and all distinct)—and one or more constants or forms or both—the *operands*—to produce a new constant or form. In this new constant or form, however, the operator variables are at certain determinate places not free variables, though they may have been free variables at those places in the operands.

To be more explicit, we remark that, in any application of an operator, the operator variables may (and commonly will) occur as free variables in some of the operands. In the new constant or form produced we distinguish three possible kinds of occurrences of the operator variables, viz.: an occurrence in one of the operands which, when considered as an occurrence in that operand alone, is an occurrence as a free variable; an occurrence in one of the operands, not of this kind; and an occurrence which is an occurrence *as* an operator variable, therefore not in any of the operands. In the new constant or form, an occurrence of one of the two latter kinds is never an occurrence as a free variable, and each occurrence of the first kind is an occurrence as a free variable or not, according to some rule associated with the particular operator.[94] The simplest case is that, in the new constant or form, none of the occurrences of the operator variables are occurrences as free variables. And this is the only case with which we shall meet in the following chapters

[92]We adopt this term from *Principia Mathematica*, giving it substantially the meaning which it acquires through changes in that work that were made (or rather, proposed) by Russell in his introduction to the second edition of it.

[93]For example, the associated function of the propositional form mentioned in footnote 91.

[94]We do require in the case of each operator variable that all occurrences of the first kind shall be occurrences as free variables or else all not, *in any one occurrence of a particular operand* in the new constant or form produced. For operators violating this requirement are not found among existing standard mathematical and logical notations, and it is clear that they would involve certain anomalies of meaning which it is preferable to avoid.

(though many operators which are familiar as standard mathematical notation fail to fall under this simplest case).

Variables thus having occurrences in a constant or form which are not occurrences as free variables of it are called *bound variables* of the constant or form.[95] The difference is that a form containing a particular variable, say x, *as a free variable* has values for various values of the variable, but a constant or form which contains x *as a bound variable only* has a meaning which is independent of x—not in the sense of having the same value for every value of x, but in the sense that the assignment of particular values to x is not a relevant procedure.[96]

It may happen that a form contains both free and bound occurrences of the same variable. This case will arise, for example, if a form containing a particular variable as a free variable and a form or constant containing that same variable as a bound variable are united by means of a binary connective.[97]

As in the case of connectives, we require that operators be such as to conform to the principles (1)–(3) at the end of §01; also that they conform to the conventions about meaning and values of variables as these were described in §02, and in particular to the principle (4) of §02.[98]

An operator is called *m-ary-n-ary* if it is used with m distinct operator variables and n operands.[99] The most common case is that of a singulary-singulary operator—or, as we shall also call it, a *simple* operator.

In particular, the notation for singulary functional abstraction, which

[95]Cf. footnote 28.

[96]Therefore a constant or form which contains a particular variable as a bound variable is unaltered in meaning by alphabetic change of that variable, at all of its bound occurrences, to a new variable (not previously occurring) which has the same range. The condition in parentheses is included only as a precaution against identifying two variables which should be kept distinct, and indeed it may be weakened somewhat—cf. the remark in §03 about alphabetic change of free variables.

E.g., the constant $\int_0^2 x^x dx$ (see footnote 36) is unaltered in meaning by alphabetic change of the variable x to the variable y: it has not only the same denotation but also the same sense as $\int_0^2 y^y dy$.

[97]See illustrations in the second paragraph of footnote 36.

[98]And also to the principle (5) of footnote 30.

[99]Thus, in the theory of real numbers, the usual notation for definite integration is a singulary-ternary operator. And in, e.g., the form $\int_0^x x^x dx$ (see footnote 36) the operator variable is x and the three operands are the constant 0, the form x, and the form x^x.

Again, the large $\prod$ (product sign), as used in the third example at the beginning of footnote 36, is part of a singulary-ternary operator. The signs = above and below the $\prod$ are not to be taken as equality signs in the ordinary sense (namely that of footnote 43) but as improper symbols, and also part of the operator. In the particular application of the operator, as it appears in this example, the operator variable is m and the operands are 1, n, and

$$\frac{x - m + 1}{m\pi}.$$

was introduced in §03, is a simple operator (the variable which is placed immediately after the letter λ being the operator variable). We shall call this the *abstraction operator* or, more explicitly, the *singulary functional abstraction operator*. In appropriate context, as we shall see in Chapter X, all other operators can in fact be reduced to this one.[100]

Another operator which we shall use—also a simple operator—is the *description operator*, (ι). To illustrate, let the operator variable be x. Then the notation (ιx) is to have as its approximate reading in words, "the x such that"; or more fully, the notation is explained as follows. It may happen that a singulary propositional form whose free variable is x has the value truth for one and only one value of x, and in this case a name of that value of x is produced by prefixing (ιx) to the form. In case there is no value of x or more than one for which the form has the value truth, there are various meanings which might be assigned to the name produced by prefixing (ιx) to the form: the analogy of English and other natural languages would suggest giving the name a sense which determines no denotation; but we prefer to select some fixed value of x and to assign this as the denotation of the name in all such cases (this selection is arbitrary, but is to be made once for all for each range of variables which is used).

Of especial importance for our purposes are the *quantifiers*. These are namely operators for which both the operands and the new constant or form produced by application of the operator are sentences or propositional forms.

As the *universal quantifier* (when, e.g., the operator variable is x) we use

---

As another example of application of the same operator, showing both bound and free occurrences of m, we cite

$$\prod_{m=m+1}^{m=m+n+1} \frac{x-m+1}{m\pi}.$$

Examples of operators taking more than one operator variable are found in familiar notations for double and multiple limits, double and multiple integrals.

It should also be noted that n-ary connectives may, if we wish, be regarded as 0-ary-n-ary operators.

[100]In the combinatory logic of H. B. Curry (based on an idea due to M. Schönfinkel) a more drastic reduction is attempted, namely the complete elimination of operators, of variables, and of all connectives, except a notation for application of a singulary function to its argument, so as to obtain a formalized language in which, with the exception of the one connective, all single symbols are constants, and which is nevertheless adequate for some or all of the purposes for which variables are ordinarily used. This is a matter beyond the scope of this book, and the present status of the undertaking is too complex for brief statement. The reader may be referred to a monograph by the present writer, *The Calculi of Lambda-Conversion* (1941), which is concerned with a related topic; also to papers by Schönfinkel, Curry, and J. B. Rosser which are there cited, to several papers by Curry and by Rosser in *The Journal of Symbolic Logic* in 1941 and 1942, to an expository paper by Robert Feys in *Revue Philosophique de Louvain*, vol. 44 (1946), pp. 74–103, 237–270, and to a paper by Curry in *Synthese*, vol. 7 (1949), pp. 391–399.

the notation $(\forall x)$ or (x), prefixing this to the operand. The universal quantifier is thus a simple operator, and we may explain its meaning as follows (still using the particular variable x as an example). (x)____ is true if the value of ____ is truth for all values of x, and (x)____ is false if there is any value of x for which the value of ____ is falsehood. Here the blank is to be filled by a singulary propositional form containing x as a free variable, the same one at all four places. Or if as a special case we fill the blank with a sentence, then (x)____ is true if and only if ____ is true. (The meaning in case the blank is filled by a propositional form containing other variables besides x as free variables now follows by the discussion of variables in §02, and may be supplied by the reader.)

Likewise the existential quantifier is a simple operator for which we shall use the notation $(\exists\ )$, filling the blank space with the operator variable and prefixing the whole to the operand. To take the particular operator variable x as an example, $(\exists x)$____ is true if the value of ____ is truth for at least one value of x, and $(\exists x)$____ is false if the value of ____ is falsehood for all values of x. Here again the blank is to be filled by a singulary propositional form containing x as a free variable. Or if as a special case we fill the blank with a sentence, then $(\exists x)$____ is true if and only if ____ is true.

In words, the notations "(x)" and "$(\exists x)$" may be read respectively as "for all x" (or "for every x") and "there is an x such that."

To illustrate the use of the universal and existential quantifiers, and in particular their iterated application, consider the binary propositional form,

$$[xy > 0],$$

where x and y are real variables, i.e., variables whose range is the real numbers. This form expresses about two real numbers x and y that their product is positive, and thus it comes to express a particular proposition as soon as values are given to x and y. If we apply to it the existential quantifier with y as operator variable, we obtain the singulary propositional form,

$$(\exists y)[xy > 0],$$

or as we may also write it, using the device (which we shall find frequently convenient later) of writing a heavy dot to stand for a bracket extending, from the place where the dot occurs, forward,

$$(\exists y) \blacksquare\, xy > 0.$$

This singulary form expresses about a real number x that there is some real number with which its product is positive; and it comes to express a particular proposition as soon as a value is given to x. If we apply to it the universal quantifier with x as operator variable, we obtain the sentence,

$$(x)(\exists y) \centerdot xy > 0.$$

This sentence expresses the proposition that for every real number there is some real number such that the product of the two is positive. It must be distinguished from the sentence,

$$(\exists y)(x) \centerdot xy > 0,$$

expressing the proposition that there is a real number whose product with every real number is positive, though it happens that both are false.[101] To bring out more sharply the difference which is made by the different order of the quantifiers, let us replace product by sum and consider the two sentences:

$$(x)(\exists y) \centerdot x + y > 0$$
$$(\exists y)(x) \centerdot x + y > 0$$

Of these sentences, the first one is true and the second one false.[102]

It should be informally clear to the reader that not both the universal and the existential quantifier are actually necessary in a formalized language, if negation is available. For it would be possible, in place of $(\exists x)$____, to write always $\sim(x)\sim$____; or alternatively, in place of (x)____, to write always $\sim(\exists x)\sim$____. And of course likewise with any other variable in place of the particular variable x.

In most treatments the universal and existential quantifiers, one or both,

---

[101]The single counterexample, of the value 0 for x, is of course sufficient to render the first sentence false.

The reader is warned against saying that the sentence $(x)(\exists y) \centerdot xy > 0$ is "nearly always true" or that it is "true with one exception" or the like. These expressions are appropriate rather to the propositional form $(\exists y) \centerdot xy > 0$, and of the sentence it must be said simply that it is false.

[102]A somewhat more complex example of the difference made by the order in which the quantifiers are applied is found in the familiar distinction between continuity and uniform continuity. Using x and y as variables whose range is the real numbers, and ε and δ as variables whose range is the positive real numbers, we may express as follows that the real function f is continuous, on the class F of real numbers (assumed to be an open or a closed interval):

$$(y)(\varepsilon)(\exists\delta)(x) \centerdot F(y) \supset \centerdot F(x) \supset \centerdot |x - y| < \delta \supset \centerdot |f(x) - f(y)| < \varepsilon$$

And we may express as follows that f is uniformly continuous on F:

$$(\varepsilon)(\exists\delta)(x)(y) \centerdot F(y) \supset \centerdot F(x) \supset \centerdot |x - y| < \delta \supset \centerdot |f(x) - f(y)| < \varepsilon$$

To avoid complications that are not relevant to the point being illustrated, we have here assumed not only that the class F is an open or closed interval but also that the range of the function f is all real numbers. (A function with more restricted range may always have its range extended by some arbitrary assignment of values; and indeed it is a common simplifying device in the construction of a formalized language to restrict attention to functions having certain standard ranges (cf. footnote 19).)

are made fundamental, notations being provided for them directly in setting up a formalized language, and other quantifiers are explained in terms of them (in a way similar to that in which, as we have just seen in the preceding paragraph, the universal and existential quantifiers may be explained, either one in terms of the other). No definite or compelling reason can be given for such a preference of these two quantifiers above others that might equally be made fundamental. But it is often convenient.

The application of one or more quantifiers to an operand (especially universal and existential quantifiers) is spoken of as *quantification.*[103]

Another quantifier is a singulary-binary quantifier for which we shall use the notation [____ $\supset$ ____], with the operands in the two blanks, and the operator variable as a subscript after the sign $\supset$. It may be explained by saying that [____ $\supset_x$ ____] is to mean the same as (x)[____ $\supset$ ____], the two blanks being filled with two propositional forms or sentences, the same two in each case (and in the same order); and of course likewise with any other variable in place of the particular variable x. The name *formal implication*[104] is given to this quantifier—or to the associated binary propositional function, i.e., to an appropriate one of the two associated functions of (say) the form $[F(u) \supset_u G(u)]$, where u is a variable with some assigned range, and F and G are variables whose range is all classes (singulary propositional functions) that have a range coinciding with the range of u.

Another quantifier is that which (or its associated propositional function) is called *formal equivalence.*[104] For this we shall use the notation [____ $\equiv$ ____], with the two operands in the two blanks, and the operator variable as a subscript after the sign $\equiv$. It may be explained by saying that [____ $\equiv_x$ ____] is to mean the same as (x)[____ $\equiv$ ____], the two blanks being filled in each case with the two operands in order; and of course likewise with any other variable in place of x.

We shall also make use of quantifiers similar in character to those just explained but having two or more operator variables. These (or their associated propositional functions) we call *binary formal implication, binary formal equivalence, ternary formal implication*, etc. E.g., binary formal implication may be explained by saying that [____ $\supset_{xy}$ ____] is to mean the

[103]The use of quantifiers originated with Frege in 1879. And independently of Frege the same idea was introduced somewhat later by Mitchell and Peirce. (See the historical account in §49.)

[104]The names *formal implication* and *formal equivalence* are those used by Whitehead and Russell in *Principia Mathematica*, and have become sufficiently well established that it seems best not to change them—though the adjective *formal* is perhaps not very well chosen, and must not be understood here in the same sense that we shall give it elsewhere.

same as $(x)(y)[\_\_\_\_ \supset \_\_\_\_]$, the two blanks being filled in each case with the two operands in order; and likewise with any two distinct variables in place of x and y as operator variables. Similarly binary formal equivalence $[\_\_\_\_ \equiv_{xy} \_\_\_\_]$, ternary formal implication $[\_\_\_\_ \supset_{xyz} \_\_\_\_]$, and so on.[105]

Besides the assertion of a sentence, as contemplated in §04, it is usual also to allow assertion of a propositional form, and to treat such an assertion as a particular fixed assertion (in spite of the presence of free variables in the expression asserted). This is common especially in mathematical contexts; where, for instance, the assertion of the equation $\sin(x + 2\pi) = \sin x$ may

---

[105]With the aid of the notations that have now been explained, we may return to §00 and rewrite the examples I–IV of that section as they might appear in some appropriate formalized language.

For this purpose let a and b be variables whose range is human beings. Let v be a variable whose range is words (taking, let us say for definiteness, any finite sequence of letters of the English alphabet as a word). Let B denote the relation of being a brother of. Let S denote the relation of having as surname. Let ϱ and σ denote the human beings Richard and Stanley respectively, and let τ denote the word "Thompson." Then the three premisses and the conclusion of I may be expressed as follows:

$$B(a, b) \supset_{ab} . S(a, v) \equiv_v S(b, v)$$
$$B(\varrho, \sigma)$$
$$S(\sigma, \tau)$$
$$S(\varrho, \tau)$$

Further, let z and w be variables whose range is complex numbers, and x a variable whose range is real numbers. Let R denote the relation of having real positive ratio, and let A denote the relation of having as amplitude. Then the premisses and conclusion of II may be expressed as follows:

$$R(z, w) \supset_{zw} . A(z, x) \equiv_x A(w, x)$$
$$R(i - \sqrt{3}/3, \omega)$$
$$A(\omega, 2\pi/3)$$
$$A(i - \sqrt{3}/3, 2\pi/3)$$

Here it is obvious that the relation of having real positive ratio is capable of being analyzed, so that instead of $R(z, w)$ we might have written, e.g.:

$$(\exists x)[x > 0][z = xw]$$

Likewise the relation of having as amplitude or (in I) the relation of being a brother of might have received some analysis. But these analyses are not relevant to the validity of the reasoning in these particular examples. And they are, moreover, in no way final or absolute; e.g., instead of analyzing the relation of having real positive ratio, we might with equal right take it as fundamental and analyze instead the relation of being greater than, in such a way that, in place of $x > y$ would be written $R(x - y, 1)$.

In the same way, for III and IV, we make no analysis of the singular propositional functions of having a portrait seen by me, of having assassinated Abraham Lincoln, and of having invented the wheeled vehicle, but let them be denoted just by P, L and W respectively. Then if β denotes John Wilkes Booth, the premisses and conclusion of III may be expressed thus:

$$P(\beta) \qquad L(\beta) \qquad (\exists a)[P(a)L(a)]$$

And the premisses and fallacious conclusion of IV thus:

$$(\exists a)P(a) \qquad (\exists a)W(a) \qquad (\exists a)[P(a)W(a)]$$

When so rewritten, the false appearance of analogy between III and IV disappears. It was due to the logically irregular feature of English grammar by which "somebody" is construed as a substantive.

be used as a means to assert this for all real numbers x; or the assertion of the inequality $x^2 + y^2 \geqq 2xy$ may be used as a means to assert that for any real numbers x and y the sum of the squares is greater than or equal to twice the product.

It is clear that, in a formalized language, if universal quantification is available, it is unnecessary to allow the assertion of expressions containing free variables. E.g., the assertion of the propositional form

$$x^2 + y^2 \geqq 2xy$$

could be replaced by assertion of the sentence

$$(x)(y) \mathbin{\blacksquare} x^2 + y^2 \geqq 2xy.$$

But on the other hand it is not possible to dispense with quantifiers in a formalized language merely by allowing the assertion of propositional forms, because, e.g., such assertions as that of

$$\sim(x)(y) \mathbin{\blacksquare} \sin(x + y) = \sin x + \sin y,$$ [106]

or that of

$$(y)[|x| \leqq |y|] \supset_x \mathbin{\blacksquare} x = 0,$$

could not be reproduced.

Consequently it has been urged with some force that the device of asserting propositional forms constitutes an unnecessary duplication of ways of expressing the same thing, and ought to be eliminated from a formalized language.[107] Nevertheless it appears that the retention of this device often facilitates the setting up of a formalized language by simplifying certain details; and it also renders more natural and obvious the separation of such restricted systems as propositional calculus (Chapter I) or functional calculus of first order (Chapter III) out from more comprehensive systems of which they are part. In the development which follows we shall therefore make free use of the assertion of propositional forms. However, in the case of such systems as functional calculus of order ω (Chapter VI) or Zermelo set theory (Chapter XI), after a first treatment employing the device in question we shall sketch briefly a reformulation that avoids it.

---

[106]This assertion (which is correct, and must sometimes be made to beginners in trigonometry) is of course to be distinguished from the different (and erroneous) assertion of

$$\sim \mathbin{\blacksquare} \sin(x + y) = \sin x + \sin y.$$

[107]The proposal to do this was made by Russell in his introduction to the second edition of *Principia Mathematica* (1925). The elimination was actually carried out by Quine in his *Mathematical Logic* (1940), and simplifications of Quine's method were effected in papers by F. B. Fitch and by G. D. W. Berry in *The Journal of Symbolic Logic* (vol. 6 (1941), pp. 18–22, 23–27).

07. The logistic method. In order to set up a formalized language we must of course make use of a language already known to us, say English or some portion of the English language, stating in that language the vocabulary and rules of the formalized language. This procedure is analogous to that familiar to the reader in language study—as, e.g., in the use of a Latin grammar written in English[108]—but differs in the precision with which the rules are stated, in the avoidance of irregularities and exceptions, and in the leading idea that the rules of the language embody a theory or system of logical analysis (cf. §00).

This device of employing one language in order to talk about another is one for which we shall have frequent occasion not only in setting up formalized languages but also in making theoretical statements as to what can be done in a formalized language, our interest in formalized languages being less often in their actual and practical use as languages than in the general theory of such use and in its possibilities in principle. Whenever we employ a language in order to talk about some language (itself or another[109]), we shall call the latter language the *object language*, and we shall call the former the *meta-language*.[110]

In setting up a formalized language we first employ as meta-language a certain portion of English. We shall not attempt to delimit precisely this portion of the English language, but describe it approximately by saying that it is just sufficient to enable us to give general directions for the manip-

[108]It is worth remark in passing that this same procedure also enters into the learning of a first language, being a necessary supplement to the method of learning by example and imitation. Some part of the language must first be learned approximately by the method of example and imitation; then this imprecisely known part of the language is applied in order to state rules of the language (and perhaps to correct initial misconceptions); then the known part of the language may be extended by further learning by example and imitation, and so on in alternate steps, until some precision in knowledge of the language is reached.

There is no reason in principle why a first language, learned in this way, should not be one of the formalized languages of this book, instead of one of the natural languages. (But of course there is the practical reason that these formalized languages are ill adapted to purposes of facility of communication.)

[109]The employment of a language to talk about that same language is clearly not appropriate as a method of setting up a formalized language. But once set up, a formalized language with adequate means of expression may be capable of use in order to talk about that language itself; and in particular the very setting up of the language may afterwards be capable of restatement in that language. Thus it may happen that object language and meta-language are the same, a situation which it will be important later to take into account.

[110]The distinction is due to David Hilbert, who, however, speaks of "Mathematik" (mathematics) and "Metamathematik" (metamathematics) rather than "object language" and "meta-language." The latter terms, or analogues of them in Polish or German, are due to Alfred Tarski and Rudolf Carnap, by whom especially (see footnotes 131, 140) the subjects of *syntax* and *semantics* have been developed.

ulation of concrete physical objects (each instance or occurrence of one of the symbols of the language being such a concrete physical object, e.g., a mass of ink adhering to a bit of paper). It is thus a language which deals with matters of everyday human experience, going beyond such matters only in that no finite upper limit is imposed on the number of objects that may be involved in any particular case, or on the time that may be required for their manipulation according to instructions. Those additional portions of English are excluded which would be used in order to treat of infinite classes or of various like abstract objects which are an essential part of the subject matter of mathematics.

Our procedure is not to define the new language merely by means of translations of its expressions (sentences, names, forms) into corresponding English expressions, because in this way it would hardly be possible to avoid carrying over into the new language the logically unsatisfactory features of the English language. Rather, we begin by setting up, in abstraction from all considerations of meaning, the purely formal part of the language, so obtaining an uninterpreted calculus or *logistic system*. In detail, this is done as follows.

The vocabulary of the language is specified by listing the single symbols which are to be used.[111] These are called the *primitive symbols* of the language,[112] and are to be regarded as indivisible in the double sense that (A) in

---

[111]Notice that we use the term "language" in such a sense that a given language has a given and uniquely determined vocabulary. E.g., the introduction of one additional symbol into the vocabulary is sufficient to produce a new and different language. (Thus the English of 1849 is not the same language as the English of 1949, though it is convenient to call them by the same name, and to distinguish, by specifying the date, only in cases where the distinction is essential.)

[112]The fourfold classification of the primitive notations of a formalized language into constants, variables, connectives, and operators is due in substance to J. v. Neumann in the *Mathematische Zeitschrift*, vol. 26 (1927), see pp. 4–6. He there adds a fifth category, composed of association-showing symbols such as parentheses and brackets. Our terms "connective" and "operator" correspond to his "Operation" and "Abstraktion" respectively.

Though there is a possibility of notations not falling in any of von Neumann's categories, such have seldom been used, and for nearly all formalized languages that have actually been proposed the von Neumann classification of primitive notations suffices. Many formalized languages have primitive notations of all four (or five) kinds, but it does not appear that this is indispensable, even for a language intended to be adequate for the expression of mathematical ideas generally.

As an interesting example of a (conceivable) notation not in any of the von Neumann categories, we mention the question of a notation by means of which from a name of a class would be formed an expression playing the role of a variable with that class as its range. Provision might perhaps be made for the formation from any class name of an infinite number of expressions playing the roles of different variables with the class as their range. But these expressions would have to differ from variables in the sense of §02 not only in being composite expressions rather than single symbols but also in the

setting up the language no use is made of any division of them into parts and (B) any finite linear sequence of primitive symbols can be regarded *in only one way* as such a sequence of primitive symbols.[113] A finite linear sequence of primitive symbols is called a *formula*. And among the formulas, rules are given by which certain ones are designated as *well-formed formulas* (with the intention, roughly speaking, that only the well-formed formulas are to be regarded as being genuinely expressions of the language).[114] Then certain among the well-formed formulas are laid down as *axioms*. And finally (primitive) *rules of inference* (or *rules of procedure*) are laid down, rules according to which, from appropriate well-formed formulas as *premisses*, a well-formed formula is *immediately inferred*[115] as *conclusion*. (So long as we are dealing only with a logistic system that remains uninterpreted, the terms *premiss, immediately infer, conclusion* have only such meaning as is conferred upon them by the rules of inference themselves.)

A finite sequence of one or more well-formed formulas is called a *proof* if each of the well-formed formulas in the sequence either is an axiom or is immediately inferred from preceding well-formed formulas in the sequence by means of one of the rules of inference. A proof is called a proof *of* the last well-formed formula in the sequence, and the *theorems* of the logistic system

---

possibility that the range might be empty. A language containing such a notation has never been set up and studied in detail and it is therefore not certain just what is feasible. (A suggestion which seems to be in this direction was made by Beppo Levi in *Universidad Nacional de Tucumán, Revista*, ser. A vol. 3 no. 1 (1942), pp. 13–78.)

The use in Chapter X of variables with subscripts indicating the range of the variable (the type) is not an example of a notation of the kind just described. For the variable, letter and subscript together, is always treated as a single primitive symbol.

[113]In practice, condition (B) usually makes no difficulty. Though the (written) symbols adopted as primitive symbols may not all consist of a single connected piece, it is ordinarily possible to satisfy (B), if not otherwise, by providing that a sequence of primitive symbols shall be written with spaces between the primitive symbols of fixed width and wider than the space at any place within a primitive symbol.

The necessity for (B), and its possible failure, were brought out by a criticism by Stanisław Leśniewski against the paper of von Neumann cited in the preceding footnote. See von Neumann's reply in *Fundamenta Mathematicae*, vol. 17 (1931), pp. 331–334, and Leśniewski's final word in the matter in an offprint published in 1938 as from *Collectanea Logica*, vol. 1 (cf. *The Journal of Symbolic Logic*, vol. 5, p. 83).

[114]The restriction to one dimension in combining the primitive symbols into expressions of the language is convenient, and non-essential. Two-dimensional arrangements are of course possible, and are familiar especially in mathematical notations, but they may always be reduced to one dimension by a change of notation. In particular the notation of Frege's *Begriffsschrift* relies heavily on a two-dimensional arrangement; but because of the difficulty of printing it this notation was never adopted by any one else and has long since been replaced by a one-dimensional equivalent.

[115]No reference to the so-called immediate inferences of traditional logic is intended. We term the inferences *immediate* in the sense of requiring only one application of a rule of inference—not in the traditional sense of (among other things) having only one premiss.

are those well-formed formulas of which proofs exist.[116] As a special case, each axiom of the system is a theorem, that finite sequence being a proof which consists of a single well-formed formula, the axiom alone.

The scheme just described—viz. the primitive symbols of a logistic system, the rules by which certain formulas are determined as well-formed (following Carnap let us call them the *formation rules* of the system), the rules of inference, and the axioms of the system—is called the *primitive basis* of the logistic system.[117]

In defining a logistic system by laying down a primitive basis, we employ as meta-language the restricted portion of English described above. In addition to this restriction, or perhaps better as part of it, we impose requirements of *effectiveness* as follows: (I) the specification of the primitive symbols shall be effective in the sense that there is a method by which, whenever a symbol is given, it can always be determined effectively whether or not it is one of the primitive symbols; (II) the definition of a well-formed formula

---

[116]Following Carnap and others, we use the term "language" in such a sense that for any given language there is one fixed notion of a proof in that language. Thus the introduction of one additional axiom or rule of inference, or a change in an axiom or rule of inference, is sufficient to produce a new and different language.

(An alternative, which might be thought to accord better with the everyday use of the word "language," would be to define a "language" as consisting of primitive symbols and a definition of well-formed formula, together with an *interpretation* (see below), and to take the axioms and rules of inference as constituting a "logic" for the language. Instead of speaking of an interpretation as *sound* or *unsound* for a logistic system (see below), we would then speak of a logic as being sound or unsound for a language. Indeed this alternative may have some considerations in its favor. But we reject it here, partly because of reluctance to change a terminology already fairly well established, partly because the alternative terminology leads to a twofold division in each of the subjects of syntax and semantics (§§08, 09)—according as they treat of the object language alone or of the object language together with a logic for it — which, especially in the case of semantics, seems unnatural, and of little use so far as can now be seen.)

[117]Besides these minimum essentials, the primitive basis may also include other notions introduced in order to use them in defining a well-formed formula or in stating the rules of inference. In particular the primitive symbols may be divided in some way into different categories: e.g., they may be classified as *primitive constants*, *variables*, and *improper symbols*, or various categories may be distinguished of primitive constants, of variables, or of improper symbols. The variables and the primitive constants together are usually called *proper symbols*. Rules may be given for distinguishing an occurrence of a variable in a well-formed formula as being a *free occurrence* or a *bound occurrence*, well-formed formulas being then classified as *forms* or *constants* according as they do or do not contain a free occurrence of a variable. Also rules may be given for distinguishing certain of the forms as *propositional forms*, and certain of the constants as *sentences*. In doing all this, the terminology often is so selected that, when the logistic system becomes a language by adoption of one of the intended principal interpretations (see below), the terms *primitive constant*, *variable*, *improper symbol*, *proper symbol*, *free*, *bound*, *form*, *constant*, *propositional form*, *sentence* come to have meanings in accord with the informal semantical explanations of §§02–06.

The *primitive basis* of a formalized language, or interpreted logistic system, is obtained by adding the semantical rules (see below) to the primitive basis of the logistic system.

shall be effective in the sense that there is a method by which, whenever a formula is given, it can always be determined effectively whether or not it is well-formed; (III) the specification of the axioms shall be effective in the sense that there is a method by which, whenever a well-formed formula is given, it can always be determined effectively whether or not it is one of the axioms; (IV) the rules of inference, taken together, shall be effective in the strong sense that there is a method by which, whenever a proposed immediate inference is given of one well-formed formula as conclusion from others as premisses, it can always be determined effectively whether or not this proposed immediate inference is in accordance with the rules of inference.

(From these requirements it follows that the notion of a proof is effective in the sense that there is a method by which, whenever a finite sequence of well-formed formulas is given, it can always be determined effectively whether or not it is a proof. But the notion of a theorem is not necessarily effective in the sense of existence of a method by which, whenever a well-formed formula is given, it can always be determined whether or not it is a theorem—for there may be no certain method by which we can always either find a proof or determine that none exists. This last is a point to which we shall return later.)

As to requirement (I), we suppose that we are able always to determine about two given symbol-occurrences whether or not they are occurrences of the same symbol (thus ruling out by assumption such difficulties as that of illegibility). Therefore, if the number of primitive symbols is finite, the requirement may be satisfied just by giving the complete list of primitive symbols, written out in full. Frequently, however, the number of primitive symbols is infinite. In particular, if there are variables, it is desirable that there should be an infinite number of different variables of each kind because, although in any one well-formed formula the number of different variables is always finite, there is hardly a way to determine a finite upper limit of the number of different variables that may be required for some particular purpose in the actual use of the logistic system. When the number of primitive symbols is infinite, the list cannot be written out in full, but the primitive symbols must rather be fixed in some way by a statement of finite length in the meta-language. And this statement must be such as to conform to (I).

A like remark applies to (III). If the number of axioms is finite, the requirement can be satisfied by writing them out in full. Otherwise the axioms must be specified in some less direct way by means of a statement of finite length in the meta-language, and this must be such as to conform to (III). It may be thought more elegant or otherwise more satisfactory that the number of axioms be finite; but we shall see that it is sometimes convenient to make use of an infinite number of axioms, and no conclusive objections appear to doing so if requirements of effectiveness are obeyed.

We have assumed the reader's understanding of the general notion of effectiveness, and indeed it must be considered as an informally familiar mathematical notion, since it is involved in mathematical problems of a frequently occurring kind, namely, problems to find a method of computation, i.e., a method by which to determine a number, or other thing, effectively.[118] We shall not try to give here a rigorous definition of effectiveness, the informal notion being sufficient to enable us, in cases we shall meet, to distinguish given methods as effective or non-effective.[119]

The requirements of effectiveness are (of course) not meant in the sense that a structure which is analogous to a logistic system except that it fails to satisfy these requirements may not be useful for some purposes or that it is forbidden to consider such—but only that a structure of this kind is unsuitable for use or interpretation as a language. For, however indefinite or imprecisely fixed the common idea of a language may be, it is at least fundamental to it that a language shall serve the purpose of communication. And to the extent that requirements of effectiveness fail, the purpose of communication is defeated.

Consider, in particular, the situation which arises if the definition of well-

---

[118]A well-known example from topology is the problem (still unsolved even for elementary manifolds of dimensionalities above 2) to find a method of calculating about any two closed simplicial manifolds, given by means of a set of incidence relations, whether or not they are homeomorphic—or, as it is often phrased, the problem to find a complete classification of such manifolds, or to find a complete set of invariants.

As another example, Euclid's algorithm, in the domain of rational integers, or in certain other integral domains, provides an effective method of calculating for any two elements of the domain their greatest common divisor (or highest common factor).

In general, an effective method of calculating, especially if it consists of a sequence of steps with later steps depending on results of earlier ones, is called an *algorithm*. (This is the long established spelling of this word, and should be preserved in spite of any considerations of etymology.)

[119]For a discussion of the question and proposal of a rigorous definition see a paper by the present writer in the *American Journal of Mathematics*, vol. 58 (1936), pp. 345–363, especially §7 thereof. The notion of effectiveness may also be described by saying that an effective method of computation, or algorithm, is one for which it would be possible to build a computing machine. This idea is developed into a rigorous definition by A. M. Turing in the *Proceedings of the London Mathematical Society*, vol. 42 (1936–1937), pp. 230–265 (and vol. 43 (1937), pp. 544–546). See further: S. C. Kleene in the *Mathematische Annalen*, vol. 112 (1936), pp. 727–742; E. L. Post in *The Journal of Symbolic Logic*, vol. 1 (1936), pp. 103–105; A. M. Turing in *The Journal of Symbolic Logic*, vol. 2 (1937), pp. 153–163; Hilbert and Bernays, *Grundlagen der Mathematik*, vol. 2 (1939), Supplement II.

formedness is non-effective. There is then no certain means by which, when an alleged expression of the language is uttered (spoken or written), say as an asserted sentence, the auditor (hearer or reader) may determine whether it is well-formed, and thus whether any actual assertion has been made.[120] Therefore the auditor may fairly demand a proof that the utterance is well-formed, and until such proof is provided may refuse to treat it as constituting an assertion. This proof, which must be added to the original utterance in order to establish its status, ought to be regarded, it seems, as part of the utterance, and the definition of well-formedness ought to be modified to provide this, or its equivalent. When such modification is made, no doubt the non-effectiveness of the definition will disappear; otherwise it would be open to the auditor to make further demand for proof of well-formedness.

Again, consider the situation which arises if the notion of a proof is non-effective. There is then no certain means by which, when a sequence of formulas has been put forward as a proof, the auditor may determine whether it is in fact a proof. Therefore he may fairly demand a proof, in any given case, that the sequence of formulas put forward is a proof; and until this supplementary proof is provided, he may refuse to be convinced that the alleged theorem is proved. This supplementary proof ought to be regarded, it seems, as part of the whole proof of the theorem, and the primitive basis of the logistic system ought to be so modified as to provide this, or its equivalent.[121] Indeed it is essential to the idea of a proof that, to any one who admits the presuppositions on which it is based, a proof carries final

---

[120]To say that an assertion has been made if there is a meaning evades the issue unless an effective criterion is provided for the presence of meaning. An understanding of the language, however reached, must include effective ability to recognize meaningfulness (in some appropriate sense), and in the purely formal aspect of the language, the logistic system, this appears as an effective criterion of well-formedness.

[121]Perhaps at first sight it will be thought that the proof as so modified might consist of something more than merely a sequence of well-formed formulas. For instance there might be put in at various places indications in the meta-language as to which rule of inference justifies the inclusion of a particular formula as immediately inferred from preceding formulas, or as to which preceding formulas are the premisses of the immediate inference.

But as a matter of fact we consider this inadmissible. For our program is to express proofs (as well as theorems) in a fully formalized object language, and as long as any part of the proof remains in an unformalized meta-language the logical analysis must be held to be incomplete. A statement in the meta-language, e.g., that a particular formula is immediately inferred from particular preceding formulas—if it is not superfluous and therefore simply omissible—must always be replaced in some way by one or more sentences of the object language.

Though we use a meta-language to set up the object language, we require that, once set up, the object language shall be an independent language capable, without continued support and supplementation from the meta-language, of expressing those things for which it was designed.

conviction. And the requirements of effectiveness (I)–(IV) may be thought of as intended just to preserve this essential characteristic of proof.

After setting up the logistic system as described, we still do not have a formalized language until an *interpretation* is provided. This will require a more extensive meta-language than the restricted portion of English used in setting up the logistic system. However, it will proceed not by translations of the well-formed formulas into English phrases but rather by *semantical rules* which, in general, *use* rather than *mention* English phrases (cf. §08), and which shall prescribe for every well-formed formula either how it denotes[122] (so making it a proper name in the sense of §01) or else how it has values[122] (so making it a form in the sense of §02).

In view of our postulation of two truth-values (§04), we impose the requirement that the semantical rules, if they are to be said to provide an interpretation, must be such that the axioms denote truth-values (if they are names) or have always truth-values as values (if they are forms), and the same must hold of the conclusion of any immediate inference if it holds of the premisses. In using the formalized language, only those well-formed formulas shall be capable of being asserted which denote truth-values (if

---

[122]Because of the possibility of misunderstanding, we avoid the wordings "what it denotes" and "what values it has."

For example, in one of the logistic systems of Chapter X we may find a well-formed formula which, under a principal interpretation of the system, is interpreted as denoting: the greatest positive integer n such that $1 + n^r$ is prime, r being chosen as the least even positive integer corresponding to which there is such a greatest positive integer n. Thus the semantical rules do in a sense determine what this formula denotes, but the remoteness of this determination is measured by the difficulty of the mathematical problem which must be solved in order to identify in some more familiar manner the positive integer which the formula denotes, or even to say whether or not the formula denotes 1.

Again in the logistic system F^{1h} of Chapter III (or A^0 of Chapter V) taken with its principal interpretation, there is a well-formed formula which, according to the semantical rules, denotes the truth-value thereof that every even number greater than 2 is the sum of two prime numbers. To say that the semantical rules determine what this formula denotes seems to anticipate the solution of a famous problem, and it may be better to think of the rules as determining indirectly what the formula expresses.

In assigning how (rather than what) a name denotes we are in effect fixing its sense, and in assigning how a form has values we fix the correspondence of sense values of the form (see footnote 27) to concepts of values of its variables. (This statement of the matter will be sufficiently precise for our present purposes, though it remains vague to the extent that we have left the meaning of "sense" uncertain—see footnotes 15, 37.)

It will be seen in particular examples below (such as rules a–g of §10, or rules a–f of §30, or rules α–ζ of §30) that in most of our semantical rules the explicit assertion is that certain well-formed formulas, usually on certain conditions, are to denote certain things or to have certain values. However, as just explained, this explicit assertion is so chosen as to give implicitly also the sense or the sense values. No doubt a fuller treatment of semantics must have additional rules stating the sense or the sense values explicitly, but this would take us into territory still unexplored.

they are names) or have always truth-values as values (if they are forms); and only those shall be capable of being rightly asserted which denote truth (if they are names) or have always the value truth (if they are forms). Since it is intended that proof of a theorem shall justify its assertion, we call an interpretation of a logistic system *sound* if, under it, all the axioms either denote truth or have always the value truth, and if further the same thing holds of the conclusion of any immediate inference if it holds of the premisses. In the contrary case we call the interpretation *unsound*. A formalized language is called sound or unsound according as the interpretation by which it is obtained from a logistic system is sound or unsound. And an unsound interpretation or an unsound language is to be rejected.

(The requirements, and the definition of soundness, in the foregoing paragraph are based on two truth-values. They are satisfactory for every formalized language which will receive substantial consideration in this book. But they must be modified correspondingly, in case the scheme of two truth-values is modified—cf. the remark in §19.)

The semantical rules must in the first instance be stated in a presupposed and therefore unformalized meta-language, here taken to be ordinary English. Subsequently, for their more exact study, we may formalize the meta-language (using a presupposed meta-meta-language and following the method already described for formalizing the object language) and restate the semantical rules in this formalized language. (This leads to the subject of *semantics* (§09).)

As a condition of rigor, we require that the proof of a theorem (of the object language) shall make no reference to or use of any interpretation, but shall proceed purely by the rules of the logistic system, i.e., shall be a *proof* in the sense defined above for logistic systems. Motivation for this is threefold, three rather different approaches issuing in the same criterion. In the first place this may be considered a more precise formulation of the traditional distinction between form and matter (§00) and of the principle that the validity of an argument depends only on the form—the form of a proof in a logistic system being thought of as something common to its meanings under various interpretations of the logistic system. In the second place this represents the standard mathematical requirement of rigor that a proof must proceed purely from the axioms without use of anything (however supposedly obvious) which is not stated in the axioms; but this requirement is here modified and extended as follows: that a proof must proceed purely from the axioms by the rules of inference, without use of anything not stated in the axioms or any method of inference not validated by the rules. Thirdly there is the motivation that the logistic system is relatively secure and definite, as compared to interpretations which we may wish to adopt, since it is based on a portion of English as meta-language so elementary and restricted that its essential reliability can hardly be doubted if mathematics is to be possible at all.

It is also important that a proof which satisfies our foregoing condition of rigor must then hold under any interpretation of the logistic system, so that there is a resulting economy in proving many things under one process.[123] The extent of the economy is just this, that proofs identical in form but different in matter need not be repeated indefinitely but may be summarized once for all.[124]

Though retaining our freedom to employ any interpretation that may be found useful, we shall indicate, for logistic systems set up in the following chapters, one or more interpretations which we have especially in mind for the system and which shall be called the *principal interpretations*.

The subject of formal logic, when treated by the method of setting up a formalized language, is called *symbolic logic*, or *mathematical logic*, or *logistic*.[125] The method itself we shall call the *logistic method*.

---

[123]This remark has now long been familiar in connection with the axiomatic method in mathematics (see below).

[124]The summarizing of a proof according to its form may indeed be represented to a certain extent, by the use of variables, within one particular formalized language. But, because of restricted ranges of the variables, such summarizing is less comprehensive in its scope than is obtained by formalizing in a logistic system whose interpretation is left open.

The procedure of formalizing a proof in a logistic system and then employing the formalized proof under various different interpretations of the system may be thought of as a mere device for brevity and convenience of presentation, since it would be possible instead to repeat the proof in full each time it were used with a new interpretation. From this point of view such use of the meta-language may be allowed as being in principle dispensable and therefore not violating the demand (footnote 121) for an independent object language.

(If on the other hand we wish to deal rigorously with the notion of logical form of proofs, this must be in a particular formalized language, namely a formalized meta-language of the language of the proofs. Under the program of §02 each variable of this meta-language will have a fixed range assigned in advance, according, perhaps, with the theory of types. And the notion of form which is dealt with must therefore be correspondingly restricted, it would seem, to proofs of a fixed class, taking no account of sameness of form between proofs of this class and others (in the same or a different language). Presumably our informal references to logical form in the text are to be modified in this way before they can be made rigorous—cf. §09.)

[125]The writer prefers the term "mathematical logic," understood as meaning logic treated by the mathematical method, especially the formal axiomatic or logistic method. But both this term and the term "symbolic logic" are often applied also to logic as treated by a less fully formalized mathematical method, in particular to the "algebra of logic," which had its beginning in the publications of George Boole and Augustus De Morgan in 1847, and received a comprehensive treatment in Ernst Schröder's *Vorlesungen über die Algebra der Logik* (1890–1905). The term "logistic" is more defi-

Familiar in mathematics is the *axiomatic method*, according to which a branch of mathematics begins with a list of *undefined terms* and a list of assumptions, or *postulates* involving these terms, and the theorems are to be derived from the postulates by the methods of formal logic.[126] If the last phrase is left unanalyzed, formal logic being presupposed as already known, we shall say that the development is by the *informal axiomatic method*.[127] And in the opposite case we shall speak of the *formal axiomatic method*.

The formal axiomatic method thus differs from the logistic method only in the following two ways:

(1) In the logistic system the primitive symbols are given in two categories: the *logical primitive symbols*, thought of as pertaining to the underlying logic, and the *undefined terms*, thought of as pertaining to the particular branch of mathematics. Correspondingly the axioms are divided into two categories: the *logical axioms*, which are well-formed formulas containing only logical primitive symbols, and the *postulates*,[128] which involve also the undefined terms and are thought of as determining the special branch of mathematics. The rules of inference, to accord with the usual conception of

---

nitely restricted to the method described in this section, and has also the advantage that it is more easily made an adjective. (Sometimes "logistic" has been used with special reference to the school of Russell or to the Frege-Russell doctrine that mathematics is a branch of logic—cf. footnote 545. But we shall follow the more common usage which attaches no such special meaning to this word.)

"Logica mathematica" and "logistica" were both used by G. W. v. Leibniz along with "calculus ratiocinator," and many other synonyms, for the calculus of reasoning which he proposed but never developed beyond some brief and inadequate (though significant) fragments. Boole used the expressions "mathematical analysis of logic," "mathematical theory of logic." "Mathematische Logik" was used by Schröder in 1877, "matématičéskaá logika" (Russian) by Platon Poretsky in 1884, "logica matematica" (Italian) by Giuseppe Peano in 1891. "Symbolic logic" seems to have been first used by John Venn (in *The Princeton Review*, 1880), though Boole speaks of "symbolical reasoning." The word "logistic" and its analogues in other languages originally meant the art of calculation or common arithmetic. Its modern use for mathematical logic dates from the International Congress of Philosophy of 1904, where it was proposed independently by Itelson, Lalande, and Couturat. Other terms found in the literature are "logischer Calcul" (Gottfried Ploucquet 1766), "algorithme logique" (G. F. Castillon 1805), "calculus of logic" (Boole 1847), "calculus of inference" (De Morgan 1847), "logique algorithmique" (J. R. L. Delboeuf 1876), "Logikkalkul" (Schröder 1877), "theoretische Logik" (Hilbert and Ackermann 1928). Also "Boole's logical algebra" (C. S. Peirce 1870), "logique algébrique de Boole" (Louis Liard 1877), "algebra of logic" (Alexander Macfarlane 1879, C. S. Peirce 1880).

[126]Accounts of the axiomatic method may of course be found in many mathematical textbooks and other publications. An especially good exposition is in the Introduction to Veblen and Young's *Projective Geometry*, vol. 1 (1910).

[127]This is the method of most mathematical treatises, which proceed axiomatically but are not specifically about logic—in particular of Veblen and Young (preceding footnote).

[128]The words "axiom" and "postulate" have been variously used, either as synonymous or with varying distinctions between them, by the present writer among others. In this book, however, the terminology here set forth will be followed closely.

the axiomatic method, must all be taken as belonging to the underlying logic. And, though they may make reference to particular undefined terms or to classes of primitive symbols which include undefined terms, they must not involve anything which, subjectively, we are unwilling to assign to the underlying logic rather than to the special branch of mathematics.[129]

(2) In the interpretation the semantical rules are given in two categories. Those of the first category fix those general aspects of the interpretation which may be assigned, or which we are willing to assign, to the underlying logic. And the rules of the second category determine the remainder of the interpretation. The consideration of different representations or interpretations of the system of postulates, in the sense of the informal axiomatic method, corresponds here to varying the semantical rules of the second category while those of the first category remain fixed.

08. Syntax. The study of the purely formal part of a formalized language in abstraction from the interpretation, i.e., of the logistic system, is called *syntax*, or, to distinguish it from the narrower sense of "syntax" as concerned with the formation rules alone,[130] *logical syntax*.[131] The meta-language used in order to study the logistic system in this way is called the *syntax language*.[131]

We shall distinguish between *elementary syntax* and *theoretical syntax*.

The elementary syntax of a language is concerned with setting up the logistic system and with the verification of particular well-formed formulas,

---

[129]Ordinarily, e.g., it would be allowed that the rules of inference should treat differently two undefined terms intended one to denote an individual and one to denote a class of individuals, or two undefined terms intended to denote a class of individuals and a relation between individuals; but not that the rules should treat differently two undefined terms intended both to denote a class of individuals. But no definitive controlling principle can be given.

The subjective and essentially arbitrary character of the distinction between what pertains to the underlying logic and what to the special branch of mathematics is illustrated by the uncertainty which sometimes arises, in treating a branch of mathematics by the informal axiomatic method, as to whether the sign of equality is to be considered as an undefined term (for which it is necessary to state postulates). Again it is illustrated by Zermelo's treatment of axiomatic set theory in his paper of 1908 (cf. Chapter XI) in which, following the informal axiomatic method, he introduces the relation ϵ of membership in a set as an undefined term, though this same relation is usually assigned to the underlying logic when a branch of mathematics is developed by the informal axiomatic method.

[130]Cf. footnote 116.

[131]The terminology is due to Carnap in his *Logische Syntax der Sprache* (1934), translated into English (with some additions) as *The Logical Syntax of Language* (1937). In connection with this book see also reviews of it by Saunders MacLane in the *Bulletin of the American Mathematical Society*, vol. 44 (1938), pp. 171–176, and by S. C. Kleene in *The Journal of Symbolic Logic*, vol. 4 (1939), pp. 82–87.

axioms, immediate inferences, and proofs as being such. The syntax language is the restricted portion of English which was described in the foregoing section, or a correspondingly restricted formalized meta-language, and the requirements of effectiveness, (I)–(IV), must be observed. The demonstration of derived rules and theorem schemata, in the sense of §§12, 33, and their application in particular cases are also considered to belong to elementary syntax, provided that the requirement of effectiveness holds which is explained in §12.

Theoretical syntax, on the other hand, is the general mathematical theory of a logistic system or systems and is concerned with all the consequences of their formal structure (in abstraction from the interpretation). There is no restriction imposed as to what is available in the syntax language, and requirements of effectiveness are or may be abandoned. Indeed the syntax language may be capable of expressing the whole of extant mathematics. But it may also sometimes be desirable to use a weaker syntax language in order to exhibit results as obtained on this weaker basis.

Like any branch of mathematics, theoretical syntax may, and ultimately must, be studied by the axiomatic method. Here the informal and the formal axiomatic method share the important advantage that the particular character of the symbols and formulas of the object language, as marks upon paper, sounds, or the like, is abstracted from, and the pure theory of the structure of the logistic system is developed. But the formal axiomatic method—the syntax language being itself formalized according to the program of §07, by employing a meta-meta-language—has the additional advantage of exhibiting more definitely the basis on which results are obtained, and of clarifying the way and the extent to which certain results may be obtained on a relatively weaker basis.

In this book we shall be concerned with the task of formalizing an object language, and theoretical syntax will be treated informally, presupposing in any connection such general knowledge of mathematics as is necessary for the work at hand. Thus we do not apply even the informal axiomatic method to our treatment of syntax. But the reader must always understand that syntactical discussions are carried out in a syntax language whose formalization is ultimately contemplated, and distinctions based upon such formalization may be relevant to the discussion.

In such informal development of syntax, we shall think of the syntax language as being a different language from the object language. But the possibility is important that a sufficiently adequate object language may be capable of expressing its own syntax, so that in this case the ultimate formalization of the syntax language may if desired consist in identifying it with the object language.[132]

We shall distinguish between theorems *of* the object language and theorems of the syntax language (which often are theorems *about* the object language) by calling the latter *syntactical theorems*. Though we demonstrate syntactical theorems informally, it is contemplated that the ultimate formalization of the syntax language shall make them theorems in the sense of §07, i.e., theorems of the syntax language in the same sense as that in which we speak of theorems of the object language.

We shall require, as belonging to the syntax language: first, names of the various symbols and formulas of the object language; and secondly, variables which have these symbols and formulas as their values. The former will be called *syntactical constants*, and the latter, *syntactical variables*.[133]

As syntactical variables we shall use the following: as variables whose range is the primitive symbols of the object language, bold Greek small letters ($\boldsymbol{\alpha}$, $\boldsymbol{\beta}$, $\boldsymbol{\gamma}$, etc.); as variables whose range is the primitive constants and variables of the object language—see footnote 117—bold roman small letters (**a**, **b**, **c**, etc.); as variables whose range is the formulas of the object language, bold Greek capitals ($\boldsymbol{\Gamma}$, $\boldsymbol{\Delta}$, etc.); and as variables whose range is the well-formed formulas of the object language, bold roman capitals (**A**, **B**, **C**, etc.). Wherever these bold letters are used in the following chapters the reader must bear in mind that they are not part of the symbolic apparatus of the object language but that they belong to the syntax language and serve the purpose of talking about the object language. In use of the object language as an independent language, bold letters do not appear (just as English words never appear in the pure text of a Latin author though they do appear in a Latin grammar written in English).

As a preliminary to explaining the device to which we resort for syntactical constants, it is desirable first to consider the situation in ordinary

[132]Cf. footnote 109. In particular the developments of Chapter VIII show that the logistic system of Chapter VII is capable of expressing its own syntax if given a suitable interpretation different from the principal interpretation of Chapter VII, namely, an interpretation in which the symbols and formulas of the logistic system itself are counted among the individuals, as well as all finite sequences of such formulas, and the functional constant S is given an appropriate (quite complicated) interpretation, details of which may be made out by following the scheme of Gödel numbers that is set forth in Chapter VIII.

[133]Given the apparatus of syntactical variables, we could actually avoid the use of syntactical constants by resorting to appropriate circumlocutions in cases where syntactical constants would otherwise seem to be demanded. Indeed the example of the preceding footnote illustrates this, as will become clear in connection with the cited chapters. But it is more natural and convenient, especially in an informal treatment of syntax, to allow free use of syntactical constants.

English, with no formalized object language specially in question. We must take into account the fact that English is not a formalized language and the consequent uncertainty as to what are its formation rules, rules of inference, and semantical rules, the contents of ordinary English grammars and dictionaries providing only some incomplete and rather vague approximations to such rules. But, with such reservations as this remark implies, we go on to consider the use of English in making syntactical statements about the English language itself.

Frequently found in practice is the use of English words *autonymously* (to adopt a terminology due to Carnap), i.e., as names of those same words.[134] Examples are such statements as "The second letter of man is a vowel," "Man is monosyllabic," "Man is a noun with an irregular plural." Of course it is equivocal to use the same word, man, both as a proper name of the English word which is spelled by the thirteenth, first, fourteenth letters of the alphabet in that order, and as a common name (see footnote 6) of featherless plantigrade biped mammals[135]—but an equivocacy which, like many others in the natural languages, is often both convenient and harmless. Whenever there would otherwise be real doubt of the meaning, it may be removed by the use of added words in the sentence, or by the use of quotation marks, or of italics, as in: "The word man is monosyllabic"; " 'Man' is monosyllabic"; "*Man* is monosyllabic."

Following the convenient and natural phraseology of Quine, we may distinguish between *use* and *mention* of a word or symbol. In "Man is a rational animal" the word "man" is used but not mentioned. In "The English translation of the French word *homme* has three letters" the word "man" is mentioned but not used. In "Man is a monosyllable" the word "man" is both mentioned and used, though used in an anomalous manner, namely autonymously.

Frege introduced the device of systematically indicating autonymy by quotation marks, and in his later publications (though not in the *Begriffsschrift*) words and symbols used autonymously are enclosed in single quotation marks in all cases. This has the effect that a word enclosed in single

---

[134]In the terminology of the Scholastics, use of a word as a name of itself, i.e., to denote itself as a word, was called *suppositio materialis*. Opposed to this as *suppositio formalis* was the use of a noun in its proper or ordinary meaning. This terminology is sometimes still convenient.

The various further distinctions of *suppositiones* are too cumbrous, and too uncertain, to be usable. All of them, like that between *suppositio materialis* and *formalis*, refer to peculiarities and irregularities of meaning which are found in many natural languages but which have to be eliminated in setting up a formalized language.

[135]To follow a definition found in *The Century Dictionary*.

quotation marks is to be treated as a different word from that without the quotation marks—as if the quotation marks were two additional letters in the spelling of the word—and equivocacy is thus removed by providing two different words to correspond to the different meanings. Many recent writers follow Frege in this systematic use of quotation marks, some using double quotation marks in this way, and others following Frege in using single quotation marks for the purpose, in order to reserve double quotation marks for their regular use as punctuation. As the reader has long since observed, Frege's systematic use of quotation marks is not adopted in this book.[136] But we may employ quotation marks or other devices from time to time, especially in cases in which there might otherwise be real doubt of the meaning.

To return to the question of syntactical constants for use in developing the syntax of a formalized object language, we find that there is in this case

[136]Besides being rather awkward in practice, such systematic use of quotation marks is open to some unfortunate abuses and misunderstandings. One of these is the misuse of quotation marks as if they denoted a function from things (of some category) to names of such things, or as if such a function might be employed at all without some more definite account of it. Related to this is the temptation to use in the role of a syntactical variable the expression obtained by enclosing a variable of an object language in quotation marks, though such an expression, correctly used, is not a variable of any kind, and not a form but a constant.

Also not uncommon is the false impression that trivial or self-evident propositions are expressed in such statements as the following: ' 'Snow is white' is true if and only if snow is white' (Tarski's example); ' 'Snow is white' means that snow is white'; ' 'Cape Town' is the [or a] name of Cape Town.'

This last misunderstanding may arise also in connection with autonymy. A useful method of combatting it is that of translation into another language (cf. a remark by C. H. Langford in *The Journal of Symbolic Logic*, vol. 2 (1937), p. 53). For example, the proposition that 'Cape Town' is the name of Cape Town would be conveyed thus to an Italian (whom we may suppose to have no knowledge of English): ' 'Cape Town' è il nome di Città del Capo.' Assuming, as we may, that the Italian words have exactly the same sense as the English words of which we use them as translations—in particular that 'Città del Capo' has the same sense as 'Cape Town' and that ' 'Cape Town' ' has the same sense in Italian as in English—we see that the Italian sentence and its English translation must express the very same proposition, which can no more be a triviality when conveyed in one language than it can in another.

The foregoing example may be clarified by recalling the remark of footnote 8 that the name relation is properly a ternary relation, and may be reduced to a binary relation only by fixing the language in a particular context. Thus we have the more explicit English sentences: ' 'Cape Town' is the English name of Cape Town'; ' 'Città del Capo' is the Italian name of Cape Town.' The Italian translations are: ' 'Cape Town' è il nome inglese di Città del Capo'; ' 'Città del Capo' è il nome italiano di Città del Capo.' Of the two propositions in question, the first one has a false appearance of obviousness when expressed in English, the illusion being dispelled on translation into Italian; the second one contrariwise does not seem obvious or trivial when expressed in English, but on translation into Italian acquires the appearance of being so.

(In the three preceding paragraphs of this footnote, we have followed Frege's systematic use of single quotation marks, and the paragraphs are to be read with that understanding. As explained, we do not follow this usage elsewhere.)

nothing equivocal in using the symbols and formulas of the object language autonymously in the syntax language, provided that care is taken that no formula of the object language is also a formula of the syntax language in any other wise than as an autonym. Therefore we adopt the following practice:

The primitive symbols of the object language will be used in the syntax language as names of themselves, and juxtaposition will be used for juxtaposition.[137]

This is the ordinary usage in mathematical writing, and has the advantage of being self-explanatory. Though we employ it only informally, it is also readily adapted to incorporation in a formalized syntax language[138] (and in fact more so than the convention of quotation marks).

As a precaution against equivocation, we shall hereafter avoid the practice—which might otherwise sometimes be convenient—of borrowing formulas of the object language for use in the syntax language (or other meta-language) with the same meaning that they have in the object language. Thus in all cases where a single symbol or a formula of the object language is found as a constituent in an English sentence, it is to be understood in accordance with the italicized rule above, i.e., autonymously.

Since we shall later often introduce conventions for abbreviating well-formed formulas of an object language, some additional explanations will be necessary concerning the use of syntactical variables and syntactical constants (and concerning autonymy) in connection with such abbreviations. These will be indicated in §11, where such abbreviations first appear. But, as explained in that section, the abbreviations themselves and therefore any special usages in connection with them are dispensable in principle, however necessary practically. In theoretical discussions of syntax and in particular in formalizing the syntax language, the matter of abbreviations of well-formed formulas may be ignored.

---

[137]I.e., juxtaposition will be used in the syntax language as a binary connective having the operation of juxtaposition as its associated function. Technically, some added notation is needed to show association, or some convention about the matter, such as that of association to the left (as in §11). But in practice, because of the associativity of juxtaposition, there is no difficulty in this respect.

[138]This is, of course, on the assumption that the syntax language is a different language from the object language.

If on the contrary a formalized language is to contain names of its own formulas, then a name of a formula must ordinarily not be that formula. E.g., a variable of a language must not be, in that same language, also a name of itself; for a proper name of a variable is no variable but a constant (as already remarked, in another connection, in footnote 136).

09. Semantics. Let us imagine the users of a formalized language, say a written language, engaged in writing down well-formed formulas of the language, and in assembling sequences of formulas which constitute chains of immediate inferences or, in particular, proofs. And let us imagine an observer of this activity who not only does not understand the language but refuses to believe that it is a language, i.e., that the formulas have meanings. He recognizes, let us say, the syntactical criteria by which formulas are accepted as well-formed, and those by which sequences of well-formed formulas are accepted as immediate inferences or as proofs; but he supposes that the activity is merely a game—analogous to a game of chess or, better, to a chess problem or a game of solitaire at cards—the point of the game being to discover unexpected theorems or ingenious chains of inferences, and to solve puzzles as to whether and how some given formula can be proved or can be inferred from other given formulas.[139]

To this observer the symbols have only such meaning as is given to them by the rules of the game—only such meaning as belongs, for example, to the various pieces at chess. A formula is for him like a position on a chess-board, significant only as a step in the game, which leads in accordance with the rules to various other steps.

All those things about the language which can be said to and understood by such an observer while he continues to regard the use of the language as merely a game constitute the (theoretical) syntax of the language. But those things which are intelligible only through an understanding that the well-formed formulas have meaning in the proper sense, e.g., that certain of them express propositions or that they denote or have values in certain ways, belong to the semantics of the language.

Thus the study of the interpretation of the language as an interpretation is called *semantics*.[140] The name is applied especially when the treatment is

---

[139]A comparison of the rules of arithmetic to those of a game of chess was made by J. Thomae (1898) and figures in the controversy between Thomae and Frege (1903–1908). The same comparison was used by Hermann Weyl (1924) in order to describe Hilbert's program of *metamathematics* or syntax of a mathematical object language.

[140]The name (or its analogue in Polish) was introduced by Tarski in a paper in *Przegląd Filozoficzny*, vol. 39 (1936), pp. 50–57, translated into German as "Grundlegung der wissenschaftlichen Semantik" in *Actes du Congrès International de Philosophie Scientifique* (1936). Other important publications in the field of semantics are: Tarski's *Pojęcie Prawdy w Językach Nauk Dedukcyjnych* (1933), afterwards translated into German (and an important appendix added) as "Der Wahrheitsbegriff in den formalisierten Sprachen" in *Studia Philosophica*, vol. 1 (1936) pp. 261–405; and Carnap's *Introduction to Semantics* (1942). Concerning Carnap's book see a review by the present writer in *The Philosophical Review*, vol. 52 (1943), pp. 298–304.

The word *semantics* has various other meanings, most of them older than that in question here. Care must be taken to avoid confusion on this account. But in this book the word will have always the one meaning, intended to be the same (or substantially

in a formalized meta-language. But in this book we shall not go beyond some unformalized semantical discussion, in ordinary English.

Theorems of the semantical meta-language will be called *semantical theorems*, and both semantical and syntactical theorems will be called *metatheorems*, in order to distinguish them from theorems of the object language.

As appears from the work of Tarski, there is a sense in which semantics can be reduced to syntax. Tarski has emphasized especially the possibility of finding, for a given formalized language, a purely syntactical property of the well-formed formulas which coincides in extension with the semantical property of being a true sentence. And in Tarski's *Wahrheitsbegriff*[141] the problem of finding such a syntactical property is solved for various particular formalized languages.[142] But like methods apply to the two semantical concepts of denoting and having values, so that syntactical concepts may be found which coincide with them in extension.[143] Therefore, if names expressing

---

so) as that in which it is used by Tarski, C. W. Morris (*Foundations of the Theory of Signs*, 1938), Carnap, G. D. W. Berry (*Harvard University, Summaries of Theses 1942*, pp. 330–334).

[141]Cited in the preceding footnote.

[142]Tarski solves also, for various particular formalized languages, the problem of finding a syntactical relation which coincides in extension with the semantical relation of satisfying a propositional form.

In a paper published in *Monatshefte für Mathematik und Physik*, vol. 42, no. 1 (1935), therefore later than Tarski's *Pojęcie Prawdy* but earlier than the German translation and its appendix, Carnap also solves both problems (of finding syntactical equivalents of being a true sentence and of satisfying a propositional form) for a particular formalized language and in fact for a stronger language than any for which this had previously been done by Tarski. Carnap's procedure can be simplified in the light of Tarski's appendix or as suggested by Kleene in his review cited in footnote 131.

On the theory of meaning which we are here adopting, the semantical concepts of being a true sentence and of satisfying a propositional form are reducible to those of denoting and having values, and these results of Tarski and Carnap are therefore implicit in the statement of the following footnote.

[143]More explicitly, this may be done as follows. In §07, in discussing the semantical rules of a formalized language, we thought of the concepts of denoting and of having values as being known in advance, and we used the semantical rules for the purpose of giving meaning to the previously uninterpreted logistic system. But instead of this it would be possible to give no meaning in advance to the words "denote" and "have values" as they occur in the semantical rules, and then to regard the semantical rules, taken together, as constituting definitions of "denote" and "have values" (in the same way that the formation rules of a logistic system constitute a definition of "well-formed"). The concepts expressed by "denote" and "have values" as thus defined belong to theoretical syntax, nothing semantical having been used in their definition. But they coincide in extension with the semantical concepts of denoting and having values, as applied to the particular formalized language.

The situation may be clarified by recalling that a particular logistic system may be expected to have many sound interpretations, leading to many different assignments of denotations and values to its well-formed formulas. These assignments of denotations and values to the well-formed formulas may be made as abstract correspondences, so that their treatment belongs to theoretical syntax. Semantics begins when we decide the meaning of the well-formed formulas by fixing a particular interpretation of the system. The distinction between semantics and syntax is found in the different signif-

these two concepts are the only specifically semantical (non-syntactical) primitive symbols of a semantical meta-language, it is possible to transform the semantical meta-language into a syntax language by a change of interpretation which consists only in altering the sense of those names without changing their denotations.

However, a sound syntax language capable of expressing such syntactical equivalents of the semantical concepts of denoting and having values—or even only a syntactical equivalent of the semantical property of truth—must ordinarily be stronger than the object language (assumed sound), in the sense that there will be theorems of the syntax language of which no translation (i.e., sentence expressing the same proposition) is a theorem of the object language. Else there will be simple elementary propositions about the semantical concepts such that the sentences expressing the corresponding propositions about the syntactical equivalents of the semantical concepts are not theorems of the syntax language.[144]

For various particular formalized languages this was proved (in effect) by Tarski in his *Wahrheitsbegriff*. And Tarski's methods[145] are such that they can be applied to obtain the same result in many other cases—in particular in the case of each of the object languages studied in this book, when a formalized syntax language of it is set up in a straightforward manner. No doubt Tarski's result is capable of precise formulation and proof as a result about a very general class of languages, but we shall not attempt this.

The significance of Tarski's result should be noticed as it affects the question of the use of a formalized language as semantical meta-language of itself. A sound and sufficiently adequate language may indeed be capable

---

icance given to one particular interpretation and to its assignment of denotations and values to the well-formed formulas; but within the domain of formal logic, including pure syntax and pure semantics, nothing can be said about this different significance except to postulate it as different.

Many similar situations are familiar in mathematics. For instance, the distinction between plane Euclidean metric geometry and plane projective geometry may be found in the different significance given to one particular straight line and one particular elliptic involution on it. And it seems not unjustified to say that the sense in which semantics can be reduced to syntax is like that in which Euclidean metric geometry can be reduced to projective geometry.

All this suggests that, in order to maintain the distinction of semantics from syntax, "denote" and "have values" should be introduced as undefined terms and treated by the axiomatic method. Our use of semantical rules is intended as a step towards this. And in fact Tarski's *Wahrheitsbegriff* already contains the proposal of an axiomatic theory of truth as an alternative to that of finding a syntactical equivalent of the concept of truth.

[144]A more precise statement of this will be found in Chapter VIII, as it applies to the special case of the logistic system of Chapter VII when interpreted, in the manner indicated in footnote 132, so as to be capable of expressing its own syntax.

[145]Related to those used by Kurt Gödel in the proof of his incompleteness theorems, set forth in Chapter VIII.

of expressing its own syntax (cf. footnote 132) and its own semantics, in the sense of containing sentences which express at least a very comprehensive class of the propositions of its syntax and its semantics. But among these sentences, if certain very general conditions are satisfied, there will always be true sentences of a very elementary semantical character which are not theorems—sentences to the effect, roughly speaking, that such and such a particular sentence is true if and only if ____ , the blank being filled by that particular sentence.[146] Hence, on the assumption that the language satisfies ordinary conditions of adequacy in other respects, not all the semantical rules (in the sense of §07), when written as sentences of the language, are theorems.

On account of this situation, the distinction between object language and meta-language, which first arises in formalizing the object language, remains of importance even after the task of formalization is complete for both the object language and the meta-language.

In concluding this Introduction, let us observe that much of what we have been saying has been concerned with the relation between linguistic expressions and their meaning, and therefore belongs to semantics. However, our interest has been less in the semantics of this or that particular language than in general features common to the semantics of many languages. And very general semantical principles, imposed as a demand upon any language that we wish to consider at all, have been put forward in some cases, notably assumptions (1), (2), (3) of §01 and assumption (4) of §02.[147]

We have not, however, attempted to formalize this semantical discussion, or even to put the material into such preliminary order as would constitute a first step toward formalization. Our purpose has been introductory and explanatory, and it is hoped that ideas to which the reader has thus been informally introduced will be held subject to revision or more precise formulation as the development continues.

From time to time in the following chapters we shall interrupt the rigorous treatment of a logistic system in order to make an informal semantical aside. Though in studying a logistic system we shall wish to hold its interpretation open, such semantical explanations about a system may serve in

[146]A more careful statement is given by Tarski.

By the results of Gödel referred to in the preceding footnote (or alternatively by Tarski's reduction of semantics to syntax), true syntactical sentences which are not theorems must also be expected. But these are of not quite so elementary a character. And the fundamental syntactical rules described in §07 may nevertheless all be theorems when written as sentences of the language.

[147]And assumption (5) of footnote 30.

particular to show a motivation for consideration of it by indicating its principal interpretations (cf. §07). *Except in this Introduction, semantical passages will be distinguished from others by being printed in smaller type, the small type serving as a warning that the material is not part of the formal logistic development and must not be used as such.*

As we have already indicated, it is contemplated that semantics itself should ultimately be studied by the logistic method.

But if semantical passages in this Introduction and in later chapters are to be rewritten in a formalized semantical language, certain refinements become necessary. Thus if the semantical language is to be a functional calculus of order ω in the sense of Chapter VI, or a language like that of Chapter X, then various semantical terms, such as the term "denote" introduced in §01, must give way to a multiplicity of terms of different types,[148] and statements which we have made using these terms must be replaced by axiom schemata[149] or theorem schemata[149] with typical ambiguity.[149] Or if the semantical language should conform to some alternative to the theory of types, changes of a different character would be required. In particular, following the Zermelo set theory (Chapter XI), we would have to weaken substantially the assumption made in §03 that every singulary form has an associated function, and explanations regarding the notation λ would have to be modified in some way in consequence.

---

[148]All the expressions of the language—formulas, or well-formed formulas—may be treated as values of (syntactical) variables of one type. But terms "denote" of different types are nevertheless necessary, because in "____ denotes ____," after filling the first blank with a syntactical variable or syntactical constant, we may still fill the second blank with a variable or constant of any type.

Analogously, various other terms that we have used have to be replaced each by a multiplicity of terms of different types. This applies in particular to "thing," and the consequent weakening is especially striking in the case of footnote 9—which must become a schema with typical ambiguity.

See also the remark in the last paragraph of footnote 87.

[149]The terminology is explained in §§27, 30, 33, and Chapter VI. (The typical ambiguity required here is ambiguity with respect to *type* in the sense described in footnote 578, and is therefore not the same as the typical ambiguity mentioned in footnote 585, which is ambiguity rather with respect to *level*.)

I. The Propositional Calculus

The name *propositional calculus*[150] is given to any one of various logistic systems—which, however, are all equivalent to one another in a sense which will be made clear later. When we are engaged in developing a particular one of these systems, or when (as often happens) it is unnecessary for the purpose in hand to distinguish among the different systems, we speak of *the* propositional calculus. Otherwise the various logistic systems are distinguished as various *formulations of* the propositional calculus.

The importance of the propositional calculus in one or another of its formulations arises from its frequent occurrence as a part of more extensive logistic systems which are considered in this book or have been considered elsewhere, the variables of the propositional calculus (propositional variables) being replaceable by sentences of the more extensive system. Because of its greater simplicity in many ways than other logistic systems which we consider, the propositional calculus also serves the purposes of introduction and illustration, many of the things which we do in connection with it being afterwards extended, with greater or less modification, to other systems.

In this chapter we develop in detail a particular formulation of the propositional calculus, the logistic system P_1. Some other formulations will be considered in the next chapter.

10. The primitive basis of P_1.[150] The primitive symbols of P_1 are three improper symbols

$$[\quad \supset \quad]$$

(of which the first and third are called *brackets*) and one primitive constant

$$f$$

and an infinite list of variables

$$p \quad q \quad r \quad s \quad p_1 \quad q_1 \quad r_1 \quad s_1 \quad p_2 \quad q_2 \quad \ldots$$

(the order here indicated being called the *alphabetic order* of the variables). The variables and the primitive constant are called *proper symbols*.[151]

---

[150]Historical questions in connection with the propositional calculus will be treated briefly in the concluding section of Chapter II.

[151]Regarding the terminology, see explanations in §07 and in footnote 117.

We shall hereafter use the abbreviations "wf" for "well-formed," "wff" for "well-formed formula," "wffs" for "well-formed formulas." The formation rules of P_1 are:[152]

10i. The primitive constant f standing alone is a wff.
10ii. A variable standing alone is a wff.
10iii. If $\mathbf{\Gamma}$ and $\mathbf{\Delta}$ are wf, then $[\mathbf{\Gamma} \supset \mathbf{\Delta}]$ is wf.[153]

To complete the definition of a wff of P_1 we add that a formula is wf if and only if its being so follows from the three formation rules. In other words, the wffs of P_1 are the least class of formulas which contains all the formulas stated in 10i and 10ii and is closed under the operation of 10iii.

Though not given explicitly in the definition, an effective test of well-formedness follows from it. If a wff is not of too great length, it may often be recognized as such at a glance. Otherwise we may employ a counting procedure in the following way. If a given formula consists of more than one symbol, it cannot be wf unless it ends with] and begins with [. Then we may start counting brackets at the beginning (or left) of the formula, pro-

---

[152]Systematic methods of numbering theorems, axioms, etc. so as to indicate the section in which each occurs were perhaps first introduced by Peano. The method adopted here has some features in common with one that has been used by Quine.

We shall number sections by numbers of two or more digits in such a way that the number of the chapter in which the section occurs may be obtained by deleting the last digit. However, chapter numbers are given by Roman numerals. Some chapters have at the beginning a brief introductory statement not in any numbered section.

We shall number axioms, rules of inference, theorems, and metatheorems by numbers of three or more digits in such a way that the number of the section in which they occur may be obtained by deleting the last digit. We place a dagger, †, before the number of an axiom or theorem of the logistic system; an asterisk, *, before the number of a (primitive) rule of inference, axiom schema, derived rule of inference, or theorem schema; and a double asterisk, **, before the number of other metatheorems. (The terminology, so far as not already explained in the introduction, will be explained in this and following chapters.) The numbers of axioms, axiom schemata, and primitive rules of inference have 0 as the next-to-last digit, and are thus distinguished from the numbers of theorems, theorem schemata, derived rules of inference, and other metatheorems.

In numbering formation rules we use the number of the section in which they occur and a small Roman numeral. Thus 10i, 10ii, 10iii are the formation rules in §10.

A collection of exercises has the same number as the section which it follows. For individual exercises in the collection we use the number of the collection of exercises, followed by a period, then another digit or digits. Thus 12.0, 12.1, and so on are the exercises which follow §12.

Definitions and definition schemata are numbered D1, D2, and so on, without regard to the section in which they occur.

[153]Concerning the use of bold letters as syntactical variables see §08. Also see the italicized statement near the end of §08.

Without the convention that juxtaposition is used for juxtaposition (see footnote 137), we would have to state 10iii more lengthily as follows: If $\mathbf{\Gamma}$ and $\mathbf{\Delta}$ are wffs, then the formula consisting of [, followed by the symbols of $\mathbf{\Gamma}$ in order, followed by $\supset$, followed by the symbols of $\mathbf{\Delta}$ in order, followed by], is wf.

ceeding from left to right, counting each left bracket, [, as +1 and each right bracket,], as −1, and adding as we go.[154] When the count of 1 first falls either on a right bracket or on a left bracket with a proper symbol immediately after it,[155] the next symbol must be the implication sign, ⊃, if the formula is wf.[156] This is called the *principal implication sign* of the formula, the part of the formula between the initial left bracket and the principal implication sign is called the *antecedent*, and the part of the formula between the principal implication sign and the final right bracket is called the *consequent*. Then the given formula is wf if and only if both antecedent and consequent are wf. Thus the question of the well-formedness of the given formula is reduced to the same question about two shorter formulas, the antecedent and the consequent. We may then repeat, applying the same procedure to the antecedent and to the consequent that we did to the given formula. After a finite number of repetitions, either we will reach the verdict that the given formula is not wf because one of the required conditions fails (or because we count all the way to the end of some formula without finding a principal implication sign), else the question of the well-formedness of the given formula will be reduced to the same question about each of a finite number of formulas consisting of no more than one symbol apiece. A formula in which the number of symbols is zero—the *null formula*—is of course not wf. And a formula consisting of just one symbol is wf if and only if that symbol is a proper symbol.[157]

Hereafter we shall speak of the *principal implication sign*, the *antecedent*, and the *consequent*, only of a wff, as indeed we shall seldom have occasion to refer to formulas that are not wf. If a formula is wf and consists of more

---

[154]The choice of left-to-right order is a concession to the habit of the eye. A like procedure could be described, equally good, in which the counting would proceed from right to left.

[155]More explicitly, if the given formula is wf and consists of more than one symbol, then the first symbol must be [, and the second symbol must be either a proper symbol or [. If the second symbol is a proper symbol, the third symbol must be ⊃, and this third symbol is the principal implication sign. If the second symbol is [, then the count is made from left to right as described; when the count of 1 falls on a], the next symbol must be ⊃, and this is the principal implication sign.

[156]Here [,], and ⊃ are being used autonymously. Notice that in addition to the use of symbols and formulas of the object language, in the syntax language, as *autonymous proper names*, i.e., as proper names of themselves in the way explained in §08, we shall also sometimes find it convenient to use them as *autonymous common names*, i.e., as common names (see footnotes 4, 6) of occurrences of themselves.

When used autonymously, [may be read orally as "left bracket,"] as "right bracket," and ⊃ as "implication sign" (or more fully, "sign of material implication").

In oral reading of wffs of P_1 (or other logistic system) the readings "left bracket" and "right bracket" may also occasionally be convenient for [and] respectively. But in this case, ⊃ should be read as "implies" or "if . . . . . then," as in the table in §05.

[157]In practice, the entire procedure may often be shortened by obvious devices.

than one symbol, it has the form $[\mathbf{A} \supset \mathbf{B}]$ in one and only one way.[158] And **A** is the antecedent, **B** is the consequent, and the $\supset$ between **A** and **B** is the principal implication sign.

By the *converse* of a wff $[\mathbf{A} \supset \mathbf{B}]$ we shall mean the wff $[\mathbf{B} \supset \mathbf{A}]$.

In P_1 all occurrences of a variable in a wff are *free occurrences*; a wff is an *n-ary form* if it contains (free) occurrences of exactly n different variables, a *constant* if it contains (free) occurrences of no variables; all forms are *propositional forms*; and all constants are propositional constants or *sentences*. (Cf. footnote 117.)

In order to state the rules of inference of P_1, we introduce the notation "S |" for the operation of substitution, so that $\mathrm{S}^{\mathbf{b}}_{\mathbf{B}}\mathbf{A}|$ is the formula which results by substitution of **B** for each occurrence of **b** throughout **A**.[159] This is a notation for which we shall have frequent use in this and later chapters. It is, of course, not a notation of P_1 (or of any logistic system studied in this book) but belongs to the syntax language, just as the apparatus of syntactical variables does: its use could always be avoided, though at the cost of some inconvenience, by employing English phrases containing the words "substitute," "substitution," or the like.

The rules of inference are the two following:[160]

*100. From $[\mathbf{A} \supset \mathbf{B}]$ and **A** to infer **B**. (*Rule of modus ponens.*)

*101. From **A**, if **b** is a variable, to infer $\mathrm{S}^{\mathbf{b}}_{\mathbf{B}}\mathbf{A}|$.[161] (*Rule of substitution.*)

In the rule of *modus ponens* (*100) the premiss $[\mathbf{A} \supset \mathbf{B}]$ is called the *major premiss*, and **A** is called the *minor premiss*. Notice the condition that the antecedent of the major premiss must be identical with the minor premiss; the conclusion is then the consequent of the major premiss.[162]

The axioms of P_1 are the three following:

†102. $[p \supset [q \supset p]]$

†103. $[[s \supset [p \supset q]] \supset [[s \supset p] \supset [s \supset q]]]$

†104. $[[[p \supset f] \supset f] \supset p]$

---

[158] As explained in §08, the bold roman capitals have as values formulas (of the logistic system under consideration) which are wf. This makes it unnecessary to put in an explicit condition, "where **A** and **B** are wf."

[159] The bold roman small letters, as "**b**" here, have proper symbols as values—see §08. When **b** does not occur in **A**, the result of the substitution is **A** itself.

[160] For brevity, we say simply "infer" in stating the rules rather than "immediately infer," which is the full expression as introduced in §07.

[161] It is meant, of course, that **B** may be any wff. The result of the substitution is wf, as may be proved by mathematical induction with respect to the number of occurrences of $\supset$ in **A**.

[162] The terms *modus ponens, major premiss, minor premiss, antecedent, consequent* are from Scholastic logic (as are, of course, also *premiss, conclusion*).

The first of these axioms (†102), or its equivalent in other formulations of the propositional calculus (whether or not it is an axiom), is called the *law of affirmation of the consequent*. Similarly, the second axiom is called the *self-distributive law of (material) implication*. And the third axiom is called the *law of double negation*.[163]

In accordance with the explanation in §07, a *proof* in P_1 is a finite sequence of one or more wffs each of which either is one of the three axioms, or is (immediately) inferred from two preceding wffs in the sequence by *modus ponens*, or is (immediately) inferred from one preceding wff in the sequence by substitution. A proof is called a proof *of* the last wff in the sequence,[164] and a wff is called a *theorem* if it has a proof.

In addition to the abbreviations "wf," "wff," we shall also use, in this and later chapters, the abbreviations "t" for "the truth-value truth" and "f" for "the truth-value falsehood."

The intended principal interpretation of the logistic system P_1 has already been indicated implicitly by discussion in §05 and in the present chapter. We now make an explicit statement of the semantical rules (in the sense of §07). These are:

a. *f* denotes f.

b. The variables are variables having the range t and f.

c. A form which consists of a variable **a** standing alone has the value t for the value t of **a**, and the value f for the value f of **a**.

d. Let **A** and **B** be constants. Then [**A** ⊃ **B**] denotes t if either **B** denotes t or **A** denotes f. Otherwise [**A** ⊃ **B**] denotes f.

e. Let **A** be a form and **B** a constant. If **B** denotes t, then [**A** ⊃ **B**] has the value t for all assignments of values to its variables. If **B** denotes f, then [**A** ⊃ **B**], for a given assignment of values to its variables, has the value f in case **A** has the

[163]The uncertainty as to whether the name *law of double negation* shall be applied to †104 or to the wff which results from †104 by interchanging antecedent and consequent is here resolved in favor of the former, and to the latter we shall therefore give the name *converse law of double negation*. Thus the law of double negation is the one which (through substitution and *modus ponens*) allows the cancellation of a double negation.

Also the self-distributive law of implication, †103, (through substitution and *modus ponens*) allows the distribution of an implication over an implication. And the *converse self-distributive law of (material) implication* is the converse of †103, which allows the inverse process to this.

If in †103 and †104 we replace the principal implication sign by the ≡ of D6 below (or by an equivalent ≡ in some other formulation of the propositional calculus), there result the *complete self-distributive law of (material) implication* and the *complete law of double negation*, as we shall call them respectively.

[164]Observe that the definition of a proof allows any number of digressions and irrelevancies, as there is no requirement that every wff appearing in a proof must actually contribute to obtaining the theorem proved or that the shortest way must always be adopted of obtaining a required wff from other wffs by a series of immediate inferences.

value t for that assignment of values to the variables, and has the value t in case **A** has the value f for that assignment of values to the variables.

f. Let **A** be a constant and **B** a form. If **A** denotes f, then [**A** ⊃ **B**] has the value t for all assignments of values to its variables. If **A** denotes t, then [**A** ⊃ **B**], for a given assignment of values to its variables, has the same value that **B** has for that assignment of values to the variables.

g. Let **A** and **B** be forms, and consider a given assignment of values to the variables of [**A** ⊃ **B**]. Then the value of [**A** ⊃ **B**] is t if, for that assignment of values to the variables, either the value of **B** is t or the value of **A** is f. Otherwise the value of [**A** ⊃ **B**] is f.

This has been written out at tedious length for the sake of illustration. Of course the last three rules may be condensed into a single statement by introducing a convention according to which a constant has a value, namely, its denotation, for any assignment of values to any variables. And rule d may be included in the same statement by a further convention according to which having a value for a null class of variables is the same as denoting.

The reader should see that these rules have the effect of assigning a unique denotation to every constant (of P_1), and a unique system of values to every form.

As to the motivation of the rules, observe that rule d just corresponds to the account of material implication as given in §05, namely, that everything implies truth, and falsehood implies everything, but truth does not imply falsehood. Rules c, e, f, g are then just what they have to be in view of the account of variables and forms in §02.

Besides this principal interpretation of P_1 other interpretations are also possible, and some of them will be mentioned in exercises later.

The reader must bear in mind that, in the formal development of the system, no use may be made of any intended interpretation, principal or other (cf. §§07, 09).

11. Definitions. As a practical matter in presenting and discussing the system, we shall make use of certain abbreviations of wffs of P_1.

In particular the outermost brackets of a wff may be omitted, so that we write, e.g.,

$$p \supset [q \supset p]$$

as an abbreviation of the wff †102. (Of course the expression $p \supset [q \supset p]$ is a formula as it stands, but not wf; since we shall hereafter be concerned with wffs only, no confusion will arise by using this expression as an abbreviation of $[p \supset [q \supset p]]$.)

We shall also omit further brackets under the convention that, in restoring omitted brackets, association shall be to the left. Thus

$$p \supset f \supset f \supset p$$

is an abbreviation of †104, while

$$p \supset [f \supset f] \supset p$$

is an abbreviation of the wff

$$[[p \supset [f \supset f]] \supset p].$$

Where, however, in omitting a pair of brackets we insert a heavy dot, ∎, the convention in restoring brackets is (instead of association to the left) that the left bracket, [, shall go in in place of the heavy dot, and the right bracket,], shall go in immediately before the next right bracket which is already present to the right of the heavy dot and has no mate to the right of the heavy dot; or, failing that, at the end of the expression.[165] (Here a left bracket is considered to be the *mate* of the first right bracket to the right of it such that an equal number, possibly zero, of left and right brackets occur between.) Thus we shall use

$$p \supset \centerdot q \supset p$$

as an abbreviation of †102, and

$$[p \supset \centerdot f \supset f] \supset p$$

as an alternative abbreviation of the same wff for which the abbreviation

$$p \supset [f \supset f] \supset p$$

was just given.

The convention regarding heavy dots may be used together with the previous convention, namely, that of association to the left when an omitted left bracket is not replaced by a heavy dot. Thus for †103 we may employ either one of the two alternative abbreviations:

$$s \supset [p \supset q] \supset \centerdot s \supset p \supset \centerdot s \supset q$$

$$[s \supset \centerdot p \supset q] \supset \centerdot s \supset p \supset \centerdot s \supset q$$

Similarly,

$$s_1 \supset [s_2 \supset \centerdot p \supset [q_1 \supset q_2] \supset r_1] \supset r_2$$

is an abbreviation of

$$[[s_1 \supset [s_2 \supset [[p \supset [q_1 \supset q_2]] \supset r_1]]] \supset r_2].$$

As we have said, these abbreviations and others to follow are not part of the logistic system P_1 but are mere devices for the presentation of it. They

[165]Compare the use of heavy dots in §06.
The use of dots to replace brackets was introduced by Peano and was adopted by Whitehead and Russell in *Principia Mathematica*. The convention here described for use of dots as brackets is not the same as that of Peano and Whitehead and Russell, but is an modification of it which the writer has found simpler and more convenient in practice.

are concessions in practice to the shortness of human life and patience, such as in theory we disdain to make. The reader is asked, whenever we write an abbreviation of a wff, to pretend that the wff has been written in full and to understand us accordingly.[166] Indeed we must actually write wffs in full whenever ambiguity or unclearness might result from abbreviating. And if any one finds it a defect that devices of abbreviation, not part of the logistic system, are resorted to at all, he is invited to rewrite this entire book without use of abbreviations, a lengthy but purely mechanical task.

Besides abbreviations by omission of brackets, we employ also abbreviations of another kind which are laid down in what we call *definitions*. Such a definition introduces a new symbol or expression (which is neither present in the logistic system itself nor introduced by any previous definition) and prescribes that it shall stand as an abbreviation[167] for a particular wff, the understanding being (unless otherwise prescribed in a special case) that the same abbreviation is used for this wff whether it stands alone or as a constituent in a longer wff.[168]

[166]There will be a few exceptions, especially in this section and in §16, where abbreviations are used as autonyms in the strict sense, i.e., as names of the abbreviations themselves rather than to denote or to abbreviate the wffs. But we shall take care that this is always clear from the context.

[167]In a few cases we may make definitions which fail to provide an abbreviation in the literal sense that the new expression introduced is actually shorter than the wff for which it stands. Use of such a definition may nevertheless sometimes serve a purpose either of increasing perspicuity or of bringing out more sharply some particular feature of a wff.

[168]Definitions in this sense we shall call *abbreviative definitions*, in order to distinguish them from various other things which also are called or may be called definitions (in connection with a formalized language). These latter include:

(1) *Explicative definitions*, which are intended to explain the meaning of a notation (symbol, wff, connective, or operator) already present in a given language, and which are expressed in a sentence of that same language. Such an explicative definition may often involve a sign of equality or of material or other equivalence, placed between the notation to be explicated, or some wff involving it, and another wff of the language. (We do not employ here the traditional term *real definition* because it carries associations and presuppositions which we wish to avoid.)

(2) Statements in a semantical meta-language, giving or explaining the meaning of a notation already present in the object language. These may be either (*primitive*) *semantical rules* in the sense of §07, or what we may call, in an obvious sense, *derived semantical rules*.

(3) Definitions which are like those of (1) in form, except that they are intended to extend the language by introducing a new notation not formerly present in it. Since definitions in this sense are as much a part of the object language in which they occur as are axioms or theorems of it, the writer agrees with Leśniewski that, if such definitions are allowed, it must be on the basis of *rules of definition*, included as a part of the primitive basis of the language and as precisely formulated as we have required in the case of the formation and transformation rules (in particular, appropriate conditions of effectiveness must be satisfied—cf. §07). Unfortunately, authors who use definitions in this sense have not always stated rules of definition with sufficient care. And even Hilbert and Bernays's *Grundlagen der Mathematik* may be criticized in this regard (their account of the matter in vol. 1, pp. 292–293 and 391–392, is much nearer than many to

In order to state definitions conveniently, we make use of an arrow, "→", to be read "stands as an abbreviation for" (or briefly, "stands for"). This arrow, therefore, belongs to the syntax language, like the term "wff" or the notation "S |" of §10. At the base (left) of the arrow we write the *definiendum*, the new symbol or expression which is being introduced by the definition. At the head (right) of the arrow we write the *definiens*, the wff for which the definiendum is to stand. And in so writing the definiens we allow ourselves to abbreviate it in accordance with any previous definitions or other conventions of abbreviation.

Our first definition is:

D1. $t \rightarrow f \supset f$

This means namely that the wff $[f \supset f]$, the definiens, may be abbreviated as t, whether it stands alone or as a part of a longer wff. In particular, then, the wff which we previously abbreviated as

$$p \supset [f \supset f] \supset p$$

may now be further abbreviated as

$$p \supset t \supset p.$$

complete rigor, but fails to allow full freedom of definition, as provision is lacking for many kinds of notation that one might wish for some purpose to introduce by definition). On the other hand, once the rules of definition have been precisely formulated, they become at least theoretically superfluous, because it would always be possible to oversee in advance everything that could be introduced by definition, and to provide for it instead by primitive notations included in the primitive basis of the language. This remains true even when the rules of definition are broad enough to allow direct introduction of new notations for functions of positive integers or of non-negative integers by means of recursion equations, as is pointed out in effect, though not in these words, by Carnap in *The Logical Syntax of Language*, §22 (compare also Hilbert and Bernays, vol. 2, pp. 293–297).

Because of the theoretical dispensability of definitions in sense (3), we prefer not to use them, and in defining a logistic system in §07 we therefore did not provide for the inclusion of rules of definition in the primitive basis. Thus we avoid such puzzling questions as whether definitions of this kind should be expressed by means of the same sign of equality or equivalence that is used elsewhere in the object language or by means of some special sign of *equality by definition* such as "$=_{\mathrm{df}}$"; and indeed whether these definitions do not after all (being about notations) belong to a meta-language rather than the object language.

Not properly in the domain of formal logic at all is the heuristic process of deciding upon a more precise meaning for a notation (often a word or an expression of a natural language) for which a vague or a partial meaning is already known, though the result of this process may be expressed in or may motivate a definition of one kind or another. Also not in the domain of formal logic is the procedure of *ostensive definition* by which a proper name, or a common name, is assigned to a concrete object by physically showing or pointing to the object.

(In connection with an unformalized meta-language we shall continue to speak of "definitions" in the usual, informal, way. It is intended that, upon formalization of the meta-language, these shall become abbreviative definitions.)

In stating definitions we shall often resort to *definition schemata*, which serve the purpose of condensing a large number (commonly an infinite number) of definitions into a single statement. For example, if **A** is any wff whatever, $[\mathbf{A} \supset f]$ is to be abbreviated by the expression which consists of the symbol $\sim$ followed by the symbols of **A** in order. This infinite list of definitions is summed up in the definition schema:

D2. $\sim\mathbf{A} \rightarrow \mathbf{A} \supset f$

Notice here, as in other examples below, that *we use the same abbreviations and the same methods of abbreviation for expressions which contain syntactical variables and have wffs as values that we do for wffs proper.* This is perhaps self-explanatory as an informal device for abbreviating expressions of the syntax language, and when so understood it need not be regarded as a departure from the program of §08 (cf. the last paragraph of that section).

We add also the following definition schemata:[169]

D3. $[\mathbf{A} \not\subset \mathbf{B}] \rightarrow \sim . \mathbf{B} \supset \mathbf{A}$
D4. $[\mathbf{A} \vee \mathbf{B}] \rightarrow \mathbf{A} \supset \mathbf{B} \supset \mathbf{B}$
D5. $[\mathbf{AB}] \rightarrow \mathbf{A} \not\subset \mathbf{B} \not\subset \mathbf{B}$
D6. $[\mathbf{A} \equiv \mathbf{B}] \rightarrow [\mathbf{A} \supset \mathbf{B}][\mathbf{B} \supset \mathbf{A}]$
D7. $[\mathbf{A} \not\equiv \mathbf{B}] \rightarrow [\mathbf{A} \not\subset \mathbf{B}] \vee [\mathbf{B} \not\subset \mathbf{A}]$
D8. $[\mathbf{A} \subset \mathbf{B}] \rightarrow \mathbf{B} \supset \mathbf{A}$
D9. $[\mathbf{A} \not\supset \mathbf{B}] \rightarrow \mathbf{B} \not\subset \mathbf{A}$
D10. $[\mathbf{A} \bar{\vee} \mathbf{B}] \rightarrow \sim\mathbf{A} \sim\mathbf{B}$
D11. $[\mathbf{A} \mid \mathbf{B}] \rightarrow \sim\mathbf{A} \vee \sim\mathbf{B}$

Of course it is understood that a wff may be abbreviated at several places simultaneously by the application of definitions. E.g.,

$$pt \vee \sim p$$

is an abbreviation of

$$[[[[[f \supset f] \supset [[[f \supset f] \supset p] \supset f]] \supset f] \supset [p \supset f]] \supset [p \supset f]].$$

Also the conventions about omission of brackets which were introduced at the beginning of this section, for wffs not otherwise abbreviated, are to be extended to the case in which abbreviations according to D1–11 are already present. (In fact we have done this several times above already; e.g., in D5 we have omitted, under the convention of association to the left,

---

[169] Some of these receive little actual use in this book, but are included so as to be available if needed. The character $\bar{\vee}$ is employed, in place of the $\vee$ with a vertical line across it, only in consequence of typographical difficulties.

two pairs of brackets belonging with the two signs $\not\subset$ according to D3, and in D6 we have omitted an outermost pair of brackets which would be present according to D5.)

Here the convention about restoring an omitted pair of brackets represented by a heavy dot remains the same as given before. For example,

$$p \,\centerdot\, q \supset r$$

becomes, on restoring brackets,

$$[p\ [q \supset r]],$$

which in turn is an abbreviation of the wff,

$$[[[q \supset r] \supset [[[q \supset r] \supset p] \supset f]] \supset f].$$

The convention of association to the left is, however, modified as follows. The bracket-pairs appearing in wffs and in expressions abbreviating wffs are divided into three categories. In the highest category are bracket-pairs which belong with the sign $\supset$ according to 10iii or which belong with one of the signs $\not\subset$, $\equiv$, $\not\equiv$, $\subset$, $\not\supset$ according to D3, D6, D7, D8, D9. In the second category are bracket-pairs belonging with one of the signs $\vee$, $\bar{\vee}$, $|$ according to D4, D10, D11. And in the third category are the bracket-pairs of D5. Among bracket-pairs of the same category, the convention of association to the left applies as before in restoring brackets. But bracket-pairs of higher category are to be restored first, without regard to those of lower category, and are to enclose those of lower category to the extent that results from this.[170] The sign $\sim$ has no brackets belonging with it, but it is of a fourth and lowest category in the sense that a restored left bracket (not represented by a heavy dot), if it falls adjacent to an occurrence of $\sim$ or a series of successive occurrences of $\sim$, must be placed to the left thereof rather than the right.

For example, upon restoring brackets in $p \vee qr$, the result is $[p \vee [qr]]$ rather than $[[p \vee q]\, r]$. Upon restoring brackets in

$$p \supset q \vee \sim rs \equiv \sim p \vee \sim q \vee s,$$

the result is $$[[p \supset [q \vee [\sim rs]]] \equiv [[\sim p \vee \sim q] \vee s]].$$

When the convention regarding categories of bracket-pairs is used in conjunction with the convention regarding heavy dots, the procedure in

[170]A similar convention about restoring brackets or parentheses is familiar in reading equations of elementary algebra, where the brackets or parentheses with the sign of equality are in a highest category, those with the signs of addition and subtraction in a second category, and those with multiplication in a third category, and where otherwise the convention of association to the left applies. For example, $xy - 3x + 2y = x - y - 4$ is to be read as $[(((xy) - (3x)) + (2y)) = ((x - y) - 4)]$, and not (e.g.) as $[(((xy) - 3)(x + (2y))) = (x - (y - 4))]$.

restoring brackets is as follows. In the case that there are no heavy dots occurring between a pair of brackets already present, we take the expression as broken up into parts by the heavy dots, restore the brackets in each of these parts separately (using the convention regarding categories of bracket-pairs, and among bracket-pairs of the same category the convention of association to the left), and then finally restore all remaining brackets as represented by the heavy dots. In the contrary case we first take a portion of the expression which occurs between a pair of brackets already present, and which contains heavy dots but contains no heavy dots between any pair of brackets already present within it; we treat this portion of the expression in the way just explained, so restoring all the brackets in it; and then we take another such portion of the resulting expression, and so continue until all brackets are restored. For example, upon restoring brackets in

$$p \supset q \,\blacksquare\, rs, \qquad p \supset qr \,\blacksquare\, r \supset s,$$
$$p \supset \,\blacksquare\, q \,\blacksquare\, rs, \qquad p \supset \,\blacksquare\, qr \,\blacksquare\, r \supset s,$$
$$s \supset [p \supset \,\blacksquare\, q \supset r \vee \,\blacksquare\, s \supset \sim q] \supset \,\blacksquare\, s \supset \sim p,$$

the results are respectively

$$[[p \supset q][rs]], \qquad [[p \supset [qr]][r \supset s]],$$
$$[p \supset [q[rs]]], \qquad [p \supset [[qr][r \supset s]]],$$
$$[[s \supset [p \supset [[q \supset r] \vee [s \supset \sim q]]]] \supset [s \supset \sim p]].$$

Finally we also allow ourselves, for convenience in abbreviating a wff, first to introduce extra brackets enclosing any wf part of it. Thus, for example, we use

$$p \supset q \,\blacksquare\, r, \qquad p \supset q \,\blacksquare\, \sim r, \qquad p \equiv q \vee \,\blacksquare\, \sim r$$

as abbreviations of wffs which would be written more fully as

$$[[p \supset q]r], \qquad [[p \supset q] \sim r], \qquad [[p \equiv q] \vee \sim r].$$

The fact that the definienda in D2–11 agree notationally with sentence connectives introduced in §05 is of course intended to show a certain agreement in meaning. Indeed in each definition schema the convention of abbreviation which is introduced corresponds to and is motivated by the recognition that a certain connective is already provided for, in the sense that there is a notation already present in P_1 (though a complex one) which, under the principal interpretation of P_1, has the same effect as the required connective.

For example, giving P_1 its principal interpretation, we need not add the connective $\sim$ to P_1 because we may always use the notation $[\mathbf{A} \supset f]$ for the negation of a sentence **A** (or of a propositional form **A**). All the purposes of the notation $\sim\mathbf{A}$, except that of brevity, are equally served by the notation $[\mathbf{A} \supset f]$, and we may therefore use the latter to the exclusion of the former.

In the same manner, D4 corresponds to the recognition that $[[\mathbf{A} \supset \mathbf{B}] \supset \mathbf{B}]$ may be used as the (inclusive) disjunction of **A** and **B**, so that it is unnecessary to provide separately for disjunction.[171] The reader may see this by observing that, for fixed values of the variables (if any), $[[\mathbf{A} \supset \mathbf{B}] \supset \mathbf{B}]$ is false[172] if and only if $[\mathbf{A} \supset \mathbf{B}]$ is true[172] and at the same time **B** is false; but, **B** being false, $[\mathbf{A} \supset \mathbf{B}]$ is true if and only if **A** is false; thus $[[\mathbf{A} \supset \mathbf{B}] \supset \mathbf{B}]$ is false if and only if both **A** and **B** are false (and of course is true otherwise); but this last is exactly what we should have for the disjunction of **A** and **B**.

Similarly, the motivation of the definition D1 is that the wff abbreviated as t is a name of the truth-value truth (according to the semantical rule d of §10).

12. Theorems of P_1.

As a first example of a theorem of P_1 we prove:

†120. $p \supset p$ *(Reflexive law of (material) implication.)*

The reader who has in mind the principal interpretation of P_1, as given in §10, may be led to remark that this proposed theorem is not only obvious but more obvious than any of the axioms. This is quite true, but it does not make unnecessary a proof of the theorem. For we wish to ascertain not merely that the proposed theorem is true but that it follows from our axioms by our rules; and not merely that it is true under the one interpretation but under all sound interpretations.[173]

A proof of †120 is the following sequence of nine formulas:

$$s \supset [p \supset q] \supset . s \supset p \supset . s \supset q$$
$$s \supset [r \supset q] \supset . s \supset r \supset . s \supset q$$
$$s \supset [r \supset p] \supset . s \supset r \supset . s \supset p$$
$$p \supset [r \supset p] \supset . p \supset r \supset . p \supset p$$
$$p \supset [q \supset p] \supset . p \supset q \supset . p \supset p$$
$$p \supset . q \supset p$$
$$p \supset q \supset . p \supset p$$
$$p \supset [q \supset p] \supset . p \supset p$$
$$p \supset p$$

The wffs have here been abbreviated by conventions introduced in the preceding section, and in verifying the proof the reader must imagine them

---

[171]It would also be possible and perhaps more natural to use $\sim\mathbf{A} \supset \mathbf{B}$ (i.e., $[[\mathbf{A} \supset f] \supset \mathbf{B}]$) as the disjunction of **A** and **B**. We have chosen $\mathbf{A} \supset \mathbf{B} \supset \mathbf{B}$ instead because there is some interest in the fact that use of the constant f (or of negation) can be avoided for this particular purpose. The definition of $\mathbf{A} \vee \mathbf{B}$ as $\mathbf{A} \supset \mathbf{B} \supset \mathbf{B}$ is given by Russell in *The Principles of Mathematics* (1903), and again, more formally, in the *American Journal of Mathematics*, vol. 28 (1906), p. 201.

[172]As explained in §04, a sentence is true or false according as it denotes t or f. And for fixed values of the free variables we call a propositional form true or false according as its value is t or f.

[173]Including interpretations that are sound in the generalized sense of §19.

rewritten in unabbreviated form (or, if necessary, must explicitly so rewrite them).

It is sufficient theoretically just to write the proof itself as above, without added explanation, since there are effective means of verification. But for the practical assistance of the reader we may explain in full, as follows. The first wff of the nine is †103. The second wff we obtain from the first by *101, substituting r for p. Again the third wff is obtained from the second by substituting p for q. The fourth one is obtained from the third by substituting p for s. The fifth is obtained from the fourth by substituting q for r. The sixth wff is †102. The seventh one results by *modus ponens* from the fifth one as major premiss and the sixth as minor premiss. Then the eighth wff results from the seventh by another application of *101, $q \supset p$ being substituted for q. Finally $p \supset p$ results by *modus ponens* from the eighth and the sixth wffs as major premiss and minor premiss respectively.

The fifth wff in the proof may conveniently be looked upon as obtained from the first by a *simultaneous substitution*, namely the substitution of p, q, p for s, p, q respectively. And the proof exhibits in detail how the effect of this simultaneous substitution may be obtained by means of four successive single substitutions.

We extend the notation for substitution introduced in §10, so that

$$\mathrm{S}^{\mathbf{b}_1 \mathbf{b}_2 \ldots \mathbf{b}_n}_{\mathbf{B}_1 \mathbf{B}_2 \ldots \mathbf{B}_n} \mathbf{A}|$$

shall be the formula which results by simultaneous substitution of $\mathbf{B}_1$, $\mathbf{B}_2$, . . ., $\mathbf{B}_n$ for $\mathbf{b}_1$, $\mathbf{b}_2$, . . ., $\mathbf{b}_n$ in $\mathbf{A}$. The substitution is to be for all occurrences of $\mathbf{b}_1$, $\mathbf{b}_2$, . . ., $\mathbf{b}_n$ throughout $\mathbf{A}$. It is required that $\mathbf{b}_1$, $\mathbf{b}_2$, . . ., $\mathbf{b}_n$ be all different (else there is no result of the substitution). But of course it is not required that all, or even any, of $\mathbf{b}_1$, $\mathbf{b}_2$, . . ., $\mathbf{b}_n$ actually occur in $\mathbf{A}$.

The effect of the simultaneous substitution,

$$\mathrm{S}^{\mathbf{b}_1 \mathbf{b}_2 \ldots \mathbf{b}_n}_{\mathbf{B}_1 \mathbf{B}_2 \ldots \mathbf{B}_n} \mathbf{A}|,$$

where $\mathbf{b}_1$, $\mathbf{b}_2$, . . ., $\mathbf{b}_n$ are variables and all different, may always be obtained by means of $2n$ successive single substitutions, i.e., $2n$ successive applications of *101. In some cases it may be possible with fewer than $2n$ single substitutions, but it is always possible with $2n$, as follows. Let $\mathbf{c}_1$, $\mathbf{c}_2$, . . ., $\mathbf{c}_n$ be the first n variables in alphabetic order not occurring in any of the wffs $\mathbf{B}_1$, $\mathbf{B}_2$, . . ., $\mathbf{B}_n$, $\mathbf{A}$ (such will always exist, because of the availability of an infinite list of variables). Then in $\mathbf{A}$ substitute successively $\mathbf{c}_1$ for $\mathbf{b}_1$, $\mathbf{c}_2$ for $\mathbf{b}_2$, . . ., $\mathbf{c}_n$ for $\mathbf{b}_n$, $\mathbf{B}_1$ for $\mathbf{c}_1$, $\mathbf{B}_2$ for $\mathbf{c}_2$, . . ., $\mathbf{B}_n$ for $\mathbf{c}_n$.

We shall use the sign $\vdash$ as a syntactical notation to express that a wff is a theorem (of P_1, or, later, of other logistic system). Thus "$\vdash p \supset p$" may be read as an abbreviation of "$p \supset p$ is a theorem," etc. (Cf. footnote 65.)

With the aid of this notation we may state as follows the metatheorem about simultaneous substitution of which we have just sketched a proof:

*121. If $\vdash$ **A**, then $\vdash \mathrm{S}^{\mathbf{b}_1\mathbf{b}_2\ldots\mathbf{b}_n}_{\mathbf{B}_1\mathbf{B}_2\ldots\mathbf{B}_n}\mathbf{A}|$.

We shall make use of this metatheorem as a *derived rule of inference.* I.e., in presenting proofs we shall pass from **A** immediately to

$$\mathrm{S}^{\mathbf{b}_1\mathbf{b}_2\ldots\mathbf{b}_n}_{\mathbf{B}_1\mathbf{B}_2\ldots\mathbf{B}_n}\mathbf{A}|,$$

not giving details of intermediate steps but referring simply to *121 (or to "simultaneous substitution" or to "substitution").

Justification for such use of derived rules of inference is similar to that for the use of definitions and other abbreviations (§11), namely, as a mere device of presentation which is fully dispensable in principle. On this account, however, it is essential that the proof of a derived rule of inference be *effective* (cf. §§07, 08) in the sense that an effective method is provided according to which from a given proof of the premisses of the derived rule it is always possible to obtain a proof of the conclusion of the derived rule.[174] For we must be sure that, whenever a proof presented by means of derived rules is challenged, we can meet the challenge by actually supplying the unabridged proof. In other words we take care that there is a mechanical procedure for supplying the unabridged proof whenever called for, and on this basis, when a proof of a particular theorem of a logistic system is presented with the aid of derived rules, we ask the reader to imagine that the proof has been written in full (and, on occasion, actually to supply it in full for himself).

The proof of *121 is clearly effective, as the reader will see on reviewing it.

With the aid of *121 as a derived rule, and some other obvious devices of abbreviation, we may now present the proof of †120 as follows:

By simultaneous substitution in †103:

$$\vdash p \supset [q \supset p] \supset . p \supset q \supset . p \supset p$$

By †102 and *modus ponens*:

$$\vdash p \supset q \supset . p \supset p$$

By substitution of $q \supset p$ for q:

$$\vdash p \supset [q \supset p] \supset . p \supset p$$

---

[174]We shall later—in particular in Chapters IV and V—consider some metatheorems whose proof is non-effective. But such metatheorems must not be used in the role of derived rules.

Finally by †102 and *modus ponens*:

$$\vdash p \supset p$$

(Thus a presentation of the proof and some practical explanation of it are condensed into about the same space as occupied above by the unabridged proof alone.)

We now go on to proofs of two further theorems of P_1.

†122. $f \supset p$

By simultaneous substitution in †102:

$$\vdash p \supset f \supset f \supset p \supset .\, f \supset .\, p \supset f \supset f \supset p$$

By †104 and *modus ponens*:

$$\vdash f \supset .\, p \supset f \supset f \supset p$$

By simultaneous substitution in †103:

$$\vdash f \supset [p \supset f \supset f \supset p] \supset .\, f \supset [p \supset f \supset f] \supset .\, f \supset p$$

By *modus ponens*:

$$\vdash f \supset [p \supset f \supset f] \supset .\, f \supset p$$

By simultaneous substitution in †102:

$$\vdash f \supset .\, p \supset f \supset f$$

Hence by *modus ponens*:

$$\vdash f \supset p$$

†123. $p \supset f \supset .\, p \supset q$

By simultaneous substitution in †102:

$$\vdash f \supset q \supset .\, p \supset .\, f \supset q$$

By substitution in †122:[175]

$$\vdash f \supset q$$

By *modus ponens*:

$$\vdash p \supset .\, f \supset q$$

By simultaneous substitution in †103:

$$\vdash p \supset [f \supset q] \supset .\, p \supset f \supset .\, p \supset q$$

Hence by *modus ponens*:

$$\vdash p \supset f \supset .\, p \supset q$$

(†123 is known as the *law of denial of the antecedent*. Notice that it may be abbreviated by D2 as $\sim p \supset .\, p \supset q$.)

---

[175]The entire proof of †122 therefore enters at this point as part of the proof of †123 when written in full.

EXERCISES 12

12.0. Prove (as a metatheorem) that the effective test of well-formedness given in §10 does in fact constitute a necessary and sufficient condition that a formula be wf according to the formation rules 10i–iii. (Use mathematical induction with respect to the number of occurrences of $\supset$ in the formula.)

12.1. Prove the assertion made in §10 that, if a formula is wf and consists of more than one symbol, it has the form $[\mathbf{A} \supset \mathbf{B}]$ in one and only one way. Also that any wf (consecutive) part of the formula is either the entire formula or a wf part of **A** or a wf part of **B**.[176] (For the proof, employ the same method of counting brackets as in the effective test of well-formedness, and again proceed by mathematical induction with respect to the number of occurrences of $\supset$ in the formula.)

12.2. Let $\mathrm{P}_{1\mathrm{L}}$ be the logistic system which we obtain from P_1 by a change of notation, writing systematically C ____ ____ in place of [____ $\supset$ ____], in the way described in footnote 91, and leaving everything else unaltered. State the primitive basis of $\mathrm{P}_{1\mathrm{L}}$. State and prove the metatheorems about $\mathrm{P}_{1\mathrm{L}}$ which are analogues of those of 12.0 and 12.1.[177]

The following proofs are to be presented with the aid of *121 *and in the same manner as is done in the latter part of* §12. *Do not use methods of later sections.*

12.3. Prove $q \supset r \supset . p \supset q \supset . p \supset r$ as a theorem of P_1. Use this theorem in order to give proofs of †122 and †123 which are briefer than those above, in the sense that they can be more briefly presented.[178]

12.4. Use the result of 12.3 in order to prove the *transitive law of (material) implication*, $p \supset q \supset . q \supset r \supset . p \supset r$, as a theorem of P_1. (One method is to apply the self-distributive law to the result of 12.3, then to use $p \supset q \supset . q \supset r \supset . p \supset q$.)

---

[176]By a different method than that indicated here, proved by S. C. Kleene in the *Annals of Mathematics*, vol. 35 (1934), pp. 531–532. Kleene's proof is carried out for a different logistic system than that of the text, but the question involved is the same in all essentials.

The first of the two metatheorems of this exercise is proved below as **143, and analogous metatheorems for the logistic system P_2 are proved in the next chapter. However, the reader should carry through the present exercise without looking forward at these later proofs. Or, alternatively, if the proofs given below are followed, they should be written out more fully, and in particular, details should be given of the proof of the lemma which is used below in proving **143.

[177]Cf. Karl Menger in *Ergebnisse eines mathematischen Kolloquiums*, no. 3 (1932), pp. 22–23; Łukasiewicz in footnote 5 (credited to Jaśkowski) of a paper in *Comptes Rendus des Séances de la Société des Sciences et des Lettres de Varsovie*, Classe III, vol. 24 (1932), pp. 153–183; Karl Schröter in *Axiomatisierung der Fregeschen Aussagenkalküle* (1943).

[178]Not necessarily in the sense that the proof written in full consists of a shorter sequence of wffs.

12.5. Prove $p \supset q \supset p \supset . p \supset f \supset p$ as a theorem of P_1. (Use †123, 12.4.)

12.6. Prove *Peirce's law*, $p \supset q \supset p \supset p$ as a theorem of P_1. (Apply the self-distributive law to $p \supset f \supset . p \supset f$, and use the result of 12.5.)

12.7. Let P_W be the logistic system which has the same primitive symbols, formation rules, and rules of inference as P_1 and which has as its axioms the transitive law of implication, Peirce's law, †102, and †122. Prove the following in order, as theorems of P_W, and hence show that P_1 and P_W are equivalent systems in the sense that they have the same theorems:

$$[p \supset . p \supset q] \supset . p \supset q$$

$$p \supset . p \supset q \supset q \qquad \textit{(Law of assertion.)}$$

$$[p \supset . q \supset r] \supset . q \supset . p \supset r \quad \textit{(Law of commutation.)}$$

$$q \supset r \supset . p \supset q \supset . p \supset r$$

$$s \supset [p \supset q] \supset . s \supset p \supset . s \supset q$$

$$p \supset f \supset f \supset p$$

Carry out the proofs in such a way that no use is made of the fourth axiom, †122, except in the proof of the last theorem, $p \supset f \supset f \supset p$.

12.8. Prove as theorems of P_W, without making use of the fourth axiom, †122: $q \supset r \supset r \supset . p \supset q \supset r \supset r$; $r \supset . p \supset . q \supset r$; $p \supset r \supset r \supset . q \supset r \supset . p \supset q \supset r$; $p \supset r \supset . p \supset q \supset r \supset r$.

12.9. For each of the three following interpretations of P_1 (cf. §10), state the remaining semantical rules, and discuss the *soundness* of the interpretation in the sense of §07: (1) Rules a, b, c are retained, but $[\mathbf{A} \supset \mathbf{B}]$ denotes t if **A** and **B** are any constants. (2) Rules a, b, c are retained, but $[\mathbf{A} \supset \mathbf{B}]$ denotes t if **A** and **B** denote the same truth-value, $[\mathbf{A} \supset \mathbf{B}]$ denotes f if **A** and **B** denote different truth-values. (3) Rules a, d are retained, but the variables (so-called) are interpreted as constants denoting t.

13. The deduction theorem. A *variant* of a wff **A** of P_1 is a wff obtained from **A** by alphabetic changes of the variables of such a sort that two occurrences of the same variable in **A** remain occurrences of the same variable, and two occurrences of distinct variables in **A** remain occurrences of distinct variables. Thus if $\mathbf{a}_1, \mathbf{a}_2, \ldots, \mathbf{a}_n$ are distinct variables, and $\mathbf{b}_1, \mathbf{b}_2, \ldots, \mathbf{b}_n$ are distinct variables, and there is no variable among $\mathbf{b}_1, \mathbf{b}_2, \ldots, \mathbf{b}_n$ which occurs in **A** and does not occur among $\mathbf{a}_1, \mathbf{a}_2, \ldots, \mathbf{a}_n$, then

$$\mathrm{S}^{\mathbf{a}_1\mathbf{a}_2\ldots\mathbf{a}_n}_{\mathbf{b}_1\mathbf{b}_2\ldots\mathbf{b}_n}\mathbf{A}|$$

is a variant of **A**. (Variants of †102, for example, are $r \supset . s \supset r$ and $q \supset . p \supset q$, but not $p \supset . r \supset r$ or $p \supset . p \supset p$.)

It is clear that if **B** is a variant of **A**, then **A** is a variant of **B**. And any variant of a variant of **A** is a variant of **A**. Also of course any wff **A** is a variant of itself.

In many ways, two wffs which are variants of one another serve the same purposes. In particular, in view of *101, every variant of a theorem is a theorem. Also, if we alter the system P_1 by replacing one or more axioms by variants of them, the theorems remain the same. In the case of theorems to which verbal names have been assigned (e.g., "the self-distributive law of implication," "Peirce's law," etc.), we shall use the same name also for any variant of the theorem.

A finite sequence of wffs is called a *variant proof* if each wff is either a variant of an axiom or is immediately inferred from preceding wffs in the sequence by one of the rules of inference. Evidently the final wff in a variant proof is always a theorem, since every variant of an axiom is a theorem; and we shall call the variant proof a variant proof *of* its final wff.

A finite sequence of wffs, $\mathbf{B}_1, \mathbf{B}_2, \ldots, \mathbf{B}_m$, is called a *proof from the hypotheses* $\mathbf{A}_1, \mathbf{A}_2, \ldots, \mathbf{A}_n$ if for each i either: (1) $\mathbf{B}_i$ is one of $\mathbf{A}_1, \mathbf{A}_2, \ldots, \mathbf{A}_n$; or (2) $\mathbf{B}_i$ is a variant of an axiom; or (3) $\mathbf{B}_i$ is inferred according to *100 from major premiss $\mathbf{B}_j$ and minor premiss $\mathbf{B}_k$, where $j < i$, $k < i$; or (4) $\mathbf{B}_i$ is inferred, according to *101, by substitution in the premiss $\mathbf{B}_j$, where $j < i$, and where the variable substituted for does not occur in $\mathbf{A}_1, \mathbf{A}_2, \ldots, \mathbf{A}_n$. Such a finite sequence of wffs, $\mathbf{B}_m$ being the final formula of the sequence, is called more explicitly a proof *of* $\mathbf{B}_m$ from the hypotheses $\mathbf{A}_1, \mathbf{A}_2, \ldots, \mathbf{A}_n$; and we use the notation

$$\mathbf{A}_1, \mathbf{A}_2, \ldots, \mathbf{A}_n \vdash \mathbf{B}_m$$

to mean: there is a proof of $\mathbf{B}_m$ from the hypotheses $\mathbf{A}_1, \mathbf{A}_2, \ldots, \mathbf{A}_n$.

Observe that the sign $\vdash$ is not a symbol belonging to the logistic system P_1 nor is it part of any schema of abbreviation of wffs of P_1, but rather it belongs to the syntax language (like the notation "S |" or the abbreviation "wff") and is used in making statements about the wffs of P_1.

The use of the sign $\vdash$ which was introduced in §12 may be regarded as amounting to a special case of the foregoing, namely the special case that $n = 0$. For a proof of $\mathbf{B}_m$ from no hypotheses is the same as a variant proof of $\mathbf{B}_m$; and we may now read the notation $\vdash \mathbf{B}_m$ either as meaning that there exists a variant proof of $\mathbf{B}_m$ or as meaning that $\mathbf{B}_m$ is a theorem (the two being trivially equivalent).

In the definition of proof from hypotheses, the condition attached to (4) should be especially noted, that the variable substituted for must not be one of the variables occurring in $\mathbf{A}_1, \mathbf{A}_2, \ldots, \mathbf{A}_n$. For example, although $q \supset f \supset f \supset f$ results from $q \supset f \supset f$ by substitution of $q \supset f$ for q, it is false that $q \supset f \supset f \vdash q \supset f \supset f \supset f$. On the other hand it is true that $q \supset f \supset f \vdash q$, by *100 and an appropriate variant of †104.

After these preliminaries, we are ready to state and prove the metatheorem which constitutes the principal topic of this section:

*130. If $\mathbf{A}_1, \mathbf{A}_2, \ldots, \mathbf{A}_n \vdash \mathbf{B}$, then $\mathbf{A}_1, \mathbf{A}_2, \ldots, \mathbf{A}_{n-1} \vdash \mathbf{A}_n \supset \mathbf{B}$.
(*The deduction theorem.*)

Proof. Let $\mathbf{B}_1, \mathbf{B}_2, \ldots, \mathbf{B}_m$ be a proof of $\mathbf{B}$ from the hypotheses $\mathbf{A}_1, \mathbf{A}_2, \ldots, \mathbf{A}_n$ ($\mathbf{B}_m$ being therefore the same as $\mathbf{B}$). And construct first the finite sequence of wffs, $\mathbf{A}_n \supset \mathbf{B}_1, \mathbf{A}_n \supset \mathbf{B}_2, \ldots, \mathbf{A}_n \supset \mathbf{B}_m$. We shall show how to insert a finite number of additional wffs in this sequence so that the resulting sequence is a proof of $\mathbf{A}_n \supset \mathbf{B}_m$, i.e., of $\mathbf{A}_n \supset \mathbf{B}$, from the hypotheses $\mathbf{A}_1, \mathbf{A}_2, \ldots, \mathbf{A}_{n-1}$. The inserted wffs will be put in before each of the wffs $\mathbf{A}_n \supset \mathbf{B}_i$ in order in such a way that, after completing the insertions as far as a particular $\mathbf{A}_n \supset \mathbf{B}_i$, the whole sequence of wffs up to that point is a proof of $\mathbf{A}_n \supset \mathbf{B}_i$ from the hypotheses $\mathbf{A}_1, \mathbf{A}_2, \ldots, \mathbf{A}_{n-1}$.[179]

In fact consider a particular $\mathbf{A}_n \supset \mathbf{B}_i$, and, if $i > 1$, suppose that the insertions have been completed as far as $\mathbf{A}_n \supset \mathbf{B}_{i-1}$. The following five cases arise:

Case 1a: $\mathbf{B}_i$ is $\mathbf{A}_n$. Then $\mathbf{A}_n \supset \mathbf{B}_i$ is $\mathbf{A}_n \supset \mathbf{A}_n$. Insert nine wffs before $\mathbf{A}_n \supset \mathbf{B}_i$, constituting namely a variant proof of an appropriate variant of †120 from which $\mathbf{A}_n \supset \mathbf{B}_i$ can be inferred by substitution.

Case 1b: $\mathbf{B}_i$ is one of $\mathbf{A}_1, \mathbf{A}_2, \ldots, \mathbf{A}_{n-1}$, say $\mathbf{A}_r$. Then $\mathbf{A}_r \supset . \mathbf{A}_n \supset \mathbf{B}_i$ is $\mathbf{A}_r \supset . \mathbf{A}_n \supset \mathbf{A}_r$. From an appropriate variant of †102, $\mathbf{A}_r \supset . \mathbf{A}_n \supset \mathbf{B}_i$ can be inferred in two steps by substitution (*101). Before $\mathbf{A}_n \supset \mathbf{B}_i$ insert first the three wffs which show this, then $\mathbf{A}_r$. From the last two of these four wffs, namely $\mathbf{A}_r \supset . \mathbf{A}_n \supset \mathbf{B}_i$ and $\mathbf{A}_r$, $\mathbf{A}_n \supset \mathbf{B}_i$ can be inferred by *modus ponens* (*100).

Case 2: $\mathbf{B}_i$ is a variant of an axiom. Following the same plan as in case 1b, insert four wffs before $\mathbf{A}_n \supset \mathbf{B}_i$, namely first a variant proof of $\mathbf{B}_i \supset . \mathbf{A}_n \supset \mathbf{B}_i$ (in two steps by substitution from a variant of †102), then $\mathbf{B}_i$ (a variant of an axiom).

Case 3: $\mathbf{B}_i$ is inferred by *modus ponens* from major premiss $\mathbf{B}_j$ and minor premiss $\mathbf{B}_k$, where $j < i$, $k < i$. Then $\mathbf{B}_j$ is $\mathbf{B}_k \supset \mathbf{B}_i$. Before $\mathbf{A}_n \supset \mathbf{B}_i$ insert first the four wffs which show the inference of $\mathbf{A}_n \supset \mathbf{B}_j \supset . \mathbf{A}_n \supset \mathbf{B}_k \supset .$

[179]Thus in effect the method of the proof is that of mathematical induction with respect to m (or i).

$\mathbf{A}_n \supset \mathbf{B}_i$, by three successive substitutions, from a variant of †103; then after these the wff $\mathbf{A}_n \supset \mathbf{B}_k \supset . \mathbf{A}_n \supset \mathbf{B}_i$ (which can be inferred by *modus ponens*, and from which then $\mathbf{A}_n \supset \mathbf{B}_i$ can be inferred by *modus ponens*, since the necessary minor premisses, $\mathbf{A}_n \supset \mathbf{B}_j$ and $\mathbf{A}_n \supset \mathbf{B}_k$, are among the earlier wffs already present in the sequence being constructed).

Case 4: $\mathbf{B}_i$ is inferred, according to *101, by substitution in $\mathbf{B}_j$, where $j < i$ and where the variable substituted for does not occur in $\mathbf{A}_1, \mathbf{A}_2, \ldots, \mathbf{A}_n$. No wffs need be inserted before $\mathbf{A}_n \supset \mathbf{B}_i$, as the same substitution suffices to infer $\mathbf{A}_n \supset \mathbf{B}_i$ from $\mathbf{A}_n \supset \mathbf{B}_j$ (here, of course, it is essential that the variable substituted for does not occur in $\mathbf{A}_n$).

As the special case of the deduction theorem in which $n = 1$ we have the following corollary:

*131. If $\mathbf{A} \vdash \mathbf{B}$, then $\vdash \mathbf{A} \supset \mathbf{B}$.

In connection with the deduction theorem we shall need also the three following metatheorems:

*132. If $\mathbf{A}_1, \mathbf{A}_2, \ldots, \mathbf{A}_n \vdash \mathbf{B}$, then $\mathbf{C}_1, \mathbf{C}_2, \ldots, \mathbf{C}_r, \mathbf{A}_1, \mathbf{A}_2, \ldots, \mathbf{A}_n \vdash \mathbf{B}$.

Proof. Let $\mathbf{a}_1, \mathbf{a}_2, \ldots, \mathbf{a}_l$ be the complete list (in alphabetic order) of those variables which occur in $\mathbf{C}_1, \mathbf{C}_2, \ldots, \mathbf{C}_r$ but not in $\mathbf{A}_1, \mathbf{A}_2, \ldots, \mathbf{A}_n$. If the given proof of $\mathbf{B}$ from the hypotheses $\mathbf{A}_1, \mathbf{A}_2, \ldots, \mathbf{A}_n$ is not also a proof of $\mathbf{B}$ from the hypotheses $\mathbf{C}_1, \mathbf{C}_2, \ldots, \mathbf{C}_r, \mathbf{A}_1, \mathbf{A}_2, \ldots, \mathbf{A}_n$, it can only be because it involves substitutions for some of the variables $\mathbf{a}_1, \mathbf{a}_2, \ldots, \mathbf{a}_l$. Therefore let $\mathbf{c}_1, \mathbf{c}_2, \ldots, \mathbf{c}_l$ be variables which are all distinct and which do not occur in $\mathbf{C}_1, \mathbf{C}_2, \ldots, \mathbf{C}_r, \mathbf{A}_1, \mathbf{A}_2, \ldots, \mathbf{A}_n$ or in the given proof of $\mathbf{B}$ from the hypotheses $\mathbf{A}_1, \mathbf{A}_2, \ldots, \mathbf{A}_n$ (to be specific, say that $\mathbf{c}_1, \mathbf{c}_2, \ldots, \mathbf{c}_l$ are the first l such variables in the alphabetic order of the variables). And throughout the given proof of $\mathbf{B}$ from the hypotheses $\mathbf{A}_1, \mathbf{A}_2, \ldots, \mathbf{A}_n$ replace $\mathbf{a}_1, \mathbf{a}_2, \ldots, \mathbf{a}_l$ by $\mathbf{c}_1, \mathbf{c}_2, \ldots, \mathbf{c}_l$ respectively. The result is a proof from the hypotheses $\mathbf{C}_1, \mathbf{C}_2, \ldots, \mathbf{C}_r, \mathbf{A}_1, \mathbf{A}_2, \ldots, \mathbf{A}_n$ of

$$\mathrm{S}^{\mathbf{a}_1\mathbf{a}_2\ldots\mathbf{a}_l}_{\mathbf{c}_1\mathbf{c}_2\ldots\mathbf{c}_l}\, \mathbf{B}|.$$

To obtain a proof of $\mathbf{B}$ from the same hypotheses, it is then necessary only to add l additional steps, substituting successively $\mathbf{a}_1$ for $\mathbf{c}_1$, $\mathbf{a}_2$ for $\mathbf{c}_2$, ..., $\mathbf{a}_l$ for $\mathbf{c}_l$.

*133. If $\vdash \mathbf{B}$, then $\mathbf{C}_1, \mathbf{C}_2, \ldots, \mathbf{C}_r \vdash \mathbf{B}$.

Proof. This is the special case of *132 in which $n = 0$.

***134.** If every wff which occurs at least once in the list $\mathbf{A}_1, \mathbf{A}_2, \ldots, \mathbf{A}_n$ also occurs at least once in the list $\mathbf{C}_1, \mathbf{C}_2, \ldots, \mathbf{C}_r$, and if $\mathbf{A}_1, \mathbf{A}_2, \ldots, \mathbf{A}_n \vdash \mathbf{B}$, then $\mathbf{C}_1, \mathbf{C}_2, \ldots, \mathbf{C}_r \vdash \mathbf{B}$.

Proof. Since it is clearly indifferent, in connection with proof from hypotheses, in what order the hypotheses are arranged, or how many times a particular hypothesis is repeated, this is a corollary of *132.

Importance of the deduction theorem to the metatheory (syntax and semantics) of the system P is clear—as a matter of showing the adequacy of the system, in a certain direction, for the purposes for which it is intended, namely for formalization of the use of sentence connectives (see §05) and of inferences involving them.

It is also possible to make use of the deduction theorem in the role of a derived rule of inference (cf. §12), since the proof of the deduction theorem provides an effective method according to which, whenever a proof of $\mathbf{B}$ from the hypotheses $\mathbf{A}_1, \mathbf{A}_2, \ldots, \mathbf{A}_n$ is given, it is possible to obtain a proof of $\mathbf{A}_n \supset \mathbf{B}$ from the hypotheses $\mathbf{A}_1, \mathbf{A}_2, \ldots, \mathbf{A}_{n-1}$—hence by repetitions of the method to obtain a proof of $\mathbf{A}_1 \supset \centerdot\, \mathbf{A}_2 \supset \centerdot \ldots \mathbf{A}_{n-1} \supset \centerdot\, \mathbf{A}_n \supset \mathbf{B}$.

As examples of this use of the deduction theorem as a derived rule, we present the following alternative proofs of the last two theorems of §12:

Proof of †122. By simultaneous substitution in †102:

$$\vdash f \supset \centerdot\, p \supset f \supset f$$

Hence by *modus ponens*:

$$f \vdash p \supset f \supset f$$

Hence by †104 and *modus ponens*:

$$f \vdash p$$

Hence by *131:

$$\vdash f \supset p$$

Proof of †123. By *modus ponens*:

$$p \supset f,\ p \vdash f$$

By the variant, $f \supset q$, of †122 and *modus ponens*:

$$p \supset f,\ p \vdash q$$

Hence by *130:

$$p \supset f \vdash p \supset q$$

Hence by *130 again (or, what comes to the same thing, by *131):

$$\vdash p \supset f \supset \centerdot\, p \supset q$$

14. Some further theorems and metatheorems of P_1. We go on to prove three additional theorems of P_1, using the deduction theorem in order to present proofs more briefly.

†140.[180] $p \supset . q \supset f \supset . p \supset q \supset f$

By two applications of *modus ponens*:

$$p, q \supset f, p \supset q \vdash f$$

By three applications of the deduction theorem:

$$\vdash p \supset . q \supset f \supset . p \supset q \supset f$$

†141. $p \supset q \supset . q \supset r \supset . p \supset r$

By two applications of *modus ponens*:

$$p \supset q, q \supset r, p \vdash r$$

Hence by the deduction theorem:

$$\vdash p \supset q \supset . q \supset r \supset . p \supset r$$

(As already indicated in 12.4, †141 is known as the *transitive law of implication.*)

†142. $p \supset f \supset r \supset . p \supset r \supset r$

By the transitive law of implication (i.e., by substitution in †141 or a suitable variant, and *modus ponens*):

$$p \supset r, r \supset f \vdash p \supset f$$

Hence by two applications of *modus ponens*:

$$p \supset f \supset r, p \supset r, r \supset f \vdash f$$

Hence by the deduction theorem:

$$p \supset f \supset r, p \supset r \vdash r \supset f \supset f$$

Hence by a variant of †104 and *modus ponens*:

$$p \supset f \supset r, p \supset r \vdash r$$

[180] $q \supset f$ may be read in words either as "not q" or as "q is false" (where "is false" is of course not the semantical term but merely a synonym of "not" or "implies falsehood"). †140 may be read in words: "If p, if not q, then p does not imply q." Similarly we may read †141 in words thus: "If p implies q, if q implies r, p implies r." And †142 thus: "If not p implies r, if p implies r, then r."

Hence by the deduction theorem:

$$\vdash p \supset f \supset r \supset . p \supset r \supset r$$

We add also the following metatheorem, which will be needed in the next section:

**143. If a formula is wf and consists of more than one symbol, it has the form $[\mathbf{A} \supset \mathbf{B}]$ in one and only one way.

Proof. It is immediate, from the definition of a wff, that a wff of more than one symbol has the form $[\mathbf{A} \supset \mathbf{B}]$ in at least one way. We must show that it cannot have this form in more than one way.

We use the same process of counting brackets which is described in §10. Namely we start at the beginning (or left) of a formula and proceed from left to right, counting each occurrence of [as $+1$ and each occurrence of] as -1, and adding as we go. The number which we thus assign to an occurrence of a bracket will be called *the number of* that occurrence of a bracket in the formula.

It follows from the definition of a wff that, if a wff contains the symbol $\supset$, it must begin with an occurrence of [and end with an occurrence of]; these we shall call respectively the *initial bracket* and the *final bracket* of the wff. By mathematical induction with respect to the total number of occurrences of $\supset$ we establish the following lemma: *The number of an occurrence of a bracket in a wff is positive, except in the case of the final bracket, which has the number* 0.

Now suppose that $[\mathbf{A} \supset \mathbf{B}]$ and $[\mathbf{C} \supset \mathbf{D}]$ are the same wff. Case 1: If **A** contains no occurrence of $\supset$, it must consist of a single symbol, either a variable or f; since **C** begins with the same symbol as **A**, it follows that **C** has no initial bracket and therefore cannot contain the symbol $\supset$; therefore **C** must be identical with **A**. Case 2: If **C** contains no occurrence of $\supset$, it follows by the same argument that **A** must be identical with **C**. Case 3: If **A** and **C** both contain the symbol $\supset$, then the final bracket of **A** is the first occurrence of a bracket with the number 0 in **A**, and therefore is the second occurrence of a bracket with the number 1 in $[\mathbf{A} \supset \mathbf{B}]$; and the final bracket of **C** is the first occurrence of a bracket with the number 0 in **C**, and therefore is the second occurrence of a bracket with the number 1 in $[\mathbf{C} \supset \mathbf{D}]$; this makes the final bracket of **A** and the final bracket of **C** coincide, and so makes **A** and **C** identical. Finally, since it follows in all three cases that **A** and **C** are identical, it is then obvious that **B** and **D** must be identical.

We do not continue further with proofs of particular theorems of P_1, although there are many more theorems of the propositional calculus which will be of importance in later chapters. For all such theorems can be obtained by the more powerful method of the next section, to establish which the theorems and metatheorems that we already have are sufficient. Indeed in the next section we shall make direct use only of *100, *101, †102, †120, †123, *130, †140, †142,**143— other axioms, theorems, and metatheorems being used only so far as they contribute to the proof of these.

EXERCISES 14

14.0. Rewrite †140 and †142 in abbreviated form, using D2, D4, and D9.

14.1. From the hypotheses p and $p \supset q$ there is a proof of q, in one step by *modus ponens*. Hence by using the method provided in the proof of *130 we may obtain a proof of the *law of assertion*, $p \supset . p \supset q \supset q$. Simplify this proof by deleting all unnecessary repetitions of the same wff or variants of it, also by using †120 and †102 in order to prove $p \supset . r \supset r$ in a more direct manner. Present the resulting proof of the law of assertion in the style of §12, without making use of the deduction theorem or of theorems whose proof has been presented only by means of the deduction theorem.

14.2. Present a proof of †140 without making use of the deduction theorem or of theorems whose proof has been presented only by means of the deduction theorem. (The proof of §14 is impracticably cumbrous when presented without the aid of the deduction theorem; nevertheless we may make heuristic use of the idea of applying, to the proof of †140 as presented in §14, the method provided in the proof of *130.)

Present proofs of the following theorems in the style of §14, *making use of the deduction theorem and of any theorems and metatheorems which have been previously proved, either in the text or as exercises*:

14.3. $p \supset . q \supset r \supset . p \supset q \supset r$

14.4. $p \supset q \supset r \supset . p \supset r \supset r$

14.5. $p \supset r \supset r \supset . p \supset f \supset r$

14.6. $p \supset q \supset [r_1 \supset s] \supset . p \supset [r_2 \supset s] \supset . r_1 \supset . r_2 \supset s$

14.7. $p \dot{\vee} q \supset q \vee p$

14.8. $[p \supset q] \vee [q \supset p]$

14.9. Establish the following four derived rules of P_1 directly—without use of *130 or of the notion of a proof from hypotheses:

(1) If $\vdash \mathbf{B}$, then $\vdash \mathbf{A}_1 \supset . \mathbf{A}_2 \supset . \ldots \mathbf{A}_n \supset \mathbf{B}$.

(2) If every wff which occurs at least once in the list $\mathbf{A}_1, \mathbf{A}_2, \ldots, \mathbf{A}_n$

also occurs at least once in the list $\mathbf{C}_1, \mathbf{C}_2, \ldots \mathbf{C}_r$, and if $\vdash \mathbf{A}_1 \supset \centerdot \mathbf{A}_2 \supset \centerdot \ldots \mathbf{A}_n \supset \mathbf{B}$, then $\vdash \mathbf{C}_1 \supset \centerdot \mathbf{C}_2 \supset \centerdot \ldots \mathbf{C}_r \supset \mathbf{B}$.

(3) If $\mathbf{B}$ is one of $\mathbf{A}_1, \mathbf{A}_2, \ldots, \mathbf{A}_n$, then $\vdash \mathbf{A}_1 \supset \centerdot \mathbf{A}_2 \supset \centerdot \ldots \mathbf{A}_n \supset \mathbf{B}$.

(4) If every wff which occurs at least once in the list $\mathbf{A}_1, \mathbf{A}_2, \ldots, \mathbf{A}_n, \mathbf{B}_1, \mathbf{B}_2, \ldots, \mathbf{B}_m$ also occurs at least once in the list $\mathbf{C}_1, \mathbf{C}_2, \ldots, \mathbf{C}_r$, if $\vdash \mathbf{A}_1 \supset \centerdot \mathbf{A}_2 \supset \centerdot \ldots \mathbf{A}_n \supset \mathbf{A}$ and $\vdash \mathbf{B}_1 \supset \centerdot \mathbf{B}_2 \supset \centerdot \ldots \mathbf{B}_m \supset \centerdot \mathbf{A} \supset \mathbf{B}$, then $\vdash \mathbf{C}_1 \supset \centerdot \mathbf{C}_2 \supset \centerdot \ldots \mathbf{C}_r \supset \mathbf{B}$.

Explain in detail how these derived rules may be used as a substitute for the deduction theorem in presenting proofs of theorems of P_1. Illustrate by presenting proofs of the three theorems of §14 with the aid of these derived rules (and without the deduction theorem).[181]

15. Tautologies, the decision problem. Let $\mathbf{B}$ be a wff of P_1, let $\mathbf{a}_1, \mathbf{a}_2, \ldots, \mathbf{a}_n$ be distinct variables among which are all the variables occurring in $\mathbf{B}$, and let $a_1, a_2, \ldots, a_n$ be truth-values (each one either t or f). We define *the value of* $\mathbf{B}$ *for the values* $a_1, a_2, \ldots, a_n$ *of* $\mathbf{a}_1, \mathbf{a}_2, \ldots, \mathbf{a}_n$ by a recursion process which assigns values to the wf parts $\mathbf{C}$ of $\mathbf{B}$, in order of increasing number of occurrences of $\supset$ in $\mathbf{C}$, as follows. If $\mathbf{C}$ is f, the value of $\mathbf{C}$ is f; if $\mathbf{C}$ is $\mathbf{a}_i$, the value of $\mathbf{C}$ is a_i; if $\mathbf{C}$ is $[\mathbf{C}_1 \supset \mathbf{C}_2]$, the value of $\mathbf{C}$ is t in case either the value of $\mathbf{C}_2$ is t or the value of $\mathbf{C}_1$ is f, and the value of $\mathbf{C}$ is f in case the values of $\mathbf{C}_1$ and $\mathbf{C}_2$ are t and f respectively. By repetitions of this process a value, t or f, is ultimately assigned to $\mathbf{B}$, and this we call the value of $\mathbf{B}$ for the values $a_1, a_2, \ldots, a_n$ of $\mathbf{a}_1, \mathbf{a}_2, \ldots, \mathbf{a}_n$.

The uniqueness of the value of $\mathbf{B}$ for a given system of values of its variables follows as a consequence of **143.

A wff $\mathbf{B}$ of P_1 is called a *tautology* if its value is t for every system of values of its variables (the values being truth-values), a *contradiction* if its value is f for every system of values of its variables.

It will be seen that the foregoing recursion process, by which we obtain the value of $\mathbf{B}$ for a system of values of its variables, just follows the semantical rules given in §10 for the principal interpretation of P_1. But in §10 we understood "denoting" and "having values" as known kinds of meaning, and we used the

[181]These derived rules have a simpler character than that of the deduction theorem in the role of a derived rule. For, like our primitive rules of inference, they require as premisses only certain asserted wffs, and, when these are given, the check of the inference is effective. But when the deduction theorem is used as a derived rule, it is necessary to submit a finite sequence of wffs not as asserted but as constituting a proof from hypotheses, and only then is an effective check available.

On the other hand these derived rules, 14.9 (1)–(4), may easily be made an efficient substitute for the deduction theorem as a means of abbreviating the presentation of proofs. Advantages of the deduction theorem in this role are largely psychological and heuristic.

semantical rules in order to assign an interpretation to P_1 as a language designed for meaningful communication. On the other hand in the present section we use the same rules, otherwise substantially unchanged, in order to define abstractly a correspondence called "having *values*," between wffs (with given *values* of their variables) and truth-values. The word "values" at its two italicized occurrences is meant as a newly introduced technical term, with no reference to the idea of meaning, and the correspondence is defined abstractly, or syntactically, in the sense that it may be used independently of what interpretation (if any) is assigned to the logistic system P_1. Compare footnote 143.

The process provided in the definition for obtaining the value of **B** for a given system of values of its variables is effective (see the discussion of the notion of effectiveness in §07, and footnotes 118, 119). Since a wff **B** can have only a finite number of variables, and hence only a finite number of systems of values of its variables, this leads to an effective process for deciding whether **B** is a tautology or a contradiction or neither. As an illustration of this algorithm, we show the following verification that †103 is a tautology, adopting a convenient arrangement of the work that is due to Quine:

s	$\supset$	$[p$	$\supset$	$q]$	$\supset .$	s	$\supset$	p	$\supset .$	s	$\supset$	q
t	t	t	t	t	t	t	t	t	t	t	t	t
t	f	t	f	f	t	t	t	t	f	t	f	f
t	t	f	t	t	t	t	f	f	t	t	t	t
t	t	f	t	f	t	t	f	f	t	t	f	f
f	t	t	t	t	t	f	t	t	t	f	t	t
f	t	t	f	f	t	f	t	t	t	f	t	f
f	t	f	t	t	t	f	t	f	t	f	t	t
f	t	f	t	f	t	f	t	f	t	f	t	f

In detail the work is as follows. First the wff †103 is written on one line. The three variables occurring in it are s, p, q; all possible systems of values of these variables are written down in the form of three columns of t's and f's, one column below the first occurrence of each of these variables in the wff. Then below each remaining occurrence of a variable in the wff is copied the same column of t's and f's that stands below its first occurrence. Then systems of values are assigned to the various wf parts of the entire wff, in order of increasing length of the parts, the system of values assigned to each part being written as a column of t's and f's below the principal implication sign of that part. For example, the values assigned to $[p \supset q]$ appear in the column below the second implication sign of the entire wff; the t at the top of this column is obtained from the values t, t of p, q, in accordance with the rule given in the first paragraph of this section; the f next to the top of the column is obtained from the values t, f of p, q, in accordance with the same rule; and so on. Again, the values assigned to $s \supset [p \supset q]$ appear in the column below the first implication sign of the entire wff; the t at the top of this column is obtained from the values t, t of p, $[p \supset q]$; and so on. The reader should carry out the work in full and compare his result with that shown above. At the end of the work, the system of values of the entire wff appears in the column below its principal implication sign, and the fact that this column consists wholly of t's shows the wff to be a tautology.

A table showing the value of $p \supset q$ for every system of values of p, q is called a *truth-table* of $\supset$. Like *truth-tables* may be calculated also for each of the notations introduced in D2–11; e.g., in accordance with D4, the truth-table of $\vee$ will show the value of $p \supset q \supset q$ for every system of values of p, q, as this is worked out by the rule given in the first paragraph of this section. The complete list of truth-tables, including that of $\supset$, is as follows:

p	$\sim p$
t	f
f	t

p	q	$p \supset q$	$p \not\subset q$	$p \vee q$	pq	$p \equiv q$	$p \not\equiv q$	$p \subset q$	$p \not\supset q$	$p \bar{\vee} q$	$p \mid q$
t	t	t	f	t	t	t	f	t	f	f	f
t	f	f	f	t	f	f	t	t	t	f	t
f	t	t	t	t	f	f	t	f	f	f	t
f	f	t	f	f	f	t	f	t	f	t	t

Though these truth-tables show explicitly, e.g., the value of $p \vee q$ for given values of p and q, of course it is understood that they may be used with arbitrary wffs replacing the variables, e.g., to find the value of $\mathbf{C}_1 \vee \mathbf{C}_2$ for given values of $\mathbf{C}_1$ and $\mathbf{C}_2$.

When the above described algorithm for calculating the system of truth-values of a wff is to be applied to a wff abbreviated by means of D1–11, the wff may be first rewritten in unabbreviated form and the algorithm then applied. In theoretical discussions we shall assume that this is done. But in practice it is more efficient to leave the wff in abbreviated form and to use the complete foregoing list of truth-tables. For example, the verification that

$$t \supset p \bar{\vee} q \equiv . p \not\equiv q \not\equiv pq \supset f$$

is a tautology, using the abbreviated form of this wff, is arranged as follows:

t	$\supset$	p	$\bar{\vee}$	q	$\equiv$	$\cdot p$	$\not\equiv$	q	$\not\equiv$	p		q	$\supset$	f
t	f	t	f	t	t	t	f	t	t	t	t	t	f	f
t	f	t	f	f	t	t	t	f	t	t	f	f	f	f
t	f	f	f	t	t	f	t	t	t	f	f	t	f	f
t	t	f	t	f	t	f	f	f	f	f	f	f	t	f

We now prove the metatheorem:

****150.** Every theorem of P_1 is a tautology.

Proof. We first establish the following lemma: *If* $\mathbf{a}_1, \mathbf{a}_2, \ldots, \mathbf{a}_n, \mathbf{b}$ *are distinct variables among which are all the variables occurring in* **A** *and all those occurring in* **B**, *and if, for the values* $a_1, a_2, \ldots, a_n, a$ *of* $\mathbf{a}_1, \mathbf{a}_2, \ldots, \mathbf{a}_n, \mathbf{b}$, *the value of* **B** *is* b *and the value of* $\mathrm{S}^{\mathbf{b}}_{\mathbf{B}}\mathbf{A}|$ *is* c, *then the value of* **A** *for the values* $a_1, a_2, \ldots, a_n, b$ *of* $\mathbf{a}_1, \mathbf{a}_2, \ldots, \mathbf{a}_n, \mathbf{b}$ *is* c.

The lemma is obvious if **A** consists of a single symbol. And we then proceed by mathematical induction with respect to the total number of occurrences of $\supset$ in **A**. If **A** is $\mathbf{A}_1 \supset \mathbf{A}_2$, then $\mathrm{S}^{\mathbf{b}}_{\mathbf{B}}\mathbf{A}|$ is $\mathrm{S}^{\mathbf{b}}_{\mathbf{B}}\mathbf{A}_1| \supset \mathrm{S}^{\mathbf{b}}_{\mathbf{B}}\mathbf{A}_2|$. Suppose that, for the values $a_1, a_2, \ldots, a_n, a$ of $\mathbf{a}_1, \mathbf{a}_2, \ldots, \mathbf{a}_n, \mathbf{b}$, the value of **B** is b, and the value of $\mathrm{S}^{\mathbf{b}}_{\mathbf{B}}\mathbf{A}|$ is c, and the value of $\mathrm{S}^{\mathbf{b}}_{\mathbf{B}}\mathbf{A}_1|$ is c_1, and the value of $\mathrm{S}^{\mathbf{b}}_{\mathbf{B}}\mathbf{A}_2|$ is c_2. Then c is f, if c_1 is t and c_2 is f; and c is t in all other cases. By the hypothesis of induction we have, for the values $a_1, a_2, \ldots, a_n, b$ of $\mathbf{a}_1, \mathbf{a}_2, \ldots, \mathbf{a}_n, \mathbf{b}$, that the value of $\mathbf{A}_1$ is c_1 and the value of $\mathbf{A}_2$ is c_2; hence the value of **A** is f if c_1 is t and c_2 is f, and the value of **A** is t in all other cases; i.e., the value of **A** is c.

The lemma then follows by mathematical induction.—For the rule of substitution, *101, we have as an immediate consequence of the lemma that, if the conclusion $\mathrm{S}^{\mathbf{b}}_{\mathbf{B}}\mathbf{A}|$ has the value f for some system of values of the variables, then the premiss **A** must have the value f for some system of values of the variables. Therefore if the premiss **A** of *101 is a tautology, the conclusion $\mathrm{S}^{\mathbf{b}}_{\mathbf{B}}\mathbf{A}|$ must be a tautology.

For the rule of *modus ponens*, *100, if the minor premiss **A** is a tautology, and if the conclusion **B** has the value f for some system of values of the variables (of **A** and **B**), then for the same system of values of the variables it follows directly from the definition of the value of a wff that the value of the major premiss $\mathbf{A} \supset \mathbf{B}$ is f. Therefore if both premisses of *100 are tautologies, the conclusion must be a tautology.

We have thus shown that the two rules of inference of P_1 preserve tautologies, in the sense that, if the premiss or premisses are tautologies, the conclusion is a tautology. We leave it to the reader to verify that the three axioms of P_1 are tautologies. **150 then follows.

*151. Let **B** be a wff of P_1, let $\mathbf{a}_1, \mathbf{a}_2, \ldots, \mathbf{a}_n$ be distinct variables among which are all the variables occurring in **B**, and let $a_1, a_2, \ldots, a_n$ be truth-values. Further, let $\mathbf{A}_i$ be $\mathbf{a}_i$ or $\mathbf{a}_i \supset f$ according as a_i is t or f; and let $\mathbf{B}'$ be **B** or $\mathbf{B} \supset f$ according as the value of **B** for the values $a_1, a_2, \ldots, a_n$ of $\mathbf{a}_1, \mathbf{a}_2, \ldots, \mathbf{a}_n$ is t or f. Then $\mathbf{A}_1, \mathbf{A}_2, \ldots, \mathbf{A}_n \vdash \mathbf{B}'$.

In order to prove that

$$(1) \qquad \mathbf{A}_1, \mathbf{A}_2, \ldots, \mathbf{A}_n \vdash \mathbf{B}'$$

we proceed by mathematical induction with respect to the number of occurrences of $\supset$ in **B**.

If there are no occurrences of $\supset$ in **B**, then **B** is either f or one of the variables $\mathbf{a}_i$. In case **B** is f, we have that $\mathbf{B}'$ is $f \supset f$, and hence (1) follows by substitution in an appropriate variant of †120. In case **B** is $\mathbf{a}_i$, we have that $\mathbf{B}'$ is the same wff as $\mathbf{A}_i$, and (1) follows trivially, the proof of $\mathbf{B}'$ from the hypotheses $\mathbf{A}_1, \mathbf{A}_2, \ldots, \mathbf{A}_n$ consisting of the single wff $\mathbf{B}'$.

Suppose that there are occurrences of $\supset$ in **B**. Then **B** is $\mathbf{B}_1 \supset \mathbf{B}_2$. By the hypothesis of induction,

$$(2) \qquad \mathbf{A}_1, \mathbf{A}_2, \ldots, \mathbf{A}_n \vdash \mathbf{B}_1',$$

$$(3) \qquad \mathbf{A}_1, \mathbf{A}_2, \ldots, \mathbf{A}_n \vdash \mathbf{B}_2',$$

where $\mathbf{B}_1'$ is $\mathbf{B}_1$ or $\mathbf{B}_1 \supset f$ according as the value of $\mathbf{B}_1$ for the values $a_1, a_2, \ldots, a_n$ of $\mathbf{a}_1, \mathbf{a}_2, \ldots, \mathbf{a}_n$ is t or f, and $\mathbf{B}_2'$ is $\mathbf{B}_2$ or $\mathbf{B}_2 \supset f$ according as the value of $\mathbf{B}_2$ for the values $a_1, a_2, \ldots, a_n$ of $\mathbf{a}_1, \mathbf{a}_2, \ldots, \mathbf{a}_n$ is t or f. In case $\mathbf{B}_2'$ is $\mathbf{B}_2$, we have that $\mathbf{B}'$ is $\mathbf{B}_1 \supset \mathbf{B}_2$, and (1) follows from (3) by substitution in an appropriate variant of †102 and *modus ponens*. In case $\mathbf{B}_1'$ is $\mathbf{B}_1 \supset f$, we have again that $\mathbf{B}'$ is $\mathbf{B}_1 \supset \mathbf{B}_2$, and (1) follows from (2) by substitution in an appropriate variant of †123 and *modus ponens*. There remains only the case that $\mathbf{B}_1'$ is $\mathbf{B}_1$ and $\mathbf{B}_2'$ is $\mathbf{B}_2 \supset f$, and in this case $\mathbf{B}'$ is $\mathbf{B}_1 \supset \mathbf{B}_2 \supset f$, and (1) follows from (2) and (3) by substitution in an appropriate variant of †140 and two uses of *modus ponens*.

Therefore *151 is proved by mathematical induction.

The proof of *151 is effective in the sense that it provides an effective method for finding a proof of $\mathbf{B}'$ from the hypotheses $\mathbf{A}_1, \mathbf{A}_2, \ldots, \mathbf{A}_n$. If **B** has no occurrences of $\supset$, this is provided directly. If **B** has occurrences of $\supset$, the proof provides directly an effective reduction of the problem of finding a proof of $\mathbf{B}'$ from the hypotheses $\mathbf{A}_1, \mathbf{A}_2, \ldots, \mathbf{A}_n$ to the two problems of finding proofs of $\mathbf{B}_1'$ and $\mathbf{B}_2'$ from the hypothesis $\mathbf{A}_1, \mathbf{A}_2, \ldots, \mathbf{A}_n$; the same reduction may then be repeated upon the two latter problems, and so on;

after a finite number of repetitions the process of reduction must terminate, yielding effectively a proof of $\mathbf{B}'$ from the hypotheses $\mathbf{A}_1, \mathbf{A}_2, \ldots, \mathbf{A}_n$.

We now go on to proof of the converse of **150, which will also be effective.

*152. If **B** is a tautology, $\vdash \mathbf{B}$.

Proof. Let $\mathbf{a}_1, \mathbf{a}_2, \ldots, \mathbf{a}_n$ be the variables of **B**, and for any system of values $a_1, a_2, \ldots, a_n$ of $\mathbf{a}_1, \mathbf{a}_2, \ldots, \mathbf{a}_n$ let $\mathbf{A}_1, \mathbf{A}_2, \ldots, \mathbf{A}_n$ be as in *151. The $\mathbf{B}'$ of *151 is **B**, because **B** is a tautology. Therefore, by *151,

$$\mathbf{A}_1, \mathbf{A}_2, \ldots, \mathbf{A}_n \vdash \mathbf{B}.$$

This holds for either choice of a_n, i.e., whether a_n is f or t, and so we have both

$$\mathbf{A}_1, \mathbf{A}_2, \ldots, \mathbf{A}_{n-1}, \mathbf{a}_n \supset f \vdash \mathbf{B}$$

and

$$\mathbf{A}_1, \mathbf{A}_2, \ldots, \mathbf{A}_{n-1}, \mathbf{a}_n \vdash \mathbf{B}.$$

By the deduction theorem,

$$\mathbf{A}_1, \mathbf{A}_2, \ldots, \mathbf{A}_{n-1} \vdash \mathbf{a}_n \supset f \supset \mathbf{B},$$

$$\mathbf{A}_1, \mathbf{A}_2, \ldots, \mathbf{A}_{n-1} \vdash \mathbf{a}_n \supset \mathbf{B}.$$

Hence, by substitution in an appropriate variant of †142 and two uses of *modus ponens*,

$$\mathbf{A}_1, \mathbf{A}_2, \ldots, \mathbf{A}_{n-1} \vdash \mathbf{B}.$$

This shows the elimination of the hypothesis $\mathbf{A}_n$. The same process may be repeated to eliminate the hypothesis $\mathbf{A}_{n-1}$, and so on, until all the hypotheses are eliminated.[182] Finally we obtain $\vdash \mathbf{B}$.

The decision problem of a logistic system is the problem to find an effective procedure or algorithm, a *decision procedure*, by which, for an arbitrary wff of the system, it is possible to determine whether or not it is a theorem (and if it is a theorem to obtain a proof of it[183]).

---

[182] Implicitly, therefore, the method of the proof is that of mathematical induction with respect to n. Cf. footnote 179.

[183] This parenthetic part of the problem will, in this book, always be included explicitly in solutions or partial solutions of the decision problem of a system. However, it might be dismissed as theoretically superfluous, on the ground that there always exists an effective enumeration of the proofs of a logistic system—as will appear in Chapter VIII—and that once a proof of a wff is known to exist, it may be found by searching through in order such an effective enumeration of all proofs. Possible objections to this are (1) that a procedure for finding something should not be called effective unless there is a predictable upper bound of the number of steps that will be required,

The effective procedure for recognizing tautologies, as described at the beginning of this section, and the effective proofs which have been given of **150 and *152, together constitute a solution of the decision problem of the logistic system P_1.

This solution of the decision problem of P_1 does not depend on any particular interpretation of P_1. Being purely syntactical in character it may be used under any interpretation of P_1, or even if no interpretation at all is adopted.

The decision problem in this sense we call more fully, the *decision problem for provability*, in a logistic system, or in a formalized language obtained by interpretation of the logistic system.

In the case of a formalized language there is also the *semantical decision problem*, as we shall call it, namely, to find an effective procedure for determining of an arbitrary sentence whether it is true in the semantical sense (§§04, 09), and of an arbitrary propositional form whether it is true for all values of its variables.[184] For the formalized language which is obtained by adopting the principal interpretation of P_1, the semantical decision problem is trivial, because the semantical rules, given at the end of §10, directly provide the required effective procedure. This triviality of the semantical decision problem, however, by no means holds for formalized languages in general, as the definition of truth contained in the semantical rules is often non-effective.

The decision problem for provability, as we have seen, is non-trivial even in the relatively simple case of the system P_1.

In view of the solution of the decision problem of P_1, the explicit presentation of proofs of particular theorems of P_1 is now no longer necessary. Whenever we require a particular theorem of P_1, it will be sufficient that we just write it down, leaving it to the reader to verify that it is a tautology and hence to find a proof of it by applying the procedure which is given in the proofs of *152 and *151. In particular we now add, on this basis, the five following theorems of P_1:

---

and (2) that not only the decision procedure itself ought to be effective, but also the demonstration of it ought to be effective in the sense that it proceeds by effectively producing the proof of the wff (when the proof exists). But these objections are not easy to maintain. Indeed the restriction on the notion of effectiveness, as proposed in (1), is vague, and the writer does not know how to make it definite without excluding procedures that must obviously be considered effective by common (informal) standards. The requirement proposed in (2) is in the direction of mathematical intuitionism—see Chapter XII—and must be regarded as radical from the point of view of classical mathematics.

[184]The writer once proposed the name "deducibility problem" for what is here called the decision problem for provability, the intention being to reserve the name "decision problem" either for the semantical decision problem or for what is called in §46 the *decision problem for validity*. It seems better, however, to use "decision problem" as a general name for problems to find an effective criterion (a decision procedure) for something, and to distinguish different decision problems by means of qualifying adjectives or phrases.

†153. $t \supset p \equiv p$

†154. $\sim\sim p \equiv p$

†155. $p \equiv q \supset . q \equiv p$

†156. $p \equiv q \supset . p \supset q$

†157. $p \equiv q \supset . q \equiv r \supset . p \equiv r$

†154 is the *complete law of double negation* (cf. footnote 163). †155 is the *commutative law of (material) equivalence*, and †157 is the *transitive law of (material) equivalence.*

Proofs of metatheorems of P_1 are also often greatly simplified by the solution of the decision problem. This is true, for example, in the case of the following:

*158. If **B** results from **A** by substitution of **N** for **M** at one or more places in **A** (not necessarily for all occurrences of **M** in **A**), and if ⊢ **M** ≡ **N**, then ⊢ **A** ≡ **B**.

Proof. Let $\mathbf{a}_1, \mathbf{a}_2, \ldots, \mathbf{a}_n$ be the complete list of variables occurring in **A** and **B** together. Since **M** ≡ **N** is a theorem, it is a tautology. Therefore, by the truth-table of ≡, **M** and **N** have the same value for every system of values of $\mathbf{a}_1, \mathbf{a}_2, \ldots, \mathbf{a}_n$. Since **B** is obtained from **A** by substitution of **N** for **M** at certain places, it follows that **A** and **B** have the same value for every system of values of $\mathbf{a}_1, \mathbf{a}_2, \ldots, \mathbf{a}_n$ (details of the proof of this, by mathematical induction with respect to the number of occurrences of ⊃ in **A**, are left to be supplied by the reader, using the result of exercise 12.1). Therefore, by the truth-table of ≡, we have that **A** ≡ **B** is a tautology. Therefore by *152, ⊢ **A** ≡ **B**.

As a corollary we have also:

*159. If **B** results from **A** by substitution of **N** for **M** at one or more places in **A** (not necessarily for all occurrences of **M** in **A**), if ⊢ **M** ≡ **N** and ⊢ **A**, then ⊢ **B**. (*Rule of substitutivity of equivalence.*)

Proof. By *158, ⊢ **A** ≡ **B**. Therefore by †156, substitution, and *modus ponens*, ⊢ **A** ⊃ **B**. Since ⊢ **A**, we have by another use of *modus ponens* that ⊢ **B**.

EXERCISES 15

15.0. Verify the following tautologies:

(1) The wffs of 14.4 and 14.5.

(2) The wff of 14.6.

(3) $pq \supset r \supset . p \supset . q \supset r$ (*Law of exportation.*)

(4) $[p \supset . q \supset r] \supset . pq \supset r$ (*Law of importation.*)

(5) $[p \supset q][p \supset r] \supset . p \supset qr$ (*Law of composition.*)

(6) $p \supset q \supset . \sim q \supset \sim p$ (*Law of contraposition.*)

(7) $p \equiv q \equiv . q \equiv p$ (*Complete commutative law of equivalence.*)

(8) The transitive law of equivalence, †157.

(9) $\sim . p \sim p$ (*Law of contradiction.*)

(10) $p \vee \sim p$ (*Law of excluded middle.*)

15.1. Determine of each of the following wffs whether it is a tautology or a contradiction or neither:

(1) $p \supset r \supset r \supset . p \supset q \supset r$

(2) $p \supset q \supset [r \supset s] \supset . p \supset r \supset . q \supset s$

(3) $f \supset f \supset . f \supset f \supset f$

(4) $p \equiv q \equiv p \vee q \equiv \sim p \vee \sim q$

15.2. Prove: If **B** results from **A** by substitution of **N** for **M** at one or more places in **A** (not necessarily for all occurrences of **M** in **A**), then $\vdash \mathbf{M} \equiv \mathbf{N} \supset . \mathbf{A} \equiv \mathbf{B}$.

15.3. Present proofs of †154 and †156 in the style of §14, not using methods or results of §15.

15.4. A wff **B** which contains n different variables is said to be in *implicative normal form* if the following conditions are satisfied: (i) **B** has the form $\mathbf{C}_1 \supset . \mathbf{C}_2 \supset . \ldots \mathbf{C}_m \supset f$; (ii) each $\mathbf{C}_i$ ($i = 1, 2, \ldots, m$) has the form $\mathbf{C}_{i1} \supset . \mathbf{C}_{i2} \supset . \ldots \mathbf{C}_{in} \supset f$; (iii) each $\mathbf{C}_{ik}$ ($i = 1, 2, \ldots, m$ and $k = 1, 2, \ldots, n$) is either $\mathbf{b}_k$ or $\sim \mathbf{b}_k$, where $\mathbf{b}_k$ is the kth of the variables occurring in **B**, according to the alphabetic order of the variables (§10); (iv) the antecedents $\mathbf{C}_i$ are all different and are arranged among themselves according to the rule that, if $\mathbf{C}_{i1}, \mathbf{C}_{i2}, \ldots, \mathbf{C}_{i(k-1)}$ are the same as $\mathbf{C}_{j1}, \mathbf{C}_{j2}, \ldots, \mathbf{C}_{j(k-1)}$ respectively, and $\mathbf{C}_{ik}$ is $\mathbf{b}_k$, and $\mathbf{C}_{jk}$ is $\sim \mathbf{b}_k$, then $i < j$. Show that for every wff **A** there is a unique corresponding wff **B** (the *implicative normal form of* **A**) such that **B** is in implicative normal form, and each $\mathbf{C}_i$ contains the same variables that **A** does, and $\vdash \mathbf{A} \equiv \mathbf{B}$. (Make use of the values of the given wff **A** for the various systems of values of its variables, in order to determine **B** in such a way that $\mathbf{A} \equiv \mathbf{B}$ is a tautology.)

What is the implicative normal form of a tautology containing the n

different variables $\mathbf{b}_1, \mathbf{b}_2, \ldots, \mathbf{b}_n$, and no other variables? Of a contradiction containing these variables and no others?

What are the possible implicative normal forms of a wff containing no variables? Of a wff containing just one variable? Of a wff containing just two (different) variables?

15.5. Show that P_1 is a commutative ring, with equivalence as the ring equality, non-equivalence as the ring addition, and conjunction as the ring multiplication, in the sense that the following analogues of the ring laws are tautologies and therefore theorems of P_1:

$$p \not\equiv q \equiv . q \not\equiv p \qquad pq \equiv qp$$
$$p \not\equiv [q \not\equiv r] \equiv . p \not\equiv q \not\equiv r \qquad p[qr] \equiv pqr$$
$$p \not\equiv q \equiv r \supset . q \equiv . p \not\equiv r \qquad p[q \not\equiv r] \equiv . pq \not\equiv pr$$

Identify the ring subtraction (cf. the third law of those above). Also identify the zero element and the unit element of the ring.

15.6. In a like sense, show further that P_1 is a Boolean ring by verifying the tautologies:[185]

$$p \not\equiv p \equiv f$$
$$pp \equiv p$$

15.7. In a like sense, show that P_1 is a Boolean ring with equivalence as the ring equality, equivalence as the ring addition, and disjunction as the

[185]The reader must be careful not to misunderstand the assertions made in 15.5–15.8. In 15.6, e.g., it is meant that, with equivalence in the role of equality, non-equivalence in the role of addition, and conjunction in the role of multiplication, the defining laws for a Boolean ring appear as theorems of P_1, and hence also all laws of Boolean rings which are derivable from these by methods of the propositional calculus (including the rule of substitution, *101, and the rule of substitutivity of equivalence, *159). There is no question of a ring in the sense of a particular system of elements and operations on them obeying the ring laws, until we deal with a particular interpretation of P_1. If we allow interpretations that are sound in the generalized sense of §19, then many sound interpretations of P_1 do turn out to be Boolean rings (with equivalence in the role of equality, etc.) in the sense of a particular system of elements and operations; but it is not true that *every sound interpretation of* P_1 *is a Boolean ring* in this sense—or better, in view of cases like those of exercises 19.11 and 19.12, it is not easy to decide on a generally satisfactory meaning for the italicized statement.

Under its principal interpretation, P_1 is not merely a Boolean ring, but a two-element field, with addition and multiplication identified in the way described in 15.5. This remark, and its application to formal work in the propositional calculus, is due to J. J. Gégalkine in *Recueil Mathématique de la Société Mathématique de Moscou*, vol. 34 (1927), pp. 9–28. To any one familiar with the procedures of elementary algebra, it is indeed very convenient to rewrite all expressions of the propositional calculus in terms of non-equivalence and conjunction as fundamental connectives, using also 0 and 1 as propositional constants, and writing the sign $+$ instead of $\not\equiv$.

The term *Boolean ring*, now standard, is due to M. H. Stone (1936).

ring multiplication. Identify the ring subtraction, and the zero element and the unit element of the ring.[186]

15.8. In a like sense, P_1 is also a Boolean algebra, again with equivalence in the role of equality, and with disjunction and conjunction identified as the Boolean sum and Boolean product respectively.[187] Verify the following tautologies that are implied in this statement: the *complete distributive law of conjunction over disjunction*; the *complete distributive law of disjunction over conjunction*; the two *laws of absorption*,

$$p \vee pq \equiv p,$$
$$p[p \vee q] \equiv p,$$

and the two *laws of De Morgan*,[188]

$$\sim[p \vee q] \equiv \sim p \sim q,$$
$$\sim[pq] \equiv \sim p \vee \sim q.$$

15.9. Various works on traditional logic treat of certain kinds of inferences, known as hypothetical syllogisms, disjunctive syllogisms, and dilemmas. These are stated verbally, and include:[189]

Hypothetical Syllogism

Modus ponens: If A then B. A. Therefore, B.
Modus tollens: If A then B. Not B. Therefore, not A.

---

[186]This is the dual (in the sense of §16) of the remark of 15.5, 15.6. It was used by Jacques Herbrand in his dissertation of 1930, independently of Gégalkine, and again by Stone in 1937. It provides another method, dual to that of the preceding footnote, by which procedures of elementary algebra may be utilized for propositional calculus; namely, all expressions of the propositional calculus are rewritten in terms of equivalence and disjunction as fundamental connectives, together with the constants 0 and 1, and the usual signs of addition and multiplication are employed instead of $\equiv$ and $\vee$ respectively. (Compare exercise 24.3.)

The laws $pp \equiv p$ and its dual are known as the *laws of tautology*, though this is quite a different sense of the word "tautology" from that introduced in the text. From the point of view of ring theory they might also be called *idempotent laws*.

[187]This remark is implicit already in Peirce's paper of 1885, cited in footnote 67. ("Peirce's law" of 12.6 is also found in this paper.)

[188]Not these laws but the corresponding laws of the class calculus were enunciated by Augustus De Morgan in his *Formal Logic* of 1847.

In verbal formulation these laws of the propositional calculus were known already to the Scholastics, perhaps first to Ockham. Cf. a paper by Łukasiewicz in *Erkenntnis*, vol. 5 (1935), pp. 111–131, in which some rudiments of the propositional calculus are traced back not only to the Scholastics but to antiquity—material implication in particular, to Philo of Megara. And concerning the history of the De Morgan laws among the Scholastics, see further Philotheus Boehner in *Archiv für Philosophie*, vol. 4 (1951), pp. 113–146.

[189]We make no attempt to enter into the history of the matter, but have merely compiled a representative list from a number of comparatively recent works of traditional character. Some discrepancies will be disclosed if parts (1) and (2) of the exercise are carried through. These may be attributed partly to uncertainties of meaning, partly to disagreements among different writers.

Disjunctive Syllogism

Modus tollendo ponens: A or B. Not A. Therefore, B.
Modus ponendo tollens: A or B. A. Therefore, not B.

Dilemma

Simple constructive: If A then C. If B then C. A or B. Therefore, C.
Simple destructive: If A then B. If A then C. Not B, or not C. Therefore, not A.
Complex constructive: If A then B. If C then D. A or C. Therefore, B or D.
Complex destructive: If A then B. If C then D. Not B, or not D. Therefore, not A, or not C.

The letters A, B, C, D are here replaceable by sentences[190]—indeed we might have used bold letters (under the conventions of §08) except for the lack of a definite object language to which they could be understood to refer. Some writers are in disagreement among themselves, and others are unclear, (a) as to whether the words "if . . . then" mean material implication or some other kind of implication, and (b) as to whether the word "or" means exclusive disjunction or inclusive disjunction. (Cf. §05.)

(1) On the assumption that "if . . . then" means material implication and "or" means exclusive disjunction, the *leading principle* of, e.g., the simple destructive dilemma is

$$p \supset q \supset . p \supset r \supset . \sim q \not\equiv \sim r \supset \sim p$$

On this assumption, write in the same manner the leading principle of each of the kinds of inference listed. Check each of the kinds of inference by ascertaining whether its leading principle is a tautology. (Wherever possible, of course, make use of known theorems of P_1 in order to shorten the work.)

(2) On the assumption that "if . . . then" means material implication and "or" means inclusive disjunction, again write the leading principle of each of the kinds of inference, and check in the same way.[191]

15.10. When Sancho Panza was governor of Barataria, the following case came before him for decision. A certain manor was divided by a river upon which was a bridge. The lord of the manor had erected a gallows at one end of the bridge and had enacted a law that whoever would cross the bridge must first swear whither he were going and on what business; if he swore truly he should be allowed to pass freely; but if he swore falsely and did then cross the bridge he should be hanged forthwith upon the gallows. One man, coming up to the other end of the bridge from the gallows, when his oath was required swore, "I go to be hanged on yonder gallows," and thereupon crossed the bridge. The vexed question whether the man shall be hanged is brought to Sancho Panza, who is holding court in the immediate vicinity, and who is of course obligated to uphold the law as validly enacted by the lord of the manor.[192]

---

[190]The traditional assumption that the sentences must have the subject-predicate form is omitted as irrelevant.

[191]For the hypothetical and disjunctive syllogisms, the question of reproduction in the notation of the propositional calculus is discussed by S. K. Langer in an appendix to her *Introduction to Symbolic Logic* (1937).

[192]The story is here only very slightly modified from the original as given by Miguel de Cervantes (1615).

Let P, Q, R, S be constants expressing the propositions, respectively, that he [the man in the story] crosses the bridge, that he is hanged on the gallows, that the oath to which he swears is true, and that the law is obeyed. Use a formulation of the propositional calculus containing these four propositional constants, as well as propositional variables. Then the given data are expressed in the three wffs: $R \equiv PQ$, P, $S \supset . Q \equiv P \sim R$. (Notice in particular that to replace the third wff by $S \supset . P \sim R \supset Q$ would not sufficiently represent the data, since we must suppose that it is as much a violation of the law to hang an innocent man as it is to let a guilty one go free.)

Verify the tautology,

$$r \equiv pq \supset . p \supset . s \supset [q \equiv p \sim r] \supset \sim s,$$

and hence by substitution and *modus ponens* demonstrate that the law cannot be obeyed in this instance.

16. Duality. The process of dualization is most conveniently applied, not to wffs of P_1 but to expressions which are abbreviations of wffs of P_1 in accordance with D1–11 (but without any omissions of brackets). *The dual* of such an expression is obtained by interchanging simultaneously, wherever they occur, the letters t and f, and each of the following pairs of connectives: $\supset$ and $\not\subset$, disjunction and conjunction, $\equiv$ and $\not\equiv$, $\subset$ and $\not\supset$, $\bar{\vee}$ and $|$. The symbol (connective), $\sim$, is left unchanged by dualization, and is therefore called *self-dual*. The letters t and f are called *duals* of each other; likewise the connectives conjunction and disjunction; likewise $\supset$ and $\not\subset$; etc.

Thus, e.g., the dual of the expression

$$[[[p \vee t] \supset [qr]] \equiv [r \vee \sim p]]$$

is the expression

$$[[[pf] \not\subset [q \vee r]] \not\equiv [r \sim p]].$$

A dual of a wff of P_1 is obtained by writing any expression of the foregoing kind which abbreviates the wff, dualizing this expression, and then finally writing the wff which the resulting expression abbreviates. It is not excluded that the wff itself may be used in the role of the expression which abbreviates it, and when this is done the *principal dual* of the wff is obtained. For example, the wff

$$[[p \supset q] \supset f]$$

has as its principal dual the wff

$$[[p \not\subset q] \not\subset t],$$

i.e., the wff

$$[[[f \supset f] \supset [[q \supset p] \supset f]] \supset f];$$

but because the same wff,

$$[[p \supset q] \supset f],$$

may also be abbreviated as $[q \not\subset p]$, it has also the wff

$$[q \supset p]$$

as a dual.

Except in the case of a wff consisting of a variable alone, the principal dual of the principal dual of a wff is not the same as the wff itself. However, of course the wff itself is always included among the various duals of any one of its duals. And any dual of a dual of a wff is equivalent to the wff in the sense of *160 below.

In order to minimize the variety of different duals of a given wff, D1–11 have been arranged as far as possible in pairs dual to each other. But this could not be done in the case of D1–3, and it is from these three that the possibility arises of different duals of the same wff. By examining D1–3, it may be seen that any two duals of the same wff can be transformed one into the other by a series of steps of the four following kinds: replacing a wf part $t \supset \mathbf{N}$ by $\mathbf{N}$, replacing a wf part $\mathbf{N}$ by $t \supset \mathbf{N}$, replacing a wf part $\sim\sim\mathbf{N}$ by $\mathbf{N}$, and replacing a wf part $\mathbf{N}$ by $\sim\sim\mathbf{N}$. By †153, †154, †155, *158, †157 (together with substitution and *modus ponens*) it therefore follows that any two duals of the same wff are equivalent in the following sense:

***160.** If $\mathbf{B}$ and $\mathbf{C}$ are duals of $\mathbf{A}$, $\vdash \mathbf{B} \equiv \mathbf{C}$.

In the truth-tables in §15 it will be seen that the truth-table for $\supset$ is transformed into that for $\not\subset$ if t and f are interchanged throughout (in all three columns of the table). In fact, if t and f are interchanged, the truth-tables for $\supset$ and $\not\subset$ are interchanged; likewise those for disjunction and conjunction; likewise those for $\equiv$ and $\not\equiv$; likewise those for $\subset$ and $\not\supset$; likewise those for $\bar{\vee}$ and $|$; and the truth-table for $\sim$ is transformed into itself. From this it follows that the dual of a tautology is a contradiction. Hence, in view of the truth-table for negation, there follows the metatheorem:

***161.** If $\vdash \mathbf{A}$, if $\mathbf{A}_1$ is a dual of $\mathbf{A}$, then $\vdash \sim\mathbf{A}_1$. (*Principle of duality.*)

Two corollaries of *161, *special principles of duality*, are obtained by means of the tautologies:

†162. $\sim[p \not\subset q] \supset \centerdot q \supset p$

†163. $\sim[p \not\equiv q] \supset \centerdot p \equiv q$

These corollaries of *161 are:

***164.** If $\vdash \mathbf{A} \supset \mathbf{B}$, if $\mathbf{A}_1$ and $\mathbf{B}_1$ are duals of $\mathbf{A}$ and $\mathbf{B}$ respectively, then $\vdash \mathbf{B}_1 \supset \mathbf{A}_1$. (*Special principle of duality for implications.*)

***165.** If $\vdash \mathbf{A} \equiv \mathbf{B}$, if $\mathbf{A}_1$ and $\mathbf{B}_1$ are duals of $\mathbf{A}$ and $\mathbf{B}$ respectively, then $\vdash \mathbf{A}_1 \equiv \mathbf{B}_1$. (*Special principle of duality for equivalences.*)

17. Consistency.

The notion of *consistency* of a logistic system is semantical in motivation, arising from the requirement that nothing which is logically absurd or self-contradictory in meaning shall be a theorem, or that there shall not be two theorems of which one is the negation of the other. But we seek to modify this originally semantical notion in such a way as to make it syntactical in character (and therefore applicable to a logistic system independently of the interpretation adopted for it). This may be done by defining *relative consistency with respect to* any transformation by which each sentence or propositional form $\mathbf{A}$ is transformed into a sentence or propositional form $\mathbf{A}'$, the definition (given below) being such that relative consistency reduces to the semantical notion of consistency under an interpretation that makes $\mathbf{A}'$ the negation of $\mathbf{A}$. Or we may define *absolute consistency* by the condition that not every sentence or propositional form shall be a theorem, since in the case of nearly all the systems with which we shall deal it is easy to see that, once we had two theorems which were negations of each other, every sentence and propositional form whatever could be proved (e.g., in the case of P_1 this follows by †123, substitution, and *modus ponens*). Or, following Hilbert, we might in the case of a particular system select an appropriate particular sentence and define the system as being consistent if that particular sentence is not a theorem (e.g., we might call P_1 consistent on condition that f is not a theorem). Or if the system has propositional variables, we may define it as being *consistent in the sense of Post*[193] if a wff consisting of a propositional variable alone is not a theorem.

Turning now to the purely syntactical statement of the matter, we have the following:

(a) A logistic system is *consistent with respect to* a given transformation by which each sentence or propositional form $\mathbf{A}$ is transformed into a sentence or propositional form $\mathbf{A}'$, if there is no sentence or propositional form $\mathbf{A}$ such that $\vdash \mathbf{A}$ and $\vdash \mathbf{A}'$.

(b) A logistic system is *absolutely consistent* if not all its sentences and propositional forms are theorems.

(c) A logistic system is *consistent in the sense of Post* (*with respect to* a

[193]E. L. Post in the *American Journal of Mathematics*, vol. 43 (1921), see p. 177. The notion of absolute consistency is, in view of the rule of substitution, closely related to that of consistency in the sense of Post; it seems to have been first used explicitly as a general definition of consistency by Tarski (*Monatshefte für Mathematik und Physik*, vol. 37 (1930), see pp. 387–388). A similar remark applies to the notion of absolute completeness (cf. Tarski, *ibid.*, pp. 390–391).

certain category of primitive symbols designated as "propositional variables") if a wff consisting of a propositional variable alone is not a theorem.

****170.** P_1 is consistent with respect to the transformation of **A** into **A** $\supset f$.

Proof. By the definition of a tautology (and the truth-table of $\supset$), not both **A** and **A** $\supset f$ can be tautologies. In fact, if **A** is a tautology, then **A** $\supset f$ is a contradiction. Therefore by **150, not both **A** and **A** $\supset f$ can be theorems of P_1.

****171.** P_1 is absolutely consistent.

Proof. The wff f is not a tautology, and therefore by **150 it is not a theorem of P_1.

****172.** P_1 is consistent in the special sense that f is not a theorem.

Proof. The same as for **171.

****173.** P_1 is consistent in the sense of Post.

Proof. A wff consisting of a propositional variable alone is not a tautology, because its value is f for the value f of the variable. Therefore by **150, it is not a theorem of P_1.

18. Completeness.

As in the case of consistency, the notion of *completeness* of a logistic system has a semantical motivation, consisting roughly in the intention that the system shall have all possible theorems not in conflict with the interpretation. As a first attempt to fix the notion more precisely, we might demand of every sentence that either it or its negation shall be a theorem; but since we allow the assertion of propositional forms (see the concluding paragraphs of §06), this may prove insufficient. Therefore, following Post,[193] we are led to define a logistic system as being complete if, for every sentence or propositional form **B**, either $\vdash$ **B** or the system would become inconsistent upon adding **B** to it as an axiom (without other change). This leads to several purely syntactical definitions of completeness, corresponding to the different syntactical definitions of consistency of a system as given in the preceding section.

Another approach starts from the idea that a system is complete if there is a sound interpretation under which every sentence that denotes truth is a theorem and every propositional form that has always the value truth is a theorem—then seeks to replace the notion of an interpretation by some suitable syntactical notion. This approach, however, requires certain restrictions on the character of the interpretation allowed, and thus leads to the introduction of *models* in the sense of Kemeny. It will be discussed briefly in Chapter X.

As syntactical definitions of completeness we have, for the present, the following:

(a) A logistic system is *complete with respect to* a given transformation by which each sentence or propositional form **A** is transformed into a sentence or propositional form **A**′, if, for every sentence or propositional form **B**, either ⊢ **B** or the system, upon addition of **B** to it as an axiom, becomes inconsistent with respect to the given transformation.

(b) A logistic system is *absolutely complete* if, for every sentence or propositional form **B**, either ⊢ **B** or the system, upon addition of **B** to it as an axiom, becomes absolutely inconsistent.

(c) A logistic system is *complete in the sense of Post* if, for every sentence or propositional form **B**, either ⊢ **B** or the system, upon addition of **B** to it as an axiom, becomes inconsistent in the sense of Post.

Let **B** be a wff of P_1 which is not a theorem. Then by *152, **B** is not a tautology. I.e., there is a system of values of the variables of **B** for which the value of **B** is f.

If **B** is added to P_1 as an axiom, it becomes possible by *121 to infer the result of any simultaneous substitution for the variables of **B**. In particular, we may take one of those systems of values of the variables of **B** for which the value of **B** is f, and substitute for each variable $\mathbf{a}_i$ either t or f according as the value a_i of that variable is t or f. Let **E** be the wff which is inferred in this way.

Since **E** contains no variables, the definition at the beginning of §15 assigns one value to **E**, and because of the way in which **E** was obtained from **B** it follows that this value is f (the explicit proof of this by mathematical induction is left to the reader). Therefore by the truth-table of ⊃, $\mathbf{E} \supset f$ is a tautology. Thus by *152, we have that $\mathbf{E} \supset f$ is a theorem of P_1, therefore also a theorem of the system which is obtained by adding **B** to P_1 as an axiom.

In the system obtained by adding **B** to P_1 as an axiom we now have that both **E** and $\mathbf{E} \supset f$ are theorems. Therefore by *modus ponens* we have that f is a theorem. Therefore by †122 and *modus ponens*, p is a theorem. Thence by substitution we may obtain any wff whatever as a theorem, including, of course, with every wff **A**, also the wff $\mathbf{A} \supset f$.

Thus we have proved the completeness of P_1 in each of the three senses:

180. P_1 is complete with respect to the transformation of **A into $\mathbf{A} \supset f$.

**181. P_1 is absolutely complete.

**182. P_1 is complete in the sense of Post.

EXERCISES 18

Discuss the consistency and completeness of each of the following logistic systems, in each of the senses of **170–**173, **180–**182:

18.0. The system obtained from P_1 by deleting the axiom †104. (Show that a wff **A** containing occurrences of f is a theorem if and only if $S^{f}_{\mathbf{a}}\mathbf{A}|$ is a theorem, where **a** is a variable not occurring in **A**.)

18.1. The primitive symbols and the formation rules are the same as those of P_1. There is one axiom, namely p. There is one rule of inference, namely *101 with the restriction added that **B** must not be f.

18.2. The primitive symbols and the formation rules are the same as those of P_1. There is one axiom, $p \supset q$, and one rule of inference, *101.

18.3. The system $P^{\mathrm{I}}_{\mathrm{B}}$ obtained from the system P_{W} of 12.7 by deleting f from among the primitive symbols, and making only such further changes as this deletion compels, namely, omitting the formation rule 10i and the fourth axiom, †122. (Make use of the results of 12.7 and 12.8; prove an analogue of *151 in which a variable **r** is selected, different from $\mathbf{a}_1, \mathbf{a}_2, \ldots, \mathbf{a}_n$, and $\mathbf{A}_i$ is defined to be $\mathbf{a}_i$ or $\mathbf{a}_i \supset \mathbf{r}$ according as a_i is t or f, and $\mathbf{B}'$ is defined to be $\mathbf{B} \supset \mathbf{r} \supset \mathbf{r}$ or $\mathbf{B} \supset \mathbf{r}$ according as the value of **B** for the values $a_1, a_2, \ldots, a_n$ of $\mathbf{a}_1, \mathbf{a}_2, \ldots, \mathbf{a}_n$ is t or f; and hence prove that *152 holds for $P^{\mathrm{I}}_{\mathrm{B}}$. In place of **170 and **180, show that $P^{\mathrm{I}}_{\mathrm{B}}$ is consistent and complete with respect to the transformation of **A** into $\mathbf{A} \supset \mathbf{a}$, where **a** is the first variable in alphabetic order not occurring in **A**.)

18.4. The system $P^{\mathrm{I}}_{\mathrm{Ł}}$ having the same primitive symbols and wffs as $P^{\mathrm{I}}_{\mathrm{B}}$, the same rules of inference, and the single following axiom:

$$p \supset q \supset r \supset . r \supset p \supset . s \supset p$$

(After verifying that this axiom is a tautology, we may prove the axioms of $P^{\mathrm{I}}_{\mathrm{B}}$ as theorems of $P^{\mathrm{I}}_{\mathrm{Ł}}$, and then use the results of 18.3. For this purpose, first establish the derived rules, that if $\vdash \mathbf{A} \supset \mathbf{B} \supset \mathbf{C}$, then $\vdash \mathbf{C} \supset \mathbf{A} \supset . \mathbf{a} \supset \mathbf{A}$, $\vdash \mathbf{a} \supset \mathbf{A} \supset \mathbf{C} \supset . \mathbf{b} \supset \mathbf{C}$, $\vdash \mathbf{B} \supset \mathbf{C}$; then following Łukasiewicz, prove

$$r \supset q \supset . r \supset q \supset p \supset . s \supset p,$$
$$p_1 \supset r \supset [s \supset p] \supset . r \supset q \supset p \supset . s \supset p,$$
$$r \supset q \supset [s \supset p] \supset . r \supset p \supset . s \supset p,$$
$$r \supset p \supset [s \supset p] \supset [p \supset q \supset r \supset p_1] \supset . q_1 \supset . p \supset q \supset r \supset p_1,$$
$$r \supset p \supset p \supset [s \supset p] \supset . p \supset q \supset r \supset . s \supset p,$$
$$p \supset r \supset q \supset q \supset . q \supset r \supset . p \supset r,$$
$$p \supset q \supset . p \supset r \supset q \supset q,$$

and the transitive law of implication, in order, as theorems of $P^{\mathrm{I}}_{\mathrm{Ł}}$.)

18.5. By means of semantical rules similar in character to a–g of §10, supply sound interpretations of the systems of 18.0–18.3, and discuss for each system the possible variety of sound interpretations of this sort.

19. Independence.[194] An axiom A of a logistic system is called *independent* if, in the logistic system obtained by omitting A from among the axioms, A is not a theorem. A primitive rule of inference R of a logistic system is called *independent* if, in the logistic system obtained by omitting R from among the primitive rules, R is not a derived rule. Or equivalently, we may define an axiom or rule of inference to be independent if there is some theorem which cannot be proved without that axiom or rule.[195]

It should not be regarded as obligatory that the axioms and rules of inference of a logistic system be independent. On the contrary there are cases in which important purposes are served by allowing non-independence. And if the requirement of independence is imposed, this is as a matter of elegance and only a part of the more general (and somewhat vague) requirement of economy of assumption.[196]

In this book we shall often ignore questions of independence of the axioms and rules of a logistic system. But for the sake of illustration we treat the matter at length in the case of P_1.

In the propositional calculus a standard device for establishing the independence of axioms and rules is to generalize the method of §15 as follows. Instead of two truth-values, a system of two or more *truth-values*,

$$0, 1, \ldots, \nu,$$

is introduced,[197] the first μ of these,

$$0, 1, \ldots, \mu,$$

[194]The reader who wishes to get on rapidly to logistic systems of more substantial character than propositional calculus may omit §19 and all of Chapter II except §§20–23, 27. Especially §§26, 28 and the accompanying exercises may well be postponed for study in connection with later chapters.

[195]In the case of rules of inference, the equivalence of the two definitions of independence depends on considerations like those adduced in footnote 183—to show that the conditions of effectiveness which we demand of a primitive rule of inference are sufficient to ensure that, when the same rule is demonstrated as a metatheorem of some other system, the required conditions of effectiveness for derived rules of inference (§12) will therefore be satisfied. In what follows, we shall make use only of the second definition of independence of a rule of inference, viz., that the rule is independent if there is at least one theorem which cannot be proved without it.

The possibility should be noticed that a rule of inference not previously independent may become so when additional axioms are adjoined to a logistic system.

[196]The requirement of economy of assumption is usually understood to concern also the length and complication (or perhaps the strength, in some sense) of individual rules and axioms—in addition to merely the number of them.

[197]It is convenient in practice to use numerals in this way to denote the truth-values, though analogy with the notation used in the case of two truth-values would suggest

(where $1 \leqq \mu < \nu$) being called *designated* truth-values.[198] To each of the primitive constants (if any) is assigned one of these truth-values as value, and to each primitive connective is assigned a truth-table in these truth-values. Analogously to the first paragraph of §15 is defined the value of a wff for given values of its variables, the possible values of the variables being the truth-values 0, 1, . . ., ν, and a wff is called a *tautology* if, for every system of values of its variables, it has one of the designated truth-values as its value. If then every rule of inference has the property of preserving tautologies (i.e., that the conclusion must be a tautology when the premisses are tautologies) and every axiom but one is a tautology, it follows that the one axiom which is not a tautology is independent. Or if every axiom is a tautology and every rule of inference except one has the property of preserving tautologies, and if further there is a theorem of the logistic system that is not a tautology, it follows that the exceptional rule of inference is independent.

In the case of P_1, it happens that we may establish the independence of each of the axioms and rules of inference, with the exception of the rule of substitution, by means of a system of three truth-values, 0, 1, 2, of which 0 is the only designated truth-value, and 2 is assigned to the primitive constant f as a value. The required truth-tables of $\supset$ are as follows (the number at the head of each column indicating the axiom or rule whose independence is established by the table in that column).

		*100	†102	†103	†104
p	q	$p \supset q$	$p \supset q$	$p \supset q$	$p \supset q$
0	0	0	0	0	0
0	1	0	2	1	1
0	2	2	2	2	2
1	0	0	2	0	0
1	1	0	2	0	0
1	2	2	0	1	2
2	0	0	0	0	0
2	1	1	0	0	0
2	2	0	0	0	0

rather $t_1, t_2, \ldots, t_\mu$ for the designated truth-values and $f_1, f_2, \ldots, f_{\nu-\mu}$ for the non-designated truth-values.

Also in the work of verifying tautologies in the manner of §15, the numerals 0 and 1 are often used instead of the letters t and f.

[198]An infinite number of truth-values may also be used, with either a finite or an infinite number of designated truth-values, and likewise of non-designated truth-values. In this case the direct process of verifying tautologies (in a manner analogous to that of §15) is no longer effective, but the notion of a tautology may nevertheless still be useful.

For the proof of independence of *100, it is necessary to supply also an example of a theorem of P_1 which is not a tautology according to the truth-table used. One such example is $f \supset p$; another is $p \supset [q \supset f] \supset f \supset q$.

The rule of substitution *101 is necessarily tautology-preserving for any system of truth-values and truth-tables, and hence its independence cannot be established by this method. However, the independence of *101 follows from the fact that without it no wff longer than the longest of the axioms could be proved. And in fact a like proof of the independence of the rule of substitution will continue to hold after the adjunction of any finite number of additional axioms (since examples are easily found of wffs of arbitrarily great length which are theorems of P_1).

The foregoing method of finding independence examples by means of a generalized system of truth-values suggests also a generalization of the propositional calculus itself. Namely, we may fix upon a generalized system of truth-values as above, then introduce a number of connectives with assigned truth-tables, and possibly also a number of constants to each of which a particular truth-value is assigned as value. The wffs of a logistic system may be constructed by using variables and these connectives and constants, and we may supply a list of axioms which are tautologies (in the generalized system of truth-values) and rules of inference which preserve tautologies. Especially if this is done in such a way that every tautology is a theorem, the resulting logistic system is called a *many-valued propositional calculus* in the sense of Łukasiewicz.

The same considerations lead also to a generalization of the requirements imposed in §07 on an interpretation of a logistic system, these requirements being modified as follows when a generalized system of truth-values is used. The semantical rules must be such that the axioms either denote truth-values or have always truth-values as values and the rules of inference preserve this property. Only those wffs are capable of being asserted which denote truth-values or have always truth-values as values; and only those are capable of being rightly asserted which denote a designated truth-value or have only designated truth-values as values. An interpretation of a logistic system is called *sound* if, under it, all the axioms either denote designated truth-values or have only designated truth-values as values, and the rules of inference preserve this property (in the sense that, if all the premisses of an immediate inference either denote designated truth-values or have only designated truth-values as values, then the same holds of the conclusion).

EXERCISES 19

19.0. Carry out in full detail the proof of independence of the axioms and rules of P_1 which is outlined in the text. (In showing that particular wffs are or are not tautologies in the generalized system of truth-values, use an arrangement analogous to that described in §15.)

19.1. Consider the possibility of demonstrating the independence of the axioms and rules of P_1 by means of a system of only two truth-values. I.e., for each axiom and rule, either supply the required demonstration or show it to be impossible.

19.2. Similarly consider the possibility of demonstrating the independence of the axioms and rules of P_1 by means of a system of three truth-values of which two are designated.

19.3. The truth-table given in the text for the independence of *100 shows that there are theorems containing the symbol *f* which cannot be proved without use of *100, but is insufficient to show that there are any such theorems not containing *f*. Prove this statement. Devise another truth-table for the independence of *100, not having this defect.

19.4. Consider a logistic system whose wffs are the same as those of P_1, whose rules of inference are *modus ponens* and substitution, which has a finite number of axioms, and for which the metatheorem *152 holds. Prove that the rules of *modus ponens* and substitution are necessarily both independent. (In the case of *modus ponens*, this can be done by exhibiting an infinite list of tautologies (in the sense of §15) no two of which are variants of each other, and proving that no one of them is obtainable by substitution from any tautology other than a variant of itself.)

19.5. Prove the independence of the axioms and rules of P_W (see exercise 12.7). Except in the case of the rule of substitution, use the method of truth-tables.

19.6. Let P^+ be the system obtained from P_1 by deleting *f* from among the primitive symbols, and making only such further changes as this deletion compels, namely, omitting the formation rule 10i and the axiom †104. Prove that the system P^+ is not complete. Determine which of the axioms of P_B^I (exercise 18.3) are theorems of P^+ and which not.

19.7. Discuss the independence of the rule of *modus ponens* in the system P^+. Does this independence follow trivially from any result already established (in text or exercises)? If not, how can it be shown?

19.8. Using *modus ponens* and substitution as rules of inference, find axioms for the following many-valued propositional calculus (due to

Łukasiewicz — cf. footnote 276). There are three truth-values, 0, 1, 2, of which 0 is designated. There are two primitive constants f_1 and f_2, to which 1 and 2 are assigned as values respectively. And there is one primitive connective, $\supset$, which is binary and to which the following truth-table is assigned:

p	q	$p \supset q$
0	0	0
0	1	1
0	2	2
1	0	0
1	1	0
1	2	1
2	0	0
2	1	0
2	2	0

Prove a modified deduction theorem for this system, that if $\mathbf{A}_1, \mathbf{A}_2, \ldots, \mathbf{A}_n \vdash \mathbf{B}$, then $\mathbf{A}_1, \mathbf{A}_2, \ldots, \mathbf{A}_{n-1} \vdash \mathbf{A}_n \supset \centerdot \mathbf{A}_n \supset \mathbf{B}$; and hence prove analogues of **150 and *152.

19.9. Consider an interpretation of P_1 by means of four truth-values, 0, 1, 2, 3, of which 0 is the only designated truth-value, and 3 is assigned to the constant f as value. For each of the following different truth-tables of $\supset$, discuss the soundness of the interpretation:

p	q	(1) $p \supset q$	(2) $p \supset q$	(3) $p \supset q$	(4) $p \supset q$	(5) $p \supset q$	(6) $p \supset q$
0	0	0	0	0	0	0	0
0	1	0	1	2	3	0	0
0	2	0	2	3	0	0	0
0	3	0	3	1	1	0	0
1	0	0	0	0	0	0	0
1	1	0	0	0	0	0	2
1	2	0	2	0	0	0	0
1	3	0	2	0	0	0	0
2	0	0	0	0	0	0	0
2	1	0	1	0	1	2	2
2	2	0	0	0	0	0	0
2	3	0	1	0	3	0	0
3	0	0	0	0	0	0	0
3	1	0	0	0	0	0	0
3	2	0	0	0	0	0	0
3	3	0	0	0	0	0	0

19.10. It may happen that a sound interpretation of P_1 by means of a system of truth-values and truth-tables is reducible to the principal interpretation (§10) by replacing the designated truth-values everywhere by t and the non-designated truth-values everywhere by f. Following Carnap, let us call such a sound interpretation of P_1 a *normal interpretation*, and other interpretations of P_1, *non-normal interpretations*.[199] Then a normal interpretation of P_1 may be thought of as differing from the principal interpretation only in that, after division of propositions into true and false in ordinary fashion, some further subdivision is made of one or both categories. But a sound non-normal interpretation differs from the principal interpretation in some more drastic way.

Of the sound interpretations of P_1 found in 19.9, determine which are normal interpretations and which are non-normal interpretations. Also determine which can be rendered normal without loss of soundness, by changing only the way in which the truth-values are divided into designated and non-designated truth-values.

19.11. Consider an interpretation of P_1 by means of six truth-values, 0, 1, 2, 3, 4, 5, of which 0 and 1 are the designated truth-values, and 5 is assigned to the constant f as value, the truth-table of $\supset$ being as follows:

p	q	$p \supset q$	p	q	$p \supset q$	p	q	$p \supset q$
0	0	0	2	0	0	4	0	0
0	1	0	2	1	1	4	1	0
0	2	0	2	2	0	4	2	0
0	3	4	2	3	4	4	3	0
0	4	4	2	4	4	4	4	0
0	5	4	2	5	3	4	5	0
1	0	0	3	0	0	5	0	0
1	1	0	3	1	1	5	1	0
1	2	2	3	2	0	5	2	0
1	3	3	3	3	0	5	3	0
1	4	4	3	4	0	5	4	0
1	5	4	3	5	2	5	5	0

(1) Show that the interpretation is sound. (*Suggestion:* Let $\mathbf{A} \supset \mathbf{B}$ and $\mathbf{A}$ be tautologies (in the six truth-values). It follows immediately from the fourth, fifth, sixth, tenth, eleventh, and twelfth entries in the table that $\mathbf{B}$ cannot have the value 3, 4, or 5 for any system of values of its variables. Hence it follows that $\mathbf{B}$ cannot have the value 2 for a system of values of its variables, because if it did,

[199]This is not Carnap's terminology but an adaptation of it to the present context, the word "interpretation" being used by Carnap in a somewhat different sense from ours. The possibility of sound interpretations of the propositional calculus (in any one of its formulations) which are not normal was pointed out, in effect, by B. A. Bernstein in the *Bulletin of the American Mathematical Society*, vol. 38 (1932) pp. 390, 592; also independently by Carnap in his book, *Formalization of Logic* (1943). See further a review of the latter by the present writer in *The Philosophical Review*, vol. 53 (1944), pp. 493–498.

the value 3 would be obtained for **B** upon interchanging the values 2 and 3 in this system of values of its variables. Hence **B** is a tautology.)

(2) Use this interpretation as a counterexample to show that the following statement is false: "In a sound interpretation of P_1 the truth-table of $\equiv$ is symmetric in the sense that, if $p \equiv q$ has a designated value for given values of p and q, it also has a designated value upon interchanging the values of p and q."

(3) Discuss the question in what way or ways it is possible to weaken this statement so as to obtain a true metatheorem of interest.

19.12. Using an interpretation of P_1 by means of the three truth-values 0, 1, 2, of which 0 and 1 are designated truth-values, and 2 is assigned to the constant f as value, show that the following statement is false: "In a sound interpretation of P_1, and for given values of p, q, and r, if $p \equiv q$ and $q \equiv r$ have designated truth-values, then $p \equiv r$ must have a designated truth-value." (*Suggestion*: The most obvious method is to use for $\supset$ the truth-table of §15, with 0 and 2 in the roles of t and f respectively, and to give to $p \supset q$ the value 1 whenever either p or q has the value 1. A different method is suggested by a remark of Church and Rescher in their review of a paper by Z. P. Dienes, in *The Journal of Symbolic Logic*, vol. 15 (1950), pp. 69–70.)

II. The Propositional Calculus (Continued)

20. The primitive basis of P_2. Another formulation of the propositional calculus is the logistic system P_2, which differs from P_1 primarily in the lack of (propositional) constants.

The primitive symbols of P_2 are the four improper symbols

$$[\quad \supset \quad] \quad \sim$$

and the infinite list of (propositional) variables

$$p \quad q \quad r \quad s \quad p_1 \quad q_1 \quad r_1 \quad s_1 \quad p_2 \quad \cdots$$

(the order here indicated being called the *alphabetic order* of the variables).

The formation rules of P_2 are:

20i. A variable standing alone is a wff.

20ii. If $\mathbf{\Gamma}$ is wf, then $\sim\mathbf{\Gamma}$ is wf.

20iii. If $\mathbf{\Gamma}$ and $\mathbf{\Delta}$ are wf, then $[\mathbf{\Gamma} \supset \mathbf{\Delta}]$ is wf.

A formula of P_2 is wf if and only if its being so follows from the three formation rules. As in the case of P_1, an effective test of well-formedness follows. In §22 we shall prove also that every wff of P_2, other than a variable standing alone, is of one and only one of the forms $\sim\mathbf{A}$ and $[\mathbf{A} \supset \mathbf{B}]$ (where $\mathbf{A}$ and $\mathbf{B}$ are wf) and in each case is of that form in only one way. In a wff $[\mathbf{A} \supset \mathbf{B}]$, the wf parts $\mathbf{A}$ and $\mathbf{B}$ are the *antecedent* and *consequent* respectively and the occurrence of $\supset$ between them is the *principal implication sign*.

The rules of inference of P_2 are the same as those of P_1:

*200. From $[\mathbf{A} \supset \mathbf{B}]$ and $\mathbf{A}$ to infer $\mathbf{B}$. (*Rule of modus ponens.*)

*201. From $\mathbf{A}$ to infer $\mathrm{S}^{\mathbf{b}}_{\mathbf{B}}\mathbf{A}|$. (*Rule of substitution.*)

The axioms of P_2 are the three following:

†202. $p \supset . q \supset p$

†203. $s \supset [p \supset q] \supset . s \supset p \supset . s \supset q$

†204. $\sim p \supset \sim q \supset . q \supset p$

The axioms are in order the *law of affirmation of the consequent*, the *self-distributive law of implication*, and the *converse law of contraposition*. In stating them we have used the same conventions about omission of brackets and use of heavy dots that were explained in §11, and we shall use these hereafter without remark in connection with any formulation of the propositional calculus. For P_2 we shall use also the definition schemata D3–11 of §11, however understanding the "$\sim$" which appears in them to be the primitive symbol $\sim$ of P_2.

The principal interpretation of P_2 is given by the following semantical rules:

a. The variables are variables having the range t and f.

b. A wff consisting of a variable **a** standing alone has the value t for the value t of **a**, and the value f for the value f of **a**.

c. For a given assignment of values to the variables of **A**, the value of $\sim$**A** is f if the value of **A** is t; and the value of $\sim$**A** is t if the value of **A** is f.

d. For a given assignment of values to the variables of **A** and **B**, the value of [**A** ⊃ **B**] is t if either the value of **B** is t or the value of **A** is f; and the value of [**A** ⊃ **B**] is f if the value of **B** is f and at the same time the value of **A** is t.

21. The deduction theorem for P_2. As in the case of P_1, we have at once the *rule of simultaneous substitution* as a derived rule:

*210. If ⊢ **A** and if $\mathbf{b}_1, \mathbf{b}_2, \ldots, \mathbf{b}_n$ are distinct variables, then ⊢ $\mathrm{S}^{\mathbf{b}_1 \mathbf{b}_2 \ldots \mathbf{b}_n}_{\mathbf{B}_1 \mathbf{B}_2 \ldots \mathbf{B}_n}\mathbf{A}|$.

We have also, as theorem of P_2:

†211. $p \supset p$ (*Reflexive law of implication.*)

Proof. By simultaneous substitution in †203:

$$\vdash p \supset [q \supset p] \supset .\, p \supset q \supset .\, p \supset p$$

Hence by †202 and *modus ponens*:

$$\vdash p \supset q \supset .\, p \supset p$$

Hence by substituting $q \supset p$ for q, and using †202 and *modus ponens* again:

$$\vdash p \supset p$$

Now the proof of the deduction theorem in §13 required only $p \supset p$, the law of affirmation of the consequent, and the self-distributive law of implication (together with the rules of *modus ponens* and simultaneous substitution). Hence by the same proof we obtain the deduction theorem, and its corollary, as metatheorems of P_2:

*212. If $\mathbf{A}_1, \mathbf{A}_2, \ldots, \mathbf{A}_n \vdash \mathbf{B}$, then $\mathbf{A}_1, \mathbf{A}_2, \ldots, \mathbf{A}_{n-1} \vdash \mathbf{A}_n \supset \mathbf{B}$.

*213. If **A** ⊢ **B**, then ⊢ **A** ⊃ **B**.

Also analogues of *133 and *134 are proved as before:

*214. If every wff which occurs at least once in the list $\mathbf{A}_1, \mathbf{A}_2, \ldots, \mathbf{A}_n$ also occurs at least once in the list $\mathbf{C}_1, \mathbf{C}_2, \ldots, \mathbf{C}_r$, and if $\mathbf{A}_1, \mathbf{A}_2, \ldots, \mathbf{A}_n \vdash \mathbf{B}$, then $\mathbf{C}_1, \mathbf{C}_2, \ldots, \mathbf{C}_r \vdash \mathbf{B}$.

*215. If $\vdash \mathbf{B}$ then $\mathbf{C}_1, \mathbf{C}_2, \ldots, \mathbf{C}_r \vdash \mathbf{B}$.

22. Some further theorems and metatheorems of P_2. Hereafter we shall adopt a more condensed arrangement in exhibiting proofs of P_2, and later of other logistic systems. In particular we shall often omit explicit references to uses of substitution, *modus ponens*, or deduction theorem.

†220. $\sim p \supset . p \supset q$ (*Law of denial of the antecedent.*)

Proof. By †202, $\sim p \vdash \sim q \supset \sim p$.
Hence by †204, $\sim p \vdash p \supset q$.
Then use the deduction theorem.

†221. $\sim\sim p \supset p$ (*Law of double negation.*)

Proof. By †220, $\sim\sim p \vdash \sim p \supset \sim\sim\sim p$.
Hence by †204, $\sim\sim p \vdash \sim\sim p \supset p$
Use *modus ponens*, then the deduction theorem.

†222. $p \supset \sim\sim p$ (*Converse law of double negation.*)

Proof. By †221, $\vdash \sim\sim\sim p \supset \sim p$. Hence use †204.

†223. $p \supset q \supset . \sim q \supset \sim p$ (*Law of contraposition.*)

Proof. By †221, $p \supset q, \sim\sim p \vdash q$.
Hence by †222, $p \supset q, \sim\sim p \vdash \sim\sim q$.
Hence $p \supset q \vdash \sim\sim p \supset \sim\sim q$.
Use †204. Then use the deduction theorem.

†224. $p \supset [r \not\subset r] \supset \sim p$

Proof. By †221, $p \supset [r \not\subset r], \sim\sim p \vdash r \not\subset r$.
Hence $p \supset [r \not\subset r] \vdash \sim\sim p \supset . r \not\subset r$.
Hence by †204, $p \supset [r \not\subset r] \vdash r \supset r \supset \sim p$.
Hence by †211, $p \supset [r \not\subset r] \vdash \sim p$.
Then use the deduction theorem.

For the effective test of well-formedness referred to in §20, and also for a number of other metatheorems which now follow, we make use of the same process of counting brackets which is described in §10. Namely, we start at the beginning (or left) of a formula and proceed from left to right, counting each occurrence of [as $+1$ and each occurrence of] as -1, and adding as we go. The number which we assign, by this counting process, to an occurrence of a bracket will be called *the number of* that occurrence of a bracket in the formula.

It follows from the definition of a wff that, if a wff contains brackets, it must end with an occurrence of] as its final symbol; this we shall call the *final bracket* of the wff. By mathematical induction with respect to the total number of occurrences of $\supset$ and $\sim$, the following lemma is readily established: *The number of an occurrence of a bracket in a wff is positive, except in the case of the final bracket, which has the number* 0, *and the number of an occurrence of* [*in a wff is greater than* 1, *except in the case of the first occurrence of* [.

**225. Every wff, other than a variable standing alone, is of one and only one of the forms $\sim\mathbf{A}$ and $[\mathbf{A} \supset \mathbf{B}]$, and in each case it is of that form in one and only one way.

Proof. The first half of the theorem is obvious, by the definition of a wff.

Again it is obvious that, if a wff has the form $\sim\mathbf{A}$, it has that form in only one way (i.e., **A** is uniquely determined), for we may obtain **A** by just deleting $\sim$ from the beginning of the wff.

It remains to show that, if a wff has the form $[\mathbf{A} \supset \mathbf{B}]$, it has that form in only one way. Suppose then that $[\mathbf{A} \supset \mathbf{B}]$ and $[\mathbf{C} \supset \mathbf{D}]$ are the same wff. If **A** contains no brackets; then—because it is evident, from the definition of a wff, that the first occurrence of $\supset$ in a wff must be preceded by an occurrence somewhere of [—it follows that the first occurrence of $\supset$ in $[\mathbf{A} \supset \mathbf{B}]$ is immediately after **A**, and hence—for the same reason—that **C** is identical with **A**. By the same argument, if **C** contains no brackets, **C** and **A** are identical. If **A** and **C** both contain brackets, then the final bracket of **A** is the first occurrence of a bracket with the number 0 in **A**, and therefore is the second occurrence of a bracket with the number 1 in $[\mathbf{A} \supset \mathbf{B}]$; and the final bracket of **C** is the first occurrence of a bracket with the number 0 in **C**, therefore the second occurrence of a bracket with the number 1 in $[\mathbf{C} \supset \mathbf{D}]$; this makes the final bracket of **A** and the final bracket of **C** coincide, and so makes **A** and **C** identical. Thus we have in every case that **A** and **C** are identical, and it then follows obviously that **B** and **D** are identical.

226. A wf part[200] of $\sim\mathbf{A}$ either coincides with $\sim\mathbf{A}$ or is a wf part of **A.

Proof. The case to be excluded is that of a wf part **M** of $\sim\mathbf{A}$, obtained by deleting one or more symbols at the end (or right) of $\sim$**A** and none at the beginning (or left). If **M** contains brackets, the impossibility of this follows because the number of the final bracket of **M** would be 0 in **A** although it is not the final bracket of **A**. If **M** contains no brackets, the impossibility follows quickly by mathematical induction with respect to the number of consecutive occurrences of $\sim$ at the beginning of **A**.

227. A wf part of [A** ⊃ **B**] either coincides with [**A** ⊃ **B**] or is a wf part of **A** or is a wf part of **B**.[201]

Proof. The case to be excluded is that of a wf part of [**A** ⊃ **B**] which, without coinciding with [**A** ⊃ **B**], includes the principal implication sign of [**A** ⊃ **B**] or the final bracket of [**A** ⊃ **B**] or the occurrence of [at the beginning of [**A** ⊃ **B**].

Suppose that **M** is such a wf part of [**A** ⊃ **B**]. Then **M** contains brackets. Either the final bracket of **M** precedes the final bracket of [**A** ⊃ **B**], and therefore has the number 0 in **M** but a positive number in [**A** ⊃ **B**]; or else the first occurrence of [in **M** is later than the occurrence of [at the beginning of [**A** ⊃ **B**], and therefore has the number 1 in **M** but a greater number in [**A** ⊃ **B**]. It follows in either case that every occurrence of a bracket in **M** has a number in **M** less than its number in [**A** ⊃ **B**]. Hence the final bracket of **M** must indeed precede the final bracket of [**A** ⊃ **B**], and the first occurrence of [in **M** must also be later than the occurrence of [at the beginning of [**A** ⊃ **B**].

Since we now have, as the only remaining possibility, that **M** includes the principle implication sign of [**A** ⊃ **B**], it must include somewhere at least one bracket which precedes this principal implication sign and is therefore in **A**. Thus **A** contains brackets. The final bracket of **A** has the number 0 in **A**, therefore the number 1 in [**A** ⊃ **B**], therefore a number less than 1 in **M**; but this is impossible because it is not the final bracket of **M**.

228. If **A, **M**, **N** are wf and **Γ** results from **A** by substitution of **N** for **M** at zero or more places (not necessarily at all occurrences of **M** in **A**), then **Γ** is wf.

Proof. For the two special cases, (a) that the substitution of **N** for **M** is at zero places in **A**, and (b) that **M** coincides with **A** and the substitution of **N** for **M** is at this one place in **A**, the result is immediate. For we have in case (a) that **Γ** is **A**, and in case (b) that **Γ** is **N**.

---

[200]By a "wf part" of a formula we shall always mean a wf consecutive part of it—as indeed is the natural and obvious terminology.

[201]As to the metatheorems **225–**227, compare 12.1 and footnote 176.

In order to prove **228 generally, we proceed by mathematical induction with respect to the total number of occurrences of the symbols $\supset$ and $\sim$ in **A**. If this total number is 0, we must have either case (a) or case (b), and the well-formedness of $\mathbf{\Gamma}$ is then immediate, as we have just seen. Consider then a wff **A** in which this total number is greater than 0; the only possible cases are the two following:

Case 1: **A** is of the form $\sim\mathbf{A}_1$. Then by **226 (unless we have the special case (b) already considered) $\mathbf{\Gamma}$ is $\sim\mathbf{\Gamma}_1$, where $\mathbf{\Gamma}_1$ results from $\mathbf{A}_1$ by substitution of **N** for **M** at zero or more places. By hypothesis of induction, $\mathbf{\Gamma}_1$ is wf. Hence by 20ii it follows that $\mathbf{\Gamma}$ is wf.

Case 2: **A** is of the form $[\mathbf{A}_1 \supset \mathbf{A}_2]$. Then by **227 (unless we have the special case (b) already considered) $\mathbf{\Gamma}$ is $[\mathbf{\Gamma}_1 \supset \mathbf{\Gamma}_2]$, where $\mathbf{\Gamma}_1$ and $\mathbf{\Gamma}_2$ result from $\mathbf{A}_1$ and $\mathbf{A}_2$ respectively by substitution of **N** for **M** at zero or more places. By hypothesis of induction, $\mathbf{\Gamma}_1$ and $\mathbf{\Gamma}_2$ are wf. Hence by 20iii it follows that $\mathbf{\Gamma}$ is wf.

The proof by mathematical induction is then complete.

*229. If **B** results from **A** by substitution of **N** for **M** at zero or more places (not necessarily at all occurrences of **M** in **A**), then

$$\mathbf{M} \supset \mathbf{N},\ \mathbf{N} \supset \mathbf{M} \vdash \mathbf{A} \supset \mathbf{B}$$

and

$$\mathbf{M} \supset \mathbf{N},\ \mathbf{N} \supset \mathbf{M} \vdash \mathbf{B} \supset \mathbf{A}.$$

Proof. For the two special cases, (a) that the substitution of **N** for **M** is at zero places in **A**, and (b) that **M** coincides with **A**, and the substitution of **N** for **M** is at this one place in **A**, the result is immediate; namely, in case (a) by substitution in †211, and in case (b) because $\mathbf{A} \supset \mathbf{B}$ and $\mathbf{B} \supset \mathbf{A}$ are the same as $\mathbf{M} \supset \mathbf{N}$ and $\mathbf{N} \supset \mathbf{M}$ respectively.

In order to prove *229 generally, we proceed by mathematical induction with respect to the total number of occurrences of the symbols $\supset$ and $\sim$ in **A**. If this total number is 0, we must have one of the special cases (a) and (b), and the result of *229 then follows immediately, as we have just seen. Consider then a wff **A** in which this total number is greater than 0; the only possible cases are the two following:

Case 1: **A** is of the form $\sim\mathbf{A}_1$. Then by **226 (unless we have the special case (b) already considered) **B** is of the form $\sim\mathbf{B}_1$, where $\mathbf{B}_1$ results from $\mathbf{A}_1$ by substitution of **N** for **M** at zero or more places. By hypothesis of induction,

$$\mathbf{M} \supset \mathbf{N},\ \mathbf{N} \supset \mathbf{M} \vdash \mathbf{A}_1 \supset \mathbf{B}_1,$$
$$\mathbf{M} \supset \mathbf{N},\ \mathbf{N} \supset \mathbf{M} \vdash \mathbf{B}_1 \supset \mathbf{A}_1.$$

Hence we get the result of *229 by substitution in †223 and *modus ponens.*

Case 2: **A** is of the form $\mathbf{A}_1 \supset \mathbf{A}_2$. Then by **227 (unless we have the special case (b) already considered) **B** is of the form $\mathbf{B}_1 \supset \mathbf{B}_2$, where $\mathbf{B}_1$ and $\mathbf{B}_2$ result from $\mathbf{A}_1$ and $\mathbf{A}_2$ respectively, by substitution of **N** for **M** at zero or more places. By hypothesis of induction,

$$\mathbf{M} \supset \mathbf{N},\ \mathbf{N} \supset \mathbf{M} \vdash \mathbf{A}_1 \supset \mathbf{B}_1,$$
$$\mathbf{M} \supset \mathbf{N},\ \mathbf{N} \supset \mathbf{M} \vdash \mathbf{B}_1 \supset \mathbf{A}_1,$$
$$\mathbf{M} \supset \mathbf{N},\ \mathbf{N} \supset \mathbf{M} \vdash \mathbf{A}_2 \supset \mathbf{B}_2,$$
$$\mathbf{M} \supset \mathbf{N},\ \mathbf{N} \supset \mathbf{M} \vdash \mathbf{B}_2 \supset \mathbf{A}_2.$$

By *modus ponens,*

$$\mathbf{B}_1 \supset \mathbf{A}_1,\ \mathbf{A}_2 \supset \mathbf{B}_2,\ \mathbf{A}_1 \supset \mathbf{A}_2,\ \mathbf{B}_1 \vdash \mathbf{B}_2,$$
$$\mathbf{A}_1 \supset \mathbf{B}_1,\ \mathbf{B}_2 \supset \mathbf{A}_2,\ \mathbf{B}_1 \supset \mathbf{B}_2,\ \mathbf{A}_1 \vdash \mathbf{A}_2.$$

Hence we get the result of *229 by use of the deduction theorem.

Thus the proof of *229 by mathematical induction is complete.

23. Relationship of P_2 to P_1. Though the constant f is absent from the system P_2, we shall nevertheless be able to show the equivalence of the systems P_1 and P_2 in a sense which involves using in P_2 the wff $r \not\subset r$ (i.e., $\sim[r \supset r]$) to replace the constant f of P_1.

Under the principal interpretations of P_1 and P_2, the constant f of P_1 and the wff $r \not\subset r$ of P_2 in fact do not have the same meaning. For the former is a constant denoting f, while the latter is a singulary form which has the value f for every value of its variable r. Nevertheless the two meanings sufficiently resemble each other that $r \not\subset r$ can be used in P_2 to serve many of the same purposes as might a constant denoting f.

If **A** is any wff of P_2, then by a process of one-by-one replacement of the various wf parts $\sim$**C** each in turn by $\mathbf{C} \supset [r \not\subset r]$ we may obtain from **A** a wff $\mathbf{A}_0$ of P_2 in which $\sim$ does not occur otherwise than as a constituent in $r \not\subset r$. We may impose the restriction that, if $\sim$**C** is the special wff $r \not\subset r$, the replacement of $\sim$**C** by $\mathbf{C} \supset [r \not\subset r]$ shall not be made. Then from a given wff **A** we obtain, by the process described, a unique wff $\mathbf{A}_0$, which we shall call *the expansion of* **A** *with respect to negation.* If then in $\mathbf{A}_0$ we replace $r \not\subset r$ everywhere by f, we obtain a unique wff $\mathbf{A}_f$ of P_1 which we shall call *the representative of* **A** *in* P_1.

We have the following metatheorems (proofs omitted when obvious):

*230. If **B** results from **A** by replacement of $\sim$**C** by $\mathbf{C} \supset [r \not\subset r]$ at one place in **A**, then $\mathbf{A} \vdash \mathbf{B}$ and $\mathbf{B} \vdash \mathbf{A}$.

Proof. This follows from *229 and *modus ponens* because, by substitution in †220 and †224,

$$\vdash \sim\mathbf{C} \supset . \mathbf{C} \supset [r \not\subset r],$$
$$\vdash \mathbf{C} \supset [r \not\subset r] \supset \sim\mathbf{C}.$$

***231**. If $\mathbf{A}_0$ is the expansion of **A** with respect to negation, $\mathbf{A} \vdash \mathbf{A}_0$ and $\mathbf{A}_0 \vdash \mathbf{A}$.

***232**. If two wffs **A** and **B** of P_2 have the same representative in P_1, then $\mathbf{A} \vdash \mathbf{B}$ and $\mathbf{B} \vdash \mathbf{A}$.

***233**. If two wffs **A** and **B** of P_2 have the same representative in P_1, then $\vdash \mathbf{A} \supset \mathbf{B}$ and $\vdash \mathbf{B} \supset \mathbf{A}$.

***234**. A wff **A** of P_2 is a theorem of P_2 if its representative $\mathbf{A}_f$ in P_1 is a theorem of P_1.

Proof. Let $\mathbf{A}_0$ be the expansion of **A** with respect to negation. By *231, it is sufficient to prove that $\mathbf{A}_0$ is a theorem of P_2 if $\mathbf{A}_f$ is a theorem of P_1. Since $\mathbf{A}_0$ is

$$\mathrm{S}^{f}_{\sim[r \supset r]}\mathbf{A}_f|,$$

we proceed as follows.

We first observe that, if **X** is an axiom of P_1, then

$$\mathrm{S}^{f}_{\sim[r \supset r]}\mathbf{X}|$$

is a theorem of P_2. In fact, if **X** is †102 or †103, this is immediate by †202 and †203; and if **X** is †104, this follows by †221 and *231.

If a proof of $\mathbf{A}_f$ in P_1 is given in which the variable r does not occur, we replace f everywhere in this proof by $\sim[r \supset r]$. The resulting sequence of wffs of P_2 becomes a proof of $\mathbf{A}_0$ in P_2 upon supplying the proof of

$$\mathrm{S}^{f}_{\sim[r \supset r]}\mathbf{X}|$$

whenever necessary (this will be, as a matter of fact, whenever **X** is †104).

If the variable r does occur in the given proof of $\mathbf{A}_f$ in P_1, we begin by selecting a variable **a** which does not occur and replacing r by **a** throughout. After that we proceed as before, i.e., we replace f everywhere by $\sim[r \supset r]$, and then supply proof (in P_2) of

$$\mathrm{S}^{f}_{\sim[r \supset r]}\mathbf{X}|$$

wherever necessary. Then finally we use *201 to substitute r for **a**.

Employing the same truth-tables of $\supset$ and $\sim$ as those given in §15, we may define the *value* of a wff of P_2 for a system of values of its variables, in the same way that we did in the case of a wff of P_1. It is also possible, in the same way as before, to carry out the actual computation of the values of a wff for all systems of values of its variables. And a wff of P_2 is called a *tautology* if its value is t for all systems of values of its variables, a *contradiction* if its value is f for all systems of values of its variables.

****235.** Every theorem of P_2 is a tautology.

Proof. The three axioms of P_2 are tautologies and the two rules of inference have the property of preserving tautologies. (Cf. the proof of **150.)

****236.** If two wffs of P_2 have the same representative in P_1, then they have the same value for any system of values of the variables occurring in them.

Proof. By *233, **235, and the truth-table of $\supset$.

****237.** A wff $\mathbf{A}$ of P_2 is a tautology if and only if its representative $\mathbf{A}_f$ in P_1 is a tautology.

Proof. Let $\mathbf{A}_0$ be the expansion of $\mathbf{A}$ with respect to negation. Because the wff $\sim[r \supset r]$ of P_2 has always the value f, it follows that $\mathbf{A}_0$ is a tautology if and only if $\mathbf{A}_f$ is a tautology. (For the full proof of this, the reader must supply an analogue of the lemma which was used in the proof of **150.) Therefore **237 follows by **236.

****238.** A wff $\mathbf{A}$ of P_2 is a theorem of P_2 only if its representative $\mathbf{A}_f$ in P_1 is a theorem of P_1.

Proof. If $\mathbf{A}$ is a theorem of P_2, then by **235 it is a tautology. Therefore, by **237, $\mathbf{A}_f$ is a tautology. Therefore, by *152, $\mathbf{A}_f$ is a theorem of P_1.

The sense in which we have equivalence of the systems P_2 and P_1 now appears in *234 and **238. Namely, in the correspondence of each wff of P_2 to its representative in P_1 we have a many-one correspondence between the wffs of P_2 and of P_1 such that theorems correspond to theorems and non-theorems to non-theorems. And this many-one correspondence satisfies the structural conditions that, if $\mathbf{A}_f$ and $\mathbf{B}_f$ are the representatives of $\mathbf{A}$ and $\mathbf{B}$ respectively, then $\mathbf{A}_f \supset \mathbf{B}_f$ is the representative of $\mathbf{A} \supset \mathbf{B}$, and $\mathbf{A}_f \supset f$ is the representative of $\sim\mathbf{A}$ [202] (unless $\mathbf{A}$ is $[r \supset r]$).

[202]The bare existence of a many-one correspondence, or even of a one-to-one correspondence, between the wffs of P_1 and of P_2, such that theorems correspond to theorems and non-theorems to non-theorems, might be demonstrated just on the ground that the theorems and the non-theorems are denumerably infinite classes both in the case of P_1 and in the case of P_2. But without some added conditions, such as the structural conditions here stated, the bare existence of such a correspondence could hardly be said to constitute a significant equivalence of the two systems. We shall return in Chapter X to the question, what meaning is best given, in general, to the *equivalence* of two logistic systems.

From this equivalence between the two systems, together with **237, we have also the converse of **235:

*239. If a wff **A** of P_2 is a tautology, $\vdash$ **A**.

In **235 and *239, together with the algorithm for determining whether a wff is a tautology, we have a solution of the decision problem of P_2. In this connection the reader should satisfy himself that the proof of *239 (together with the proofs of preceding metatheorems on which it depends) directly provides an effective procedure to construct a proof of **A** in P_2 if it has been verified that **A** is a tautology.

As in the case of P_1, the consistency and completeness of P_2, in the various senses discussed in Chapter I, now follow as corollaries of this solution of the decision problem.

Principles of duality for P_2, analogues of *161, *164, *165, also follow in the same way as for P_1.

EXERCISES 23

23.0. Let P_{2L} be the logistic system which is identical with P_2 except that the wffs are translated into the parenthesis-free notation of Łukasiewicz. State the primitive basis of P_{2L}, and state and prove the analogues of **225–**227 for P_{2L}. (Compare exercise 12.2, and footnote 91.)

23.1. By the methods of §19, discuss the independence of the axioms and rules of P_2.

23.2. As a corollary of *229, prove the analogue of *159 (substitutivity of equivalence) for P_2.

23.3. Prove for P_2 that every tautology is a theorem, by a method which parallels the proof of the corresponding metatheorem of P_1 as this is contained in §§12–15, and which therefore avoids use of **226–*229.

23.4. According to our conventions, the expression "$\sim p \supset . q \not\subset p$" abbreviates a certain wff of P_1 and a certain (different) wff of P_2. Write each of these wffs without abbreviations other than omissions of brackets. For each wff, as thus written, carry out the computation of its values for all systems of values of its variables. Verify that corresponding values of the two wffs are always the same; and explain why this must be so in all such cases.

23.5. By analogy to §16, treat in detail the matter of duality in P_2.

23.6. Let P_F be the logistic system having the same primitive symbols and wffs as P_2, *200 and *201 as its rules of inference, and the six following axioms:

$$p \supset . q \supset p$$
$$s \supset [p \supset q] \supset . s \supset p \supset . s \supset q$$
$$p \supset [q \supset r] \supset . q \supset . p \supset r$$
$$p \supset q \supset . \sim q \supset \sim p$$
$$\sim\sim p \supset p$$
$$p \supset \sim\sim p$$

(1) Prove †204 as a theorem of P_F, and thus establish that the theorems of P_F are the same as those of P_2. (2) Discuss the independence of the axioms of P_F.

23.7. Let $P_Ł$ be the logistic system having the same primitive symbols and wffs as P_2, *200 and *201 as its rules of inference, and the single following axiom:

$$p_1 \supset [q_1 \supset p_1] \supset [\sim r \supset [p \supset \sim s] \supset [r \supset [p \supset q] \supset . s \supset p \supset . s \supset q] \supset p_2] \supset . q_2 \supset p_2$$

Establish that the theorems of $P_Ł$ are the same as those of P_2 by showing (1) that the single axiom of $P_Ł$ is a tautology, and (2) that the axioms of P_2 are theorems of $P_Ł$.

23.8. Establish the same result also for the logistic system $P_ł$ which is like $P_Ł$ except that the following (somewhat shorter) single axiom is used:

$$[p \supset . \sim q \supset [r \supset s] \supset . s \supset q \supset . p_1 \supset . r \supset q] \supset [\sim q_1 \supset [q_1 \supset r_1] \supset s_1] \supset s_1$$

23.9. Establish the same result also for the logistic system P_S, which is like $P_Ł$ except that the single axiom is the following:

$$r_1 \supset [p \supset q \supset s \supset q_1] \supset [s_1 \supset . s \supset p_1 \supset . \sim p \supset \sim r] \supset . s_1 \supset . s \supset p \supset . r \supset p$$

(Make use of the result of exercise 18.4.)

24. Primitive connectives for the propositional calculus.

In P_2 we used implication and negation as primitive connectives for the propositional calculus, and in P_1 we used implication and the constant f. We go on now to consider some other choices of primitive connectives (including, for convenience of expression, the constants as 0-ary connectives).

With one exception, we shall not consider connectives which take more than two operands. The exception is a ternary connective for which, when applied to operands **A**, **B**, **C**, we shall use the notation [**A**, **B**, **C**]. We call this connective *conditioned disjunction*, and assign to it the following truth-table:[203]

---

[203]A convenient oral reading of "$[p, q, r]$" is "p or r according as q or not q."

p	q	r	$[p, q, r]$
t	t	t	t
t	t	f	t
t	f	t	t
t	f	f	f
f	t	t	f
f	t	f	f
f	f	t	t
f	f	f	f

(It follows that the dual of $[p, q, r]$ is $[r, q, p]$ in the sense that the truth-table of $[p, q, r]$ becomes the truth-table of $[r, q, p]$ upon interchanging t and f.)

The singulary and binary connectives which we shall consider are those of §05,[204] with truth-tables as in §15. The constants are t and f, with values assigned as t and f respectively. If we include the truth-table

p	p
t	t
f	f

it will be seen that all possible truth-tables are then covered (for connectives that are no more than binary), except those truth-tables in which the value in the last column is independent of one of the earlier columns.

When a number of primitive connectives are given, together with the usual infinite list of propositional variables, the definition of wff is then immediate by analogy with §§10, 20. And the value of a wff for each system of values of its variables is then given by a definition analogous to that of §15. It is clear that the values of a particular wff for all systems of values of its variables may be given completely in a finite table like the truth-table of a connective; we shall call this the *truth-table of* the wff.

A system of primitive connectives for propositional calculus will be called *complete* if all possible truth-tables of two or more columns[205] are found among the truth-tables of the resulting wffs. And a particular one of the connectives

[204] I.e., the sentence connectives of §05.

[205] Notice that a wff with one variable has a truth-table of two columns, a wff with two variables a truth-table of three columns, and so on. A wff with no variables has just one (fixed) value and may therefore be said to have a truth-table of just one column (and one row). In the definition of completeness we have purposely excluded one-column truth-tables because we wish to allow as complete not only such a system of connectives as implication and f but also, e.g., such a system as implication and negation.

will be called *independent* if, upon suppression of that particular connective, its truth-table is no longer among the truth-tables of the resulting wffs—or in the case of a constant, *t*, or *f*, if upon suppression of it there is no longer among the resulting wffs any one which has the value t, or the value f, respectively, for all systems of values of its variables.

The problem of complete systems of independent primitive connectives for the propositional calculus has been treated systematically by Post.[206] We shall not make an exhaustive treatment here, but shall consider only certain special cases. We begin with the following:

Conditioned disjunction, t, and f constitute a complete system of independent primitive connectives for the propositional calculus.

The completeness is proved by mathematical induction with respect to the number of different variables in a wff constructed from these connectives. Among the wffs containing no variables, it is clear that all possible systems of values are found, since the wff *f* has the value f, and the wff *t* has the value t. Suppose that among the wffs containing *n* variables all possible systems of values, i.e., all possible truth-tables, are found. And consider a proposed system of values for a wff containing $n + 1$ variables, i.e., a proposed $(n + 2)$-column truth-table, T. Let the first column in the truth-table T be for the variable **b**. Let the truth-table T_1 be obtained from T by deleting all the rows which have f in the first column and then deleting the first column. Let T_2 be obtained from T by deleting all the rows which have t in the first column and then deleting the first column. By hypothesis of induction, wffs **A** and **C** exist whose truth-tables are T_1 and T_2 respectively. Then the wff [**A**, **b**, **C**] has the truth-table T, as may be seen by reference to the truth-table of conditioned disjunction. (Thus the completeness of the given system of primitive connectives follows by mathematical induction.)

The independence of conditioned disjunction is obvious, since without conditioned disjunction there would be no wffs except those consisting of a single symbol (*f* or *t* or a variable).

The independence of *t* may be proved by reference to the last row in the truth-table of conditioned disjunction, since it follows from this last row that, if a wff is constructed from conditioned disjunction and *f* (without

[206]In his monograph, *The Two-Valued Iterative Systems of Mathematical Logic* (1941). For connectives which are no more than binary, there is a different treatment by William Wernick in the *Transactions of the American Mathematical Society*, vol. 51 (1942), pp. 117–132.

Post deals also with the problem of characterizing the truth-tables which result from an arbitrary system of primitive connectives (with assigned truth-tables).

use of t), then it must have the value f when the value f is given to all its variables. In the same way the independence of f may be proved by reference to the first row in the truth-table of conditioned disjunction.

We may now prove completeness of other systems of primitive connectives by defining by means of the given primitive connectives (in the manner of §11) the three connectives, conditional disjunction, t, and f, and showing that the definitions give the value t to t, the value f to f, and the required truth-table to conditioned disjunction. Thus the completeness of implication and f follows by D1 (see §11) and the definition:

D12. $$[\mathbf{A}, \mathbf{B}, \mathbf{C}] \rightarrow [\mathbf{B} \supset \mathbf{A}][\sim\mathbf{B} \supset \mathbf{C}]$$

In the case of systems of primitive connectives which do not include constants, it is not possible to give definitions of t and f. But in proving completeness it is sufficient instead to give an example of a wff which has the value t for all systems of values of its variables, i.e., which is a tautology, and an example of a wff which has the value f for all systems of values of its variables, i.e., which is a contradiction. Thus the completeness of implication and negation follows by D12 above (as reconstrued when implication and negation are the primitive connectives) together with any example of a tautology, say $r \supset r$, and any example of a contradiction, say $r \not\subset r$.

In each of the systems, implication and f, implication and negation, the independence of the second connective follows because no wff constructed from implication alone can be a contradiction (as we may prove by mathematical induction, using the truth-table of implication). The independence of implication is in each case obvious because of the very restricted class of wffs which could be constructed without implication. Thus:

Each of the systems, implication and f, implication and negation, is a complete system of independent primitive connectives for the propositional calculus.

Having this, we may now also prove completeness of a system of primitive connectives by defining either implication and f or implication and negation, and showing that the definitions give the required truth-tables (and, if f is defined, that they give the value f to f).

In particular the completeness of negation and disjunction follows by the definition of $[\mathbf{A} \supset \mathbf{B}]$ as $\sim\mathbf{A} \vee \mathbf{B}$. And the completeness of negation and conjunction follows by the definition of $[\mathbf{A} \supset \mathbf{B}]$ as $\sim[\mathbf{A} \sim\mathbf{B}]$. Independence may be proved in each case in a manner analogous to that in which the independence of implication and negation was proved. Thus:

Each of the systems negation and disjunction, negation and conjunction, is a complete system of independent primitive connectives for the propositional calculus.

We leave to the reader the proof of the following:

Implication and converse non-implication constitute a complete system of independent primitive connectives for the propositional calculus.

This last system of primitive connectives has, like the system consisting of conditioned disjunction, *t*, and *f*, the substantial advantage of being *self-dual* in the sense that the dual of each primitive connective either is itself a primitive connective or is obtained from a primitive connective by permuting the operands. Indeed it is clear that a treatment of duality, like that of §16 for P_1, would be much simpler in the case of a formulation of the propositional calculus based on a self-dual system of primitive connectives.

Of the two self-dual systems of primitive connectives suggested, that consisting of implication and converse non-implication has the disadvantage that it is impossible to make a definition of negation which is *self-dual* in the sense that the dual of $\sim\mathbf{A}$ is $\sim\mathbf{A}_1$, where $\mathbf{A}_1$ is the dual of $\mathbf{A}$. Therefore it becomes necessary (for convenience in dualizing) to make two definitions of negation which are duals of each other. Then, if the symbols $\sim$ and $-$ are used for the two negations, the dual of $\sim\mathbf{A}$ will be $-\mathbf{A}_1$, where $\mathbf{A}_1$ is the dual of $\mathbf{A}$.

The self-dual system consisting of conditioned disjunction, *t*, and *f*, does not have this disadvantage. But it does have the obvious disadvantages associated with the use, as primitive, of a connective which takes more than two operands.

For this reason it has sometimes been suggested that the requirement of independence be abandoned in the interest of admitting a more convenient self-dual system of primitive connectives. In particular the system consisting of negation, conjunction, and disjunction has been proposed. Another possibility, of course, is negation, implication, and converse non-implication.

For certain purposes there are advantages in a complete system of primitive connectives which consists of one connective only, although to obtain such a system it is necessary to make a rather artificial choice of the primitive connective (and also to abandon any requirement of self-duality if the primitive connective is to be no more than binary). We leave to the reader the proof of the following:[207]

---

[207]The possibility of a single primitive connective for the propositional calculus was known to C. S. Peirce, as appears from a fragment, written about 1880, and from

Non-conjunction, taken alone, constitutes a complete system of primitive connectives for the propositional calculus. Likewise non-disjunction alone. These are the only connectives which are no more than binary and which have the property of constituting a complete system of primitive connectives for the propositional calculus when taken alone.

EXERCISES 24

24.0. Taking conditioned disjunction, t, and f as primitive, give definitions of all the singulary and binary connectives. Select the simplest possible definitions, subject to the conditions that definitions of mutually dual connectives shall be dual to each other and that the definition of negation shall be self-dual.

24.1. Taking implication and converse non-implication as primitive, give definitions of the singulary and remaining binary connectives. Select the simplest possible definitions, subject to the condition that definitions of mutually dual connectives shall be dual to each other. As indicated in the text, supply definitions of two negations dual to each other.

24.2. With conditioned disjunction, t, and f as primitive, assume that a formulation of the propositional calculus has been given such that every theorem is a tautology and every tautology is a theorem. Supply for this formulation of the propositional calculus a treatment of duality, analogous to that of §16 for P_1. Discuss first the dualization of wffs proper, and then

---

Chapter 3 of his unfinished *Minute Logic*, dated January-February 1902. These were unpublished during Peirce's lifetime, but appeared in 1933 in the fourth volume of his *Collected Papers* (see pp. 13–18, 215–216 thereof). First publication of the remark that the propositional calculus may be based on a single primitive connective was by H. M. Sheffer in a paper in the *Transactions of the American Mathematical Society*, vol. 14 (1913), pp. 481–488.

The analogous remark for Boolean algebra is that the usual Boolean operations may be based on a single primitive operation, for which two choices, dual to each other, are possible. Both of these operations are used together by Edward Stamm as a self-dual system of primitive operations for a postulational treatment of Boolean algebra, in a paper in *Monatshefte für Mathematik und Physik*, vol. 22 (1911), pp. 137–149. The explicit basing of Boolean algebra on a single primitive connective first appears in Sheffer's paper of 1913 (there is also some suggestion of this in Peirce's unpublished fragment of 1880).

Peirce's notations, which are his two alternative single primitive connectives for the propositional calculus, have not been used by others and need not be reproduced here. Sheffer uses the sign of disjunction, $\vee$, inverted as a sign of non-disjunction; he introduces non-conjunction only in a footnote and uses no special sign for it. The vertical line, since called Sheffer's stroke, was used by Sheffer only in connection with Boolean algebra; its use as a sign of non-conjunction was introduced by J. G. P. Nicod in the paper which has his single axiom for the propositional calculus, discussed in the next section (cf. *Proceedings of the Cambridge Philosophical Society*, vol. 19 (1917–1920), pp. 32–41).

also the dualization of expressions that are abbreviations of wffs under the definitions of 24.0.

24.3. Show that equivalence, disjunction, and *f* constitute a complete system of independent primitive connectives for the propositional calculus. (Compare exercise 15.7, and footnote 186.)

24.4. In each of the following systems of primitive connectives for the propositional calculus, demonstrate the independence of the connectives: (1) negation and disjunction; (2) negation and conjunction; (3) implication and converse non-implication.

24.5. Taking non-conjunction as the only primitive connective, give definitions of the singulary and remaining binary connectives. Neglecting considerations of duality, select the definitions in which the definiens is shortest (when written out in full, in terms of the primitive connective).

24.6. Show that equivalence and non-equivalence do not constitute a complete system of primitive connectives for the propositional calculus. Determine all possible ways of adding to the list one or more connectives which are no more than binary, so as to obtain a complete system of independent primitive connectives for the propositional calculus.

24.7. It may happen that a complete system of primitive connectives for the propositional calculus, though all are independent, is capable of being reduced without loss of completeness by replacing one of the connectives by a connective which can be defined from it alone and takes fewer operands than it does. For this purpose, any tautology is to be treated as if it supplied a definition of *t*, and any contradiction a definition of *f*, though these are not definitions in the proper sense (cf. the remark in the text on this point). When a complete system of independent primitive connectives for the propositional calculus is not capable of being reduced in this way, it is a *specialized* system of primitive connectives for the propositional calculus, in the sense of Post. Of the various complete systems of independent primitive connectives for the propositional calculus which are mentioned in the text, determine which are specialized systems. Of those which are not, supply all possible reductions to specialized systems.

24.8. Do the same thing for each of the complete systems of independent primitive connectives found in 24.6.

24.9. Consider a formulation of the propositional calculus in which the primitive connectives are negation, conjunction, and disjunction. A wff **B** which contains *n* different variables is said to be in *full disjunctive normal form* if the following conditions are satisfied: (i) **B** has the form of a disjunction $\mathbf{C}_1 \vee \mathbf{C}_2 \vee \ldots \vee \mathbf{C}_m$; (ii) each term $\mathbf{C}_i$ of this disjunction has the form of a conjunction, $\mathbf{C}_{i1}\mathbf{C}_{i2}\ldots\mathbf{C}_{in}$; (iii) each $\mathbf{C}_{ik}$ ($i = 1, 2, \ldots, m$ and $k = 1, 2, \ldots, n$) is either $\mathbf{b}_k$ or $\sim\mathbf{b}_k$, where $\mathbf{b}_k$ is the kth of the variables occurring in **B**, according to the alphabetic order of the variables (§20); (iv) the terms $\mathbf{C}_i$ are all different and are arranged among themselves according to the rule that, if $\mathbf{C}_{i1}, \mathbf{C}_{i2}, \ldots, \mathbf{C}_{i(k-1)}$ are the same as $\mathbf{C}_{j1}, \mathbf{C}_{j2}, \ldots, \mathbf{C}_{j(k-1)}$ respectively, and $\mathbf{C}_{ik}$ is $\mathbf{b}_k$, and $\mathbf{C}_{jk}$ is $\sim\mathbf{b}_k$, then $i < j$. Introduce material equivalence by an appropriate definition, and let a wff **B** be called a *full disjunctive normal form of* a wff **A** if **B** is in full disjunctive normal form and contains the same variables as **A** does and $\mathbf{A} \equiv \mathbf{B}$ is a tautology. Show that every wff not a contradiction has a unique full disjunctive normal form; and by means of the two laws of De Morgan (cf. 15.8) and commutative, associative, and distributive laws involving conjunction and disjunction (cf. 15.5, 15.7, 15.8), show how to reduce the wff to full disjunctive normal form. Show that a wff is a tautology if and only if it has a full disjunctive normal form in which $m = 2^n$.

24.10. For a formulation of the propositional calculus in which the primitive connectives are conditioned disjunction, t, and f, we may define *normal form* as follows, by recursion with respect to the number of different variables which a wff contains: a wff containing no variables is in normal form if and only if it is one of the two wffs, t, f; a wff in which the distinct variables contained are, in alphabetic order, $\mathbf{b}_1, \mathbf{b}_2, \ldots, \mathbf{b}_n$ is in normal form if and only if it has the form $[\mathbf{A}, \mathbf{b}_n, \mathbf{C}]$, where each of the wffs **A** and **C** contains all of the variables $\mathbf{b}_1, \mathbf{b}_2, \ldots, \mathbf{b}_{n-1}$, does not contain $\mathbf{b}_n$, and is in normal form. For this case establish a result about reduction to normal form, analogous to that of exercise 24.9 for reduction to full disjunctive normal form in the case of negation, conjunction, and disjunction as primitive connectives.

25. Other formulations of the propositional calculus. Formulations of the propositional calculus so far considered, in the text and in exercises, have been based either on implication and f or on implication and negation as primitive connectives. We go on now to describe briefly some formulations based on other primitive connectives.

Very well known is the formulation, $\mathrm{P_R}$, of the propositional calculus which is used in *Principia Mathematica*. In this the primitive connectives are negation and disjunction. The rules of inference are substitution and *modus ponens* (the latter in the form, from $\sim\mathbf{A} \vee \mathbf{B}$ and **A** to infer **B**). The axioms are the five following, in which $\mathbf{A} \supset \mathbf{B}$ is to be understood as an abbreviation of $\sim\mathbf{A} \vee \mathbf{B}$:

$$p \vee p \supset p$$
$$q \supset p \vee q$$
$$p \vee q \supset q \vee p$$
$$p \vee [q \vee r] \supset q \vee [p \vee r]$$
$$q \supset r \supset . p \vee q \supset p \vee r$$

Several reductions of this sytem have been proposed. Of these the most immediate is the system P_B obtained by just deleting the fourth axiom, which is, in fact, non-independent. Another is the system P_N in which the five axioms are replaced by the following four axioms:

$$p \vee p \supset p$$
$$p \supset p \vee q$$
$$p \vee [q \vee r] \supset q \vee [p \vee r]$$
$$q \supset r \supset . p \vee q \supset p \vee r$$

Still another is the system P_G in which the five axioms are replaced by the following three:

$$p \vee p \supset p$$
$$p \supset p \vee q$$
$$q \supset r \supset . p \vee q \supset r \vee p$$

For some purposes there may be advantages in a formulation of the propositional calculus which is based on only one primitive connective, only one axiom, and besides substitution only one rule of inference, although in order to accomplish this it is necessary to make a rather artificial selection of the primitive connective and to allow the single axiom to be relatively long. As long as the procedure in constructing a logistic system is regarded as tentative, with the choice held open as to what assumptions (in the form of axioms or rules) will finally be accepted—or if the emphasis is upon fixing the ground of theorems and metatheorems in the sense of distinguishing what assumptions each one rests upon —the preference will be for naturalness in the selection of primitive connectives and for simplicity in the individual axioms and rules, rather than for reduction in their number. On the other hand the proof of desired metatheorems may well be simplified in some cases by the contrary course of reducing the number of primitive connectives, axioms, and rules, even at the expense of naturalness or simplicity; and a metatheorem, once proved for one formulation, may perhaps be extended to other formulations by establishing equivalence of the formulations (in some appropriate sense).

For the propositional calculus, a formulation of the proposed kind was first found by J. G. P. Nicod. His system, call it P_n, is based on non-conjunction as the primitive connective. The rules of inference are substitution and the rule: from **A | . B | C** and **A** to infer **C**. The single axiom is:

$$p \mid [q \mid r] \mid . p_1 \mid [p_1 \mid p_1] \mid . s \mid q \mid . p \mid s \mid . p \mid s$$

In another such formulation of the propositional calculus, P_w, the primitive connective and the rules of inference are the same as in P_n, and the single axiom is the following:

$$p \mid [q \mid r] \mid . [s \mid r \mid . p \mid s \mid . p \mid s] \mid . p \mid . p \mid q$$

(This axiom, unlike Nicod's, is *organic*, in the sense of Wajsberg and Leśniewski, i.e., no wf part shorter than the whole is a tautology.)

In still another such formulation of the propositional calculus, P_L, the primitive connective and the rules of inference are still the same as in P_n, and the single axiom is:

$$p \mid [q \mid r] \mid . p \mid [r \mid p] \mid . s \mid q \mid . p \mid s \mid . p \mid s$$

(This axiom is closer to Nicod's, and still organic.)

EXERCISES 25

25.0. Establish the sufficiency of P_R for the propositional calculus by carrying the development of the system far enough, either to prove directly that a wff is a theorem if and only if it is a tautology, or to show equivalence to P_2 in a sense analogous to that of §23. (To facilitate the development, the derived rule should be established as early as possible, that if **M**, **N**, **A**, **B** satisfy the conditions stated in *229, and if $\vdash$ **M** $\supset$ **N** and $\vdash$ **N** $\supset$ **M**, then $\vdash$ **A** $\supset$ **B** and $\vdash$ **B** $\supset$ **A**.)

25.1. Establish the sufficiency of P_B for the propositional calculus by proving the fourth axiom of P_R as a theorem.

25.2. Discuss the independence of the axioms and rules of P_B.

25.3. Establish the sufficiency of P_N for the propositional calculus by proving the second and third axioms of P_R as theorems.

25.4. Discuss the independence of the axioms and rules of P_N.

25.5. Establish the sufficiency of P_G for the propositional calculus by proving the axioms of P_R as theorems. (The chief difficulty is to prove the theorem $p \supset p$. For this purpose, following H. Rasiowa, we may first prove in order the theorems $p \supset {\sim}{\sim}p$, $q \vee p \supset . {\sim}p \supset q$, ${\sim}{\sim} . {\sim}{\sim}{\sim}p \supset {\sim}p$, ${\sim}p \vee . p \vee {\sim}{\sim}{\sim}p$, ${\sim}{\sim}[p \supset r] \supset . s \vee [q \vee p] \supset r \vee q \vee s$.)

25.6. Discuss the independence of the axioms and rules of P_G.

25.7. Establish the sufficiency of P_n for the propositional calculus by showing its equivalence to P_R in a sense analogous to that of §23.

25.8. Establish the sufficiency of P_n for the propositional calculus, without use of P_R, by showing its equivalence to P_2 in a sense analogous to that of §23.

25.9. Discuss the independence of the axioms and rules of P_n. (This question can be answered by means of immediately obvious considerations, without use of truth-tables or generalized systems of truth-values in the sense of §19.)

25.10. By the method of §19 or otherwise, determine whether P_n remains sufficient for the propositional calculus when its second rule of inference is weakened to the following: from $\mathbf{A} \mid . \mathbf{B} \mid \mathbf{B}$ and **A** to infer **B**.

25.11. Establish the sufficiency of P_w for the propositional calculus by proving the axiom of P_n as a theorem.

25.12. Establish the sufficiency of P_L for the propositional calculus.

25.13.[208] (1) Given implication and converse non-implication as primitive connectives, and substitution and *modus ponens* as rules of inference, find axioms so that the resulting system is sufficient for the propositional calculus. Seek, as far as feasible, to make the individual axioms simple, and after that to make their number small. (2) Establish the independence of the axioms and rules.

25.14.[208] Given negation, conjunction, and disjunction as primitive connectives, find axioms and rules of inference so that the resulting logistic system is sufficient for the propositional calculus. In doing so, make the system of axioms and rules of inference self-dual in such a sense that a metatheorem analogous to *161 (principle of duality) follows immediately therefrom. Subject to this condition seek, as far as feasible, to make the individual axioms and rules simple, and after that to make their number small. Can the axioms and rules be made independent without excessive complication or loss of the feature of self-duality?

25.15.[208] Answer the same questions if the primitive connectives are conditioned disjunction, t, and f.

25.16.[208] Let the primitive connectives be conditioned disjunction, t, and f. Let the rules of inference be the rule of substitution and the following rule: from [**A**, **B**, **C**] and **B** to infer **A**. Find axioms so that the resulting system is sufficient for the propositional calculus. Seek, as far as feasible, to make the individual axioms simple, and after that to make their number small. (Ignore the matter of duality.)

---

[208]This is offered as an open problem for investigation, rather than as an exercise in the ordinary sense. The writer has not attempted to find a solution.

26. Partial systems of propositional calculus. We have so far emphasized the matter of logistic systems adequate to the full propositional calculus, in the sense of being equivalent in some appropriate sense to P_1 or P_2. Studies have also been made, however, of various partial systems, not adequate to the full propositional calculus, and in this section we shall describe briefly some of these.

One sort of partial system of propositional calculus is based on an incomplete system of primitive connectives, axioms and rules being so chosen that the theorems coincide with the tautologies in those connectives. An example is the *implicational propositional calculus*, which has implication as its only primitive connective, and which is formulated by (e.g.) either the logistic system P^I_B of exercise 18.3 or the system $P^I_Ł$ of 18.4. Other examples will be found in the exercises following this section.

The chief interest of partial systems of this sort would seem to be as stepping-stones toward formulations of the full propositional calculus. For example, the result of exercise 18.4, together with that of 12.7, leads to a formulation P_λ of the propositional calculus in which the primitive connectives are implication and f, the rules of inference are *modus ponens* and substitution, and the axioms are the two following:

$$p \supset q \supset r \supset . r \supset p \supset . s \supset p$$
$$f \supset p$$

The foregoing formulation of the full propositional calculus is elegant for its brevity, and sharply separates out the role of the constant f from that of implication, but fails to separate from one another what may be regarded as different assumptions about implication. If we wished to separate from the others those properties of implication which are involved in the deduction theorem, we might begin with the logistic system P^+ of exercise 19.6 (or an equivalent)—the *positive implicational propositional calculus* of Hilbert—and add primitive connectives and axioms to obtain a formulation of the full propositional calculus.

Akin to the positive implicational propositional calculus is the *positive propositional calculus* of Hilbert, designed to embody the part of the propositional calculus which may be said to be independent in some sense of the existence of a negation. This may be formulated as a logistic system P^P, as follows. The primitive connectives are implication, conjunction, disjunction, and equivalence (which then are not independent, even as primitive connectives for this partial system of propositional calculus). The rules of inference are *modus ponens* and substitution. And the axioms are the eleven following:

$$p \supset . q \supset p$$
$$s \supset [p \supset q] \supset . s \supset p \supset . s \supset q$$
$$pq \supset p$$
$$pq \supset q$$
$$p \supset . q \supset pq$$
$$p \supset p \vee q$$
$$q \supset p \vee q$$
$$p \supset r \supset . q \supset r \supset . p \vee q \supset r$$
$$p \equiv q \supset . p \supset q$$
$$p \equiv q \supset . q \supset p$$
$$p \supset q \supset . q \supset p \supset . p \equiv q$$

The system P^P may be extended to a formulation of the full propositional calculus by adding negation as a primitive connective and one or more suitably chosen axioms involving negation. We may for example use †204 as a single additional axiom, so obtaining a formulation of the propositional calculus which we shall call P_H.

On the other hand by adding to P^P a weaker axiom or axioms involving negation we may obtain a formulation of the *intuitionistic propositional calculus* of Heyting.[209]

The *mathematical intuitionism* of L. E. J. Brouwer will be discussed in Chapter XII. On grounds to be explained in that chapter, it rejects certain principles of logic which mathematicians have traditionally accepted without question, among them certain laws of propositional calculus, especially the law of double negation and the law of excluded middle. (Of course this involves also rejection of such an interpretation of propositional calculus as that of §10 or §20, not perhaps in itself but in the light of the actual use of the propositional calculus as a part of a more extensive language.)

Heyting's logistic formalization of the ideas of Brouwer (accepted by Brouwer) gave them a precision which they otherwise lacked, and has played a major role in subsequent debate of the merits of the intuitionistic critique of classical mathematics. For the intuitionistic propositional calculus we adopt not Heyting's original formulation but the following equivalent formulation P^i_S.

The primitive connectives of P^i_S are implication, conjunction, disjunction, equivalence, and negation. The rules of inference are *modus ponens* and

[209]Arend Heyting in *Sitzungsberichte der Preussischen Akademie der Wissenschaften*, Physikalisch-mathematische Klasse, 1930, pp. 42–56.

substitution. The axioms are the eleven axioms of P^P and the two following additional axioms:

$$p \supset \sim p \supset \sim p \qquad \textit{(Special law of reductio ad absurdum.)}$$
$$\sim p \supset . p \supset q \qquad \textit{(Law of denial of the antecedent.)}$$

The *minimal propositional calculus* of Kolmogoroff and Johansson[210] makes a more drastic rejection of classical laws involving negation. A formulation of it, P_0^m, may be obtained from P_S^i by replacing the two foregoing axioms by the single axiom:

$$p \supset q \supset . p \supset \sim q \supset \sim p \qquad \textit{(Law of reductio ad absurdum.)}$$

Wajsberg has shown[211] that any theorem **A** of P_S^i can be proved from the first two of the thirteen axioms together with only those axioms which contain the connectives, other than implication, actually appearing in **A**. As a corollary the same thing may be shown also for P_0^m. It follows that those theorems of the intuitionistic propositional calculus P_S^i, or of the minimal propositional calculus P_0^m, which do not contain negation are identical with the theorems of the positive propositional calculus; further, that those theorems of any one of the three systems in which implication appears as the only connective are identical with the theorems of the positive implicational propositional calculus.

The decision problem of P_S^i has been solved by Gentzen, and again by Wajsberg.[212] By the results referred to in the preceding paragraph, solution of the decision problem follows for P^P and P^+. And by the result of exercise 26.19 (2), solution of the decision problem follows also for P_0^m.

---

[210]A. Kolmogoroff in *Recueil Mathématique de la Société Mathématique de Moscou*, vol. 32 (1924–1925), pp. 646–667, and Ingebrigt Johansson in *Compositio Mathematica*, vol. 4 (1936), pp. 119–136. See also the paper of Wajsberg cited in the next footnote.

Kolmogoroff considers primarily not the full minimal calculus but the part of it obtained by suppressing the three primitive connectives, conjunction, disjunction, equivalence, and the axioms containing them (and finds for this calculus a similar result to that later found by V. Glivenko, quoted in exercise 26.15, for the intuitionistic propositional calculus). Addition of the three axioms for disjunction which are given in the text is mentioned by Kolmogoroff in a footnote, but the full minimal calculus and the name ("minimal calculus" or "Minimalkalkül") first occur in Johansson's paper. Kolmogoroff uses for implication not the two axioms in the text but four axioms taken from Hilbert that are equivalent to these; and he takes the three axioms for disjunction from Ackermann. Johansson uses Heyting's axioms (from the paper cited in footnote 209), suppressing one of those for negation.

[211]Mordchaj Wajsberg in *Wiadomości Matematyczne*, vol. 46 (1938), pp. 45–101.

[212]Gerhard Gentzen in *Mathematische Zeitschrift*, vol. 39 (1934), pp. 176–210, 405–431. Wajsberg, *loc.cit.* Other decision procedures for the intuitionistic propositional calculus are due to J. C. C. McKinsey and Alfred Tarski in *The Journal of Symbolic Logic*, vol. 13 (1948), pp. 1–15, to Ladislav Rieger in *Acta Facultatis Rerum Naturalium Universitatis Carolinae*, no. 189 (1949), and to B. Ú. Pil'čak in the *Doklady Akadémii Nauk SSSR*, vol. 75 (1950), pp. 773–776.

EXERCISES 26

26.0. Let $\mathrm{P}^{\mathrm{E}}_{\mathrm{W}}$ be the partial system of propositional calculus based on equivalence as the only primitive connective, the rules of inference being substitution and the rule, from $\mathbf{A} \equiv \mathbf{B}$ and $\mathbf{A}$ to infer $\mathbf{B}$, and the axioms being the two following:

$$p \equiv q \equiv . q \equiv p$$
$$p \equiv [q \equiv r] \equiv . p \equiv q \equiv r$$

Prove the following theorems of $\mathrm{P}^{\mathrm{E}}_{\mathrm{W}}$:

$$p \equiv p \equiv q \equiv q$$
$$p \equiv p$$
$$p \equiv q \equiv . r \equiv . p \equiv . q \equiv r$$
$$[s \equiv . p \equiv . q \equiv r] \equiv . p \equiv q \equiv r \equiv s$$
$$r \equiv p \equiv [q \equiv r] \equiv . p \equiv q$$
$$q \equiv r \equiv . p \equiv q \equiv . p \equiv r$$
$$p_1 \equiv q_1 \equiv . p_2 \equiv q_2 \equiv . p_1 \equiv p_2 \equiv . q_1 \equiv q_2$$

(The order in which the theorems are given is one possible order in which they may be proved. Heuristically, solution of 26.0 and 26.2 may be facilitated by noticing that the given axioms are the complete commutative and associative laws of equivalence.)

26.1. Hence prove the following metatheorem of $\mathrm{P}^{\mathrm{E}}_{\mathrm{W}}$, by a method analogous to that of the proof of *229: If $\mathbf{B}$ results from $\mathbf{A}$ by substitution of $\mathbf{N}$ for $\mathbf{M}$ at zero or more places (not necessarily at all occurrences of $\mathbf{M}$ in $\mathbf{A}$), and if $\vdash \mathbf{M} \equiv \mathbf{N}$, then $\vdash \mathbf{A} \equiv \mathbf{B}$.

26.2. Hence prove that a wff of $\mathrm{P}^{\mathrm{E}}_{\mathrm{W}}$ is a theorem if and only if every variable in it occurs an even number of times.[213] Hence the theorems of $\mathrm{P}^{\mathrm{E}}_{\mathrm{W}}$ coincide with the tautologies in which equivalence appears as the only connective.

26.3. (1) Let $\mathrm{P}^{\mathrm{E}}_{\mathrm{w}}$ be the system obtained from $\mathrm{P}^{\mathrm{E}}_{\mathrm{W}}$ by replacing the two axioms by the following single axiom:

$$p \equiv [q \equiv r] \equiv [r \equiv s \equiv s] \equiv . p \equiv q$$

Prove that the theorems of $\mathrm{P}^{\mathrm{E}}_{\mathrm{w}}$ are identical with those of $\mathrm{P}^{\mathrm{E}}_{\mathrm{W}}$.

(2) Let $\mathrm{P}^{\mathrm{E}}_{\text{Ł}}$ be the system obtained from $\mathrm{P}^{\mathrm{E}}_{\mathrm{W}}$ by replacing the two axioms by the following single axiom:

---

[213]This solution of the decision problem of the equivalence calculus is due to Leśniewski. Notice its relationship to 15.7.

$$q \equiv r \equiv . p \equiv q \equiv . r \equiv p$$

Prove that the theorems of $P^E_Ł$ are identical with those of P^E_W.

(3) Following Łukasiewicz, use the result of 26.2 and the method of §19 to show that no shorter single axiom can thus replace the two axioms of P^E_W.

26.4. Let P^{EN} be the system obtained from P^E_W by adjoining negation as an additional primitive connective, and one additional axiom:

$$\sim p \equiv \sim q \equiv . p \equiv q$$

In a sense analogous to that of §23, demonstrate equivalence of P^{EN} to the system P^{Ef} obtained from P^E_W by adjoining f as an additional primitive symbol, and no additional axioms—negation being defined thus in P^{Ef}:

$$\sim \mathbf{A} \rightarrow \mathbf{A} \equiv f$$

Hence prove that a wff of P^{EN} is a theorem if and only if every variable in it and the sign $\sim$ occur each an even number of times (if at all).[214] Hence the theorems of P^{EN} and of P^{Ef} coincide with the tautologies in equivalence and negation, and in equivalence and f, respectively.

26.5. Show that the system P^{EN} is not complete in the sense of Post, since the wff $p \equiv \sim p$ can be added as an axiom without making the wff p a theorem.[215]

26.6. A partial system of propositional calculus is to have equivalence and disjunction as primitive connectives, and, besides the rule of substitution, the two following rules of inference: from **A** and $\mathbf{A} \equiv \mathbf{B}$ to infer **B**; from **A** to infer $\mathbf{A} \vee \mathbf{B}$.[216] (1) Find axioms such that the theorems coincide with the tautologies in equivalence and disjunction. (2) With the aid of any previous results proved in the text or in exercises, show that the system (as based on these axioms) is complete in the sense of Post. (3) Discuss also the independence of the axioms and rules of inference.

26.7. Making use of results already found for P_2 (so far as they apply), show that a wff of P_H is a theorem if and only if it is a tautology.

26.8. (1) Establish the independence of $p \supset . q \supset p$ as an axiom of P_H by means of the following truth-tables, in which 0 and 1 are the designated truth-values:

---

[214] This solution of the decision problem of the equivalence-negation calculus is due independently to McKinsey and Mihailescu, as a corollary of Leśniewski's solution of the decision problem of the equivalence calculus; see *The Journal of Symbolic Logic*, vol. 2, p. 175, and vol. 3, p. 55. Here again the relationship to 15.7 should be noticed.

[215] Eugen Gh. Mihailescu in *Annales Scientifiques de l'Université de Jassy*, part 1, vol. 23 (1937), pp. 369–408, iv.

[216] These two rules are used by M. H. Stone in *American Journal of Mathematics*, vol. 59 (1937), pp. 506–514.

p	q	$p \supset q$	pq	$p \vee q$	$p \equiv q$	$\sim p$
0	0	0	1	1	1	3
0	1	1	1	1	1	
0	2	2	4	1	4	
0	3	2	4	1	4	
0	4	4	4	1	4	
1	0	0	1	1	1	2
1	1	1	1	1	1	
1	2	2	4	1	4	
1	3	2	4	1	4	
1	4	4	4	1	4	
2	0	0	4	1	4	3
2	1	1	4	1	4	
2	2	0	4	4	1	
2	3	0	4	4	4	
2	4	1	4	4	1	
3	0	2	4	1	4	3
3	1	1	4	1	4	
3	2	2	4	4	4	
3	3	2	4	4	4	
3	4	1	4	4	1	
4	0	0	4	1	4	0
4	1	1	4	1	4	
4	2	0	4	4	1	
4	3	0	4	4	1	
4	4	1	4	4	1	

(2) By a modification of these truth-tables establish also the independence of $p \supset . q \supset p$ as an axiom of P^i_S. (*Suggestion*: In both parts (1) and (2), in order to minimize the labor of verifying tautologies mechanically, make use as far as possible of arguments of a general character.)

26.9. Discuss the independence of the remaining axioms (1) of the system P_H, and (2) of the system P^i_S.

26.10. Consider a system of truth-values 0, 1, . . ., ν with 0 as the only designated truth-value, and the following truth-tables of the connectives of P^i_S: the value of $p \supset q$ is 0 if the value of p is greater than or equal to the value of q, and in the contrary case it is the same as the value of q; the value of pq is the greater of the values of p and q; the value of $p \vee q$ is the lesser of the values of p and q; the value of $p \equiv q$ is 0 if the values of p and q are the same, and in the contrary case it is the greater of the values of p and q; the value of $\sim p$ is 0 if the value of p is ν, and in all other cases the value of $\sim p$ is ν. Show that all theorems of P^i_S are tautologies according to these truth-tables.

26.11. Hence show that the following are not theorems of P^i_S: the law of double negation; the law of excluded middle; the converse law of contraposition; and the *law of indirect proof*,

$$\sim p \supset q \supset . \sim p \supset \sim q \supset p.$$

26.12. Following Kurt Gödel, use these same truth-tables to show that

$$[p_1 \equiv p_2] \vee [p_1 \equiv p_3] \vee \ldots \vee [p_1 \equiv p_n]$$
$$\vee [p_2 \equiv p_3] \vee [p_2 \equiv p_4] \vee \ldots \vee [p_2 \equiv p_n] \vee \ldots\ldots \vee [p_{n-1} \equiv p_n]$$

is not a theorem of P^i_S (for any n), also that this wff becomes a theorem of P^i_S upon identifying any two of its variables (by substituting one of the variables everywhere for the other), and hence finally that *there is no system of truth-tables in finitely many truth-values* such that under it not only are all theorems of P^i_S tautologies but also all tautologies are theorems of P^i_S.[217]

26.13. With the aid of the deduction theorem (which can be demonstrated for P^i_S in the same way as for P_2), show that the following are theorems of P^i_S:

$p \supset q \supset . p \supset \sim q \supset \sim p$	(*Law of reductio ad absurdum.*)
$p \supset \sim\sim p$	(*Converse law of double negation.*)
$\sim\sim\sim p \supset \sim p$	(*Law of triple negation.*)
$p \supset q \supset . \sim q \supset \sim p$	(*Law of contraposition.*)
$\sim\sim . p \vee \sim p$	(*Weak law of excluded middle.*)
$\sim . p \sim p$	(*Law of contradiction.*)

26.14. Let P_Γ be the system obtained by adjoining the law of excluded middle, $p \vee \sim p$, to P^i_S as an additional axiom. Prove †204 as a theorem of P_Γ. Hence show that the theorems of P_Γ are the same as the theorems of P_H—therefore, by the result of 26.7, that the theorems of P_Γ are the same as the tautologies (according to the usual two-valued truth-tables, §15).

26.15. Using the results of 26.13 and 26.14, establish the following results

[217]Stanisław Jaśkowski has constructed a system of truth-tables in infinitely many truth-values which is such that under it the tautologies coincide with the theorems of the intuitionistic propositional calculus (i.e., which, in the terminology of McKinsey, is *characteristic* for the intuitionistic propositional calculus). See *Actes du Congrès International de Philosophie Scientifique,* Paris 1935 (published 1936), part VI, pp. 58–61. (*Added in proof.* See further a paper by Gene F. Rose in the *Transactions of the American Mathematical Society,* vol. 75 (1953), pp. 1-19.)

of V. Glivenko: If a wff **A** of P^i_S is a tautology according to the usual two-valued truth-tables, then $\sim\sim$**A** is a theorem of P^i_S. A wff $\sim$**A** of P^i_S is a theorem of P^i_S if and only if it is a tautology according to the usual two-valued truth-tables.

26.16. As a corollary of the results of 26.15, establish also the following result of Gödel:[218] A wff **A** of P^i_S in which conjunction and negation are the only connectives appearing is a theorem of P^i_S if and only if it is a tautology according to the usual two-valued truth-tables.

26.17. Discuss the independence of the axioms of P_r (exercise 26.14), showing in particular that the axiom $p \supset \sim p \supset \sim p$ is non-independent.

26.18. Let P^i_2 be the system obtained from P^i_S by replacing the axiom $p \supset \sim p \supset \sim p$ by the two following axioms:

$$\sim . p \sim p$$

$$p \supset q \supset . \sim q \supset \sim p$$

(1) Show that the theorems of P^i_2 are the same as those of P^i_S. (2) Show that the theorems of P^i_2 which can be proved without use of the axiom $\sim p \supset . p \supset q$ are the same as the theorems of P^m_0. (3) Discuss the independence of the axioms of P^i_2.

26.19. Let P^i_W be the system obtained from P^P by adjoining f as an additional primitive symbol and $f \supset p$ as an additional axiom. Let P^m_J be the system obtained from P^P by adjoining f as an additional primitive symbol and no additional axioms. Show that *229 is valid as a metatheorem of P^i_S, and of P^m_0. (1) Hence establish the equivalence of P^i_S and P^i_W in the sense of §23. (2) Likewise establish the equivalence of P^m_0 and P^m_J in the sense of §23.[219]

26.20. Establish the following result (substantially that of Kolmogoroff referred to in footnote 210): In any theorem of the full propositional calculus P_2—in which the connectives occurring are only implication and negation—let every variable **a** be replaced throughout by its double negation $\sim\sim$**a**. The resulting formula is a theorem of the minimal calculus P^m_0.

26.21. Establish the equivalence of P^m_0 and the system P^m_w obtained from it by replacing its last axiom (the law of *reductio ad absurdum*) by the axiom $p \supset \sim q \supset . q \supset \sim p$.

26.22. For the system P^P, show that implication is definable from con-

---

[218] See Glivenko's paper cited in footnote 271, and a paper by Gödel in *Ergebnisse eines Mathematischen Kolloquiums*, no. 4 (1933), pp. 34–38.

[219] This result regarding the minimal calculus was found by Johansson, *loc.cit.*; and later by Wajsberg, *loc.cit*, for Kolmogoroff's minimal calculus with implication and negation as only primitive connectives. (See footnotes 210, 211.)

junction and equivalence, in the sense that $p \supset q \equiv . p \equiv pq$ is a theorem of the system.

26.23. For the system P^P, show in a like sense that both implication and conjunction are definable from disjunction and equivalence. Hence disjunction, equivalence, and *f* constitute (in an appropriate sense) a complete system of independent primitive connectives for P^i_W (see 26.19).[220]

26.24. By means of the truth-table of exercise 26.10 with $\nu = 2$, show, for the system P^i_S, that equivalence is not definable from disjunction and negation. Hence, by the result of the preceding exercise, disjunction, equivalence, and negation constitute (in an appropriate sense) a complete system of independent primitive connectives for P^i_S.[220]

27. Formulations employing axiom schemata. Formulations of the propositional calculus so far considered have been based each on a finite number of axioms, although the program of §07 allows also that the number of axioms be infinite, provided there is supplied an effective method by which to recognize a given wff as being or not being an axiom.

When the axioms are infinite in number, of course they cannot be written out in full, and it is necessary rather to indicate them (or all but a finite number of them) by one or more statements in the syntax language each introducing an infinite class of axioms. Such a statement in the syntax language may always be reworded as a rule of inference with an empty class of premisses, and in this sense the distinction between an infinite and a finite number of axioms is illusory. The more significant distinction is between formulations which rely more or less heavily on syntactical statements (such as rules of inference) to take the place of separately stated axioms in the object language—but here no sharp line of division can be drawn.

Formulations of the kind which we describe as based on an infinite number of axioms have important advantages in some cases. We consider in this section a particular class of such formulations, namely, those in which the primitive basis involves *axiom schemata*.[221]

[220]Wajsberg has shown (*loc.cit.*) that implication-conjunction-disjunction-*f* and implication-conjunction-disjunction-negation constitute complete systems of independent primitive connectives for P^i_W and P^i_S respectively, i.e., for the intuitionistic propositional calculus.

[221]A formulation of a different kind having an infinite number of axioms is obtained by choosing some suitable system of connectives as primitive and then making every tautology an axiom, no rules of inference being then necessary—as pointed out, in effect, by Herbrand in 1930.

This procedure provides no deductive analysis of the propositional calculus, and no opportunity to consider the effects of making or rejecting various particular assumptions (such as, e.g., the law of excluded middle or the law of denial of the antecedent—cf. §26). Nevertheless it may be useful in a case where it is desired to deal with the prop-

An *axiom schema* represents an infinite number of axioms by means of an expression containing syntactical variables—a form, in the sense of §02—which has wffs as values.[222] Every value of the expression is to be taken as an axiom. For convenience of statement we shall indicate this by writing the expression itself in the same manner as an axiom.

An example of a formulation of the propositional calculus with axiom schemata is the following system P.

The primitive symbols and the wffs of P are the same as those of P_2 (§20). The axioms, infinite in number, are given by the three following axiom schemata:

$$\mathbf{A} \supset . \mathbf{B} \supset \mathbf{A}$$
$$\mathbf{A} \supset [\mathbf{B} \supset \mathbf{C}] \supset . \mathbf{A} \supset \mathbf{B} \supset . \mathbf{A} \supset \mathbf{C}$$
$$\sim\mathbf{A} \supset \sim\mathbf{B} \supset . \mathbf{B} \supset \mathbf{A}$$

And the only rule of inference (if we do not count the axiom schemata as such) is the rule of *modus ponens*.

*270. Every theorem of P is a theorem of P_2.

Proof. Every axiom of P either is an axiom of P_2 or is obtained from an axiom of P_2 by a substitution (*201) or a simultaneous substitution (*210). And the one rule of inference of P is also a rule of inference of P_2.

**271. Every theorem of P_2 is a theorem of P.

Proof. Since every axiom of P_2 is an axiom of P, and the rules of inference are the same except for the rule of substitution (*201), it will be sufficient to show that the rule of substitution is a derived rule of P. This is done as follows:

In an application of the rule of *modus ponens*, let the major premiss be $\mathbf{C} \supset \mathbf{D}$, the minor premiss **C**, and the conclusion **D**. If we substitute the wff **B** for the variable **b** throughout in both premisses and in the conclusion, the three resulting wffs,

$$\mathrm{S}^{\mathbf{b}}_{\mathbf{B}}\mathbf{C} \supset \mathbf{D}|, \qquad \mathrm{S}^{\mathbf{b}}_{\mathbf{B}}\mathbf{C}|, \qquad \mathrm{S}^{\mathbf{b}}_{\mathbf{B}}\mathbf{D}|,$$

are also premisses and conclusion of an application of the rule of *modus*

ositional calculus rapidly and only as a preliminary to a study of more comprehensive systems. It has been used in this way by Hilbert and Bernays (*Grundlagen der Mathematik*, vol. 1 (1934), see pp. 83, 105), by Quine (*Mathematical Logic*, 1940, see pp. 88–89) and by others.

A similar short cut is indeed possible in the treatment of any logistic system whose decision problem has been solved in such a way as to provide a practically feasible decision procedure, but of course only after the solution of the decision problem.

[222]Compare the discussion of definition schemata in §11.

ponens, in view of the fact that

$$\mathrm{S}^{\mathbf{b}}_{\mathbf{B}}\mathbf{C} \supset \mathbf{D}|$$

is the same wff as

$$\mathrm{S}^{\mathbf{b}}_{\mathbf{B}}\mathbf{C}| \supset \mathrm{S}^{\mathbf{b}}_{\mathbf{B}}\mathbf{D}|.$$

Since the only rule of inference of P is *modus ponens*, and since the result of making the substitution of **B** for **b** throughout an axiom of P is again an axiom of P, it follows that a proof of a wff **A** as a theorem of P can be transformed into a proof of

$$\mathrm{S}^{\mathbf{b}}_{\mathbf{B}}\mathbf{A}|$$

as a theorem of P by just substituting **B** for **b** throughout, in every wff in the proof.

This completes the proof of the metatheorem **271.[223] It may of course be used as a derived rule of P, but we have numbered it with a double asterisk as a metatheorem of P_2.

As a corollary of **271, we have also the following metatheorem of P_2:

272. There is an effective process by which any proof of a theorem of P_2 can be transformed into a proof of the same theorem of P_2 in which substitution is applied only to axioms (i.e., in every application of the rule of substitution the premiss **A is one of the axioms of P_2).

In the foregoing we have chosen the system P as an example, because it is closely related to the particular formulations of the functional calculi of first order that receive treatment in the next chapter. It is clear, however, that any formulation of the propositional calculus or any partial system of propositional calculus, if the rules of inference are *modus ponens* and substitution, may be reformulated in the same way, i.e., we may replace each axiom by a corresponding axiom schema and take *modus ponens* as the one rule of inference, so obtaining a new system which has the same theorems as the original one.

A like reformulation is also possible if the rules of inference are substitution and one or more rules similar in character to *modus ponens*. For example, the system P_L of §25 may be reformulated as follows, as a system $P_{L\sigma}$

[223]The remark should be made that, in spite of the equivalence of P to P_2 in the strong sense that is given by *270 and **271, and in spite of the completeness of P_2, nevertheless P is not complete in any of the three senses of §18. In fact the wff $p \supset q$, for example, could be added to P as an axiom without producing any inconsistency. Semantically, this is connected with the fact that P has a wider class of sound interpretations than P_2, the rules of P being (in consequence of the omission of the rule of substitution) no longer sufficient to distinguish between propositional variables and propositional constants.

whose theorems are the same as those of P_L. The primitive symbols and the wffs of $P_{L\sigma}$ are the same as those of P_L. The one rule of inference is: From $\mathbf{A} \mid . \mathbf{B} \mid \mathbf{C}$ and $\mathbf{A}$ to infer $\mathbf{C}$. And the axioms are all the wffs,

$$\mathbf{A} \mid [\mathbf{B} \mid \mathbf{C}] \mid . \mathbf{A} \mid [\mathbf{C} \mid \mathbf{A}] \mid . \mathbf{D} \mid \mathbf{B} \mid . \mathbf{A} \mid \mathbf{D} \mid . \mathbf{A} \mid \mathbf{D},$$

where **A**, **B**, **C**, **D** are wffs (to be taken in all possible ways).

28. Extended propositional calculus and protothetic.

The *extended propositional calculus* of Russell and Łukasiewicz-Tarski[224] has, besides notations of the propositional calculus, also the *universal quantifier* or the *existential quantifier* or both (cf. §06), with propositional variables as the operator variables.

Primitive symbols for a formulation of the extended propositional calculus may be selected in various ways, among which may seem most obvious the addition of one or both of the quantifiers to the primitive symbols of a formulation of the propositional calculus. But because this latter method, when applied to P_1 or P_2, leads to non-independence of the primitive connectives and operators, Łukasiewicz and Tarski propose instead implication and the universal quantifier as primitive connective and operator.[225] Other connectives and operators are then introduced by definition. E.g., the following definitions may be made:[226]

$$f \rightarrow (s)s$$

$$\sim\mathbf{A} \rightarrow \mathbf{A} \supset f$$

$$(\exists \mathbf{c})\mathbf{A} \rightarrow \sim(\mathbf{c})\sim\mathbf{A}$$

$$t \rightarrow (\exists s)s$$

$$[\mathbf{A} \supset_{\mathbf{c}} \mathbf{B}] \rightarrow (\mathbf{c})[\mathbf{A} \supset \mathbf{B}]$$

And other sentence connectives may then be defined as in §11.

---

[224]The extended propositional calculus was treated by Russell under the name of "theory of implication" in the *American Journal of Mathematics*, vol. 28 (1906), pp. 159–202. It was treated by Łukasiewicz and Tarski as "erweiterter Aussagenkalkül" in the *Comptes Rendus des Séances de la Société des Sciences et des Lettres de Varsovie*, Classe III, vol. 23 (1930), pp. 44–50.

[225]As appears from an informal account in *The Principles of Mathematics* (1903) and from his further discussion of the matter in 1906, Russell also intended to use these primitives for the extended propositional calculus. But in 1906 he takes negation as an additional primitive connective on the ground that it would otherwise be impossible to express the proposition that not everything is true—which, he holds, is adequately expressed by $\sim(p)p$ but not by $(p)p \supset (s)s$. (To the writer it would seem that Russell's position of 1906 involves the very doubtful thesis that there is *one* indispensable concept of negation, given *a priori*, which it is the business of the logician to reproduce: perhaps not even the extreme realism of Frege would support this.)

[226]In 1903 Russell defines $\sim p$, in effect, as $p \supset_r r$. In 1906 he considers and rejects the definition of $\sim p$ as $p \supset (s)s$. Also due to Russell (1903, 1906) is the definition of the conjunction pq (or, more generally, of **AB**), which is suggested by the third of the four displayed formulas on the next page.

The definition of $\sim p$ as $p \supset_r r$ is foreshadowed in C. S. Peirce's paper of 1885 (*American Journal of Mathematics*, vol. 7, see pp. 189–190). And it may have been from this source that Russell had the idea. However, Peirce does not explicitly use the universal quantifier in a definition of negation, but rather expresses the negation of p by $p \supset \alpha$, where α is explained verbally as an "index of no matter what token."

Adopting the Łukasiewicz-Tarski primitives and the above definitions, together with conventions about omissions of brackets parallel to those of §11, we may cite the following as some examples of wffs (sentences or propositional forms) which are true or have always the value truth in the intended interpretation and therefore ought to be theorems:

$$p \supset q \supset_q [q \supset r] \supset r$$
$$p \supset q \equiv . q \supset r \supset_r . p \supset r$$
$$pq \equiv . p \supset [q \supset r] \supset_r r$$
$$(p)(\exists q)(r) . p \supset q \supset . q \supset p \supset r$$

Using still the same primitives, and relying on the intended interpretation, we can show that the extended propositional calculus and the formulation P_1 of the propositional calculus are equivalent in a sense similar to that in which P_1 and P_2 were shown to be equivalent in §23. For it is clear that (**b**)**B** is concurrent (in the sense of §02) to the conjunction **CD**,[227] where **C** is obtained from **B** by substituting $f \supset f$ (i.e., $(s)s \supset (s)s$) for all free occurrences of **b**, and **D** is obtained from **B** by substituting f (i.e., $(s)s$) for all free occurrences of **b**. Given a wff **A** of the extended propositional calculus, we may iterate the operation of replacing a wf part (**b**)**B** by the conjunction **CD** just described, only obeying the restriction that this replacement is not be made if (**b**)**B** is the particular wff $(s)s$. After a sufficient number of iterations of this, the wff **A** will be changed to a wff $\mathbf{A}_0$ in which the universal quantifier does not appear except in wf parts $(s)s$. Upon replacing $(s)s$ everywhere by the primitive symbol f of P_1, $\mathbf{A}_0$ becomes a wff $\mathbf{A}_f$ of P_1. The correspondence between **A** and $\mathbf{A}_f$ is a many-one correspondence between wffs of the extended propositional calculus and wffs of P_1. And by assumption (4) of §02, **A** and $\mathbf{A}_f$ are concurrent. In view of the solution of the decision problem of P_1, this leads us to a solution of the semantical decision problem of the extended propositional calculus; and in formulating the extended propositional calculus as a logistic system we may be guided by the demand that solution of the decision problem for provability shall be the same as of the semantical decision problem.

The *protothetic* of Leśniewski[228] has, in addition to the notations of the extended propositional calculus, also variables whose values are truth-functions (in the sense of the last paragraph of §05), say

$$f^1, g^1, h^1, f^1_1, g^1_1, h^1_1, f^1_2, \ldots$$

as variables whose range consists of the singulary truth-functions, and

$$f^2, g^2, h^2, f^2_1, g^2_1, h^2_1, f^2_2, \ldots$$

as variables whose range is the binary truth-functions, and

$$f^3, g^3, h^3, f^3_1, g^3_1, h^3_1, f^3_2, \ldots$$

[227]In addition to the use of the word "conjunction" to denote the connective or its associated truth-function, as explained in §05, it will be convenient also to speak of a wff **CD** (formed from **C** and **D** by means of this connective) as "a conjunction." Similarly a wff $\mathbf{C} \vee \mathbf{D}$ will be called "a disjunction," a wff $\mathbf{C} \equiv \mathbf{D}$ "an equivalence," a wff $\sim\mathbf{C}$ "a negation," and so on.

[228]Stanisław Leśniewski, "Grundzüge eines neuen Systems der Grundlagen der Mathematik," in *Fundamenta Mathematicae*, vol. 14 (1929), pp. 1–81.

as variables whose range is the ternary truth-functions, and so on. Further, the notation for application of a function to its argument or arguments (see §03) is provided for among the primitive symbols. And the quantifiers are allowed to have not only propositional variables but variables of any kind as operator variables. Leśniewski allows also variables of still other types, e.g., variables whose values are propositional functions of truth-functions, but these seem to play a less important role, and we venture to change his terminology to the extent of excluding them from protothetic.[229] Finally, Leśniewski allows assertion of sentences only, and not of wffs containing free variables (cf. the end of §06); but this is from one point of view a non-essential feature, and we would propose that the name *protothetic* be applied also to systems, otherwise like Leśniewski's, in which wffs with or without free variables may be asserted.

For the primitive symbols of a formulation of protothetic, besides the various kinds of variables and the notation for application of a function to its argument or arguments, we may take implication as primitive sentence connective, and the universal quantifier as primitive operator (allowing it to have a variable of any kind as operator variable). Or, following a discovery of Tarski,[230] implication may be replaced by (material) equivalence as primitive connective.

Equivalence of protothetic to extended propositional calculus, and thus ultimately to propositional calculus, may be shown by a similar method and in a similar sense to those for the equivalence of extended propositional calculus to P_1. In lieu of an explicit statement of the many-one correspondence between wffs of protothetic and of extended propositional calculus (which would be lengthy), we shall merely indicate the correspondence by giving some examples. For the wff[231] $(f^1) \centerdot p \supset (q)f^1(q) \equiv (q) \centerdot p \supset f^1(q)$ of protothetic, the corresponding wff of extended propositional calculus is the conjunction of the four following:[232]

$$p \supset (q)t \equiv (q) \centerdot p \supset t$$
$$p \supset (q)q \equiv (q) \centerdot p \supset q$$
$$p \supset (q) \sim q \equiv (q) \centerdot p \supset \sim q$$
$$p \supset (q)f \equiv (q) \centerdot p \supset f$$

Again, for the wff $(f^2) \centerdot p \equiv q \supset (r) \centerdot f^2(p, r) \equiv f^2(q, r)$ of protothetic, the corresponding wff of extended propositional calculus is a conjunction of sixteen others, which we shall not write out in full but which include, e.g., the following:

---

[229]We may speak of *higher protothetic* when variables of such higher types are to be allowed. (*Added in proof*. Since this was written a comprehensive account of protothetic by Jerzy Słupecki has appeared in *Studia Logica* (Warsaw), vol. 1 (1953), pp. 44–112; Słupecki gives the names *propositional calculus with quantifiers*, *elementary protothetics*, and *protothetics* to what we here call *extended propositional calculus*, *protothetic*, and *higher protothetic* respectively.)

[230]*Fundamenta Mathematicae*, vol. 4 (1923), pp. 196–200.

[231]As an abbreviation, in writing particular wffs of protothetic, the superscripts after the letters *f*, *g*, *h* may simply be omitted. No confusion can result among variables whose values are functions of different numbers of arguments, or even with the letter *f* denoting the truth-value falsehood.

[232]By the conjunction of four wffs **A**, **B**, **C**, **D** we mean, of course, the conjunction **ABCD**, understood according to the convention of association to the left (§11).

$$p \equiv q \supset (r) \, . \, t \equiv t$$
$$p \equiv q \supset (r) \, . \, p \vee r \equiv q \vee r$$
$$p \equiv q \supset (r) \, . \, p \subset r \equiv . \, q \subset r$$
$$p \equiv q \supset (r) \, . \, p \equiv q$$
$$p \equiv q \supset (r) \, . \, p \supset r \equiv . \, q \supset r$$
$$p \equiv q \supset (r) \, . \, r \equiv r$$

For a wff (say) (f^3)**B** of protothetic, where **B** contains f^3 as a free variable only and contains no other truth-functional variables, the corresponding wff of extended propositional calculus would be a conjunction of 256 others (there are, namely, 256 different ternary truth-functions, and we must use some systematic method of setting down for each one a propositional form which has it as an associated function).[233] If the entire given wff (as distinguished from a wf part of it) has free truth-functional variables, it is necessary first to prefix universal quantifiers binding these variables, and then to apply the indicated method of obtaining a corresponding wff of extended propositional calculus.

Like propositional calculus, both extended propositional calculus and protothetic or a modified form of protothetic[234] will occur as parts of more extensive logistic systems to be considered later—in particular, of functional calculi of second or higher orders. However, in the treatment of these logistic systems which we shall adopt, extended propositional calculus and protothetic do not play a fundamental role in the way that the propositional calculus does. Therefore in this section we have confined ourselves to a brief sketch.

EXERCISES 28

28.0. Using the solution of the decision problem which is indicated in the text, verify the four examples which are given of wffs of extended propositional calculus that are true or have always the value truth. (Of course, where possible, make use of known results regarding propositional calculus in order to shorten the work.)

28.1. In the same way, verify the following as wffs of protothetic that have always the value truth.[231]

---

[233]For this solution of the decision problem of protothetic, cf. Leśniewski, *loc.cit.*, and references to Łukasiewicz and to Tarski which are there given.

[234]In the case of the functional calculus of fourth or higher order, the modification of protothetic (besides a change in the letters used as truth-functional variables) will consist in allowing the notation for application of a function to its arguments to be used only in such combinations as **a**(**b**), **a**(**b**, **c**), **a**(**b**, **c**, **d**), . . ., where **a** is in each case a truth-functional variable of appropriate kind and **b**, **c**, **d**, . . . must be propositional variables. For reasons which will become clear later, this may be considered a modification in the particular formulation of protothetic rather than in protothetic itself, the decision depending on what notion of equivalence between logistic systems we are willing to accept for this purpose. But, e.g., although 28.1(4) and 28.1(7) are wffs of formulations of protothetic which are contemplated in this section, they are not wffs of any functional calculus of higher order (even with change of the letters *f*, *g*).

On the other hand, protothetic occurs as a part of the logistic system of Chapter X with no modifications other than essentially trivial changes of notation.

(1)[235] $p \equiv q \supset \centerdot f(p) \equiv f(q)$

(2) $p \equiv q \equiv (f) \centerdot f(p) \supset f(q)$

(3)[236] $pq \equiv (f) \centerdot p \equiv \centerdot f(p) \equiv f(q)$

(4) $f(p, p) \supset \centerdot f(p, \sim p) \supset (q) f(p, q)$

(5) $g(p) \equiv g(t)p \vee g(f) \sim p$ (*Boole's law of development.*[237])

(6) $g(p, q) \equiv g(t, t)pq \vee g(t, f)\ p \sim q \vee g(f, t) \sim p\ q \vee g(f, f) \sim p \sim q$
(*Boole's law of development in two variables.*[237])

(7) $(\exists q)(p) \centerdot f(g(p)) \equiv g(f(p)) \equiv q$

(8)[238] $pq \equiv (f) \centerdot f(p, q) \equiv f(t, t)$

(9)[238] $pq \equiv (f) \centerdot f(p, q) \equiv f(q, p \equiv q)$

28.2. With aid of the solution of the decision problem which is indicated in the text, establish a principle of duality for a formulation of extended propositional calculus with the Łukasiewicz-Tarski primitives, analogous to the principle of duality *161 for the propositional calculus.

28.3. Likewise establish a principle of duality for a formulation of protothetic with implication as the only primitive sentence connective, and the universal quantifier as the only primitive operator.

29. Historical notes. *The algebra of logic* had its beginning in 1847,[239] in the publications of Boole and De Morgan.[240] This concerned itself at first with an algebra or calculus of classes, to which a similar algebra of

---

[235]Notice the relationship of this to the metatheorem of exercise 15.2 (or to the analogue of this metatheorem for any other formulation of the propositional calculus). Namely, all theorems given by this metatheorem are in a certain sense included in the one theorem of protothetic, being directly obtainable from it by a rule of substitution for propositional variables, like *101 or *201, and a rule of substitution for truth-functional variables, analogous to the rule of substitution for functional variables which is discussed in the next chapter. Thus the relationship is like that of, e.g., the theorem $\sim p \supset \centerdot p \supset q$ of the propositional calculus to the metatheorem that every wff $\sim \mathbf{A} \supset \centerdot \mathbf{A} \supset \mathbf{B}$ is a theorem.

[236]Cf. Tarski in the paper cited in footnote 230.

[237]Given by George Boole for the class calculus. Notice the relationship to the metatheorem of exercise 24.9 (concerning reduction to full disjunctive normal form).

[238]From a paper by Bolesław Sobociński, *Z Badań nad Prototetyką*, which was published as an offprint in 1939, but nearly all copies of which were destroyed in the war. An English translation with an added explanatory introduction was published in 1949 by the "Institut d'Études Polonaises en Belgique" under the title *An Investigation of Protothetic.*

[239]There were a number of anticipations of the idea of an algebra or a calculus of logic, especially by Leibniz, Gottfried Ploucquet, J. H. Lambert, G. F. Castillon, but these were in various ways inadequate or incomplete and never led to a connected development. See Louis Couturat's *La Logique de Leibniz* (1901) and *Opuscules et Fragments Inédits de Leibniz* (1903); C. I. Lewis's *A Survey of Symbolic Logic* (1918); Jørgen Jørgensen's *A Treatise of Formal Logic* (1931); Karl Dürr's "Die Logistik Johann Heinrich Lamberts" in *Festschrift Andreas Speiser* (1945); and the writer's "A Bibliography of Symbolic Logic" in volumes 1 and 3 of *The Journal of Symbolic Logic.*

[240]George Boole, *The Mathematical Analysis of Logic* (1847), and *An Investigation of the Laws of Thought* (1854). Augustus De Morgan, *Formal Logic* (1847).

relations[241] was later added. Though it was foreshadowed in Boole's treatment of "Secondary Propositions," a true propositional calculus perhaps first appeared from this point of view in the work of Hugh MacColl, beginning in 1877.[242]

The logistic method was first applied by Frege in his *Begriffsschrift* of 1879. And this work contains in particular the first formulation of the propositional calculus as a logistic system, the system P_F of 23.6. Due to Łukasiewicz[243] as a simplification of Frege's formulation is the system P_2, which we have used in this chapter (§20).

However, Frege's work received little recognition or understanding until long after its publication, and the propositional calculus continued development from the older point of view, as may be seen in the work of C. S. Peirce, Ernst Schröder, Giuseppe Peano, and others. The beginnings of a change (though not yet the logistic method) appear in the work of Peano and his school. And from this source A. N. Whitehead and Bertrand Russell derived much of their earlier inspiration; later they became acquainted with the more profound work of Frege and were perhaps the first to appreciate its significance.[244]

After Frege, the earliest treatments of propositional calculus by the logistic method are by Russell. Some indications of such a treatment may be found in *The Principles of Mathematics* (1903). It is extended propositional calculus (§28) which is there contemplated rather than propositional calculus; but by making certain changes in the light of later developments, it is possible to read into Russell's discussion the following axioms for a partial system P_R^{IK} of propositional calculus with implication and conjunction as primitive connectives, the rules of inference being *modus ponens* (explicitly stated by Russell) and substitution (tacit):

---

[241]De Morgan, *Syllabus of a Proposed System of Logic* (1860), and a paper in the *Transactions of the Cambridge Philosophical Society*, vol. 10 (1864), pp. 331–358; C. S. Peirce, various papers 1870–1903, reprinted in volume 3 of his *Collected Papers*; J. J. Murphy, various papers 1875–1891; Ernst Schröder, *Algebra der Logik*, vol. 3 (1895).

[242]*Mathematical Questions*, vol. 28 (1877), pp. 20–23; *Proceedings of the London Mathematical Society*, vol. 9 (1877–1878), pp. 9–20, 177–186, and vol. 10 (1878–1879), pp. 16–28; *Mind*, vol. 5 (1880), pp. 45–60; *Philosophical Magazine*, 5s. vol. 11 (1881), pp. 40–43.

[243]See Łukasiewicz and Tarski, "Untersuchungen über den Aussagenkalkül" in *Comptes Rendus des Séances de la Société des Sciences et des Lettres de Varsovie*, Classe III, vol. 23 (1930), pp. 30–50.

[244]An excellent historical and expository account of the work of Whitehead and Russell is found in Chapter 2 (by W. V. Quine) of *The Philosophy of Alfred North Whitehead*.

$$pq \supset p$$
$$[p \supset q][q \supset r] \supset . p \supset r$$
$$p \supset [q \supset r] \supset . pq \supset r$$
$$pq \supset r \supset . p \supset . q \supset r$$
$$[p \supset q][p \supset r] \supset . p \supset qr$$
$$p \supset q \supset p \supset p$$

As a part of Russell's treatment of extended propositional calculus in 1906,[224] there appears a formulation P_r of the propositional calculus, with implication and negation as primitive connectives, *modus ponens* and substitution as rules of inference, and the following axioms:

$$p \supset p$$
$$p \supset . q \supset p$$
$$p \supset q \supset . q \supset r \supset . p \supset r$$
$$p \supset [q \supset r] \supset . q \supset . p \supset r$$
$$\sim\sim p \supset p$$
$$p \supset \sim p \supset \sim p$$
$$p \supset \sim q \supset . q \supset \sim p$$

The formulation P_R of the propositional calculus (§25) was published by Russell in 1908,[245] and was afterwards used by Whitehead and Russell in *Principia Mathematica* in 1910. It may be simplified to the system P_B by just deleting the axiom whose non-independence was discovered by Paul Bernays.[246] Other simplifications of it are the system P_N, due to J. G. P. Nicod,[247] and the system P_G, due to Götlind and Rasiowa.[248]

Statement of the rule of substitution was neglected by Frege in 1879, but appears explicitly in connection with a different system in his *Grundgesetze der Arithmetik*, vol. 1 (1893). First statement of the rule of substitution specifically for the propositional calculus is by Louis Couturat in *Les Principes des Mathématiques* (1905), but his statement is perhaps insufficient as failing to make clear that the expression substituted for a (propositional) variable may itself contain variables. Russell states this rule more satisfactorily in his paper of 1906, but omits it in 1908. And in *Principia Mathematica* the authors hold that the rule of substitution cannot be stated, writing: "The recognition that a certain proposition is an instance of some general proposition previously proved . . . cannot itself be erected into a

---

[245] *American Journal of Mathematics*, vol. 30 (1908), pp. 222–262.
[246] *Mathematische Zeitschrift*, vol. 25 (1926), pp. 305–320.
[247] *Proceedings of the Cambridge Philosophical Society*, vol. 19 (1917–1920), pp. 32–41.
[248] Erik Götlind in the *Norsk Matematisk Tidsskrift*, vol. 29 (1947), pp. 1–4; H. Rasiowa, *ibid.*, vol. 31 (1949), pp. 1–3.

general rule." This seems to show that Whitehead and Russell had abandoned Frege's method of stating rules of inference syntactically, or perhaps had never fully accepted it. But C. I. Lewis, writing in immediate connection with *Principia Mathematica*, states the rule of substitution explicitly for a proposed system of Strict Implication in 1913,[249] and in his *Survey of Symbolic Logic* (1918) he supplies this rule also for the system of *Principia*. And Russell, in his *Introduction to Mathematical Philosophy* (1919), recognizes that there is an omission in *Principia* in failing to state the rule of substitution for the propositional calculus.

The device of using axiom schemata, as in §27, so that a rule of substitution becomes unnecessary was introduced by J. v. Neumann.[250]

For the propositional calculus, the name "calculus of equivalent statements" was used by MacColl. The name "Aussagenkalkul" was introduced by Schröder in German in 1890 and 1891. Perhaps as a translation of this, Russell uses "propositional calculus" in 1903 and again in 1906—but, at least in 1903, he applies the name to extended propositional calculus rather than to propositional calculus proper (according to the terminology now standard). Couturat[251] translates Russell's "propositional calculus" into French as "calcul des propositions," but at the same time he so alters Russell's method that the name comes to be applied to propositional calculus proper. The name "calculus of propositions" was used by Lewis in a series of papers beginning in 1912 and in *A Survey of Symbolic Logic* (1918). Since then the names "propositional calculus" and "calculus of propositions" have received general acceptance in the sense, or about the sense, in which we have used the former.[252]

---

[249]*The Journal of Philosophy, Psychology, and Scientific Methods*, vol. 10 (1913), pp. 428–438.

[250]*Mathematische Zeitschrift*, vol. 26 (1927), pp. 1–46. Von Neumann's device may be employed as a means of formulating a logistic system for which a rule of substitution cannot be used because of the absence of propositional variables or other variables suitable for the purpose. Thus in particular a simple applied functional calculus of first order, in the sense of the next chapter (§30), must be formulated with the aid of axiom schemata rather than a rule or rules of substitution.

See a discussion of the matter by Hilbert and Bernays in *Grundlagen der Mathematik*, vol. 1, pp. 248–249; also, specially for the propositional calculus, by Bernays in *Logical Calculus* (1935–1936), pp. 44–47; also by Bernays, *ibid.*, pp. 50–53.

[251]In *Revue de Métaphysique et de Morale*, vol. 12 (1904), pp. 25–30, and in *Les Principes des Mathématiques* (1905).

[252]Other names found in the literature are "theory of deduction" (in *Principia Mathematica* and in Russell's *Introduction to Mathematical Philosophy*) and "sentential calculus" (i.e., calculus of sentences, by a number of recent writers). But both of these names seem rather inappropriate because they refer to certain syntactical aspects of the calculus (deducibility, sentences) rather than to corresponding meanings (implication, propositions). Indeed a theory of deduction or a calculus of sentences would more naturally be a branch of logical syntax and be expressed in a meta-language.

A formulation of the propositional calculus with a single axiom was given by Nicod in January 1917,[253] the system P_n of §25. Modifications of this, having still a single axiom, are the systems P_W of Wajsberg,[254] and P_L of Łukasiewicz.[255] All three systems are based on Sheffer's stroke (non-conjunction) as primitive connective and employ a more powerful rule of inference in place of *modus ponens*.

Regarding formulations of the propositional calculus with a single axiom and with only *modus ponens* and substitution as rules of inference, see the historical account given by Łukasiewicz and Tarski in the paper cited in footnote 243. Of these, the system $P_{Ł}$ (exercise 23.7) is due to Łukasiewicz and is given in the same paper; the system $P_{ł}$ (23.8) is also due to Łukasiewicz, being credited to him in a paper by Bolesław Sobociński;[256] and the system of P_S (23.9) is obtained, by a method of Sobociński,[256] from the single axiom of exercise 18.4 for the implicational propositional calculus (the system $P^I_{Ł}$), which latter is due to Łukasiewicz.[257]

Another formulation of the implicational propositional calculus is the system P^I_B of exercise 18.3, which is due to Tarski and Bernays.[258] Wajsberg has shown that,[259] the rules of inference being always *modus ponens* and substitution, a complete formulation of the implicational propositional calculus becomes a complete formulation of the (full) propositional calculus with implication and f as primitive connectives upon adjoining $f \supset p$ as an additional axiom. Hence the system P_W of 12.7, obtained thus from P^I_B, and the system P_λ (§26), which is obtained in the same way from $P^I_{Ł}$.

Leśniewski was especially interested in equivalence as a primitive connective because he took definitions in sense (3) of footnote 168, and therefore expressed definitions, in the propositional calculus or in protothetic, as (material) equivalences.[260] The first formulation of a partial system of propositional calculus with equivalence as sole primitive connective (such that

[253]In the paper cited in footnote 247.

[254]*Monatshefte für Mathematik und Physik*, vol. 39 (1932), pp. 259–262.

[255]Given in Wajsberg's paper cited in the preceding footnote. Still another modification of Nicod's axiom, due to Łukasiewicz, is given by Leśniewski in the paper cited in footnote 228, p. 10.

[256]*Przegląd Filozoficzny*, vol. 35 (1932), pp. 171–193.

[257]*Proceedings of the Royal Irish Academy*, vol. 52 section A no. 3 (1948), pp. 25–33. (*Added in proof.* Various still shorter single axioms for the propositional calculus are given by C. A. Meredith in *The Journal of Computing Systems*, vol. 1 no. 3 (1953), pp. 155–164.)

[258]See the account of the matter by Łukasiewicz and Tarski in the paper cited in footnote 243.

[259]In *Wiadomości Matematyczne*, vol. 43 (1937), pp. 131–168.

[260]See the paper cited in footnote 228, pp. 10–11.

all tautologies in this connective are theorems) was by Leśniewski.[261] Other such formulations are the system P_W^E of 26.0, by Wajsberg;[262] and the systems with a single axiom, P_w^E of 26.3(1), by Wajsberg, and $P_Ł^E$ of 26.3(2), by Łukasiewicz.[263] The system P^{EN} of 26.4, with equivalence and negation as primitive connectives, is due to Mihailescu.[264]

Returning to formulations of the full propositional calculus, we mention also the system P_Λ of Łukasiewicz, having implication and negation as primitive connectives, *modus ponens* and substitution as rules of inference, and the three following axioms:

$$p \supset q \supset . q \supset r \supset . p \supset r$$
$$\sim p \supset p \supset p$$
$$p \supset . \sim p \supset q$$

A proof of the completeness of P_Λ (not quite the same as the original treatment of the system by Łukasiewicz[265]) is outlined in exercise 29.2 below.

Opposite in tendency to such formulations as P_λ and P_Λ, in which economy is emphasized, are formulations of the propositional calculus by Hilbert[266] which are designed to separate the roles of the various connectives, though at the cost of economy and of independence of the primitive connectives. Of this latter kind is the system P_H (§26), which is one of a number of closely related such formulations that are given by Hilbert and Bernays in the first volume of their *Grundlagen der Mathematik* (1934).[267]

---

[261]In the same paper, p. 16.

[262]In the paper cited in footnote 259, p. 163. The axioms were previously announced, without proof of their sufficiency, in the paper cited in footnote 254.

[263]The system of 26.3(1) is in Wajsberg's paper of footnote 259, p. 165, previously announced in the paper of footnote 254. The same papers have also two other formulations of the equivalence calculus by Wajsberg, one of them with a single axiom. Various other single axioms for the equivalence calculus are quoted (without proof of sufficiency) in the paper cited in footnote 256. However, the shortest single axioms for the equivalence calculus, with rules of inference as in 26.0, are due to Łukasiewicz, one of them being the axiom given in 26.3(2) (see a review by Heinrich Scholz in *Zentralblatt für Mathematik und ihre Grenzgebiete*, vol. 22 (1940), pp. 289–290).

[264]In the paper cited in footnote 215.

[265]In his mimeographed *Elementy Logiki Matematycznej*, Warsaw 1929.

[266]E.g., in *Abhandlungen aus dem Mathematischen Seminar der Hamburgischen Universität*, vol. 6 (1928), pp. 65–85.

[267]Hilbert and Bernays use the axiom $p \supset q \supset . p \supset r \supset . p \supset qr$ in place of the shorter axiom $p \supset . q \supset pq$. They mention the obvious possibility of using the shorter axiom, but point out that by doing so in the case of the formulation of the propositional calculus which they adopt primarily (and which is in some other respects not quite the same as our system P_H) the independence of the axiom $p \supset . q \supset p$ would be destroyed. In his *Logical Calculus* (1935–1936), p. 44, Bernays introduces a formulation $P_{H'}$ of the propositional calculus which differs from P_H only in having the law of *reductio ad absurdum* and the law of double negation as axioms in place of the converse law of contraposition. The independence of $p \supset . q \supset p$ as an axiom of P_H was established

By omitting from P_H negation and the one axiom containing negation, there is obtained a formulation P^P of what Hilbert and Bernays call "positive Logik," or *positive propositional calculus*. This is intended to be that part of the propositional calculus which is in some sense independent of the existence of a negation (e.g., Peirce's law is not a theorem of it). By omitting all connectives other than implication, and the axioms containing them, there is obtained a formulation P^+ of the *positive implicational propositional calculus*.

Such a system as P_H has also the advantage (indicated by Hilbert and Bernays) of exhibiting in very convenient form the relationship between the full propositional calculus and the intuitionistic propositional calculus. Namely, the systems P^i_S and P^m_0 of §26 are obtained by adding to P^P negation and appropriate axioms containing negation, or in other words by altering only the negation axioms of P_H. The formulation P^i_S of the intuitionistic propositional calculus is due to Heinrich Scholz and Karl Schröter.[268] The formulation P^i_W of the intuitionistic propositional calculus (employing f as primitive instead of negation, and adding the axiom $f \supset p$ to P^P) was given by Wajsberg,[269] but a similar observation regarding the possibility of using f in place of negation in formulating the intuitionistic propositional calculus had been made also by Gerhard Gentzen.[270]

The formulation P_Γ (26.14) of the full propositional calculus may be credited to V. Glivenko on the basis of his remark that a formulation of the full propositional calculus is obtained from a formulation of the intuitionistic propositional calculus by adjoining only the law of excluded middle as an additional axiom.[271]

Many other formulations of the propositional calculus and partial systems of propositional calculus are found in the literature. We mention in this section only those which we have actually used in text or exercises, or which seem to have some outstanding interest or historical importance.

The *truth-table decision procedure* for the propositional calculus (cf. §15) is applied in an informal way to special cases by Frege in his *Begriffsschrift*

---

above in 26.8; and its independence as an axiom of P_H, (a question left open by Bernays) may be established by a minor modification of the same method.

Additional results and remarks in connection with P_H and related systems, including the positive implicational propositional calculus, the positive propositional calculus, and the intuitionistic propositional calculus, are in Supplement III of the second volume of Hilbert and Bernays's *Grundlagen der Mathematik*.

[268] Credited to them by Wajsberg in the paper cited in footnote 211.

[269] In the paper cited in footnote 211.

[270] See *Mathematische Zeitschrift*. vol. 39, p. 189.

[271] In his paper in *Académie Royale de Belgique, Bulletins de la Classe des Sciences*, series 5 vol. 15 (1929), pp. 183–188 (which contains also, as its principal result, the metatheorems of exercise 26.15).

of 1879 (in connection with implication and negation as primitive connectives). The first statement of it as a general decision procedure is six years later by Peirce[272] (in connection with implication and non-implication as primitive connectives). Much of the recent development of the method stems from its use by Łukasiewicz[273] and by Post.[274] The term *tautology* is taken from Wittgenstein.[275]

Using three truth-values instead of two, and truth-tables in these three truth-values, Łukasiewicz first introduced a three-valued propositional calculus (cf. §19) in 1920.[276] He was led to this by ideas about modality, according to which a third truth-value—possibility, or better, contingency—has to be considered in addition to truth and falsehood; but the abstract importance of the new calculus transcends that of any particular associated ideas of this kind. Generalization to a many-valued propositional calculus, with $\nu + 1$ truth-values of which $\mu + 1$ are designated $(1 \leqq \mu < \nu)$, was made by Post in 1921,[277] independently of Lukasiewicz, and from a purely abstract point of view. Afterwards, but independently of Post, Łukasiewicz generalized his three-valued propositional calculus to obtain higher many-valued propositional calculi; this was in 1922, according to his statement, but was not published until 1929 and 1930.[278] Łukasiewicz's calculi differ from those of Post in that there is just one designated truth-value, and also in not being *full* (i.e., not every possible truth-function, in terms of the ν truth-values, is represented by a form of which it is an associated function). When, however, Łukasiewicz's many-valued propositional calculi are extended to full many-valued propositional calculi by the method of Słupecki,[279] they become special cases of those of Post.

A primitive basis (especially axioms and rules of inference) for Łukasiewicz's three-valued propositional calculus, so that it becomes a logistic system, was provided by Wajsberg;[280] and Łukasiewicz and Tarski assert that this may also

[272]In his paper cited in footnote 67, pp. 190–192 (or *Collected Papers*, vol. 3, pp 223–225).

[273]Jan Łukasiewicz in *Przegląd Filozoficzny*, vol. 23 (1921), pp. 189–205, and in later publications.

[274]E. L. Post in the *American Journal of Mathematics*, vol. 43 (1921), pp. 163–185.

[275]Ludwig Wittgenstein, "Logisch-philosophische Abhandlung" in *Annalen der Naturphilosophie*, vol. 14 (1921), pp. 185–262; reprinted in book form, with English translation in parallel, as *Tractatus Logico-philosophicus*.

[276]*Ruch Filozoficzny*, vol. 5 (1920), pp. 169–171. Łukasiewicz's truth-table for $\supset$ appears in exercise 19.8, where a three-valued propositional calculus closely related to that of Łukasiewicz is described.

[277]In the paper cited in footnote 274, which is Post's dissertation of 1920.

[278]In the publications cited in footnotes 243, 265, and in Łukasiewicz's *Philosophische Bemerkungen zu mehrwertigen Systemen des Aussagenkalküls*, which immediately follows (in the same periodical) the paper of footnote 243.

[279]Jerzy Słupecki in *Comptes Rendus des Séances de la Société des Sciences et des Lettres de Varsovie*, Classe III, vol. 29 (1936), pp. 9–11, and vol. 32 (1939), pp. 102–128.

[280]In the *Comptes Rendus des Séances de la Société des Sciences et des Lettres de Varsovie*, Classe III, vol. 24 (1931), pp. 126–148. Another system of axioms and rules of inference for Łukasiewicz's three-valued propositional calculus has been given recently by Alan Rose in *The Journal of the London Mathematical Society*, vol. 26 (1951), pp. 50–58.

be done for all the Łukasiewicz finitely many-valued propositional calculi. For *full* finitely many-valued propositional calculi primitive bases have been given by Słupecki.[279] And more recently the question of primitive bases for finitely many-valued propositional calculi (and functional calculi) has been treated by J. B. Rosser and A. R. Turquette in a series of papers in *The Journal of Symbolic Logic*. Łukasiewicz introduced also an infinitely many-valued propositional calculus, but the question of a primitive basis for this seems to be still open.

For the purpose of proving independence of the axioms of the propositional calculus, the use of many-valued truth-tables (with one or more designated values) was introduced by Bernays in his *Habilitationsschrift* of 1918, but not published until 1926.[281] This idea was also discovered independently by Łukasiewicz but not published. The remark that the method can be extended to rules of inference was made by Huntington.[282] The method employed in §19 of proving independence of the rule of substitution is due to Bernays (who suggested it to the writer in 1936).

The independence results of 26.11 and the three-valued truth-table used to obtain them are due to Heyting.[283] The result of 26.12, and the many-valued truth-tables of 26.10 (except in the cases $\nu = 1,2$) are Gödel's.[284]

Proofs of consistency and completeness of the propositional calculus, based on the truth-table method, were first made by Post.[285] Since then, a number of different proofs of completeness of the propositional calculus have been published, of which we mention here only those by Kalmár[286] and Quine.[287] Quine makes use of the particular formulation P_W of the propositional calculus, with implication and *f* as primitive connectives. The formulation P_1 of the propositional calculus, due to Wajsberg,[288] combines some of the features of P_2 and P_W: and the method by which we proved the completeness of P_1 in Chapter I is an adaptation to this case of the method

[281]In the *Mathematische Zeitschrift*, vol. 25 (1926), pp. 305–320.

[282]E. V. Huntington in the *Annals of Mathematics*, ser. 2 vol. 36 (1935), pp. 313–324.

[283]The first two of the independence results of 26.11 are proved in his paper of 1930, cited in footnote 209. All of them follow immediately by the same method, whether for Heyting's original formulation of the intuitionistic propositional calculus or any other.

[284]*Akademie der Wissenschaften in Wien, Mathematisch-naturwissenschaftliche Klasse, Anzeiger*, vol. 69 (1932), pp. 65–66.

[285]In the paper of footnote 274.

[286]László Kalmár in *Acta Scientiarum Mathematicarum*, vol. 7 no. 4 (1935), pp. 222–243.

[287]In *The Journal of Symbolic Logic*, vol. 3 (1938), pp. 37–40. References to earlier completeness proofs will be found in this paper.

[288]In a paper in *Wiadomości Matematyczne*, vol. 47 (1939), pp. 119–139. The idea of applying Kalmár's method to this formulation of the propositional calculus was suggested to the writer by Leon Henkin as yielding perhaps the briefest available completeness proof for the propositional calculus (if based on independent axioms with *modus ponens* and substitution as rules of inference).

The system P_W^m of exercise 26.21 is obtained from this same paper of Wajsberg, where it is briefly mentioned on p. 139 in indicating a correction to the paper of footnote 211.

of Kalmár. Especially the method of the proof of *152 (which is a necessary preliminary to the proof of completeness in §18), the idea of using *151 as a lemma for *152, and the method of the proof of *151, are taken from Kalmár with obvious adaptations.

For the implicational propositional calculus, a completeness proof was published by Wajsberg[289] (an earlier completeness proof by Tarski seems not to have been published). The method which was suggested in 18.3 for a completeness proof for the implicational propositional calculus follows Kalmár[286] in general plan; but the idea of using a propositional variable in place of *f* is taken from Wajsberg,[289] and the crucial idea of taking **B**′ to be **B** ⊃ **r** ⊃ **r** (rather than **B**) when the value of **B** is t, together with corresponding changes at various places in proving analogues of *151 and *152, was communicated to the writer by Leon Henkin in July 1948.

The proof of equivalence of P_2 to P_1 in §23 follows in large part a proof used by Wajsberg[289] for a similar purpose.

The deduction theorem (§13) is not a peculiarity of the propositional calculus but has analogues for many other logistic systems (in particular for functional calculi of first and higher orders, as we shall see in the chapters following). The idea of the deduction theorem and the first proof of it for a particular system must be credited to Jacques Herbrand.[290] Its formulation as a general methodological principle for logistic systems is due to Tarski.[291] The name "deduction theorem" is taken from the German "Deduktionstheorem" of Hilbert and Bernays.[292]

The idea of using the deduction theorem as a primitive rule of inference in formulations of the propositional calculus or functional calculus is due independently to Jaśkowski[293] and Gentzen.[294] Such a primitive rule of inference has a less elementary character than is otherwise usual (cf. footnote 181), and indeed it would not be admissible for a *logistic system* according

[289]In the paper cited in footnote 259.

[290]A modified form of the deduction theorem, adapted to a special formulation of the functional calculus of first order, was stated by him, without proof, in an abstract in the *Comptes Rendus des Séances de l'Académie des Sciences* (Paris), vol. 186 (1928), see p. 1275. The deduction theorem was stated and proved, for a system like the functional calculus of order ω, in his Paris dissertation, *Recherches sur la Théorie de la Démonstration*, published in 1930 as no. 33 of the *Travaux de la Société des Sciences et des Lettres de Varsovie*, see pp. 61–62. In proving the deduction theorem, in Chapter I and again in Chapter III, we employ what is substantially Herbrand's method.

[291]It is stated by him in this way in a paper in the *Comptes Rendus des Séances de la Société des Sciences et des Lettres de Varsovie*, Classe III, vol. 23 (1930)—see Axiom 8* on p. 24.

[292]*Grundlagen der Mathematik*, vol. 1, p. 155, and vol. 2, p. 387.

[293]Stanisław Jaśkowski, *On the Rules of Suppositions in Formal Logic*, 1934.

[294]In the *Mathematische Zeitschrift*, vol. 39 (1934), pp. 176–210, 405–431; see especially II. Abschnitt, pp. 183–190.

to the definition as we actually gave it in §07. But this disadvantage may be thought to be partly offset by a certain naturalness of the method; indeed to take the deduction theorem as a primitive rule is just to recognize formally the usual informal procedure (common especially in mathematical reasoning) of proving an implication by making an assumption and drawing a conclusion.

Employment of the deduction theorem as primitive or derived rule must not, however, be confused with the use of *Sequenzen* by Gentzen[295] (cf. 39.10–39.12 below). For Gentzen's arrow, $\rightarrow$, is not comparable to our syntactical notation, $\vdash$, but belongs to his object language (as is clear from the fact that expressions containing it appear as premisses and conclusions in applications of his rules of inference). And in fact we might well have introduced *Sequenzen* in connection with P_1 by means of the following definition schemata:[296]

$$\mathbf{A}_1, \mathbf{A}_2, \ldots, \mathbf{A}_n \rightarrow$$
$$\qquad \rightarrow \mathbf{A}_1 \supset . \mathbf{A}_2 \supset . \ldots \mathbf{A}_n \supset f,$$
$$\mathbf{A}_1, \mathbf{A}_2, \ldots, \mathbf{A}_n \rightarrow \mathbf{B}_1, \mathbf{B}_2, \ldots, \mathbf{B}_m$$
$$\qquad \rightarrow \mathbf{A}_1 \supset . \mathbf{A}_2 \supset . \ldots \mathbf{A}_n \supset \mathbf{B}_1 \vee \mathbf{B}_2 \vee \ldots \vee \mathbf{B}_m$$

where $n = 0, 1, 2, 3, \ldots$, and $m = 1, 2, 3, \ldots$, and where in both cases the abbreviation is to be used only for an entire asserted wff (never for a wf proper part of such). Thus we would have obtained all the formal properties of Gentzen's *Sequenzen*, and in particular would have been able to state the derived rules of **14.9** somewhat more conveniently.

For the derived rules of **14.9** themselves (as an alternative to use of the deduction theorem) credit should be given to Frege, rules very similar in nature and purpose having been employed by him in the first volume of his *Grundgesetze der Arithmetik* (1893).

The derived rule of substitutivity of equivalence (*159) and the related metatheorem of **15.2** were demonstrated for the propositional calculus—specifically, for P_R—by Post.[297] The related metatheorem *229 (where two implications take the place of an equivalence) was obtained for the implicational propositional calculus by Wajsberg.[298]

The full disjunctive normal form (**24.9**) may be traced back to Boole's law

[295]Ibid., III. Abschnitt, pp. 190–210. Gentzen's use of *Sequenzen* is taken in part from Paul Hertz—see a paper by Gentzen in the *Mathematische Annalen*, vol.107 (1932), pp. 329–350, and references to Hertz which are there given.

[296]Substantially the same equivalences as those which appear from these definition schemata are given by Gentzen in the paper just cited, p. 180 and p. 418.

[297]In the paper cited in footnote 274. See also the discussion of truth-functions and formal equivalence in the introduction to the first volume of *Principia Mathematica* (1910).

[298]In the paper cited in footnote 259.

of development (28.1(5), (6)), and it is used on this basis by Schröder in *Der Operationskreis des Logikkalkuls* (1877). Both the full disjunctive normal form and its dual, the *full conjunctive normal form*, were given in 1880 by Peirce.[299] Schröder and Peirce state the normal forms for the class calculus, but extension to the propositional calculus is immediate.

The principle of duality is to be credited to Schröder, who gives it for the class calculus in his *Operationskreis*, just cited, and again in his *Algebra der Logik*. Schröder does not extend the principle of duality to the propositional calculus, but such extension is immediate and seems to have been assumed by various later authors without special statement of it (e.g., by Whitehead, Couturat, Sheffer) Heinrich Behmann, in the paper cited in footnote 299, explicitly establishes the principle of duality in a form corresponding to *165 not only for the propositional calculus but also, in effect, for the functional calculi of first and second orders (where quantifiers are involved in addition to sentence connectives—see Chapters III–V below); and Hilbert and Ackermann in the first edition of their *Grundzüge der theoretischen Logik* (1928) establish analogues of *164 and *165 for the propositional calculus and for the functional calculus of first order. However, a principle of duality in connection with quantifiers appears already in the third volume of Schröder's *Algebra der Logik*.

EXERCISES 29

29.0. (1) Establish the completeness of P_r by proving as theorems the axioms of some formulation of the propositional calculus which has the same primitive connectives and the same rules of inference and is already known to be complete. (2) Discuss the independence of the axioms of P_r.

29.1. (1) Establish the completeness of the partial system of propositional calculus P_R^{IK}. (Compare 18.3.) (2) Discuss the independence of the axioms of P_R^{IK}.

---

[299]In the *American Journal of Mathematics*, vol. 3 (1880), pp. 37–39 (or *Collected Papers*, vol. 3, pp. 133–134).

Heinrich Behmann (in a paper in the *Mathematische Annalen*, vol. 86 (1922), pp. 163–229) gives the name "disjunktive Normalform" to a disjunction of terms $\mathbf{C}_i$ in which each term $\mathbf{C}_i$ has the form of a conjunction of propositional variables and negations of propositional variables; and the name "konjunktive Normalform" to the dual of this. For the use of these normal forms in the propositional calculus, Behmann refers to Bernays's unpublished *Habilitationsschrift* of 1918.

The full conjunctive normal form corresponds to the "ausgezeichnete konjunktive Normalform" of Hilbert and Ackermann (in *Grundzüge der theoretischen Logik*) and is thus a special case of the "konjunktive Normalform" of Bernays, Behmann, Hilbert and Ackermann. And dually the full disjunctive normal form is a special case of the disjunctive normal form.

29.2. Establish the completeness of P_Λ by proving the following theorems of P_Λ (in order) and then using the result of exercise 12.7:

$$\sim p \supset q \supset [\sim q \supset q] \supset . p \supset . \sim q \supset q$$
$$\sim q \supset \sim p \supset . p \supset . \sim q \supset q$$
$$\sim q \supset \sim p \supset . \sim q \supset q \supset q \supset . p \supset q$$
$$q \supset . \sim q \supset q \supset q \supset . p \supset q$$
$$p \supset . \sim q \supset q \supset q$$
$$\sim q \supset q \supset q \supset [p \supset q] \supset . \sim [p \supset q] \supset . p \supset q$$
$$\sim q \supset q \supset q \supset [p \supset q] \supset . p \supset q$$
$$q \supset . p \supset q$$
$$\sim q \supset \sim p \supset . p \supset q$$
$$\sim p \supset . p \supset q$$
$$p \supset q \supset p \supset p$$

29.3. In the system P_Λ, use the method of truth-tables to establish the independence of the three axioms and the rule of *modus ponens*. (Except in one case, that of the first axiom, a system of two truth-values is sufficient.)

29.4. Use the result of 29.2 to establish the completeness of the system P_j of propositional calculus in which the primitive connectives and the rules of inference are the same as for P_H (see §26) and the axioms are the nine following:

$$p \vee q \equiv . p \supset q \supset q$$
$$p \supset q \supset . q \vee . p \supset r$$
$$p \equiv q \supset . p \supset q$$
$$p \equiv q \vee . q \vee p$$
$$p \supset . q \equiv p \equiv q$$
$$p \supset . pq \equiv q$$
$$p \supset . q \supset . r \supset p$$
$$p \equiv \sim p \supset q$$
$$pq \supset p$$

(This system—minimizing the length of the separate axioms rather than the number of axioms—is due to Stanisław Jaśkowski, in *Studia Societatis Scientiarum Torunensis*, vol. 1 no. 1 (1948).)

29.5. Discuss the independence of the axioms of P_j.

III. Functional Calculi of First Order

The *functional calculus of first order* has or may have, in addition to notations of the propositional calculus, also *individual variables,* quantifiers with individual variables as operator variables (cf. §06), *individual constants, functional variables, functional constants.*

Various different functional calculi of first order are distinguished according to just which of these notations are introduced. But the individual variables are always included, and either some functional variables or some functional constants. And one or more quantifiers are always included, either the universal quantifier or one or more quantifiers which are (when taken together, and in the presence of the other primitive notations) *definitionally equivalent* to the universal quantifier in the sense that they can be obtained from the universal quantifier, and the universal quantifier can be obtained from them, by abbreviative definitions (cf. §11, and footnote 168) which reproduce the requisite formal properties. Propositional variables are not necessarily included, but there must be a complete system of primitive connectives for the propositional calculus, or something from which such a complete system of sentence connectives can be obtained by abbreviative definitions so as to reproduce the requisite formal properties.

In this chapter we study a particular formulation of each of the functional calculi of first order, the various formulations being in their development so nearly parallel to one another that they can be treated simultaneously without confusion or awkwardness. Where not necessary to distinguish the various different functional calculi of first order we speak just of "the functional calculus of first order," and the particular formulation of the functional calculus of first order studied in this chapter is then called "F^1." One of the functional calculi of first order is the *pure functional calculus of first order* (as explained in §30 below), and we call our formulation of it "F^{1p}." Thus "F^1" is ambiguous among various logistic systems, one of which is F^{1p}.

In Chapter IV we shall consider further the pure functional calculus of first order, introducing, in particular, an alternative formulation of it, F_2^{1p}.

30. The primitive basis of F^1. The primitive symbols of F^1 are the eight improper symbols

$$[\quad \supset \quad] \quad \sim \quad (\quad , \quad) \quad \forall$$

and the infinite list of individual variables

$$x \quad y \quad z \quad x_1 \quad y_1 \quad z_1 \quad x_2 \quad y_2 \quad z_2 \quad \ldots;$$

also some or all of the following, including either at least one of the infinite lists of functional variables or at least one functional constant:[300] the infinite list of propositional variables

$$p \quad q \quad r \quad s \quad p_1 \quad q_1 \quad r_1 \quad s_1 \quad p_2 \quad \ldots$$

and for each positive integer n an infinite list of n-ary functional variables, namely, the infinite list of singulary functional variables

$$F^1 \quad G^1 \quad H^1 \quad F^1_1 \quad G^1_1 \quad H^1_1 \quad F^1_2 \quad \ldots$$

and the infinite list of binary functional variables

$$F^2 \quad G^2 \quad H^2 \quad F^2_1 \quad G^2_1 \quad H^2_1 \quad F^2_2 \quad \ldots$$

and so on, further any number of individual constants, any number of singulary functional constants, any number of binary functional constants, any number of ternary functional constants, and so on. We do not specify the particular symbols to be used as functional constants, but allow them to be introduced as required, subject to conditions (B) and (I) of §07 (as to (B), cf. also footnote 113).

In the case of each category of variables, the order indicated for them is called their *alphabetic order*.

The formation rules of F^1 are:

30i. A propositional variable standing alone is a wff.[301]

30ii. If **f** is an n-ary functional variable or an n-ary functional constant, and if $\mathbf{a}_1, \mathbf{a}_2, \ldots, \mathbf{a}_n$ are individual variables or individual constants or both (not necessarily all different), then $\mathbf{f}(\mathbf{a}_1, \mathbf{a}_2, \ldots, \mathbf{a}_n)$ is a wff.

30iii. If $\mathbf{\Gamma}$ is wf then $\sim\mathbf{\Gamma}$ is wf.

30iv. If $\mathbf{\Gamma}$ and $\mathbf{\Delta}$ are wf then $[\mathbf{\Gamma} \supset \mathbf{\Delta}]$ is wf.

30v. If $\mathbf{\Gamma}$ is wf and **a** is an individual variable then $(\forall \mathbf{a})\mathbf{\Gamma}$ is wf.

---

[300]We require that either all or none of the variables in any one category be included, e.g., either the entire infinite list of binary functional variables or none of them. But where individual or functional constants are present, their number may be finite.

[301]Of course 30i may be vacuous in the case of certain systems F^1, namely, if propositional variables are not included among the primitive symbols. We might therefore have written, "A propositional variable standing alone, if the system F^1 contains such, is a wff." The added clause (between commas) increases clearness, but is not otherwise actually necessary.

A formula of F^1 is wf if and only if its being so follows from the formation rules. As in the case P_1 and P_2 (see §§10, 20) there follows, for a particular given system F^1, an effective test of well-formedness. The demonstration of this is left to the reader.

The wf parts $\mathbf{A}$ and $\mathbf{B}$ of a wff $[\mathbf{A} \supset \mathbf{B}]$ are called the *antecedent* and *consequent* respectively, and the occurrence of $\supset$ between them is called the *principal implication sign*. That the antecedent and consequent and principal implication sign of any wff $[\mathbf{A} \supset \mathbf{B}]$ are unique is part of the metatheorem **312 of the next section,

The *converse* of $[\mathbf{A} \supset \mathbf{B}]$ is the wff $[\mathbf{B} \supset \mathbf{A}]$ obtained by interchanging the antecedent and consequent. The *converse* of

$$(\forall \mathbf{a}_1)(\forall \mathbf{a}_2) \ldots (\forall \mathbf{a}_n)[\mathbf{A} \supset \mathbf{B}] \text{ is } (\forall \mathbf{a}_1)(\forall \mathbf{a}_2) \ldots (\forall \mathbf{a}_n)[\mathbf{B} \supset \mathbf{A}].$$

The *elementary parts* of a wff are the parts which are wf according to 30i or 30ii, i.e., they are those wf parts which have either the form of a propositional variable alone or the form $\mathbf{f}(\mathbf{a}_1, \mathbf{a}_2, \ldots, \mathbf{a}_n)$ where $\mathbf{f}$ is an n-ary functional variable or constant and $\mathbf{a}_1, \mathbf{a}_2, \ldots, \mathbf{a}_n$ are individual variables or constants.

An occurrence of a variable $\mathbf{a}$ in a wff $\mathbf{A}$ is called a *bound occurrence* of $\mathbf{a}$ in $\mathbf{A}$ if it is an occurrence in a wf part of $\mathbf{A}$ of the form $(\forall \mathbf{a})\mathbf{B}$; otherwise it is called a *free occurrence* of $\mathbf{a}$ in $\mathbf{A}$. The *bound variables* of $\mathbf{A}$ are the variables which have bound occurrences in $\mathbf{A}$; and the *free variables* of $\mathbf{A}$ are those which have free occurrences in $\mathbf{A}$.[302]

From the definition of a wff of F^1 it follows that all occurrences of propositional and functional variables are free occurrences. But an occurrence of an individual variable in a wff of F^1 may be either free or bound.

A wff is an *n-ary form* if it has exactly n different free variables, a *constant* if it has no free variables. In F^1, all forms are *propositional forms*, and all constants (wffs) are propositional constants or *sentences*. (Cf. note 117.)

In addition to the syntactical notations,

$$\mathrm{S}_{\mathbf{\Gamma}}^{\mathbf{b}}\mathbf{A}|, \qquad \mathrm{S}_{\mathbf{\Gamma}_1\mathbf{\Gamma}_2\ldots\mathbf{\Gamma}_n}^{\mathbf{b}_1\mathbf{b}_2\ldots\mathbf{b}_n}\mathbf{A}|,$$

explained in §§10, 12, we shall use also the syntactical notation

$$\underset{\smile}{\mathrm{S}}_{\mathbf{\Gamma}}^{\mathbf{b}}\mathbf{A}|$$

for the result of substituting $\mathbf{\Gamma}$ for all free occurrences of $\mathbf{b}$ in $\mathbf{A}$, and

$$\underset{\smile}{\mathrm{S}}_{\mathbf{\Gamma}_1\mathbf{\Gamma}_2\ldots\mathbf{\Gamma}_n}^{\mathbf{b}_1\mathbf{b}_2\ldots\mathbf{b}_n}\mathbf{A}|$$

for the result of substituting simultaneously $\mathbf{\Gamma}_1$ for all free occurrences of

[302] Compare the first four paragraphs of §06, and footnotes 28, 36, 96.

$\mathbf{b}_1$, $\mathbf{\Gamma}_2$ for all free occurrences of $\mathbf{b}_2, \ldots, \mathbf{\Gamma}_n$ for all free occurrences of $\mathbf{b}_n$, throughout **A** (the dot indicating substitution for free occurrences only).

In writing wffs of F^1 we make use of the same abbreviations by omission of brackets that were explained in §11, including the same conventions about association to the left and about the use of heavy dots. We also abbreviate by omitting superscripts on functional variables—writing, e.g., $F(x)$ instead of $F^1(x)$, and $F(x, y)$ instead of $F^2(x, y)$—since the superscript required to make the formula wf can always be supplied in only one way (cf. 30ii).

Also we adopt for use in connection with F^1 the definition schemata D3–12, understanding the "$\sim$" which appears in them to be the primitive symbol $\sim$ of F^1. However, the brackets which appear as part of the notation [**A**, **B**, **C**] introduced by D12 must (unlike other brackets) never be omitted.

And we add further the following definition schemata, in which $\mathbf{a}, \mathbf{a}_1, \mathbf{a}_2, \ldots$ must be individual variables:

D13.[303] $(\mathbf{a})\mathbf{A} \to (\forall \mathbf{a})\mathbf{A}$

D14.[304] $(\exists \mathbf{a})\mathbf{A} \to \sim(\mathbf{a})\sim\mathbf{A}$

D15.[305] $[\mathbf{A} \supset_{\mathbf{a}_1\mathbf{a}_2\ldots\mathbf{a}_n} \mathbf{B}] \to (\mathbf{a}_1)(\mathbf{a}_2)\ldots(\mathbf{a}_n) \centerdot \mathbf{A} \supset \mathbf{B}$ $\qquad n = 1, 2, 3, \ldots$

D16.[305] $[\mathbf{A} \equiv_{\mathbf{a}_1\mathbf{a}_2\ldots\mathbf{a}_n} \mathbf{B}] \to (\mathbf{a}_1)(\mathbf{a}_2)\ldots(\mathbf{a}_n) \centerdot \mathbf{A} \equiv \mathbf{B}$ $\qquad n = 1, 2, 3, \ldots$

D17. $[\mathbf{A} \mid_{\mathbf{a}} \mathbf{B}] \to (\mathbf{a}) \centerdot \mathbf{A} \mid \mathbf{B}$

In abbreviating wffs by omission of brackets, we use (as already stated) the convention of association to the left of §11. This convention is modified in the same way as in §11 by dividing bracket-pairs into categories, the same division into categories being used as before with the following additions: (1) when one of the signs $\supset$, $\equiv$, $\mid$ is used with subscripts (according to D15–17), the bracket-pair that belongs with it is considered in the same category as if there were no subscripts; (2) the combinations of symbols $(\forall \mathbf{a})$, $(\mathbf{a})$, $(\exists \mathbf{a})$ (simple quantifiers with operator variable) are placed along with the sign $\sim$, in a fourth and lowest category, in the same sense as already explained in §11 for $\sim$.

In stating the rules of inference and the axiom schemata of F^1, we make use (for convenience) of the definition schemata and other conventions of abbreviation which have just been described.

---

[303] I.e., in writing wffs we may as an abbreviation simply omit the symbol $\forall$. Compare the discussion of the universal quantifier in §06.

[304] Compare the discussion of the existential quantifier in §06.

[305] Compare the discussion of formal implication and formal equivalence in §06. The purely abbreviative definition schemata, D15 and D16, are to be distinguished from the semantical statements made in §06. The latter provide definitions of formal implication and formal equivalence rather in sense (2) of footnote 168, and they must apply to object languages in which the notations in question are truly present.

The rules of inference are the two following:

*300. From $\mathbf{A} \supset \mathbf{B}$ and $\mathbf{A}$ to infer $\mathbf{B}$. (*Rule of modus ponens.*)

*301. From $\mathbf{A}$, if $\mathbf{a}$ is an individual variable, to infer $(\mathbf{a})\mathbf{A}$. (*Rule of generalization.*)

In an application of the rule of *modus ponens,* we call $\mathbf{A} \supset \mathbf{B}$ the *major premiss,* and $\mathbf{A}$ the *minor premiss.* In an application of the rule of generalization, we say that the variable $\mathbf{a}$ has been *generalized upon.*

The axioms of F^1 are infinite in number, and are represented by means of five axiom schemata in the manner described in §27. These axiom schemata are as follows:

*302. $\mathbf{A} \supset . \mathbf{B} \supset \mathbf{A}$

*303. $\mathbf{A} \supset [\mathbf{B} \supset \mathbf{C}] \supset . \mathbf{A} \supset \mathbf{B} \supset . \mathbf{A} \supset \mathbf{C}$

*304. $\sim\mathbf{A} \supset \sim\mathbf{B} \supset . \mathbf{B} \supset \mathbf{A}$

*305. $\mathbf{A} \supset_{\mathbf{a}} \mathbf{B} \supset . \mathbf{A} \supset (\mathbf{a})\mathbf{B}$, where $\mathbf{a}$ is any individual variable which is not a free variable of $\mathbf{A}$.

*306. $(\mathbf{a})\mathbf{A} \supset \check{S}^{\mathbf{a}}_{\mathbf{b}}\mathbf{A}|$, where $\mathbf{a}$ is an individual variable, $\mathbf{b}$ is an individual variable or an individual constant, and no free occurrence of $\mathbf{a}$ in $\mathbf{A}$ is in a wf part of $\mathbf{A}$ of the form $(\mathbf{b})\mathbf{C}$.

In *305 and *306, we here meet for the first time with *axiom schemata* which, unlike those introduced in §27, have conditions attached to them (stated in the syntax language). For example, according to *305, not every wff $\mathbf{A} \supset_{\mathbf{a}} \mathbf{B} \supset . \mathbf{A} \supset (\mathbf{a})\mathbf{B}$ is an axiom, but only those wffs of this form which satisfy the further condition that $\mathbf{A}$ contains no free occurrence of $\mathbf{a}$.

The intention of *305 and *306 may be made clearer by giving some examples, for the sake of which we suppose that the propositional variables and the singulary and binary functional variables are included among the primitive symbols of F^1.

Thus one of the wffs which is an axiom according to *305 is[306]

$$p \supset_x F(x) \supset . p \supset (x)F(x),$$

or, as we may write it if we do not use the abbreviation of D15,

$$(x)[p \supset F(x)] \supset . p \supset (x)F(x).$$

(This may be called a *basic instance* of *305, in the sense that all other in-

[306]In this example, $\mathbf{A}$ is the wff p, $\mathbf{a}$ is the individual variable x, and $\mathbf{B}$ is the wff $F(x)$.

stances of *305 may be obtained from it by means of rules of substitution to be discussed in §35, and that no shorter instance of *305 has this property.) Again the wff

$$G(x) \supset_y H(y) \supset . G(x) \supset (y)H(y)$$

is an axiom, an instance of *305 (though not a basic instance). And also an axiom by *305 is

$$F(x) \supset_y [G(y, z) \supset_z H(z)] \supset . F(x) \supset . G(y, z) \supset_{yz} H(z),$$

and so on, an infinite number of axioms altogether. But the following wff is not an instance of *305 and not an axiom:

$$F(x) \supset_x G(x) \supset . F(x) \supset (x)G(x)$$

Some wffs which are instances of *306 and therefore axioms are the following (the first two are basic instances):[307]

$$(x)F(x) \supset F(y)$$
$$(x)F(x) \supset F(x)$$
$$F(x) \supset_x (y)G(y) \supset . F(y) \supset (y)G(y)$$
$$F(x) \supset_x (x)G(x) \supset . F(y) \supset (x)G(x)$$
$$F(x, y) \supset_x (z)G(x, z) \supset . F(y, y) \supset (z)G(y, z)$$

On the other hand the following wffs are not instances of *306 and not axioms:[308]

$$(x)(y)F(x, y) \supset (y)F(y, y)$$
$$F(x) \supset_x (y)G(x, y) \supset . F(y) \supset (y)G(y, y)$$

As we did in connection with formulations of the propositional calculus, we shall place the sign $\vdash$ before a wff to express that it is a theorem.

Among the various functional calculi of first order, F^1, whose primitive bases have now been stated, we distinguish certain ones by special names as follows. The *pure functional calculus of first order*, F^{1p}, is that in which the primitive symbols include all the propositional variables and all the functional variables (singulary, binary, ternary, etc.), but no individual constants and no functional constants. The *singulary functional calculus of first order*, $F^{1,1}$, is that in which the primitive symbols include all the propositional variables and all the singulary functional variables, but no other functional variables, no individual constants, and no functional constants; if to these

[307]To state quite explicitly how these five wffs are instances of *306, we have, taking them in order: in the first one, **A** is $F(x)$, **a** is x, **b** is y; in the second one, **A** is $F(x)$, **a** is x, **b** is x; in the third one, **A** is $F(x) \supset (y)G(y)$, **a** is x, **b** is y; in the fourth one, **A** is $F(x) \supset (x)G(x)$, **a** is x, **b** is y; in the last one, **A** is $F(x, y) \supset (z)G(x, z)$, **a** is x, and **b** is y.

[308]The first one of the two is, however, a theorem, as we shall see later (exercise 34.3).

primitive symbols we add the binary functional variables we have the *binary functional calculus of first order*, $F^{1,2}$; and so on. A functional calculus of first order in which the primitive symbols include individual constants or functional constants or both is an *applied functional calculus of first order*; if no propositional variables and no functional variables are included, it is a *simple applied functional calculus of first order* (in this case there must be at least one functional constant).

This terminology will be useful to us later. But in this chapter we shall often not need to distinguish different functional calculi of first order by name, because, as already explained, we treat the various functional calculi of first order simultaneously by giving a development which holds equally for any one or all of them.

The intended principal interpretations of the functional calculi of first order may be roughly indicated by saying that propositional variables and sentence connectives are to have the same meaning as in the propositional calculus, the universal quantifier, $(\forall\ )$, is to have the meaning described in §06, and the functional variables are to have propositional functions of individuals as values (e.g., the range of a singulary functional variable is the singulary functions from individuals to truth-values, and similarly for binary functional variables, etc.). For application of a function to its argument or arguments the notation described in §03 is used. For the *individuals*, i.e., the range of the individual variables, various choices may be made, so that various different principal interpretations result. Indeed if no individual or functional constants are among the primitive symbols, it is usual to allow the individuals to be any well-defined non-empty class.[309] But if there are individual or functional constants, the intended interpretation of these may lead to restrictions upon or a special choice of the class of individuals.

We illustrate by stating the semantical rules explicitly in two cases, namely that of the pure functional calculus of first order, F^{1p}, and that of a simple applied functional calculus of first order, F^{1h}, which has two ternary functional constants, Σ and Π, and no other functional constants or individual constants among its primitive symbols.

In the case of F^{1p}, some non-empty class must first be chosen as the individuals, and there is then one principal interpretation, as follows:

a. The individual variables are variables having the individuals as their range.[310]

---

[309]The term "individual" was introduced by Russell in connection with the theory of types, to be discussed below in Chapter VI. A rather special meaning was given to the term by Russell, the individuals being described as things "destitute of complexity" (Russell, 1908) or as objects which "are neither propositions nor functions" (Whitehead and Russell, 1910). But in the light of Russell's recognition that only relative types are actually relevant in any context, it is now usual to employ "individual" in the way described in the text. Cf. Carnap, *Abriss der Logistik* (1929), p. 19.

b_0. The propositional variables are variables having the range t and f.

b_1. The singulary functional variables are variables having as their range the singulary (propositional) functions from individuals to truth-values.[310]

b_2. The binary functional variables are variables having as their range the binary propositional functions whose range is the ordered pairs of individuals.[310]

b_n. The n-ary functional variables are variables having as their range the n-ary propositional functions whose range is the ordered n-tuples of individuals.[310]

c_0. A wff consisting of a propositional variable **a** standing alone has the value t for the value t of **a**, and the value f for the value f of **a**.

c_n. Let $\mathbf{f}(\mathbf{a}_1, \mathbf{a}_2, \ldots, \mathbf{a}_n)$ be a wff in which **f** is an n-ary functional variable, and $\mathbf{a}_1, \mathbf{a}_2, \ldots, \mathbf{a}_n$ are individual variables, not necessarily all different. Let $\mathbf{b}_1, \mathbf{b}_2, \ldots, \mathbf{b}_m$ be the complete list of different individual variables among $\mathbf{a}_1, \mathbf{a}_2, \ldots, \mathbf{a}_n$. Consider a system of values, b of **f**, and $b_1, b_2, \ldots, b_m$ of $\mathbf{b}_1, \mathbf{b}_2, \ldots, \mathbf{b}_m$; and let $a_1, a_2, \ldots, a_n$ be the values which are thus given to $\mathbf{a}_1, \mathbf{a}_2, \ldots, \mathbf{a}_n$ in order. Then the value of $\mathbf{f}(\mathbf{a}_1, \mathbf{a}_2, \ldots, \mathbf{a}_n)$ for the system of values $b, b_1, b_2, \ldots, b_m$ of $\mathbf{f}, \mathbf{b}_1, \mathbf{b}_2, \ldots, \mathbf{b}_m$ (in that order) is $b(a_1, a_2, \ldots, a_n)$.[311]

d. For a given system of values of the free variables of $\sim$**A**, the value of $\sim$**A** is f if the value of **A** is t; and the value of $\sim$**A** is t if the value of **A** is f.[312]

e. For a given system of values of the free variables of $[\mathbf{A} \supset \mathbf{B}]$, the value of $[\mathbf{A} \supset \mathbf{B}]$ is t if either the value of **B** is t or the value of **A** is f; and the value of $[\mathbf{A} \supset \mathbf{B}]$ is f if the value of **B** is f and at the same time the value of **A** is t.

f. Let **a** be an individual variable and let **A** be any wff. For a given system of values of the free variables of $(\forall\mathbf{a})\mathbf{A}$, the value of $(\forall\mathbf{a})\mathbf{A}$ is t if the value of **A** is t for every value of **a**; and the value of $(\forall\mathbf{a})\mathbf{A}$ is f if the value of **A** is f for at least one value of **a**.[313]

[310] Since the individual and functional variables have values as variables, it might therefore be thought more natural to consider them wffs when standing alone and to provide semantical rules giving them values as forms (as rule c_0 does in the case of the propositional variables). Also a similar remark might be thought to apply to individual and functional constants in an applied functional calculus of first order. For the logistic system of Chapter X we shall indeed follow this idea. But for the functional calculi of first (and higher) order it is practically more convenient not to call a formula wf which consists of an individual or functional variable or an individual or functional constant standing alone. In adopting this terminology for the functional calculi, we shall nevertheless regard the individual and functional variables and the individual and functional constants as proper symbols (cf. footnote 80) and as having meaning in isolation. And in particular we shall speak of the individual and functional constants as having denotations—which are given by the semantical rules.

[311] The same notation for application of a function to its arguments that we have introduced in the object language is here used also in the meta-language. We shall follow this practice generally, not considering it a violation of the next-to-last paragraph of §08. For a rare case in which this might lead to ambiguity in connection with autonymy, we hold in reserve the possibility of an appropriate circumlocution to render the meaning unmistakable.

[312] This and following rules are to be understood according to the conventions (introduced in §10) that: (1) to have a value for a null class of variables is to denote; (2) the value of a constant for any system of values of any variables is the denotation of the constant; (3) the value of a form for any system of values of *any variables which include the free variables of the form* is the same as the value of the form for the given values of the free variables of the form (values of other variables being ignored).

[313] Note should be taken of the following special cases of rule f and rule ζ, in accordance with the conventions referred to in footnote 312. If **A** contains the individual variable

For convenience of reference we have here indicated an infinite list of rules $b_0, b_1, b_2, \ldots$ and an infinite list of rules $c_0, c_1, c_2, \ldots$. These may, however, be condensed in statement into just two rules, b and c.[314]

In the case of F^{1h}, the individuals shall be the *natural numbers*, i.e., the positive integers and 0. There is one principal interpretation, as follows:

α_0. The individual variables are variables having a non-empty range $\mathfrak{J}$.

α_1. The range $\mathfrak{J}$ of the individual variables is the natural numbers.

β_0. Each of Σ and Π denotes a ternary propositional function of the members of $\mathfrak{J}$.

β_1. Σ denotes the ternary propositional function of natural numbers whose value, for any natural numbers a_1, a_2, a_3 as arguments (in that order), is t or f according as a_3 is or is not the sum of a_1 and a_2.

β_2. Π denotes the ternary propositional function of natural numbers whose value, for any natural numbers a_1, a_2, a_3 as arguments (in that order), is t or f according as a_3 is or is not the product of a_1 and a_2.

γ. Let **f** be a ternary functional constant denoting the propositional function b, and let $\mathbf{a}_1$, $\mathbf{a}_2$, $\mathbf{a}_3$ be individual variables, not necessarily all different. Then the value of $\mathbf{f}(\mathbf{a}_1, \mathbf{a}_2, \mathbf{a}_3)$ for a system of values of the individual variables is $b(a_1, a_2, a_3)$, where a_1, a_2, a_3 are the respective values of $\mathbf{a}_1$, $\mathbf{a}_2$, $\mathbf{a}_3$.

δ. Identical in wording to rule d above.

ε. Identical in wording to rule e above.

ζ. Identical in wording to rule f above.[313]

EXERCISES 30

30.0. Express the following proposition by a wf of F^{1h} (taking the principal interpretation of F^{1h}): For all natural numbers a, b, c, if $a + b = c$, then $b + a = c$.

30.1. Similarly, express in F^{1h}: For all natural numbers a, b, either $a \leqq b$ or $b \leqq a$. (An abbreviative definition should first be introduced to represent the relation $\leqq$, say $[\mathbf{a} \leqq \mathbf{b}] \rightarrow (\exists \mathbf{c})\Sigma(\mathbf{a}, \mathbf{c}, \mathbf{b})$ where **a** and **b** are any individual variables and **c** is the first individual variable in alphabetic order which is not the same as either **a** or **b**.)

---

a as its sole free variable, then $(\forall\mathbf{a})\mathbf{A}$ denotes t if the value of **A** is t for every value of **a**, and $(\forall\mathbf{a})\mathbf{A}$ denotes f if the value of **A** is f for at least one value of **a**. If **A** does not contain the individual variable **a** as a free variable, the value of $(\forall\mathbf{a})\mathbf{A}$ is the same as the value of **A**, for any system of values of the free variables. If **A** contains no free variables, and if **a** is any individual variable, $(\forall\mathbf{a})\mathbf{A}$ has the same denotation as **A**.

[314]I.e., *in an appropriate semantical meta-language* the two infinite lists of rules may be reduced to two rules, as described. It is beyond the scope of our present treatment to undertake the detailed formalization (of the meta-language) which would be necessary to answer the question, what is an appropriate semantical meta-language for this purpose. We remark, however, using a terminology to be explained in later chapters, that such a semantical meta-language might with the aid of certain artifices be based on an applied functional calculus of sufficiently high (finite) order, functional constants being introduced to represent certain syntactical and semantical notions and axioms (postulates) added concerning them.

30.2. Similarly, express in F^{1h}: if two natural numbers have a product equal to 0, then one of them is equal to 0. (A notation, say $Z_0(\mathbf{a})$, should first be introduced by abbreviative definition to represent the propositional function, equality to 0. Use may then be made of the wff $(\exists z) \centerdot \Pi(x, y, z)Z_0(z)$ to express that two natural numbers [values of x and y] have a product equal to 0.)

30.3. Similarly, find in F^{1h} wffs expressing as nearly as possible each of the following: (1) A natural number remains the same if it is multiplied by 1. (2) The sum of two odd numbers is an even number. (3) For every prime number there exists a greater prime number; (4) The one and only even prime number is 2. (5) For all natural numbers a, b, and all natural numbers c except 0, if $ac \leqq bc$, then $a \leqq b$; (6) For all natural numbers a, b, $(a + b)^2 = (a^2 + b^2) + 2ab$.

30.4. Consider an applied functional calculus of first order, F^{1a}, which has the functional constants Σ and Π, with the same meaning as in F^{1h}, and in addition has all propositional and functional variables. The principal interpretation may be taken as obvious by analogy with those given for F^{1p} and F^{1h}, the individuals being again the natural numbers. In this functional calculus of first order, express as nearly as possible, by means of a wff containing free functional variables, each of the following assertions: (1) In any non-empty class of natural numbers there is a least number.[315] (2) If a non-empty class of natural numbers contains no greatest natural number, then for any given natural number it contains a greater natural number. (3) The relation $\leqq$ between natural numbers is characterized by the three conditions;[316] that it holds between 0 and every natural number; that it does not hold between any natural number other than 0 and 0; and that (for all natural numbers a and b) it holds between $a + 1$ and $b + 1$ if and only if it holds between a and b. (4) For every natural number k, the sum of the odd numbers less than $2k$ is k^2. (*Suggestion:* Given a class C of natural numbers, the relation between a natural number k and the sum of the natural numbers of C which are less than k is characterized by the three conditions: that it holds between 0 and 0; that it does not hold between 0 and any natural number other than 0; and that, for all natural numbers a and b, it holds between $a + 1$ and b if and only if there is a natural number c such that it holds between a and c, and b is equal to $c + a$ or c according as a does or does not belong to C.) (5) Perfect numbers exist.[317]

30.5. Taking the individuals to be the odd perfect numbers,[317] let an interpretation of F^{1p} be given by the same semantical rules a–f as for a principal interpretation. (1) If there are odd perfect numbers, the interpretation is a principal interpretation; on this hypothesis, show that the interpretation is sound. (2) On the hypothesis that there are no odd perfect numbers, show that the interpretation is unsound. (On the latter hypothesis, observe that a wff containing free individual variables must be said to have the value t for all

[315] Recall that, according to §04, a class is the same thing as a singulary propositional function.

[316] Or in other words, a relation between natural numbers is the relation $\leqq$ if and only if it satisfies these three conditions.

[317] A perfect number is a natural number, other than 0, which is equal to the sum of its aliquot parts, i.e., to the sum of the natural numbers less than itself by which it is exactly divisible.

systems of values of its free variables, and also to have the value f for all systems of values of its free variables; but it must not be said that it has the value f for at least one system of values of its free variables.)

30.6. The operator, or quantifier, which is introduced by abbreviative definition in D17 may of course also be used as a primitive (singulary-binary) quantifier in a formulation of the functional calculus of first order. For a formulation of the pure functional calculus of first order in which this is the *only* primitive sentence connective or quantifier, state the formation rules; state the semantical rules for a principal interpretation; and supply definitions of the universal quantifier, implication, and negation which will provide the appropriate agreement in meaning of the kind explained in the semantical paragraphs at the end of §11.

30.7. The intended principal interpretation of the *extended propositional calculus* (in the sense of §28) is informally indicated by the semantical discussion in §§04–06, 28. For a formulation of the extended propositional calculus in which implication is the one primitive sentence connective, and the universal quantifier is the one primitive quantifier, state the formation rules; and state the semantical rules for the principal interpretation.

30.8. The intended principal interpretation of *protothetic* (in the sense of §28) is informally indicated by the semantical discussion in §§04–06, 28. For a formulation of protothetic in which equivalence is the one primitive sentence connective, and the universal quantifier is the one primitive quantifier (which takes, however, either a propositional variable or a truth-functional variable as operator variable), state the formation rules; state the semantical rules for the principal interpretation; and supply definitions of implication and negation which will provide the appropriate agreement in meaning.

31. Propositional calculus. If the primitive symbols of F^1 include the propositional variables, then every theorem of the propositional calculus, in implication and negation as primitive connectives, is a theorem also of F^1—as follows immediately from the results of §27, because the axioms of the system P of §27 are included among the axioms of F^1 (in *302, *303, and *304) and the one rule of inference of P is a rule of inference also of F^1.

Even if the propositional variables are not included among the primitive symbols of F^1, we may draw a similar conclusion regarding *substitution instances* of theorems of the propositional calculus.

By a *substitution instance* of a wff **A** of any formulation of the propositional calculus we mean, namely, an expression or formula

$$\mathrm{S}^{\mathbf{b}_1\mathbf{b}_2\ldots\mathbf{b}_n}_{\mathbf{B}_1\mathbf{B}_2\ldots\mathbf{B}_n}\,\mathbf{A}|,$$

where $\mathbf{b}_1, \mathbf{b}_2, \ldots, \mathbf{b}_n$ is the complete list of (distinct) propositional variables of **A**, and $\mathbf{B}_1, \mathbf{B}_2, \ldots, \mathbf{B}_n$ are any wffs of the logistic system under consideration in the particular context, in this chapter, the logistic system F^1.

It is clear, then, that a substitution instance of a wff of P is a wff of F^1. Moreover, a substitution instance

$$\mathrm{S}^{\mathbf{b}_1 \mathbf{b}_2 \ldots \mathbf{b}_n}_{\mathbf{B}_1 \mathbf{B}_2 \ldots \mathbf{B}_n} \mathbf{A}|$$

of a theorem **A** of P must be a theorem of F^1. For every substitution instance of an axiom of P is an axiom of F^1 (by *302–*304). And if a proof is given of **A** as a theorem of P, let $\mathbf{b}_1, \mathbf{b}_2, \ldots, \mathbf{b}_n, \mathbf{c}_1, \mathbf{c}_2, \ldots, \mathbf{c}_m$ be the complete list of (distinct) propositional variables occurring, choose arbitrary wffs $\mathbf{C}_1, \mathbf{C}_2, \ldots, \mathbf{C}_m$ of F^1, and substitute simultaneously $\mathbf{B}_1$ for $\mathbf{b}_1$, $\mathbf{B}_2$ for $\mathbf{b}_2, \ldots,$ $\mathbf{B}_n$ for $\mathbf{b}_n$, $\mathbf{C}_1$ for $\mathbf{c}_1$, $\mathbf{C}_2$ for $\mathbf{c}_2, \ldots,$ $\mathbf{C}_m$ for $\mathbf{c}_m$ throughout the given proof. The result of this substitution is a proof of

$$\mathrm{S}^{\mathbf{b}_1 \mathbf{b}_2 \ldots \mathbf{b}_n}_{\mathbf{B}_1 \mathbf{B}_2 \ldots \mathbf{B}_n} \mathbf{A}|$$

as a theorem of F^1—valid applications of the rule of *modus ponens* in P being transformed by the substitution into valid applications of the rule of *modus ponens* in F^1.

Since we know (by **235, *239, *270, **271) that the theorems of P are the same as the *tautologies* of P (in the sense of §15), we may, without loss, put the foregoing results in the following form:

*310. Every tautology of P is theorem of F^1 if it is a wff of F^1.

*311. Every substitution instance of a tautology of P is a theorem of F^1.

In the foregoing, the arbitrary choice of the wffs $\mathbf{C}_1, \mathbf{C}_2, \ldots, \mathbf{C}_m$ is easily replaced by a definite rule for their choice. For example, since $n > 0$, we might just make each of the wffs $\mathbf{C}_1, \mathbf{C}_2, \ldots, \mathbf{C}_m$ identical with $\mathbf{B}_1$. Thus the proof of *311 becomes effective, in the sense of §12, so that *311 (as well as *310) may be employed as a derived rule of inference.

The use of *311 as a derived rule is facilitated by the fact that there is an effective procedure to determine whether a given wff of F^1 is or is not a substitution instance of a tautology of P—and to find a tautology of P of which it is a substitution instance, in case there is one. Details of this are now left to the reader. We shall rely on it in order to set down, whenever required, a wff of F^1 as a substitution instance of a tautology of P, and to leave it to the reader to verify it as such.

We shall often make use of *311 as a derived rule of inference in this way. And ordinarily, as sufficient indication of such use of *311, either alone or followed by one or more applications of *modus ponens*, we shall write simply the words, "by P," or "use P," or the like.

We add here, for reference, the five following metatheorems:

**312 Every wff is of one and only one of the five following forms, and in each case it is of that form in one and only one way: a propositional variable standing alone; $\mathbf{f}(\mathbf{a}_1, \mathbf{a}_2, \ldots, \mathbf{a}_n)$, where $\mathbf{f}$ is an n-ary functional variable or n-ary functional constant and $\mathbf{a}_1, \mathbf{a}_2, \ldots, \mathbf{a}_n$ are individual variables or individual constants or both; $\sim\mathbf{A}$; $[\mathbf{A} \supset \mathbf{B}]$; $(\forall\mathbf{a})\mathbf{A}$, where $\mathbf{a}$ is an individual variable.

**313. A wf part of $\sim\mathbf{A}$ either coincides with $\sim\mathbf{A}$ or is a wf part of $\mathbf{A}$.

**314. A wf part of $[\mathbf{A} \supset \mathbf{B}]$ either coincides with $[\mathbf{A} \supset \mathbf{B}]$ or is a wf part of $\mathbf{A}$ or is a wf part of $\mathbf{B}$.

**315. A wf part of $(\forall\mathbf{a})\mathbf{A}$ either coincides with $(\forall\mathbf{a})\mathbf{A}$ or is a wf part of $\mathbf{A}$.

**316. If $\mathbf{\Gamma}$ results from $\mathbf{A}$ by substitution of $\mathbf{N}$ for $\mathbf{M}$ at zero or more places (not necessarily at all occurrences of $\mathbf{M}$ in $\mathbf{A}$), then $\mathbf{\Gamma}$ is wf.

These metatheorems are used in particular in the proof of **323, of *340, and of **390, below. Their proofs are analogous to those of **225-**228, and are left to the reader.

32. Consistency of F¹. A wff of F¹ is called *quantifier-free* if it contains no quantifiers—or, what comes to the same thing, if it contains no occurrence of the symbol $\forall$. From any wff of F¹ we obtain its *associated quantifier-free formula* (also wf) by deleting all occurrences of the universal quantifier —i.e., by deleting the four symbols $(\forall\mathbf{a})$, at every place where such four symbols occur consecutively.

From a wff of F¹ we obtain an *associated formula of the propositional calculus* (abbreviated "afp") by first forming the associated quantifier-free formula, and then, in the latter, replacing every wf part $\mathbf{f}(\mathbf{a}_1, \mathbf{a}_2, \ldots, \mathbf{a}_n)$ by a propositional variable not previously occurring, in accordance with the following rule: two wf parts $\mathbf{f}(\mathbf{a}_1, \mathbf{a}_2, \ldots, \mathbf{a}_n)$ and $\mathbf{g}(\mathbf{b}_1, \mathbf{b}_2, \ldots, \mathbf{b}_m)$ are replaced by the same propositional variable if and only if $\mathbf{f}$ and $\mathbf{g}$ are the same functional variable or functional constant (as, of course, can happen only when $m = n$).

For example, in F¹ᵖ, the wff $G(x) \supset_y H(y) \supset . G(x) \supset (y)H(y)$ has as its associated quantifier-free formula $G(x) \supset H(y) \supset . G(x) \supset H(y)$, and thus an afp is $p \supset q \supset . p \supset q$. The wff $F(x, y) \supset_x (z)G(x, z) \supset . F(y, y) \supset (z)G(y, z)$ has as its associated quantifier-free formula $F(x, y) \supset G(x, z) \supset . F(y, y) \supset G(y, z)$, and therefore again an afp is $p \supset q \supset . p \supset q$.

Clearly, all the afps of a given wff of F[1] are variants of one another in the sense of §13. Hence, if one of the afps of a given wff of F[1] is a tautology, all of them are.

Now it is easily verified that every afp of an axiom of F[1] is a tautology. In fact for any instance $\mathbf{A} \supset . \mathbf{B} \supset \mathbf{A}$ of axiom schema *302, an afp must have the form $\mathbf{A}_0 \supset . \mathbf{B}_0 \supset \mathbf{A}_0$, where $\mathbf{A}_0$ and $\mathbf{B}_0$ are afps of $\mathbf{A}$ and $\mathbf{B}$ respectively; and $\mathbf{A}_0 \supset . \mathbf{B}_0 \supset \mathbf{A}_0$ must be a tautology, because it is obtainable by substitution from the tautology $p \supset . q \supset p$. Similarly, afps of instances of axiom schemata *303–*305 must have, in order, the forms:

$$\mathbf{A}_0 \supset [\mathbf{B}_0 \supset \mathbf{C}_0] \supset . \mathbf{A}_0 \supset \mathbf{B}_0 \supset . \mathbf{A}_0 \supset \mathbf{C}_0$$
$$\sim\mathbf{A}_0 \supset \sim\mathbf{B}_0 \supset . \mathbf{B}_0 \supset \mathbf{A}_0$$
$$\mathbf{A}_0 \supset \mathbf{B}_0 \supset . \mathbf{A}_0 \supset \mathbf{B}_0$$

And each of these is obviously a tautology because obtainable by substitution from a known tautology. In the case of *306, $\mathbf{A}$ and

$$\check{\mathrm{S}}^{\mathbf{a}}_{\mathbf{b}}\, \mathbf{A}|$$

differ (if at all) only by the substitution of one individual variable or constant for another, and therefore they must have the same afp $\mathbf{A}_0$. It follows that an afp of an instance of axiom schema *306 must have the form $\mathbf{A}_0 \supset \mathbf{A}_0$, which is obviously a tautology because obtainable by substitution from the known tautology $p \supset p$.

Moreover, the rules of inference of F[1] preserve the property of having a tautology as afp, i.e., if the premiss or premisses of the rule have this property, then the conclusion does also. In the case of *301, this is immediate. In the case of *300, we must make use of the remark that, if one afp of a wff is a tautology, then all are. Let $\mathbf{A}_0 \supset \mathbf{B}_0$ be an afp of the major premiss $\mathbf{A} \supset \mathbf{B}$. Then $\mathbf{A}_0$ and $\mathbf{B}_0$ are afps of $\mathbf{A}$ and $\mathbf{B}$ respectively. Since $\mathbf{A}_0$ and $\mathbf{A}_0 \supset \mathbf{B}_0$ are tautologies, it follows that $\mathbf{B}_0$ is a tautology, in consequence of the truth-table of $\supset$ (compare the proof of **150).

Since the axioms all have the property of having a tautology as afp, and since the rules of inference preserve this property, there follows the metatheorem:

**320. Every theorem of F[1] has a tautology as afp.

Now it is clear that if a wff $\mathbf{A}$ has a tautology as afp, its negation $\sim\mathbf{A}$ has as afp, not a tautology but a contradiction. Hence by **320, not both $\mathbf{A}$

and $\sim\mathbf{A}$ can be theorems. Thus we have the consistency of F^1 in the following senses:[318]

**321. F^1 is consistent with respect to the transformation of $\mathbf{A}$ into $\sim\mathbf{A}$.

**322. F^1 is absolutely consistent.

This proof of the consistency of F^1 differs from our proof of consistency of the propositional calculus (§17) in that it is not associated in the same way with a solution of the decision problem. In fact the converse of **320 fails, as we go on to show by giving an example of a non-theorem which has a tautology as afp. For this purpose we first establish the two following metatheorems:

**323. For every quantifier-free theorem of F^1 there is a proof in which only quantifier-free formulas occur.

Proof. Let any proof be given of a quantifier-free formula $\mathbf{C}$, and let $\mathbf{c}_1, \mathbf{c}_2, \ldots, \mathbf{c}_n$ be the complete list of individual variables and individual constants occurring in the proof. Then replace every wff $\mathbf{B}$ occurring in the proof (i.e., occurring as one of the finite sequence of wffs which constitutes the proof) by a quantifier-free formula $\mathbf{B}\ddagger$, according to the following procedure.

Choose individual variables $\mathbf{b}_1, \mathbf{b}_2, \ldots, \mathbf{b}_n$ which are distinct among themselves and distinct from all of $\mathbf{c}_1, \mathbf{c}_2, \ldots, \mathbf{c}_n$; and throughout the wff $\mathbf{B}$ substitute $\mathbf{b}_1, \mathbf{b}_2, \ldots, \mathbf{b}_n$ for $\mathbf{c}_1, \mathbf{c}_2, \ldots, \mathbf{c}_n$ respectively. In the resulting wff $\mathbf{B}'$, a wf part $(\forall\mathbf{b}_r)\mathbf{A}$ is to be replaced by the conjunction $\mathbf{A}_1\mathbf{A}_2 \ldots \mathbf{A}_n$, where $\mathbf{A}_i$ is

$$\mathrm{S}^{\mathbf{b}_r}_{\mathbf{c}_i}\mathbf{A}|$$

$(i = 1, 2, \ldots, n)$. If there is more than one wf part of $\mathbf{B}'$ of the form $(\forall\mathbf{b}_r)\mathbf{A}$, then the stated replacement is to be made first for one of the wf parts

[318] If propositional variables are among the primitive symbols, it follows by the same methods that F^1 is consistent in the sense of Post.

(*Added in proof.*) The demonstration of consistency is here made in a form which could be applied also, with obvious modifications, to a formulation such as F^{1p}_2 (see §40) having a rule of substitution for functional variables as a primitive rule. However, as pointed out to me by John G. Kemeny, the argument leading up to **321 and **322 in the present section could be greatly simplified by using a more simply defined associated formula in the propositional calculus. Namely let the *singulary associated formula of the propositional calculus* (abbreviated "sfp") be obtained from the associated quantifier-free formula of any wff of F^1 by replacing every elementary part by the one propositional variable p. —It follows from this that, e.g., $F(x) \supset G(y)$ could be added as an axiom to F^{1p} without producing inconsistency, but could not be so added to F^{1p}_2.

$(\forall \mathbf{b}_r)\mathbf{A}$ in which $\mathbf{A}$ is quantifier-free (obviously there must be one such); then in the resulting wff another wf part $(\forall \mathbf{b}_r)\mathbf{A}$ is to be chosen in which $\mathbf{A}$ is quantifier-free, and the stated replacement is to be made again; and so on, the successive replacements being continued until $\mathbf{B}'$ has become a quantifier-free formula $\mathbf{B}\dagger$. Then $\mathbf{B}\ddagger$ is to be the conjunction (in some specified order) of all the n^n wffs

$$\mathrm{S}^{\mathbf{b}_1\mathbf{b}_2\ldots\mathbf{b}_n}_{\mathbf{d}_1\mathbf{d}_2\ldots\mathbf{d}_n}\mathbf{B}\dagger\,|$$

where $\mathbf{d}_1, \mathbf{d}_2, \ldots, \mathbf{d}_n$ are any among the variables and constants $\mathbf{c}_1, \mathbf{c}_2, \ldots, \mathbf{c}_n$, taken in any order, and not necessarily all different.

If $\mathbf{B}$ is an axiom, then $\mathbf{B}\ddagger$ is a substitution instance of a tautology (of P) as the reader may verify by considering separately each of the schemata *302–*306. By *311, $\mathbf{B}\ddagger$ is therefore a theorem of F^1, and in fact the method which we used in establishing *311 provides without difficulty a proof of $\mathbf{B}\ddagger$ in which only quantifier-free formulas occur.

If, in the given proof of $\mathbf{C}$, $\mathbf{B}$ is inferred by *300 from premisses $\mathbf{A} \supset \mathbf{B}$ and $\mathbf{A}$, then $\mathbf{A}\ddagger$ is the conjunction of the n^n wffs

$$\mathrm{S}^{\mathbf{b}_1\mathbf{b}_2\ldots\mathbf{b}_n}_{\mathbf{d}_1\mathbf{d}_2\ldots\mathbf{d}_n}\mathbf{A}\dagger\,|,$$

and $[\mathbf{A} \supset \mathbf{B}]\ddagger$ is the conjunction of the n^n wffs

$$\mathrm{S}^{\mathbf{b}_1\mathbf{b}_2\ldots\mathbf{b}_n}_{\mathbf{d}_1\mathbf{d}_2\ldots\mathbf{d}_n}\mathbf{A}\dagger\,| \supset \mathrm{S}^{\mathbf{b}_1\mathbf{b}_2\ldots\mathbf{b}_n}_{\mathbf{d}_1\mathbf{d}_2\ldots\mathbf{d}_n}\mathbf{B}\dagger\,|;$$

and therefore it is possible by a series of steps involving methods of the propositional calculus only (cf. §31) to infer each of the n^n wffs

$$\mathrm{S}^{\mathbf{b}_1\mathbf{b}_2\ldots\mathbf{b}_n}_{\mathbf{d}_1\mathbf{d}_2\ldots\mathbf{d}_n}\mathbf{B}\dagger\,|$$

from $[\mathbf{A} \supset \mathbf{B}]\ddagger$ and $\mathbf{A}\ddagger$; and therefore by further steps involving propositional calculus only it is possible to infer $\mathbf{B}\ddagger$. Specifically, what is needed is proofs of two substitution instances each, of the n^n tautologies,

$$p_1p_2 \ldots p_{n^n} \supset p_i \qquad (i = 1, 2, \ldots, n^n),$$

and proof of an appropriate substitution instance of the tautology,

$$p_1 \supset {\scriptstyle\blacksquare}\, p_2 \supset {\scriptstyle\blacksquare} \ldots p_{n^n-1} \supset {\scriptstyle\blacksquare}\, p_{n^n} \supset p_1p_2 \ldots p_{n^n},$$

and a number of applications of *modus ponens*. By the method used in the demonstration of *311, all of this can be accomplished without use of other than quantifier-free formulas.

If, in the given proof of $\mathbf{C}$, $\mathbf{B}$ is inferred from premiss $\mathbf{A}$ by generalizing upon the individual variable c_r (thus by *301), then $\mathbf{B}\dagger$ is the conjunction of the n wffs

$$\mathrm{S}^{\mathbf{b}_r}_{\mathbf{c}_i}\mathbf{A}\dagger\,|$$

($i = 1, 2, \ldots, n$), and $\mathbf{A}\ddagger$ is the conjunction of the n^n wffs

$$\mathrm{S}^{\mathbf{b}_1\mathbf{b}_2\ldots\mathbf{b}_n}_{\mathbf{d}_1\mathbf{d}_2\ldots\mathbf{d}_n}\mathbf{A}\dagger\,|.$$

Thus

$$\mathbf{A}\ddagger \supset \mathrm{S}^{\mathbf{b}_1\mathbf{b}_2\ldots\mathbf{b}_n}_{\mathbf{d}_1\mathbf{d}_2\ldots\mathbf{d}_n}\mathbf{B}\dagger\,|$$

is a substitution instance of a tautology

$$p_1p_2\ldots p_{n^n} \supset p_{j_1}p_{j_2}\ldots p_{j_n}$$

(where the subscripts $j_1, j_2, \ldots, j_n$ are a certain n different ones among the subscripts $1, 2, \ldots, n^n$), and therefore is a theorem of F^1 by *311. Hence from $\mathbf{A}\ddagger$ we may infer by *modus ponens* each of the wffs

$$\mathrm{S}^{\mathbf{b}_1\mathbf{b}_2\ldots\mathbf{b}_n}_{\mathbf{d}_1\mathbf{d}_2\ldots\mathbf{d}_n}\mathbf{B}\dagger\,|,$$

and hence finally by a suitable substitution instance of the tautology

$$p_1 \supset {\scriptstyle\blacksquare}\, p_2 \supset {\scriptstyle\blacksquare} \ldots p_{n^n} \supset p_1p_2\ldots p_{n^n}$$

and *modus ponens* we may infer $\mathbf{B}\ddagger$. Again by the method used in the demonstration of *311, this can all be accomplished without use of other than quantifier-free formulas.

To sum up, we have now shown how the given proof of **C** can be transformed into a proof of $\mathbf{C}\ddagger$ in which only quantifier-free formulas occur. But by hypothesis **C** is quantifier-free. Therefore $\mathbf{C}\dagger$ is

$$\mathrm{S}^{\mathbf{c}_1\mathbf{c}_2\ldots\mathbf{c}_n}_{\mathbf{b}_1\mathbf{b}_2\ldots\mathbf{b}_n}\mathbf{C}\,|,$$

and $\mathbf{C}\ddagger$ is a conjunction of wffs one of which is **C**. By a further application of propositional calculus we can therefore go on to prove **C**, and by the method of *311 this can be done still without use of other than quantifier-free formulas.

**324. Every quantifier-free theorem of F^1 is a substitution instance of a tautology of P.

Proof. Given a quantifier-free theorem **C** of F^1 we can find, by **323, proof of **C** in which only quantifier-free formulas occur. In this proof of **C**, the only axioms used must be instances of the schemata *302, *303, *304, and therefore substitution instances of axioms of P; and the only rule of inference used must be *modus ponens*. Thus we have for each successive wff in the proof of **C**, as it is obtained, that it is a substitution instance of a theorem of P. Ultimately we have that **C** is a substitution instance of a theorem of P, and therefore a substitution instance of a tautology of P.

We remark, in passing, that from **324 we have a new proof of the consistency of F^1. Indeed the absolute consistency of F^1 follows from **324 upon giving one example of a quantifier-free formula which is not a substitution instance of a tautology. And the consistency of F^1 with respect to the transformation of **A** into **~A** then follows because, by *311 and the law of denial of the antecedent

$$\sim p \supset . p \supset q,$$

if **A** and **~A** were both theorems in any instance, then every wff would be a theorem. (This proof of the consistency of F^1 again is not associated with any general solution of the decision problem of F^1, but it does involve a solution of the decision problem for the special case of quantifier-free formulas.)

Now in particular, $F(x) \supset F(y)$ is a quantifier-free formula of F^{1p} which is not a substitution instance of a tautology, therefore it is a non-theorem, although it has a tautology as afp. Or, more generally, if **f** is an n-ary functional variable or an n-ary functional constant, and if $\mathbf{a}_1, \mathbf{a}_2, \ldots, \mathbf{a}_n$, $\mathbf{b}_1, \mathbf{b}_2, \ldots, \mathbf{b}_n$ are individual variables or individual constants, then

$$\mathbf{f}(\mathbf{a}_1, \mathbf{a}_2, \ldots, \mathbf{a}_n) \supset \mathbf{f}(\mathbf{b}_1, \mathbf{b}_2, \ldots, \mathbf{b}_n)$$

has a tautology as afp, but is not a theorem of F^1 unless $\mathbf{a}_1, \mathbf{a}_2, \ldots, \mathbf{a}_n$ are in order the same as $\mathbf{b}_1, \mathbf{b}_2, \ldots, \mathbf{b}_n$.

The proof of **320 makes use of no property of the axioms of F^1 except that every axiom has a tautology as afp. In consequence, the addition to F^1 of another axiom having a tautology as afp would not alter the property of the system that every theorem has a tautology as afp, and therefore would not destroy the consistency of the system. It follows that F^1 is not complete in any of the senses of §18, and especially:

325. F^1 is not complete with respect to the transformation of **A into **~A**, and is not absolutely complete.

However, in §44 we shall prove a completeness theorem for F^{1p}, and for the equivalent system F^{1p}_2, establishing their completeness in a weaker sense.

An explanation of the incompleteness of F^{1p} may quickly be seen from the point of view of the interpretation. The wff $F(x) \supset F(y)$, for example, has the value t for all values of its free variables, in the case of a principal interpretation in which there is just one individual; also, regardless of the number of individuals, in an interpretation which is like a principal interpretation except that the range of some or all of the singulary functional variables, including the variable F, is restricted to two particular singulary propositional functions of individuals, namely, the propositional function whose value is t for all arguments and the propositional function whose value is f for all arguments. (An interpretation of this latter kind, though not principal, is sound, as may readily be verified.) If only such interpretations as these were contemplated, it would be natural to expect $F(x) \supset F(y)$ as a theorem, and to add it as an axiom if it were not otherwise a theorem. But in other principal interpretations of F^{1p}, in which the number of individuals is greater than one, $F(x) \supset F(y)$ does not have the value t for all values of its free variables and therefore, for the sake of the soundness of the interpretation, must not be a theorem.

The completeness theorem of §44 will mean, semantically, that all those wffs of F^{1p} are theorems which have, in every principal interpretation, the value t for all values of their free variables. Hence the theorems of any functional calculus of first order may be described by saying that they are the wffs which, under the intended way of interpreting the connectives and quantifier, are true (1) for all values of the free variables, (2) regardless of the denotations assigned to the constants, and (3) independently of the nature and number of the individuals—provided only that there are individuals, that the values of the individual variables and the denotations of the individual constants are restricted to be individuals, that the values of the propositional variables are restricted to truth-values, and that the values of the n-ary functional variables and the denotations of the n-ary functional constants are restricted to be n-ary propositional functions of individuals.

33. Some theorem schemata of F^1. A *theorem schema* is a syntactical expression which represents many theorems (commonly an infinite number of different theorems) of a logistic system, in the same way that an axiom schema represents many different axioms. In the treatment of F^1 we shall deal with theorem schemata rather than with particular theorems, and shall supply for each theorem schema, by means of a *schema of proof*, an effective demonstration that each particular theorem which it represents can be proved. As in the case of derived rules of inference (discussed in Chapter I), justification of this lies in the effectiveness of the demonstration, whereby for any particular theorem represented by a given theorem schema the particular proof can always be supplied on demand. Thus our procedure amounts not to an actual formal development of the system F^1 but rather to giving effective instructions which might guide such an actual development. It is important to remember that the theorem schemata are in fact syntactical theorems about F^1, and only their *instances*, the particular theorems which they represent, are the theorems of F^1.

*330. $\vdash \underset{\cdot}{\mathrm{S}}^{\mathbf{a}}_{\mathbf{b}}\mathbf{A}| \supset (\exists \mathbf{a})\mathbf{A}$, where **a** is an individual variable, **b** is an individual variable or an individual constant, and no free occurrence of **a** in **A** is in a wf part of **A** of the form $(\mathbf{b})\mathbf{C}$.

Proof. By *306, $\vdash$ **(a)~A** $\supset$ **~Ṣ$^{\mathbf{a}}_{\mathbf{b}}$A|**.
Hence by P,[319] $\vdash$ **Ṣ$^{\mathbf{a}}_{\mathbf{b}}$A|** $\supset$ **~(a)~A**.[320]

***331.** $\vdash$ **(a)A** $\supset$ **(∃a)A**.[321]

Proof. By *306, $\vdash$ **(a)A** $\supset$ **A**.[322]
By *330, $\vdash$ **A** $\supset$ **(∃a)A**.[323]
Then use the transitive law of implication.[324]

***332.** $\vdash$ **A** $\supset_{\mathbf{a}}$ **B** $\supset$ **.** **(a)A** $\supset$ **B**.

Proof. By *306, $\vdash$ **A** $\supset_{\mathbf{a}}$ **B** $\supset$ **. A** $\supset$ **B**.[325]
Also by *306, $\vdash$ **(a)A** $\supset$ **A**.[325]
Then use P.[326]

***333.** $\vdash$ **A** $\supset_{\mathbf{a}}$ **B** $\supset$ **.** **(a)A** $\supset$ **(a)B**.

Proof. By *332 and generalization,[327] $\vdash$ **A** $\supset_{\mathbf{a}}$ **B** $\supset_{\mathbf{a}}$ **. (a)A** $\supset$ **B**.
Hence by *305, $\vdash$ **A** $\supset_{\mathbf{a}}$ **B** $\supset$ **. (a)A** $\supset_{\mathbf{a}}$ **B**.[328]
By *305, $\vdash$ **(a)A** $\supset_{\mathbf{a}}$ **B** $\supset$ **. (a)A** $\supset$ **(a)B**.
Then use the transitive law of implication.[324]

***334.** $\vdash$ **A** $\equiv_{\mathbf{a}}$ **B** $\supset$ **.** **(a)A** $\equiv$ **(a)B**.

Proof. By P, $\vdash$ **A** $\equiv$ **B** $\supset$ **. A** $\supset$ **B**.

---

[319]For explanation of the phrase "by P" (i.e., by propositional calculus) see the explanation which follows *311 in §31.

[320]This final expression is identical with *330, the theorem schema to be proved (cf. D14). In such cases we shall not refer explicitly to the definitions or definition schemata involved but shall merely leave it to the reader to see that the proof is complete.

[321]The condition that **a** shall be an individual variable may be taken as obvious, since the formula would not otherwise be wf. We shall hereafter, in stating theorem schemata, systematically omit explicit statement of such conditions when obvious for this reason.

[322]This special case of *306 in which **b** is the same variable as **a** will be used frequently. In particular (by *modus ponens*) it provides the inverse of the rule of generalization, as a derived rule; and we shall later have occasion to use it in this way also.

[323]This is the special case of *330, in which **b** is the same variable as **a**.

[324]Such a reference to a particular tautology of P by name will be employed as a more explicit substitute for the words "by P" or "then use P", thus as including a reference to *311 (cf. footnote 319).

[325]Compare footnote 322.

[326]See the explanation in §31.

[327]I.e., with *332 as premiss, the rule of inference *301 is applied.

[328]More explicitly, we take as major premiss

$$\mathbf{A} \supset_{\mathbf{a}} \mathbf{B} \supset_{\mathbf{a}} [(\mathbf{a})\mathbf{A} \supset \mathbf{B}] \supset . \mathbf{A} \supset_{\mathbf{a}} \mathbf{B} \supset . (\mathbf{a})\mathbf{A} \supset_{\mathbf{a}} \mathbf{B},$$

which is an instance of *305, and use *modus ponens*.

Hence by generalization and *333,[329] $\vdash \mathbf{A} \equiv_{\mathbf{a}} \mathbf{B} \supset . \mathbf{A} \supset_{\mathbf{a}} \mathbf{B}$.
Hence by *333 and the transitive law of implication,

$$\vdash \mathbf{A} \equiv_{\mathbf{a}} \mathbf{B} \supset . (\mathbf{a})\mathbf{A} \supset (\mathbf{a})\mathbf{B}.$$

Again, by P, $\vdash \mathbf{A} \equiv \mathbf{B} \supset . \mathbf{B} \supset \mathbf{A}$.
And hence by a similar series of steps, $\vdash \mathbf{A} \equiv_{\mathbf{a}} \mathbf{B} \supset . (\mathbf{a})\mathbf{B} \supset (\mathbf{a})\mathbf{A}$.
Then use P.

*335. $\vdash \mathbf{A} \supset (\mathbf{a})\mathbf{B} \equiv . \mathbf{A} \supset_{\mathbf{a}} \mathbf{B}$, if **a** is not free in **A**.

Proof. By *306, $\vdash (\mathbf{a})\mathbf{B} \supset \mathbf{B}$.
Hence by P, $\vdash \mathbf{A} \supset (\mathbf{a})\mathbf{B} \supset . \mathbf{A} \supset \mathbf{B}$.
Hence by generalization and *305, $\vdash \mathbf{A} \supset (\mathbf{a})\mathbf{B} \supset . \mathbf{A} \supset_{\mathbf{a}} \mathbf{B}$.
Then use *305 and P.

*336. $\vdash (\mathbf{a})(\mathbf{b})\mathbf{A} \equiv (\mathbf{b})(\mathbf{a})\mathbf{A}$.

Proof. By *306, $\vdash (\mathbf{b})\mathbf{A} \supset \mathbf{A}$.
Hence by generalization and *333, $\vdash (\mathbf{a})(\mathbf{b})\mathbf{A} \supset (\mathbf{a})\mathbf{A}$.
Hence by generalization and *305, $\vdash (\mathbf{a})(\mathbf{b})\mathbf{A} \supset (\mathbf{b})(\mathbf{a})\mathbf{A}$.
Similarly, $\vdash (\mathbf{b})(\mathbf{a})\mathbf{A} \supset (\mathbf{a})(\mathbf{b})\mathbf{A}$.
Then use P.

*337. $\vdash (\mathbf{a})\mathbf{A} \equiv \mathbf{A}$, if **a** is not free in **A**.

Proof. By P, $\vdash \mathbf{A} \supset \mathbf{A}$.
Hence by generalization and *305, $\vdash \mathbf{A} \supset (\mathbf{a})\mathbf{A}$.
By *306, $\vdash (\mathbf{a})\mathbf{A} \supset \mathbf{A}$.
Then use P.

*338. $\vdash \sim(\exists\mathbf{a})\mathbf{A} \equiv (\mathbf{a})\sim\mathbf{A}$.

Proof. By P, $\vdash \sim\sim(\mathbf{a})\sim\mathbf{A} \equiv (\mathbf{a})\sim\mathbf{A}$.

*339. $\vdash (\mathbf{a})\mathbf{A} \equiv (\mathbf{b})\mathbf{B}$, if there is no free occurrence of **b** in **A**, and no free occurrence of **a** in **A** is in a wf part of **A** of the form $(\mathbf{b})\mathbf{C}$, and **B** is $\check{S}^{\mathbf{a}}_{\mathbf{b}}\mathbf{A}|$.

Proof. By *306, $\vdash (\mathbf{a})\mathbf{A} \supset \mathbf{B}$. Hence, by generalizing upon **b** and then

---

[329] I.e., more explicitly, we take

$$\mathbf{A} \equiv \mathbf{B} \supset . \mathbf{A} \supset \mathbf{B}$$

as premiss, and generalize upon **a** (*301); then we take the resulting wff as minor premiss, and an appropriate instance of *333 as major premiss, and use *modus ponens*.

using *305, we have that ⊢ (**a**)**A** ⊃ (**b**)**B**. Now the given relation between the wffs (**a**)**A** and (**b**)**B** is reciprocal, i.e., there is no free occurrence of **a** in **B**, and no free occurrence of **b** in **B** is in a wf part of **B** of the form (**a**)**D**, and **A** is $\mathrm{\dot{S}}^{\mathbf{b}}_{\mathbf{a}}\mathbf{B}|$. Therefore in the same way we have that ⊢ (**b**)**B** ⊃ (**a**)**A**. Therefore by P, ⊢ (**a**)**A** ≡ (**b**)**B**.

34. Substitutivity of equivalence. In this section we establish the rule of substitutivity of equivalence (*342) and some related derived rules of inference. (Compare *158, *159, 15.3 in the propositional calculus.)

*340. If **B** results from **A** by substitution of **N** for **M** at zero or more places (not necessarily at all occurrences of **M** in **A**), and if $\mathbf{a}_1, \mathbf{a}_2, \ldots, \mathbf{a}_n$ is a list of individual variables including at least those free variables of **M** and **N** which occur also as bound variables of **A**, then $\vdash \mathbf{M} \equiv_{\mathbf{a}_1\mathbf{a}_2\ldots\mathbf{a}_n} \mathbf{N} \supset . \mathbf{A} \equiv \mathbf{B}$.

Proof. In a manner analogous to that of the proof of *229, we proceed by mathematical induction with respect to the total number of occurrences of the symbols ⊃, ~, ∀ in **A**.

We consider first the two special cases, (a) that the substitution of **N** for **M** is at zero places in **A**, and (b) that **M** coincides with **A** and the substitution of **N** for **M** is at this one place in **A**. In case (a), **B** is the same as **A**, and therefore $\vdash \mathbf{M} \equiv_{\mathbf{a}_1\mathbf{a}_2\ldots\mathbf{a}_n} \mathbf{N} \supset . \mathbf{A} \equiv \mathbf{B}$ by P. In case (b), **A** and **B** are the same as **M** and **N** respectively, and therefore by n uses of *306,[330] and the transitive law of implication, $\vdash \mathbf{M} \equiv_{\mathbf{a}_1\mathbf{a}_2\ldots\mathbf{a}_n} \mathbf{N} \supset . \mathbf{A} \equiv \mathbf{B}$.

Now if the total number of occurrences of the symbols ⊃, ~, ∀ in **A** is 0, we must have one of the special cases (a), (b), and the result of *340 then follows quickly, as we have just seen. Consider then a wff **A** in which this total number is greater than 0; the possible cases are the three following:

Case 1: **A** is of the form $\mathbf{A}_1 \supset \mathbf{A}_2$. Then (unless we have the special case (b) already considered) **B** is of the form $\mathbf{B}_1 \supset \mathbf{B}_2$, where $\mathbf{B}_1$ and $\mathbf{B}_2$ result from $\mathbf{A}_1$ and $\mathbf{A}_2$, respectively, by substitution of **N** for **M** at zero or more places. By hypothesis of induction,

$$\vdash \mathbf{M} \equiv_{\mathbf{a}_1\mathbf{a}_2\ldots\mathbf{a}_n} \mathbf{N} \supset . \mathbf{A}_1 \equiv \mathbf{B}_1,$$

$$\vdash \mathbf{M} \equiv_{\mathbf{a}_1\mathbf{a}_2\ldots\mathbf{a}_n} \mathbf{N} \supset . \mathbf{A}_2 \equiv \mathbf{B}_2.$$

Hence we get the result of *340 by P, using an appropriate substitution instance of the tautology,

[330]Again this is the special case of *306 in which **a** and **b** are the same variable.

$$[p \supset . q_1 \equiv r_1] \supset . [p \supset . q_2 \equiv r_2] \supset . p \supset . q_1 \supset q_2 \equiv . r_1 \supset r_2.$$

Case 2: **A** is of the form $\sim\mathbf{A}_1$. Then (unless we have the special case (b) already considered) **B** is of the form $\sim\mathbf{B}_1$, where $\mathbf{B}_1$ results from $\mathbf{A}_1$ by substitution of **N** for **M** at zero or more places. By hypothesis of induction,

$$\vdash \mathbf{M} \equiv_{\mathbf{a}_1\mathbf{a}_2\ldots\mathbf{a}_n} \mathbf{N} \supset . \mathbf{A}_1 \equiv \mathbf{B}_1.$$

Hence we get the result of *340 by P, using an appropriate substitution instance of the tautology,

$$[p \supset . q \equiv r] \supset . p \supset . \sim q \equiv \sim r.$$

Case 3: **A** is of the form $(\mathbf{a})\mathbf{A}_1$. Then (unless we have the special case (b) already considered) **B** is of the form $(\mathbf{a})\mathbf{B}_1$, where $\mathbf{B}_1$ results from $\mathbf{A}_1$ by substitution of **N** for **M** at zero or more places. By hypothesis of induction,

$$\vdash \mathbf{M} \equiv_{\mathbf{a}_1\mathbf{a}_2\ldots\mathbf{a}_n} \mathbf{N} \supset . \mathbf{A}_1 \equiv \mathbf{B}_1.$$

Hence by generalizing upon **a** and then using *305,[331] we have that

$$\vdash \mathbf{M} \equiv_{\mathbf{a}_1\mathbf{a}_2\ldots\mathbf{a}_n} \mathbf{N} \supset . \mathbf{A}_1 \equiv_{\mathbf{a}} \mathbf{B}_1.$$

Hence we get the result of *340 by using *334 and the transitive law of implication.

Thus the proof of *340 by mathematical induction is complete.

The two remaining metatheorems of this section follow as corollaries:

*341. If **B** results from **A** by substitution of **N** for **M** at zero or more places (not necessarily at all occurrences of **M** in **A**), and if $\vdash \mathbf{M} \equiv \mathbf{N}$, then $\vdash \mathbf{A} \equiv \mathbf{B}$.

Proof. By *340, *301, and *300.

*342. If **B** results from **A** by substitution of **N** for **M** at zero or more places (not necessarily at all occurrences of **M** in **A**), if $\vdash \mathbf{M} \equiv \mathbf{N}$ and $\vdash \mathbf{A}$, then $\vdash \mathbf{B}$. (*Rule of substitutivity of* (*material*) *equivalence*.)

Proof. By *341 and P.

---

[331]At this step it is essential that **a** is not a free variable of

$$\mathbf{M} \equiv_{\mathbf{a}_1\mathbf{a}_2\ldots\mathbf{a}_n} \mathbf{N}.$$

This is secured by the hypothesis that among $\mathbf{a}_1, \mathbf{a}_2, \ldots, \mathbf{a}_n$ are all the free variables of **M** and **N** which have bound occurrences in **A**.

EXERCISES 34

34.0. With aid of the results of §32, show that the following are non-theorems of F^{1p}:

(1) $$F(x) \supset_x G(x) \supset (\exists x) \,.\, F(x)G(x)$$

(2) $$(\exists x)F(x) \supset (x)F(x)$$

(3) $$F(x) \supset_x \,.\, F(y) \supset_y [G(x) \supset G(y)] \vee (z)F(z)$$

34.1. Show that any proof of a wff $(\mathbf{a})\mathbf{A}$ as a theorem of F^1 must contain an application of the rule of generalization (*301) in which the variable that is generalized upon is $\mathbf{a}$.[332]

34.2. In *340, to what extent may the hypothesis be weakened that among $\mathbf{a}_1, \mathbf{a}_2, \ldots, \mathbf{a}_n$ are all the free individual variables of $\mathbf{M}$ and $\mathbf{N}$ which occur as bound variables in $\mathbf{A}$?

Establish the following theorem schemata of F^1, *using methods and results of* §§30–34, *but not those of any later section*:

34.3. The theorem schema of which $(x)(y)F(x, y) \supset (y)F(y, y)$ is a basic instance.

34.4. $\vdash \mathbf{B} \supset_{\mathbf{a}} \mathbf{A} \supset \,.\, (\exists \mathbf{a})\mathbf{B} \supset \mathbf{A}$, if $\mathbf{a}$ is not free in $\mathbf{A}$.

34.5. $\vdash \mathbf{A} \supset_{\mathbf{a}} \mathbf{B} \supset \,.\, (\exists \mathbf{a})\mathbf{A} \supset (\exists \mathbf{a})\mathbf{B}$.

34.6. $\vdash \mathbf{A} \equiv_{\mathbf{a}} \mathbf{B} \supset \,.\, (\exists \mathbf{a})\mathbf{A} \equiv (\exists \mathbf{a})\mathbf{B}$.

35. Derived rules of substitution. By taking advantage of the device of axiom schemata, as discussed in §27, we have formulated the system F^1 without use of rules of substitution as primitive rules of inference. And indeed this way of doing it seems to be the only possibility in the case of a simple applied functional calculus of first order. But if there is a sufficient apparatus of variables, an alternative formulation is possible in which there are primitive rules of substitution (in addition to the rules of *modus ponens* and generalization) and the number of axioms is finite—as we shall see in §40.

In this section, the rules of substitution in question are obtained as derived rules of F^1. In doing this no distinction need be made of different kinds of functional calculi of first order, as the rules in fact all hold even in the case of a simple applied functional calculus of first order. But in a case in which variables of a particular kind are not present, of course a rule of substitution for variables of this kind reduces to something trivial.

[332]In consequence of the metatheorem of this exercise, the treatment of the deduction theorem for F^1 is unsound as it appears in Chapter II of the 1944 edition of *Introduction to Mathematical Logic, Part I*. An amended treatment of the deduction theorem appears in §36 below.

***350.** If **a** is an individual variable which is not free in **N** and **b** is an individual variable which does not occur in **N**, if **B** results from **A** by substituting $\mathrm{S}^{\mathbf{a}}_{\mathbf{b}}\mathbf{N}|$ for a particular occurrence of **N** in **A**, if $\vdash$ **A**, then $\vdash$ **B**.

(*Rule of alphabetic change of bound (individual) variable.*)

Proof. By *339 and *342 (the various wf parts of **N** of the form $(\mathbf{a})\mathbf{A}_i$ being taken one by one in left-to-right order of their initial symbols).

***351.** If **a** is an individual variable and **b** is an individual variable or an individual constant, if no free occurrence of **a** in **A** is in a wf part of **A** of the form (**b**)**C**, if $\vdash$ **A**, then $\vdash \underset{\cdot}{\mathrm{S}}{}^{\mathbf{a}}_{\mathbf{b}}\mathbf{A}|$.

(*Rule of substitution for individual variables.*)

Proof. By *301 and *306.

In order to state rules of substitution for propositional and functional variables we introduce a new substitution notation for which we use the letter Š.

If **p** is a propositional variable, the notation

$$\check{\mathrm{S}}^{\mathbf{p}}_{\mathbf{B}}\,\mathbf{A}|$$

shall stand for[333] **A** unless the condition is satisfied that (1) no wf part of **A** of the form (**b**)**C**, where **b** is a free variable of **B**, contains a free[334] occurrence of **p**; and, if this condition is satisfied, it shall stand for[333]

$$\underset{\cdot}{\mathrm{S}}{}^{\mathbf{p}}_{\mathbf{B}}\,\mathbf{A}|.$$

If **f** is an *n*-ary functional variable and $\mathbf{x}_1, \mathbf{x}_2, \ldots, \mathbf{x}_n$ are distinct individual variables, the notation

$$\check{\mathrm{S}}^{\mathbf{f}(\mathbf{x}_1,\mathbf{x}_2,\ldots,\mathbf{x}_n)}_{\mathbf{B}}\,\mathbf{A}|$$

shall stand for [335] **A** unless the two conditions are satisfied that: (1) no wf part of **A** of the form (**b**)**C**, where **b** is a free variable of **B** other than $\mathbf{x}_1, \mathbf{x}_2, \ldots, \mathbf{x}_n$, contains a free[336] occurrence of **f**; and (2) for each ordered

[333]I.e., for any particular propositional variable **p** and any particular wffs **A** and **B**, the syntactical notation in question denotes the wff **A** if the condition (1) is not satisfied, and, if the condition (1) is satisfied, it denotes the wff which results by substituting **B** for all free occurrences **p** in **A**.

[334]In connection with F^1 the word "free" here is superfluous. It is included because we shall wish to use the same substitution notation also in connection with other systems (without changing the wording of the definition).

[335]Compare footnote 333.

[336]The restriction to free occurrences of **f** is superfluous in connection with F^1, because in a wff of F^1 every occurrence of **f** is a free occurrence. As before, the restriction is included for the sake of use of the same notation in connection with other logistic systems.

n-tuple $\mathbf{a}_1, \mathbf{a}_2, \ldots, \mathbf{a}_n$ of individual variables or individual constants (or both, not necessarily all distinct) for which $\mathbf{f}(\mathbf{a}_1, \mathbf{a}_2, \ldots, \mathbf{a}_n)$ occurs in $\mathbf{A}$ in such a way that the occurrence of $\mathbf{f}$ is a free occurrence,[336] the wf parts of $\mathbf{B}$, if any, that have the forms $(\mathbf{a}_1)\mathbf{C}, (\mathbf{a}_2)\mathbf{C}, \ldots, (\mathbf{a}_n)\mathbf{C}$ contain no free occurrences of $\mathbf{x}_1, \mathbf{x}_2, \ldots, \mathbf{x}_n$ respectively.[337] And, if these two conditions are satisfied, the notation shall stand for[335] the result of replacing $\mathbf{f}(\mathbf{a}_1, \mathbf{a}_2, \ldots, \mathbf{a}_n)$, at all of its occurrences in $\mathbf{A}$ at which $\mathbf{f}$ is free,[336] by

$$\dot{S}^{\mathbf{x}_1\mathbf{x}_2\ldots\mathbf{x}_n}_{\mathbf{a}_1\mathbf{a}_2\ldots\mathbf{a}_n}\mathbf{B}|,$$

this replacement to be carried out simultaneously for all ordered n-tuples $\mathbf{a}_1, \mathbf{a}_2, \ldots, \mathbf{a}_n$ of individual variables or individual constants (or both, not necessarily all distinct) such that $\mathbf{f}(\mathbf{a}_1, \mathbf{a}_2, \ldots, \mathbf{a}_n)$ has an occurrence in $\mathbf{A}$ at which $\mathbf{f}$ is free.

*352_0. If $\mathbf{p}$ is a propositional variable, if $\vdash \mathbf{A}$, then

$$\vdash \check{S}^{\mathbf{p}}_{\mathbf{B}}\mathbf{A}|.$$

(*Rule of substitution for propositional variables.*)

*352_n. If $\mathbf{f}$ is an n-ary functional variable and $\mathbf{x}_1, \mathbf{x}_2, \ldots, \mathbf{x}_n$ are distinct individual variables, if $\vdash \mathbf{A}$, then

$$\vdash \check{S}^{\mathbf{f}(\mathbf{x}_1, \mathbf{x}_2, \ldots, \mathbf{x}_n)}_{\mathbf{B}}\mathbf{A}|.$$

(*Rule of substitution for n-ary functional variables.*)

Proof. The proof of *352_0, *352_n is analogous to that of **271.

We make use of a wff $\mathbf{B}'$ which differs from $\mathbf{B}$ by alphabetic changes of the bound and free individual variables of $\mathbf{B}$ in such manner that: (i) the individual variables occurring in $\mathbf{B}'$ are none of them the same as individual variables occurring anywhere in the given proof of $\mathbf{A}$; and (ii) the same variable occurs at two places in $\mathbf{B}'$ if and only if the variables occurring at the two corresponding places in $\mathbf{B}$ are the same. Let $\mathbf{y}_i$ ($i = 1, 2, \ldots, n$) be the variable which occurs in $\mathbf{B}'$ in place of the variable $\mathbf{x}_i$ in $\mathbf{B}$; or if $\mathbf{x}_i$ does not occur in $\mathbf{B}$, choose $\mathbf{y}_i$ to be an individual variable not occurring otherwise.

We observe that, if $\mathbf{E}$ is any axiom occuring in the given proof of $\mathbf{A}$, then

$$\check{S}^{\mathbf{p}}_{\mathbf{B}'}\mathbf{E}| \quad \text{or} \quad \check{S}^{\mathbf{f}(\mathbf{y}_1, \mathbf{y}_2, \ldots, \mathbf{y}_n)}_{\mathbf{B}'}\mathbf{E}|$$

(as the case may be) is again an axiom.

[337] In other words, to satisfy condition (2), if $\mathbf{f}(\mathbf{a}_1, \mathbf{a}_2, \ldots, \mathbf{a}_n)$ has an occurrence in $\mathbf{A}$ at which $\mathbf{f}$ is free, and $(\mathbf{a}_1)\mathbf{C}$ is a wf part of $\mathbf{B}$, then $(\mathbf{a}_1)\mathbf{C}$ shall contain no free occurrence of $\mathbf{x}_1$; if $\mathbf{f}(\mathbf{a}_1, \mathbf{a}_2, \ldots, \mathbf{a}_n)$ has an occurrence in $\mathbf{A}$ at which $\mathbf{f}$ is free, and $(\mathbf{a}_2)\mathbf{C}$ is a wf part of $\mathbf{B}$, then $(\mathbf{a}_2)\mathbf{C}$ shall contain no free occurrence of $\mathbf{x}_2$; and so on.

Moreover, in any application of the rule of *modus ponens* in the given proof of **A**, let the premisses and conclusion be **C** ⊃ **D**, **C**, **D**. Then

$$\check{S}^{\mathbf{p}}_{\mathbf{B}'}\mathbf{C} \supset \mathbf{D}|, \quad \check{S}^{\mathbf{p}}_{\mathbf{B}'}\mathbf{C}|, \quad \check{S}^{\mathbf{p}}_{\mathbf{B}'}\mathbf{D}|$$

or

$$\check{S}^{\mathbf{f}(\mathbf{y}_1, \mathbf{y}_2, \ldots, \mathbf{y}_n)}_{\mathbf{B}'}\mathbf{C} \supset \mathbf{D}|, \quad \check{S}^{\mathbf{f}(\mathbf{y}_1, \mathbf{y}_2, \ldots, \mathbf{y}_n)}_{\mathbf{B}'}\mathbf{C}|, \quad \check{S}^{\mathbf{f}(\mathbf{y}_1, \mathbf{y}_2, \ldots, \mathbf{y}_n)}_{\mathbf{B}'}\mathbf{D}|,$$

(as the case may be) are also premisses and conclusion for an application of the rule of *modus ponens*.[338]

Again, in any application of the rule of generalization in the given proof of **A**, let the premiss and conclusion be **C**, (**b**)**C**. Then

$$\check{S}^{\mathbf{p}}_{\mathbf{B}'}\mathbf{C}|, \quad \check{S}^{\mathbf{p}}_{\mathbf{B}'}(\mathbf{b})\mathbf{C}|$$

or

$$\check{S}^{\mathbf{f}(\mathbf{y}_1, \mathbf{y}_2, \ldots, \mathbf{y}_n)}_{\mathbf{B}'}\mathbf{C}|, \quad \check{S}^{\mathbf{f}(\mathbf{y}_1, \mathbf{y}_2, \ldots, \mathbf{y}_n)}_{\mathbf{B}'}(\mathbf{b})\mathbf{C}|$$

are also premiss and conclusion for an application of the rule of generalization.[339]

If, therefore, in the given proof,

$$\mathbf{A}_1, \mathbf{A}_2, \ldots, \mathbf{A}_m,$$

of **A** we replace each wff $\mathbf{A}_i$ ($i = 1, 2, \ldots, m$) by

$$\check{S}^{\mathbf{p}}_{\mathbf{B}'}\mathbf{A}_i| \quad \text{or} \quad \check{S}^{\mathbf{f}(\mathbf{y}_1, \mathbf{y}_2, \ldots, \mathbf{y}_n)}_{\mathbf{B}'}\mathbf{A}_i|,$$

we obtain a proof of

$$\check{S}^{\mathbf{p}}_{\mathbf{B}'}\mathbf{A}| \quad \text{or} \quad \check{S}^{\mathbf{f}(\mathbf{y}_1, \mathbf{y}_2, \ldots, \mathbf{y}_n)}_{\mathbf{B}'}\mathbf{A}|$$

(as the case may be).

In order to obtain the required proof of the wff

$$\check{S}^{\mathbf{p}}_{\mathbf{B}}\mathbf{A}| \quad \text{or} \quad \check{S}^{\mathbf{f}(\mathbf{x}_1, \mathbf{x}_2, \ldots, \mathbf{x}_n)}_{\mathbf{B}}\mathbf{A}|$$

(unless this is the same wff as **A**, in which case the matter is trivial) we use the proof just found of

$$\check{S}^{\mathbf{p}}_{\mathbf{B}'}\mathbf{A}| \quad \text{or} \quad \check{S}^{\mathbf{f}(\mathbf{y}_1, \mathbf{y}_2, \ldots, \mathbf{y}_n)}_{\mathbf{B}'}\mathbf{A}|$$

and add to it a series of steps in which the required alphabetic changes of

---

[338]The reason for introducing the wff **B′** may be seen in this paragraph of the proof. Namely, it is necessary to employ **B′** instead of **B** for substitution in **C** and **C** ⊃ **D**, because of the possibility that the numbered conditions, (1), or (1) and (2), though holding for the Š-substitution of **B** in **D** may fail for the Š-substitution of **B** in **C** and in **C** ⊃ **D**.

[339]Again in this paragraph of the proof the necessity of employing **B′** instead of **B** is seen, because of the possibility that **b** might be a free variable of **B**.

bound individual variables are accomplished by means of *339 and *342 (as in the proof of *350) and the required substitutions for free individual variables are accomplished by means of *301 and *306 (as in the proof of *351).[340] To make matters definite we may specify that first the required alphabetic changes of bound variables shall be made in alphabetic order of the variables to be changed, and, for any one variable, in left-to-right order of the relevant occurrences of **B**′; and that then the required substitutions shall be made in alphabetic order of the variables to be substituted for.

This completes the proof of $*352_0$, $*352_n$, except that, in order to make it possible to use these metatheorems as derived rules, it is necessary to fix explicitly how the individual variables of **B**′ and the individual variables $\mathbf{y}_1, \mathbf{y}_2, \ldots, \mathbf{y}_n$ shall be chosen. We do this by taking the different individual variables of **B** in order of their first occurrence in **B**, then after them the remaining variables (if any) among $\mathbf{x}_1, \mathbf{x}_2, \ldots, \mathbf{x}_n$, in order. To each of these in turn, as corresponding variable (in **B**′ or among $\mathbf{y}_1, \mathbf{y}_2, \ldots, \mathbf{y}_n$), is assigned the first individual variable in alphabetic order which occurs nowhere in the given proof of **A** and which has not previously been assigned.[341]

EXERCISES 35

35.0. In *350, to what extent may the conditions be weakened that **a** is not free in **N** and that **b** does not occur in **N**: (1) if the remainder of the metatheorem is to remain unchanged; and (2) if instead of $\mathrm{S}^{\mathbf{a}}_{\mathbf{b}}\mathbf{N}|$ is used the result of substituting **b** for the bound occurrences of **a** throughout **N**?

35.1. In each of the following cases, write the result of the indicated substitution:

(1) $\check{\mathrm{S}}^{F(x,y)}_{(z)[F(y,z)\supset G(x,y)]}(\exists x)(y)F(y,x)\supset(\exists x)F(x,x)|$

(2) $\check{\mathrm{S}}^{F(x)}_{(y)F(x,y)}(x)F(x)\supset F(y)|$

(3) $\check{\mathrm{S}}^{F(x,y)}_{(x)G(x)\supset G(y)}F(x,z)\supset(\exists y)F(y,z)|$

---

[340] It is here that the numbered conditions, (1), or (1) and (2), in the definition of the notation Š are essential, in order to assure that in this added series of steps constituting the final part of the proof nothing is required except alphabetic changes of bound individual variables and substitutions for free individual variables in accordance with *350 and *351 respectively.

[341] Notice that the derived rules of *350 and *351, and our proofs of them, will continue to hold for a system obtained from F^1 by adjoining any additional axioms. But *352 will continue to hold only if the additional axioms are such as to maintain the situation that the result of an Š-substitution in any axiom is again an axiom, or at least a theorem.

(4) $\check{S}^{F(x,y)}_{G(x,y,z)}F(x,y) \supset_{xy} (\exists z)F(x,z)|$

(5) $\check{S}^{F(x,y)}_{G(x,z) \supset (y)G(y,z)}F(y,z) \supset_x G(x,y) \supset_{yz} . F(y,z) \supset (z)G(z,y)|$

(6) $\check{S}^{F(x,y)}_{F(x,z) \supset F(z,y)}(x)F(z,x) \vee (y)F(y,z) \supset F(z,z)|$

35.2. In order to verify the necessity for each of the numbered conditions, (1) and (2), in the definition of the notation

$$\check{S}^{\mathbf{f}(\mathbf{x}_1, \mathbf{x}_2, \ldots, \mathbf{x}_n)}_{\mathbf{B}}\mathbf{A}|$$

show by examples that *352$_1$ would fail if either of these numbered conditions were omitted.

36. The deduction theorem. We shall now establish a *deduction theorem* for F^1, analogous to that for the propositional calculus (cf. §13).

The notion of a *variant* of a wff, introduced in §13 for the system P$_1$, may be extended in obvious fashion to wffs of F^1 or of other formulations of the functional calculus of first order: namely, a variant **B**′ of a wff **B** differs from **B** only by alphabetic changes of the variables of **B** of all kinds (bound or free, individual, propositional, or functional), in such a way that the same variable occurs at two places in **B**′ if and only if the variables occurring at the two corresponding places in **B** are the same. In connection with the deduction theorem for F^1, we need only observe that every variant of an axiom is also an axiom. But in other cases, such as, e.g., the formulations of functional calculi of first order introduced in §40, this will not hold; and in such cases the notion of a variant must play a role in the treatment of the deduction theorem analogous to that which it had in §13.

A finite sequence of wffs, $\mathbf{B}_1, \mathbf{B}_2, \ldots, \mathbf{B}_m$, of F^1 is called a *proof from the hypotheses* $\mathbf{A}_1, \mathbf{A}_2, \ldots, \mathbf{A}_n$ if for each i either: (1) $\mathbf{B}_i$ is one of $\mathbf{A}_1, \mathbf{A}_2, \ldots, \mathbf{A}_n$; or (2) $\mathbf{B}_i$ is an axiom; or (3) $\mathbf{B}_i$ is inferred according to *300 from major premiss $\mathbf{B}_j$ and minor premiss $\mathbf{B}_k$, where $j < i$, $k < i$; or (4) $\mathbf{B}_i$ is inferred according to *301 (the rule of generalization) from the premiss $\mathbf{B}_j$, where $j < i$, and where the variable that is generalized upon does not occur as a free variable in $\mathbf{A}_1, \mathbf{A}_2, \ldots, \mathbf{A}_n$; or (5) $\mathbf{B}_i$ is inferred by an alphabetic change of bound variable, according to *350,[342] from the premiss $\mathbf{B}_j$, where $j < i$; or (6) $\mathbf{B}_i$ is inferred according to *351,[342] by substitution in the premiss $\mathbf{B}_j$, where $j < i$, and where the variable, **a**, that is substituted for does not occur as a free variable in $\mathbf{A}_1, \mathbf{A}_2, \ldots, \mathbf{A}_n$; or (7) $\mathbf{B}_i$ is inferred according to *352,[342] by substitution in the premiss $\mathbf{B}_j$, where $j < i$, and where the variable,

[342]Or, more exactly, according to *350—or *351—or *352, as these would be restated to make them read as primitive rules of inference.

p or **f**, that is substituted for does not occur as a free variable in $\mathbf{A}_1$, $\mathbf{A}_2$, . . ., $\mathbf{A}_n$.

Such a finite sequence of wffs, $\mathbf{B}_m$ being the final formula of the sequence, is called more explicitly a proof *of* $\mathbf{B}_m$ from the hypotheses $\mathbf{A}_1, \mathbf{A}_2, \ldots, \mathbf{A}_n$. And we use the (syntactical) notation

$$\mathbf{A}_1, \mathbf{A}_2, \ldots, \mathbf{A}_n \vdash \mathbf{B}_m$$

to mean: there is a proof of $\mathbf{B}_m$ from the hypotheses $\mathbf{A}_1, \mathbf{A}_2, \ldots, \mathbf{A}_n$.

The special case that $n = 0$ is not excluded. It is true that, because of clauses (5), (6), (7) in the foregoing definition, a proof from the null class of hypotheses is not the same thing as a proof. But we shall hereafter use the notation, $\vdash \mathbf{B}_m$, indifferently in the sense of §30, to mean that there is a proof of the wff in question, and in the sense of the present section, to mean that there is a proof of it from the null class of hypotheses—relying on the metatheorems *350–*352 to enable us to obtain (effectively) a proof of any wff whenever we have a proof of it from the null class of hypotheses.[343]

*360. If $\mathbf{A}_1, \mathbf{A}_2, \ldots, \mathbf{A}_n \vdash \mathbf{B}$, then $\mathbf{A}_1, \mathbf{A}_2, \ldots, \mathbf{A}_{n-1} \vdash \mathbf{A}_n \supset \mathbf{B}$.

(*The deduction theorem.*)

Proof. Let $\mathbf{B}_1, \mathbf{B}_2, \ldots, \mathbf{B}_m$ be a proof of **B** from the hypotheses $\mathbf{A}_1$, $\mathbf{A}_2$, . . ., $\mathbf{A}_n$ ($\mathbf{B}_m$ being therefore the same as **B**). And construct the finite sequence of wffs $\mathbf{A}_n \supset \mathbf{B}_1$, $\mathbf{A}_n \supset \mathbf{B}_2, \ldots, \mathbf{A}_n \supset \mathbf{B}_m$. We shall show how to insert a finite number of additional wffs in this sequence so that the resulting sequence is a proof of $\mathbf{A}_n \supset \mathbf{B}_m$, i.e., of $\mathbf{A}_n \supset \mathbf{B}$, from the hypotheses $\mathbf{A}_1, \mathbf{A}_2, \ldots, \mathbf{A}_{n-1}$.

In fact consider a particular $\mathbf{A}_n \supset \mathbf{B}_i$, and if $i > 1$ suppose that the insertions have been completed as far as $\mathbf{A}_n \supset \mathbf{B}_{i-1}$. The eight following cases arise:

Case 1a: $\mathbf{B}_i$ is $\mathbf{A}_n$. Then $\mathbf{A}_n \supset \mathbf{B}_i$ is $\mathbf{A}_n \supset \mathbf{A}_n$, a substitution instance of the tautology $p \supset p$. Therefore insert before $\mathbf{A}_n \supset \mathbf{B}_i$ the wffs needed to make up the proof of it that is obtained by the method used in the demonstration of *311. (No substitutions or generalizations appear in this.)

Case 1b: $\mathbf{B}_i$ is one of $\mathbf{A}_1, \mathbf{A}_2, \ldots, \mathbf{A}_{n-1}$, say $\mathbf{A}_r$. Then $\mathbf{A}_r \supset . \mathbf{A}_n \supset \mathbf{B}_i$ is an axiom, an instance of *302. Therefore insert before $\mathbf{A}_n \supset \mathbf{B}_i$ the two wffs

[343] The ambiguity of sense will lead to no confusion, in the contexts in which the notation will actually be used, because we know that $\vdash \mathbf{B}$ in one sense if and only if $\vdash \mathbf{B}$ in the other sense. However, the ambiguity may be removed if desired by agreeing that when hypotheses are explicitly written (as in, e.g., "$p \vdash q \supset p$" or "$\mathbf{A}_1, \mathbf{A}_2, \ldots, \mathbf{A}_n \vdash \mathbf{B}_m$") the notation shall be understood in the sense of the present section—and in such a case as "$\mathbf{A}_1, \mathbf{A}_2, \ldots, \mathbf{A}_n \vdash \mathbf{B}_m$" this shall not be affected by the possibility of assigning the value 0 to n. But when the sign $\vdash$ is written actually without hypotheses appearing before it (as, e.g., "$\vdash p \supset . q \supset p$"), it shall be understood in the sense of §30.

$\mathbf{A}_r \supset . \mathbf{A}_n \supset \mathbf{B}_i$ and $\mathbf{A}_r$, from which $\mathbf{A}_n \supset \mathbf{B}_i$ is inferred by *300 (*modus ponens*).

Case 2: $\mathbf{B}_i$ is an axiom. Then $\mathbf{B}_i \supset . \mathbf{A}_n \supset \mathbf{B}_i$ is an axiom, an instance of *302. Therefore insert before $\mathbf{A}_n \supset \mathbf{B}_i$ the two wffs $\mathbf{B}_i \supset . \mathbf{A}_n \supset \mathbf{B}_i$ and $\mathbf{B}_i$, which both are axioms, and from which $\mathbf{A}_n \supset \mathbf{B}_i$ is inferred by *modus ponens.*

Case 3: $\mathbf{B}_i$ is inferred by *modus ponens* from major premiss $\mathbf{B}_j$ and minor premiss $\mathbf{B}_k$, where $j < i$, $k < i$. Then $\mathbf{B}_j$ is $\mathbf{B}_k \supset \mathbf{B}_i$. Insert before $\mathbf{A}_n \supset \mathbf{B}_i$ first the wff $\mathbf{A}_n \supset \mathbf{B}_j \supset . \mathbf{A}_n \supset \mathbf{B}_k \supset . \mathbf{A}_n \supset \mathbf{B}_i$ (which is an axiom, an instance of *303) and then the wff $\mathbf{A}_n \supset \mathbf{B}_k \supset . \mathbf{A}_n \supset \mathbf{B}_i$ (which can be inferred by *modus ponens*, and from which then $\mathbf{A}_n \supset \mathbf{B}_i$ can be inferred by *modus ponens*, since the necessary minor premisses, $\mathbf{A}_n \supset \mathbf{B}_j$ and $\mathrm{A}_n \supset \mathbf{B}_k$, are among the earlier wffs already present in the sequence being constructed).

Case 4: $\mathbf{B}_i$ is inferred by the rule of generalization from the premiss $\mathbf{B}_j$, where $j < i$. Then $\mathbf{B}_i$ is $(\mathbf{a})\mathbf{B}_j$, where $\mathbf{a}$ is an individual variable which does not occur as a free variable in $\mathbf{A}_1, \mathbf{A}_2, \ldots, \mathbf{A}_n$. Insert before $\mathbf{A}_n \supset \mathbf{B}_i$ first the wff $\mathbf{A}_n \supset_{\mathbf{a}} \mathbf{B}_j \supset . \mathbf{A}_n \supset \mathbf{B}_i$ (since $\mathbf{a}$ is not a free variable of $\mathbf{A}_n$,[344] this is an axiom, an instance of *305), and then the wff $\mathbf{A}_n \supset_{\mathbf{a}} \mathbf{B}_j$ (which can be inferred by generalization[344] from the earlier wff $\mathbf{A}_n \supset \mathbf{B}_j$ already present in the sequence being constructed, and from which then $\mathbf{A}_n \supset \mathbf{B}_i$ can be inferred by *modus ponens*).

Case 5: $\mathbf{B}_i$ is inferred by an alphabetic change of bound variable, according to *350, from the premiss $\mathbf{B}_j$, where $j < i$. In this case a corresponding alphabetic change of bound variable suffices to infer $\mathbf{A}_n \supset \mathbf{B}_i$ from $\mathbf{A}_n \supset \mathbf{B}_j$.

Case 6: $\mathbf{B}_i$ is inferred according to *351, by substitution in the premiss $\mathbf{B}_j$, where $j < i$, and where the (individual) variable that is substituted for does not occur as a free variable in $\mathbf{A}_1, \mathbf{A}_2, \ldots, \mathbf{A}_n$. In this case the same substitution suffices to infer $\mathbf{A}_n \supset \mathbf{B}_i$ from $\mathbf{A}_n \supset \mathbf{B}_j$.[345]

Case 7: $\mathbf{B}_i$ is inferred according to *352, by substitution in the premiss $\mathbf{B}_j$, where $j < i$, and where the (propositional or functional) variable that is substituted for does not occur as a free variable in $\mathbf{A}_1, \mathbf{A}_2, \ldots, \mathbf{A}_n$. In this case the same substitution suffices to infer $\mathbf{A}_n \supset \mathbf{B}_i$ from $\mathbf{A}_n \supset \mathbf{B}_j$.[345]

This completes the proof of the deduction theorem. From the special case of it in which $n = 1$ we have the corollary:

***361.** If $\mathbf{A} \vdash \mathbf{B}$, then $\vdash \mathbf{A} \supset \mathbf{B}$.

---

[344]Notice here the role of the condition that $\mathbf{a}$ does not occur as a free variable in $\mathbf{A}_1, \mathbf{A}_2, \ldots, \mathbf{A}_n$ (clause (4) in the definition of *proof from hypotheses* at the beginning of this section).

[345]Notice, in particular, the role of the condition that the variable which is substituted for does not occur as a free variable in $\mathbf{A}_n$.

In what follows we shall often use the deduction theorem as a derived rule in establishing theorems or theorem schemata. For this purpose it is essential that our proof of it is effective.[346] It is left to the reader to verify this, after supplying a definite particular proof of the wff $\mathbf{A}_n \supset \mathbf{A}_n$ to be used for the case 1a.

The following metatheorems, *362 and *363,[347] are also needed in connection with use of the deduction theorem as a derived rule. Tacit use will often be made especially of *363.[348]

*362. If every wff which occurs at least once in the list $\mathbf{A}_1, \mathbf{A}_2, \ldots, \mathbf{A}_n$ also occurs at least once in the list $\mathbf{C}_1, \mathbf{C}_2, \ldots, \mathbf{C}_r$, and if $\mathbf{A}_1, \mathbf{A}_2, \ldots, \mathbf{A}_n \vdash \mathbf{B}$, then $\mathbf{C}_1, \mathbf{C}_2, \ldots, \mathbf{C}_r \vdash \mathbf{B}$.

Proof. Let $\mathbf{a}_1, \mathbf{a}_2, \ldots, \mathbf{a}_l$ be the complete list of those variables of all kinds (individual, propositional, functional) which occur as free variables in $\mathbf{C}_1, \mathbf{C}_2, \ldots, \mathbf{C}_r$ but do not occur as free variables in $\mathbf{A}_1, \mathbf{A}_2, \ldots, \mathbf{A}_n$ (though some of them may perhaps occur as bound variables in $\mathbf{A}_1, \mathbf{A}_2, \ldots, \mathbf{A}_n$). Then, if the given proof of $\mathbf{B}$ from the hypotheses $\mathbf{A}_1, \mathbf{A}_2, \ldots, \mathbf{A}_n$ is not also a proof of $\mathbf{B}$ from the hypotheses $\mathbf{C}_1, \mathbf{C}_2, \ldots, \mathbf{C}_r$, it can only be because it involves generalizations upon or substitutions for some of the variables $\mathbf{a}_1, \mathbf{a}_2, \ldots, \mathbf{a}_l$. Therefore let $\mathbf{c}_1, \mathbf{c}_2, \ldots, \mathbf{c}_l$ be variables which are all distinct and which do not occur in $\mathbf{C}_1, \mathbf{C}_2, \ldots, \mathbf{C}_r$ or in the given proof of $\mathbf{B}$ from the hypotheses $\mathbf{A}_1, \mathbf{A}_2, \ldots, \mathbf{A}_n$, $\mathbf{c}_1$ being a variable of the same type (individual, propositional, singulary functional, binary functional, etc.) as $\mathbf{a}_1$, $\mathbf{c}_2$ being a variable of the same type as $\mathbf{a}_2$, $\mathbf{c}_3$ a variable of the same type as $\mathbf{a}_3$, and so on. And throughout the given proof of $\mathbf{B}$ from the hypotheses $\mathbf{A}_1, \mathbf{A}_2, \ldots, \mathbf{A}_n$ replace $\mathbf{a}_1, \mathbf{a}_2, \ldots, \mathbf{a}_l$ by $\mathbf{c}_1, \mathbf{c}_2, \ldots, \mathbf{c}_l$ respectively. The result is a proof of

$$\mathrm{S}^{\mathbf{a}_1\mathbf{a}_2\ldots\mathbf{a}_l}_{\mathbf{c}_1\mathbf{c}_2\ldots\mathbf{c}_l}\mathbf{B}|$$

from the hypotheses $\mathbf{D}_1, \mathbf{D}_2, \ldots, \mathbf{D}_r$, where $\mathbf{D}_1, \mathbf{D}_2, \ldots, \mathbf{D}_r$ differ from $\mathbf{C}_1, \mathbf{C}_2, \ldots, \mathbf{C}_r$, respectively, at most by certain alphabetic changes of bound variables. This is changed into a proof of $\mathbf{B}$ from the hypotheses $\mathbf{D}_1, \mathbf{D}_2, \ldots, \mathbf{D}_r$ by adding at the end, if necessary, an appropriate series of alphabetic changes of bound variables and substitutions (under clauses (5), (6), (7) in the definition of proof from hypotheses). Then finally a proof of $\mathbf{B}$

[346]Compare the discussion of derived rules of inference in §12, and the discussion at the end of §13 of the use of the deduction theorem for the propositional calculus.

[347]Compare *132–*134.

[348]For example, when *333 is used in the proof of *365, when *364 is used in the proof of *366, when *365 is used in the proof of *367, when *333 and *392 are used in the proof of *421.

from the hypotheses $\mathbf{C}_1, \mathbf{C}_2, \ldots, \mathbf{C}_r$ is obtained by inserting, for various values of i, as necessary, wffs to constitute a proof of $\mathbf{D}_i$ from $\mathbf{C}_i$ by alphabetic changes of bound variables.

The foregoing construction has to be made more explicit at several places, in order to allow the metatheorem to be used as a derived rule. For example, definite instructions must be given as to the choice of the variables $\mathbf{c}_1, \mathbf{c}_2, \ldots, \mathbf{c}_l$, so as to make it fully determinate. Details of this are obvious but cumbersome, and may be left to the reader.[349]

By taking $n = 0$ in *362, we have as a corollary:

*363. If $\vdash \mathbf{B}$, then $\mathbf{C}_1, \mathbf{C}_2, \ldots, \mathbf{C}_r \vdash \mathbf{B}$.

We go on to establish a number of derived rules (*366–*369) that facilitate the use of the existential quantifier, in connection with the deduction theorem. As a preliminary to this, two theorem schemata (*364, *365) are demonstrated, with aid of the deduction theorem as a derived rule.

*364. $\vdash \mathbf{B} \supset_{\mathbf{a}} \mathbf{A} \supset . (\exists \mathbf{a})\mathbf{B} \supset \mathbf{A}$, if $\mathbf{a}$ is not free in $\mathbf{A}$.

Proof. By *306, $\mathbf{B} \supset_{\mathbf{a}} \mathbf{A} \vdash \mathbf{B} \supset \mathbf{A}$.
Hence by P, $\mathbf{B} \supset_{\mathbf{a}} \mathbf{A} \vdash \sim\mathbf{A} \supset \sim\mathbf{B}$.
Hence by generalization, $\mathbf{B} \supset_{\mathbf{a}} \mathbf{A} \vdash \sim\mathbf{A} \supset_{\mathbf{a}} \sim\mathbf{B}$.[350]
Hence by *305, $\mathbf{B} \supset_{\mathbf{a}} \mathbf{A} \vdash \sim\mathbf{A} \supset (\mathbf{a})\sim\mathbf{B}$.
Hence by P, $\mathbf{B} \supset_{\mathbf{a}} \mathbf{A} \vdash \sim(\mathbf{a})\sim\mathbf{B} \supset \mathbf{A}$.
Then use the deduction theorem.

*365. $\vdash \mathbf{A} \supset_{\mathbf{a}} \mathbf{B} \supset . (\exists \mathbf{a})\mathbf{A} \supset (\exists \mathbf{a})\mathbf{B}$.

Proof. By *306, $\mathbf{A} \supset_{\mathbf{a}} \mathbf{B} \vdash \mathbf{A} \supset \mathbf{B}$.
Hence by P, $\mathbf{A} \supset_{\mathbf{a}} \mathbf{B} \vdash \sim\mathbf{B} \supset \sim\mathbf{A}$.
Hence by generalization, $\mathbf{A} \supset_{\mathbf{a}} \mathbf{B} \vdash \sim\mathbf{B} \supset_{\mathbf{a}} \sim\mathbf{A}$.[350]
Hence by *333, $\mathbf{A} \supset_{\mathbf{a}} \mathbf{B} \vdash (\mathbf{a})\sim\mathbf{B} \supset (\mathbf{a})\sim\mathbf{A}$.
Hence by P, $\mathbf{A} \supset_{\mathbf{a}} \mathbf{B} \vdash \sim(\mathbf{a})\sim\mathbf{A} \supset \sim(\mathbf{a})\sim\mathbf{B}$.
Then use the deduction theorem.

*366. If $\mathbf{A}_1, \mathbf{A}_2, \ldots, \mathbf{A}_n \vdash \mathbf{B}$, and $\mathbf{a}$ is an individual variable which does not occur as a free variable in $\mathbf{A}_1, \mathbf{A}_2, \ldots, \mathbf{A}_{n-1}, \mathbf{B}$, then $\mathbf{A}_1, \mathbf{A}_2, \ldots, \mathbf{A}_{n-1}, (\exists \mathbf{a})\mathbf{A}_n \vdash \mathbf{B}$.[351]

---

[349]Compare the last two paragraphs in the proof of *352.

[350]Where generalization is thus used in connection with the deduction theorem, care must be taken that the variable **a** which is generalized upon does not occur as a free variable in any of the hypotheses. It is left to the reader to verify this in each case.

Proof. By the deduction theorem,

$$\mathbf{A}_1, \mathbf{A}_2, \ldots, \mathbf{A}_{n-1} \vdash \mathbf{A}_n \supset \mathbf{B}.$$

Hence by generalizing upon **a** and then using *364, we have that

$$\mathbf{A}_1, \mathbf{A}_2, \ldots, \mathbf{A}_{n-1} \vdash (\exists \mathbf{a})\mathbf{A}_n \supset \mathbf{B}.$$

Hence by *modus ponens*,

$$\mathbf{A}_1, \mathbf{A}_2, \ldots, \mathbf{A}_{n-1}, (\exists \mathbf{a})\mathbf{A}_n \vdash \mathbf{B}.$$

*367. If $\mathbf{A}_1, \mathbf{A}_2, \ldots, \mathbf{A}_n \vdash \mathbf{B}$, and **a** is an individual variable which does not occur as a free variable in $\mathbf{A}_1, \mathbf{A}_2, \ldots, \mathbf{A}_{n-1}$, then $\mathbf{A}_1, \mathbf{A}_2, \ldots, \mathbf{A}_{n-1}, (\exists \mathbf{a})\mathbf{A}_n \vdash (\exists \mathbf{a})\mathbf{B}$.

Proof. By the same method as the proof of *366, but with use of *365 replacing that of *364.

*368. If $\mathbf{A}_1, \mathbf{A}_2, \ldots, \mathbf{A}_n \vdash \mathbf{B}$, and **a** is an individual variable which does not occur as a free variable in $\mathbf{A}_1, \mathbf{A}_2, \ldots, \mathbf{A}_{n-r}, \mathbf{B}$, then $\mathbf{A}_1, \mathbf{A}_2, \ldots, \mathbf{A}_{n-r}, (\exists \mathbf{a}) \centerdot \mathbf{A}_{n-r+1}\mathbf{A}_{n-r+2} \ldots \mathbf{A}_n \vdash \mathbf{B}$.

Proof. By P,

$$\mathbf{A}_1, \mathbf{A}_2, \ldots, \mathbf{A}_{n-r}, \mathbf{A}_{n-r+1}\mathbf{A}_{n-r+2} \ldots \mathbf{A}_n \vdash \mathbf{B}.$$

Hence use *366.

*369. If $\mathbf{A}_1, \mathbf{A}_2, \ldots, \mathbf{A}_n \vdash \mathbf{B}$, and **a** is an individual variable which does not occur as a free variable in $\mathbf{A}_1, \mathbf{A}_2, \ldots, \mathbf{A}_{n-r}$, then $\mathbf{A}_1, \mathbf{A}_2, \ldots, \mathbf{A}_{n-r}, (\exists \mathbf{a}) \centerdot \mathbf{A}_{n-r+1}\mathbf{A}_{n-r+2} \ldots \mathbf{A}_n \vdash (\exists \mathbf{a})\mathbf{B}$.

Proof. By the same method as the proof of *368, but with use of *367 replacing that of *366.

37. Duality. As in § 16, we begin by applying the process of dualization not to wffs but to expressions which are abbreviations of wffs in accordance with certain definitions. Namely, we allow abbreviation by D3–11 and D14, but not by other definition schemata[352] and not by omissions of brackets.

---

[351]The case $n = 1$ of *366 is proved as a metatheorem of the functional calculus of first order by Hilbert and Bernays, *Grundlagen der Mathematik*, vol. 1 (1934), pp. 157–158. It may also be compared with theorems I and III of the writer's "A Set of Postulates for the Foundation of Logic" (in the *Annals of Mathematics*, vol. 33 (1932), see pp. 358, 366)—with which, however, it is far from identical, because the logistic system considered in that paper is a quite different system from any of the functional calculi of first or higher order, and indeed is one which was afterwards shown by Kleene and Rosser (in the *Annals of Mathematics*, vol. 36 (1935), pp. 630–636) to be inconsistent.

[352]Here D13 might well also be allowed, but it simplifies the statement of the matter slightly if we exclude it. In order to allow D12, D15–17, we would have to add duals

Of such an expression *the dual* is obtained by interchanging simultaneously, wherever they occur, each of the following pairs of connectives and quantifiers (or better, each of the following pairs of symbols): $\supset$ and $\not\subset$, disjunction and conjunction, $\equiv$ and $\not\equiv$, $\subset$ and $\not\supset$, $\bar{\vee}$ and $|$, $\forall$ and $\exists$.

Of a wff of F^1 *a dual* is obtained by writing any expression of the foregoing kind which abbreviates the wff, dualizing this expression, and then finally writing the wff which the resulting expression abbreviates. As a particular case, the given wff itself may be used in the role of the expression which abbreviates it, and when this is done *the principal dual* of the given wff is obtained.

By examining D3–11 and D14, it will be seen that any two duals of the same wff can be transformed one into the other by a series of steps of which each consists, either in replacing a wf part $\sim\sim\mathbf{N}$ by $\mathbf{N}$, or in replacing a wf part $\mathbf{N}$ by $\sim\sim\mathbf{N}$ (i.e., as we may say, either in deleting or in inserting a double negation). Hence by P and *341:

*370. If $\mathbf{B}$ and $\mathbf{C}$ are duals of $\mathbf{A}$, then $\vdash \mathbf{B} \equiv \mathbf{C}$.

In order to establish for F^1 a principle of duality analogous to *161, we shall show for each axiom of F^1 that the negation of any dual of it is a theorem of F^1; also for each rule of inference of F^1 that, if we replace the premisses and conclusion by negations of duals of them, the inference still holds as a derived rule. It will then follow that the negation of a dual of a theorem of F^1 is always a theorem of F^1.

To begin with the rules of inference, consider first *300. Here the premisses are $\mathbf{A} \supset \mathbf{B}$ and $\mathbf{A}$, the conclusion $\mathbf{B}$. Let $\mathbf{A}_1$ be a dual of $\mathbf{A}$, and $\mathbf{B}_1$ a dual of $\mathbf{B}$. Then one of the duals of $\mathbf{A} \supset \mathbf{B}$ is $\mathbf{A}_1 \not\subset \mathbf{B}_1$. By P, if $\vdash \sim . \mathbf{A}_1 \not\subset \mathbf{B}_1$ and $\vdash \sim\mathbf{A}_1$, then $\vdash \sim\mathbf{B}_1$. Hence by *370 and P, if the negation of any dual of $\mathbf{A} \supset \mathbf{B}$ and the negation of any dual of $\mathbf{A}$ are theorems, then the negation of every dual of $\mathbf{B}$ is a theorem.

Likewise consider *301. The premiss is $\mathbf{A}$ and the conclusion $(\mathbf{a})\mathbf{A}$. From $\sim\mathbf{A}_1$, the negation of any dual of $\mathbf{A}$, we may infer first $(\mathbf{a})\sim\mathbf{A}_1$ by *301, and thence $\sim(\exists\mathbf{a})\mathbf{A}_1$ by *338 and P. This is the negation of one of the duals of the conclusion, and from it the negation of every other dual of the conclusion follows by *370 and P.

Turning now to the axioms, we see that one of the duals of any axiom which is an instance of *302, *303, *304 must have, in corresponding order,

---

of these definition schemata, inventing suitable notations for the purpose; but this seems not worth while, as the notations so introduced would hardly be used except in connection with the treatment of duality.

the following forms (where $\mathbf{A}_1$, $\mathbf{B}_1$, $\mathbf{C}_1$ are duals of **A**, **B**, **C** respectively):

$$\mathbf{A}_1 \not\subset . \mathbf{B}_1 \not\subset \mathbf{A}_1$$

$$\mathbf{A}_1 \not\subset [\mathbf{B}_1 \not\subset \mathbf{C}_1] \not\subset . \mathbf{A}_1 \not\subset \mathbf{B}_1 \not\subset . \mathbf{A}_1 \not\subset \mathbf{C}_1$$

$$\sim\mathbf{A}_1 \not\subset \sim\mathbf{B}_1 \not\subset . \mathbf{B}_1 \not\subset \mathbf{A}_1$$

And the negation of each of these may be seen to be a substitution instance of a tautology, therefore a theorem by *311. That the negation of every other dual of the same axiom is also a theorem, then follows by *370 and P.

In the case of an axiom which is an instance of *306, one of its duals has the form

$$(\exists\mathbf{a})\mathbf{A}_1 \not\subset \check{S}_{\mathbf{b}}^{\mathbf{a}}\mathbf{A}_1|,$$

where **a** is an individual variable, **b** is an individual variable or an individual constant, and no free occurrence of **a** in $\mathbf{A}_1$ is in a wf part of $\mathbf{A}_1$ of the form $(\mathbf{b})\mathbf{C}$ [353] That the negation of this is a theorem follows by *330 and P. Hence by *370 and P, the negation of every dual of the axiom is also a theorem.

In the case of an axiom which is an instance of *305, in order to prove similarly that the negation of every dual of it is a theorem, we need only the following theorem schema of F^1:

*371. $\vdash \mathbf{A} \not\subset (\exists\mathbf{a})\mathbf{B} \supset (\exists\mathbf{a}) . \mathbf{A} \not\subset \mathbf{B}$, if **a** is not free in **A**.

Proof. By P, $\vdash \mathbf{B} \supset \mathbf{A} \equiv \sim\sim . \mathbf{B} \supset \mathbf{A}$.
Hence by *364 and *342, $\vdash (\mathbf{a})\sim\sim[\mathbf{B} \supset \mathbf{A}] \supset . (\exists\mathbf{a})\mathbf{B} \supset \mathbf{A}$.
Then use the law of contraposition (†223).

Thus we have shown that every axiom of F^1 has the property that the negation of every dual of it is a theorem of F^1, and that the rules of inference of F^1 preserve this property. Hence every theorem of F^1 has the property, i.e.:

*372. If $\vdash \mathbf{A}$, and if $\mathbf{A}_1$ is a dual of **A**, then $\vdash \sim\mathbf{A}_1$.

(*Principle of duality.*)

As in §16, two *special principles of duality* follow as corollaries, by P:

*373. If $\vdash \mathbf{A} \supset \mathbf{B}$, and if $\mathbf{A}_1$ and $\mathbf{B}_1$ are duals of **A** and **B** respectively, then $\vdash \mathbf{B}_1 \supset \mathbf{A}_1$.

(*Special principle of duality for (material) implications.*)

---

[353] In *306, the condition that *no free occurrence of* **a** *in* **A** *is in a wf part of* **A** *of the form* $(\mathbf{b})\mathbf{C}$ is clearly equivalent to the condition that *no free occurrence of* **a** *in* **A** *is in a wf part of* **A** *of either of the forms* $(\mathbf{b})\mathbf{C}$ *or* $(\exists\mathbf{b})\mathbf{C}$. In the latter form, it is obvious that the condition is unchanged by dualization.

*374. If ⊢ **A** ≡ **B**, and if $\mathbf{A}_1$ and $\mathbf{B}_1$ are duals of **A** and **B** respectively, then ⊢ $\mathbf{A}_1 \equiv \mathbf{B}_1$.

(*Special principle of duality for (material) equivalences.*)

By *the dual* of a theorem schema or axiom schema of F^1 we shall mean: (1) if the schema has the form of an implication, the theorem schema obtained from it by *373; (2) if the schema has the form of an equivalence, the theorem schema obtained from it by *374; (3) in other cases, the theorem schema obtained from it by *372. In case (1), the dualization is to be performed on the antecedent and consequent of the schema as actually written, and according to the instructions as given in the first paragraph of this section; similarly, in case (2) the dualization is to be performed on the two parts of the schema as actually written, and in case (3) it is to be performed on the schema as actually written, again according to the instructions in the first paragraph of this section. Thus the dual of a theorem schema or axiom schema may differ according to what abbreviations are used in writing the schema, but it is unique for any schema as actually written. It is understood that, before dualizing a schema, any abbreviations by omission of brackets or by D12, D13, D15–17 are first to be withdrawn, restoration towards unabbreviated form proceeding thus far but no farther.

In writing the dual of a theorem schema or axiom schema, the subscripts 1 on the bold capital letters to indicate dualization—as we used them, e.g., in proving *372—may be omitted, on the ground that every wff is a dual of some wff. If verbally stated conditions are attached to a theorem schema or axiom schema (as for instance in the case of *305, *306, *339), the conditions must be dualized in an appropriate sense; but in most cases with which we shall meet in practice the verbally stated conditions are the same as or equivalent to their duals and may therefore be left unaltered.[354]

To illustrate the dualization of theorem schemata and axiom schemata, we may cite the following examples. The dual of *302 is the theorem schema which asserts that

$$\vdash \mathbf{B} \not\subset \mathbf{A} \supset \mathbf{A}.$$

The dual of *304 is the theorem schema asserting that

$$\vdash \mathbf{B} \not\subset \mathbf{A} \supset . \sim\mathbf{A} \not\subset \sim\mathbf{B}.$$

The dual of *305 is *371. The dual of *306 is *330. The dual of *330 is *306,

[354]Besides theorem schemata, we shall sometimes speak also of *duals* of other metatheorems, in the sense of corollaries of them by *372, *373, *374. We do not attempt to make this notion more precise, but shall use the terminology in this case only heuristically or suggestively.

or, more correctly, it is the theorem schema which is an immediate corollary of *306, asserting that every instance of *306 is a theorem. The dual of *331 is *331, i.e., as we shall say, *331 is *self-dual.*

Again, the following theorem schemata are, in order, the duals of the theorem schemata *336, *337, *338, *339:

*375. $\vdash (\exists \mathbf{a})(\exists \mathbf{b})\mathbf{A} \equiv (\exists \mathbf{b})(\exists \mathbf{a})\mathbf{A}$.

*376. $\vdash (\exists \mathbf{a})\mathbf{A} \equiv \mathbf{A}$, if $\mathbf{a}$ is not free in $\mathbf{A}$.

*377. $\vdash \sim(\mathbf{a})\mathbf{A} \equiv (\exists \mathbf{a})\sim\mathbf{A}$.

*378. $\vdash (\exists \mathbf{a})\mathbf{A} \equiv (\exists \mathbf{b})\mathbf{B}$, if there is no free occurrence of $\mathbf{b}$ in $\mathbf{A}$, and no free occurrence of $\mathbf{a}$ in $\mathbf{A}$ is in a wf part of $\mathbf{A}$ of the form $(\mathbf{b})\mathbf{C}$, and $\mathbf{B}$ is $\underset{.}{S}^{\mathbf{a}}_{\mathbf{b}}\mathbf{A}|$.

38. Some further theorem schemata.

*380. $\vdash (\exists \mathbf{a})\mathbf{B} \supset \mathbf{A} \equiv . \mathbf{B} \supset_{\mathbf{a}} \mathbf{A}$, if $\mathbf{a}$ is not free in $\mathbf{A}$.

Proof. By dualizing *335, $\vdash \mathbf{A} \not\subset (\exists \mathbf{a})\mathbf{B} \equiv (\exists \mathbf{a}) . \mathbf{A} \not\subset \mathbf{B}$.
Hence by *377 and P, $\vdash \mathbf{A} \not\subset (\exists \mathbf{a})\mathbf{B} \equiv \sim(\mathbf{a}) . \mathbf{B} \supset \mathbf{A}$.
Then use P.

*381. $\vdash (\mathbf{a})\mathbf{A} \supset (\exists \mathbf{a})\mathbf{B} \equiv (\exists \mathbf{a}) . \mathbf{A} \supset \mathbf{B}$.

Proof. By P, $\vdash \sim\mathbf{A} \supset . \mathbf{A} \supset \mathbf{B}$.
Hence by generalization and *365, $\vdash (\exists \mathbf{a})\sim\mathbf{A} \supset (\exists \mathbf{a}) . \mathbf{A} \supset \mathbf{B}$.
Hence by *377 and P, $\vdash \sim(\mathbf{a})\mathbf{A} \supset (\exists \mathbf{a}) . \mathbf{A} \supset \mathbf{B}$.
Also by *302, generalization, and *365, $\vdash (\exists \mathbf{a})\mathbf{B} \supset (\exists \mathbf{a}) . \mathbf{A} \supset \mathbf{B}$
Hence by P,[355] $\vdash (\mathbf{a})\mathbf{A} \supset (\exists \mathbf{a})\mathbf{B} \supset (\exists \mathbf{a}) . \mathbf{A} \supset \mathbf{B}$.

By *306 and *modus ponens*, $\mathbf{A} \supset \mathbf{B}, (\mathbf{a})\mathbf{A} \vdash \mathbf{B}$.
Hence by *367, $(\exists \mathbf{a})[\mathbf{A} \supset \mathbf{B}], (\mathbf{a})\mathbf{A} \vdash (\exists \mathbf{a})\mathbf{B}$.
Hence by the deduction theorem, $\vdash (\exists \mathbf{a})[\mathbf{A} \supset \mathbf{B}] \supset . (\mathbf{a})\mathbf{A} \supset (\exists \mathbf{a})\mathbf{B}$.

Then use P.

*382. $\vdash \mathbf{A} \supset (\exists \mathbf{a})\mathbf{B} \equiv (\exists \mathbf{a}) . \mathbf{A} \supset \mathbf{B}$, if $\mathbf{a}$ is not free in $\mathbf{A}$.

Proof. By *381, *337, and *342.

*383. $\vdash (\mathbf{a})\mathbf{B} \supset \mathbf{A} \equiv (\exists \mathbf{a}) . \mathbf{B} \supset \mathbf{A}$, if $\mathbf{a}$ is not free in $\mathbf{A}$.

Proof. By *381, *376, and *342.

[355]The tautology used is $\sim p \supset r \supset . q \supset r \supset . p \supset q \supset r$.

*384. $\vdash [\sim\mathbf{A} \supset_{\mathbf{a}} \mathbf{C}][\mathbf{B} \supset_{\mathbf{a}} \mathbf{C}] \equiv . \mathbf{A} \supset \mathbf{B} \supset_{\mathbf{a}} \mathbf{C}$.

Proof. By P,[355] $\sim\mathbf{A} \supset \mathbf{C}$, $\mathbf{B} \supset \mathbf{C} \vdash \mathbf{A} \supset \mathbf{B} \supset \mathbf{C}$.
Hence by P and *306, $[\sim\mathbf{A} \supset_{\mathbf{a}} \mathbf{C}][\mathbf{B} \supset_{\mathbf{a}} \mathbf{C}] \vdash \mathbf{A} \supset \mathbf{B} \supset \mathbf{C}$.
Hence by generalization, $[\sim\mathbf{A} \supset_{\mathbf{a}} \mathbf{C}][\mathbf{B} \supset_{\mathbf{a}} \mathbf{C}] \vdash \mathbf{A} \supset \mathbf{B} \supset_{\mathbf{a}} \mathbf{C}$.
Hence $\vdash [\sim\mathbf{A} \supset_{\mathbf{a}} \mathbf{C}][\mathbf{B} \supset_{\mathbf{a}} \mathbf{C}] \supset . \mathbf{A} \supset \mathbf{B} \supset_{\mathbf{a}} \mathbf{C}$.

By *306 and P, $\mathbf{A} \supset \mathbf{B} \supset_{\mathbf{a}} \mathbf{C} \vdash \sim\mathbf{A} \supset \mathbf{C}$.
Hence by generalization, $\mathbf{A} \supset \mathbf{B} \supset_{\mathbf{a}} \mathbf{C} \vdash \sim\mathbf{A} \supset_{\mathbf{a}} \mathbf{C}$.
Again by *306, P, and generalization, $\mathbf{A} \supset \mathbf{B} \supset_{\mathbf{a}} \mathbf{C} \vdash \mathbf{B} \supset_{\mathbf{a}} \mathbf{C}$.
Hence by P, $\mathbf{A} \supset \mathbf{B} \supset_{\mathbf{a}} \mathbf{C} \vdash [\sim\mathbf{A} \supset_{\mathbf{a}} \mathbf{C}][\mathbf{B} \supset_{\mathbf{a}} \mathbf{C}]$.
Hence $\vdash \mathbf{A} \supset \mathbf{B} \supset_{\mathbf{a}} \mathbf{C} \supset [\sim\mathbf{A} \supset_{\mathbf{a}} \mathbf{C}][\mathbf{B} \supset_{\mathbf{a}} \mathbf{C}]$.

Then use P.

*385. $\vdash (\mathbf{a})[\mathbf{A} \vee \mathbf{B}] \equiv \mathbf{A} \vee (\mathbf{a})\mathbf{B}$, if **a** is not free in **A**.

Proof. By *335 and P, $\vdash (\mathbf{a})[\sim\mathbf{A} \supset \mathbf{B}] \equiv . \sim\mathbf{A} \supset (\mathbf{a})\mathbf{B}$.
Then use P and *342.

*386. $\vdash (\mathbf{a})[\mathbf{B} \vee \mathbf{A}] \equiv (\mathbf{a})\mathbf{B} \vee \mathbf{A}$, if **a** is not free in **A**.

Proof. By *385, P, and *342.

*387. $\vdash \mathbf{A} \equiv_{\mathbf{a}} \mathbf{B} \supset . (\exists\mathbf{a})\mathbf{A} \equiv (\exists\mathbf{a})\mathbf{B}$.

Proof. By *340 and *350.

*388. $\vdash \mathbf{A} \equiv_{\mathbf{a}} \mathbf{B} \supset . \mathbf{A} \equiv (\exists\mathbf{a})\mathbf{B}$, if **a** is not free in **A**.

Proof. By *387, *376, and *342.

EXERCISES 38

38.0. For the proof of the deduction theorem, *360, case 1a, write out explicitly the full list of wffs that are to be inserted before $\mathbf{A}_n \supset \mathbf{B}_i$.

38.1. Write the duals of the theorem schemata *383–*388.

38.2. Establish the theorem schema *365 as a corollary of *333 by dualization.

38.3. Similarly establish the theorem schema *387 as a corollary of *334 by dualization.

38.4. Similarly establish a corollary of *332 by dualization (in the expression of which, "$\sim$" shall occur only through the abbreviation "$(\exists\mathbf{a})$" for "$\sim(\mathbf{a})\sim$", and, in particular, "$\not\subset$" shall not occur).

38.5. Establish the following theorem schemata of F^1:

(1) $\vdash (\mathbf{a})\mathbf{A}(\mathbf{a})\mathbf{B} \equiv (\mathbf{a}) \centerdot \mathbf{AB}$.

(2) $\vdash (\mathbf{a})\mathbf{A}(\exists\mathbf{a})\mathbf{B} \supset (\exists\mathbf{a}) \centerdot \mathbf{AB}$.

(3) $\vdash (\mathbf{a})(\exists\mathbf{b})[\mathbf{A} \supset \mathbf{B}] \equiv (\exists\mathbf{b})(\mathbf{a}) \centerdot \mathbf{A} \supset \mathbf{B}$, if **a** is not free in **A** and **b** is not free in **B**.

(4) $\vdash (\mathbf{a})(\exists\mathbf{b})[\mathbf{B} \supset \mathbf{A}] \equiv (\exists\mathbf{b})(\mathbf{a}) \centerdot \mathbf{B} \supset \mathbf{A}$, if **a** is not free in **A** and **b** is not free in **B**.

38.6. A formulation F^{1i} of the *intuitionistic functional calculus of first order* may be given as follows. The primitive symbols are those of F^1, with such additions to them as to make the existential quantifier primitive as well as the universal quantifier, and to supply all the primitive sentence connectives of the system P_2^i (see 26.18). The formation rules are those of F^1, with the obvious added rules to correspond to the additional primitive symbols. The definition of *bound* and *free* occurrences of variables must be changed so that an occurrence of **a** in **A** is bound if it is in a wf part of **A** of either of the forms $(\forall\mathbf{a})\mathbf{B}$ or $(\exists\mathbf{a})\mathbf{B}$, otherwise free. Of the definition schemata employed in connection with F^1, only D13, D15, D16 are retained. The same abbreviations by omission of brackets are retained, including the same convention about the use of heavy dots, and also the abbreviation by omitting superscripts on functional variables. The rules of inference are *300 and *301, the same as for F^1. The axioms comprise all substitution instances of axioms of P_2^i, and all instances of four additional axiom schemata which, with obviously necessary modifications, are the same as *305, *306, *330, *364.[356]

By the same or nearly the same proofs as for F^1, the following hold (with obvious modifications) also for F^{1i}: a modified form of *311 with theorems of P_2^i taking the place of tautologies of P; *331–*337; *339. Hence show that *365 and *340–*342 hold for F^{1i}.

38.7. Show that the rule of alphabetic change of bound variable, *350, holds for F^{1i}. (Hence *351, *352, *360–*363, *366–*369 hold for F^{1i} by the same proofs as for F^1.)

[356] Specifically, the changes to be made in *305, *306, *330, *364 are as follows: *330 and *364 are to be modified to read as axiom schemata instead of theorem schemata; existential quantifiers occurring are to be read, not according to D14, but rather as involving the primitive symbol $\exists$ of F^{1i}; the terms "free occurrence" and "free variable" are to be construed according to the modified definition as just given for F^{1i}; and in *306 and *330 the last of the verbally stated conditions is to be altered to read, "and no free occurrence of **a** in **A** is in a wf part of **A** of either of the forms $(\forall\mathbf{b})\mathbf{C}$ or $(\exists\mathbf{b})\mathbf{C}$."

38.8. Assuming the results of the two preceding exercises, as well as results obtained in Exercises 26 regarding the intuitionistic propositional calculus, establish the following theorem schemata of F^{1i}:

(1) $\vdash (\exists a){\sim}A \supset {\sim}(a)A$.

(2) $\vdash {\sim}(\exists a)A \equiv (a){\sim}A$.

(3) $\vdash (\exists a){\sim}{\sim}A \supset {\sim}{\sim}(\exists a)A$.

(4) $\vdash {\sim}{\sim}(a)A \supset (a){\sim}{\sim}A$.

(5) $\vdash (\exists a)(\exists b)A \equiv (\exists b)(\exists a)A$.

(6) $\vdash (\exists a)A \equiv A$, if **a** is not free in **A**.

(7) $\vdash (\exists a)B \supset A \supset . B \supset_a A$, if **a** is not free in **A**.

(8) $\vdash (\exists a)[B \supset A] \supset . (a)B \supset A$, if **a** is not free in **A**.

(9) $\vdash (\exists a)[A \supset B] \supset . A \supset (\exists a)B$, if **a** is not free in **A**.

(10) $\vdash (\exists a)[A \vee B] \equiv A \vee (\exists a)B$, if **a** is not free in **A**.

38.9. A formulation F^1_r of the (ordinary, or non-intuitionistic) functional calculus of first order may be obtained from F^{1i} by adding the axiom schema, $A \vee {\sim}A$. State and prove an appropriate metatheorem of equivalence between F^1_r and F^1. (Compare 26.14.)

38.10. A formulation F^{1m} of the *minimal functional calculus of first order* may be obtained from F^{1i} by omitting the axiom schema ${\sim}A \supset . A \supset B$ (with no other change). Of the results of 38.6–38.8 regarding F^{1i}, extend as many as possible to F^{1m}.

38.11. Another formulation, F^{1m}_J, of the minimal functional calculus of first order may be obtained from F^{1i} by suppressing the primitive symbol ${\sim}$, introducing a new primitive symbol f (a propositional constant), and altering accordingly the formation rules, rules of inference, and axiom schemata. (The three axiom schemata which involve the symbol ${\sim}$ explicitly are thus omitted; the other axiom schemata and rules of inference remain unaltered except in that the notion of a wff has been changed by the change in the formation rules; no new axiom schemata or rules of inference are added.) Establish the equivalence of F^{1m} and F^{1m}_J in a sense like that of §23. (Compare 26.19.)

38.12. For every wff **A** of F^1_r let an associated wff **A*** be defined, by recursion as follows: if **A** is a propositional variable standing alone, or if **A** is of the form $\mathbf{f}(\mathbf{a}_1, \mathbf{a}_2, \ldots, \mathbf{a}_n)$ where **f** is an n-ary functional variable or constant and $\mathbf{a}_1, \mathbf{a}_2, \ldots, \mathbf{a}_n$ are individual variables or constants, then $\mathbf{A}^*$ is $\sim\sim\mathbf{A}$; $[\mathbf{A} \supset \mathbf{B}]^*$ is $\sim\sim[\mathbf{A}^* \supset \mathbf{B}^*]$; $[\mathbf{AB}]^*$ is $\sim\sim[\mathbf{A}^*\mathbf{B}^*]$; $[\mathbf{A} \vee \mathbf{B}]^*$ is $\sim\sim[\mathbf{A}^* \vee \mathbf{B}^*]$; $[\mathbf{A} \equiv \mathbf{B}]^*$ is $\sim\sim[\mathbf{A}^* \equiv \mathbf{B}^*]$; if **C** is $\sim\mathbf{A}$, then $\mathbf{C}^*$ is $\sim\mathbf{A}^*$; if **C** is $(\forall\mathbf{a})\mathbf{A}$, then $\mathbf{C}^*$ is $\sim\sim(\forall\mathbf{a})\mathbf{A}^*$; if **C** is $(\exists\mathbf{a})\mathbf{A}$, then $\mathbf{C}^*$ is $\sim\sim(\exists\mathbf{a})\mathbf{A}^*$. Show that **A** is a theorem of F^1_I if and only if $\mathbf{A}^*$ is a theorem of $\mathrm{F}^{1\mathrm{m}}$.[357] (This may be done by showing that the axioms of F^1_I have the property that the associated wff is a theorem of $\mathrm{F}^{1\mathrm{m}}$, and that the rules of inference preserve this property.)

38.13. In a wff of F^1 the *elementary parts*, as defined in §30, are those wf parts which have either the form of a propositional variable alone or the form $\mathbf{f}(\mathbf{a}_1, \mathbf{a}_2, \ldots, \mathbf{a}_n)$ where **f** is an n-ary functional variable or constant and $\mathbf{a}_1, \mathbf{a}_2, \ldots, \mathbf{a}_n$ are individual variables or constants. Let $\mathbf{A}_*$ be the wff obtained from **A** by replacing each elementary part **E** of **A** by $\sim\sim\mathbf{E}$. Show that **A** is a theorem of F^1 if and only if $\mathbf{A}_*$ is a theorem of $\mathrm{F}^{1\mathrm{m}}$. (Use the result of 38.12, together with that of 26.20, and 38.8(4) as a theorem schema of $\mathrm{F}^{1\mathrm{m}}$, and *342 as a metatheorem of $\mathrm{F}^{1\mathrm{m}}$.)

38.14. Extend the result of 38.13 to wffs **A** of F^1_I which do not contain either disjunction or the existential quantifier.[358]

39. Prenex normal form. If $(\forall\mathbf{a})\mathbf{C}$ or $(\exists\mathbf{a})\mathbf{C}$ appears as a wf part of a wff **A**, the *scope* of that particular occurrence of the quantifier, $(\forall\mathbf{a})$ or $(\exists\mathbf{a})$, in **A** is the particular occurrence of **C** immediately following that occurrence of $(\forall\mathbf{a})$ or $(\exists\mathbf{a})$.[359]

An occurrence of a quantifier, $(\forall\mathbf{a})$ or $(\exists\mathbf{a})$, in a wff is *initially placed* if either it is at the beginning of the wff (i.e., with no symbols preceding it,

[357]This result is due in substance to Kolmogoroff in the paper cited in footnote 210. (*Added in proof.*) A number of further results similar to those of 38.12, 38.13, 38.14, but for intuitionistic rather than minimal functional calculus of first order, are in Kleene's *Introduction to Metamathematics* (1952), see p. 495.

[358]This result, regarding the minimal functional calculus of first order, should be compared with a result of Gödel, regarding intuitionistic arithmetic, in his paper in *Ergebnisse eines Mathematischen Kolloquiums*, no. 4 (1933), pp. 34–38.

[359]On the analogy of D14, we are here using "$(\exists\mathbf{a})$" as abbreviation of "$\sim(\forall\mathbf{a})\sim$", and "$(\exists\ \ )$" as an abbreviation of "$\sim(\forall\ \ )\sim$". Also in speaking of an occurrence of $(\forall\ \ )$ or of $(\exists\ \ )$, we mean that the blank space shall be filled by a variable—so that, e.g., an occurrence of $(\forall\ \ )$ consists of one occurrence of each of the three symbols (, $\forall$,), in that order, the three occurrences being consecutive except that a single symbol, a variable, must stand between the occurrence of $\forall$ and that of).

Strictly speaking, an occurrence of the universal quantifier is an occurrence of $(\forall\ \ )$, and (in F^1) an occurrence of the existential quantifier is an occurrence of $\sim(\forall\ \ )\sim$. But we shall sometimes find it convenient to speak loosely, in a way that includes the operator variable in an occurrence of the universal or existential quantifier. Thus we speak here of occurrences of $(\forall\mathbf{a})$ and $(\exists\mathbf{a})$ as occurrences of the universal and existential quantifiers. And there is a similar intent in the first paragraph of §32, in the statement of **391 below, and elsewhere.

not even brackets) or it is preceded only by one or more occurrences of quantifiers, $(\forall\ )$ and $(\exists\ )$, each with its own operator variable.[359]

An occurrence of a quantifier, $(\forall \mathbf{a})$ or $(\exists \mathbf{a})$, in a wff is called *vacuous* if its operator variable **a** has no free occurrence in its scope. In the contrary case it is called *non-vacuous*.[359]

A wff is said to be *in prenex normal form* if it has no occurrences of quantifiers otherwise than in initially placed non-vacuous occurrences of $(\forall\ )$ and $(\exists\ )$.[359]

Thus a wff **A** is in prenex normal form if and only if it has the form

$$\mathbf{\Pi}_1\mathbf{\Pi}_2 \ldots \mathbf{\Pi}_n\mathbf{M},$$

where **M** is wf and quantifier-free, where each $\mathbf{\Pi}_i$ is either $(\forall \mathbf{a}_i)$ or $(\exists \mathbf{a}_i)$ $(i = 1, 2, \ldots, n)$, and where $\mathbf{a}_1, \mathbf{a}_2, \ldots, \mathbf{a}_n$ are variables which are all different and which all have at least one (free) occurrence in **M**. Then the formula

$$\mathbf{\Pi}_1\mathbf{\Pi}_2 \ldots \mathbf{\Pi}_n$$

is called the *prefix* of **A**, and the wff **M** is called the *matrix* of **A**. (As a special case, we may have that $n = 0$; in this case the prefix is the null formula, and the matrix **M** coincides with **A**.)

In order to obtain what we shall call *the prenex normal form of* a wff **A**, we consider the following operations of reduction, applicable to a wff containing quantifiers that are not initially placed:

(i) If (in left-to-right order) the first occurrence of a quantifier that is not initially placed is in a wf part $\sim(\forall \mathbf{a})\mathbf{C}$, where **C** does not begin with $\sim$, then this wf part (i.e., this one occurrence of it) is replaced by $(\exists \mathbf{a})\sim\mathbf{C}$. Cf. *377.

(ii) If the first occurrence of a quantifier that is not initially placed is in a wf part $\sim(\exists \mathbf{a})\mathbf{C}$, this wf part is replaced by $(\forall \mathbf{a})\sim\mathbf{C}$. Cf. *338.

(iii) If the first occurrence of a quantifier that is not initially placed is in a wf part $[(\forall \mathbf{a})\mathbf{C} \supset \mathbf{D}]$, this wf part is replaced by

$$(\exists \mathbf{b})[\mathrm{S}^{\mathbf{a}}_{\mathbf{b}}\mathbf{C}| \supset \mathbf{D}],$$

where **b** is either **a**, in case **a** has no free occurrence in **D**, or otherwise the first individual variable in alphabetic order after **a** which does not occur in **C** and has no free occurrence in **D**. Cf. *350, *383.

(iv) If the first occurrence of a quantifier that is not initially placed is in a wf part $[(\exists \mathbf{a})\mathbf{C} \supset \mathbf{D}]$, this wf part is replaced by

$$(\forall \mathbf{b})[\mathrm{S}^{\mathbf{a}}_{\mathbf{b}}\mathbf{C}| \supset \mathbf{D}],$$

where **b** is determined as in (iii). Cf. *350, *380.

(v) If the first occurrence of a quantifier that is not initially placed is in a wf part [**D** ⊃ (∀**a**)**C**], where **D** is quantifier-free, then this wf part is replaced by

$$(\forall \mathbf{b})[\mathbf{D} \supset \mathrm{S}^{\mathbf{a}}_{\mathbf{b}}\mathbf{C}|],$$

where **b** is determined as in (iii). Cf. *350, *335.

(vi) If the first occurrence of a quantifier that is not initially placed is in a wf part [**D** ⊃ (∃**a**)**C**], where **D** is quantifier-free, then this wf part is replaced by

$$(\exists \mathbf{b})[\mathbf{D} \supset \mathrm{S}^{\mathbf{a}}_{\mathbf{b}}\mathbf{C}|],$$

where **b** is determined as in (iii). Cf. *350, *382.

390. By a finite number of successive applications of the reduction steps (i)–(vi), any wff **A of F^1 can be reduced to a wff **A**′ of F^1 in which all quantifiers are initially placed. This process of reduction is effective, and the resulting wff **A**′ is uniquely determined when **A** is given.

Proof. At each stage in the process, as long as there are any quantifiers not initially placed, one and only one of the reduction steps (i)–(vi) is possible and the result of making this reduction step is determined effectively and uniquely. It remains only to show that the process must terminate in a finite number of steps.

Let a particular occurrence of one of the signs ⊃ or ~ be called *external* to a particular occurrence of a quantifier if: (1) it is not in the scope of that occurrence of the quantifier, and also (2) in case of an occurrence of (∀) with an occurrence of ~ both immediately before and immediately after it (or, in other words, in case of an occurrence of (∃)) it is not one of those two occurrences of ~ (which form part of that occurrence of (∃)).

Now each reduction step either diminishes the number of occurrences of quantifiers that are not initially placed, or else, while leaving this number unchanged, diminishes the total number of occurrences of the signs ⊃ and ~ external to the first occurrence of a quantifier that is not initially placed. Since both numbers are of course finite in the given wff **A**, it follows that the process of reduction must terminate in the required wff **A**′ whose quantifiers are all initially placed.

391. Any wff **A of F^1 can be reduced to a wff **B** of F^1 in prenex normal form, by first applying the reduction process of **390 to reduce **A** to a wff **A**′ in which all quantifiers are initially placed, and then deleting all vacuous occurrences of quantifiers, (∀**a**) or (∃**a**), in **A**′

to obtain **B**. This process of reduction is effective, and the resulting wff **B** in prenex normal form is uniquely determined when **A** is given.

Proof. Obvious as a corollary of **390.

Definition. The wff **B** which is obtained from the wff **A** by the reduction process of **391 is called the prenex normal form of **A**.

*392. If **B** is the prenex normal form of **A**, $\vdash \mathbf{A} \equiv \mathbf{B}$.

Proof. At each step of the reduction process of **390, the wff obtained is equivalent to the previous wff, in the sense that the material equivalence of the two wffs is a theorem of F^1. Thus by the transitive law of equivalence, $\vdash \mathbf{A} \equiv \mathbf{A}'$. Again, in the reduction of $\mathbf{A}'$ to **B** by deleting vacuous occurrences of quantifiers, the deletions may be performed one by one, and at each step the wff obtained is equivalent to the previous wff. Therefore, by the transitive law of equivalence, $\vdash \mathbf{A} \equiv \mathbf{B}$.

The derived rule *341 here has to be used at each step, in establishing the equivalence of the wff obtained to the previous one. Given *341, the required equivalence follows, in the case of reduction step (i), by *377; in the case of reduction step (ii), by *338; in the case of (iii), by *350 and *383; in the case of (iv), by *350 and *380; in the case of (v), by *350 and *335; in the case of (vi), by *350 and *382; in the case of deletion of a vacuous occurrence of a quantifier, by *337 or *376.

Although *the prenex normal form of* a wff is unique, as we have here defined it, and although a wff is always equivalent to its prenex normal form (in the sense that the equivalence is a theorem of F^1), it is not true in general that, if two wffs are equivalent to each other (in this sense), they therefore have the same prenex normal form. Counterexamples are obvious, and are left to the reader.

EXERCISES 39

39.0. Find the prenex normal form of each of the following:

(1) $$\sim\sim(y)F(y, z) \supset_z \sim\sim(\exists x)G(x, y, z)$$

(Answer: $(z)(\exists x_1)(\exists x) \centerdot \sim\sim F(x_1, z) \supset G(x, y, z)$.)

(2) $$(y)F(x, y) \supset (y)F(x, x)$$

(3) $$F(y) \supset_y \sim \centerdot (x)G(x, y) \supset (y)\sim F(y)$$

(4) $$F(x) \vee \sim(y)F(y)$$

(5) $$(\exists x)F(x, y, z) \equiv (y)G(x, y, z).$$

39.1. Show that the matrix of the prenex normal form of a wff differs from the associated quantifier-free formula (in the sense of §32) at most by changes in the individual variables at certain places, and deletions of double negations $\sim\sim$.

39.2. A formulation of the functional calculus of first order is to have negation, conjunction, and disjunction as its primitive sentence connectives, and the universal and existential quantifiers as its primitive quantifiers. Otherwise the primitive symbols are to be the same as for F^1. (1) Write the formation rules for this formulation of the functional calculus of first order. (2) Given that the rules of inference are generalization (*301) and *modus ponens* (in the form, from $\sim\mathbf{A} \vee \mathbf{B}$ and $\mathbf{A}$ to infer $\mathbf{B}$), supply suitable axiom schemata—making them as few and as simple as feasible—and then demonstrate equivalence of the system to F^1 in an appropriate sense.

39.3. Extend the definition of *full disjunctive normal form* (see exercise 24.9) to quantifier-free formulas of the system introduced in 39.2. Show for this system that, if $\mathbf{A}$ and $\mathbf{B}$ are quantifier-free, then $\vdash \mathbf{A} \equiv \mathbf{B}$ if and only if $\mathbf{A}$ and $\mathbf{B}$ either have the same full disjunctive normal form or both have no full disjunctive normal form.

39.4. Define *prenex normal form* in an appropriate way for the system of exercise 39.2, and demonstrate analogues of **390, **391, *392.

39.5. In the system of exercise 39.2, if the propositional variables are included among the primitive symbols, let a wff be said to be in *prenex-disjunctive normal form* if either it is $p \sim p$ or: (I) it is in prenex normal form, and (II) the variables in its prefix are, in order of their occurrence in the prefix, and for some n, the first n individual variables in alphabetic order, and (III) its matrix is in full disjunctive normal form. (1) Establish metatheorems about reduction to prenex-disjunctive normal form, analogous to those of 39.4 about reduction to prenex normal form. (2) Answer the question whether it is true in general that, if $\vdash \mathbf{A} \equiv \mathbf{B}$, then $\mathbf{A}$ and $\mathbf{B}$ have the same prenex-disjunctive normal form.

39.6. In the case of the singulary functional calculus of first order, with primitive basis as given in §30 (i.e., in the formulation $F^{1,1}$), the process of bringing quantifiers forward by means of the reduction steps (i)–(vi) of this section can be in a certain sense reversed. The following reduction steps are used: (a) to delete a vacuous occurrence of a quantifier $(\exists\mathbf{a})$ or $(\forall\mathbf{a})$; (b) to replace a wf part $(\exists\mathbf{a})\sim\mathbf{C}$ by $\sim(\forall\mathbf{a})\mathbf{C}$; (c) to replace a wf part $(\forall\mathbf{a})\sim\mathbf{C}$ by $\sim(\exists\mathbf{a})\mathbf{C}$, if it is not immediately preceded by $\sim$; (d) to replace a wf part $(\exists\mathbf{a})[\mathbf{C} \supset \mathbf{D}]$ by $[(\forall\mathbf{a})\mathbf{C} \supset (\exists\mathbf{a})\mathbf{D}]$; (e) to replace a wf part $(\forall\mathbf{a})[\mathbf{C} \supset \mathbf{D}]$ by $[(\exists\mathbf{a})\mathbf{C} \supset \mathbf{D}]$, if $\mathbf{a}$ is not free in $\mathbf{D}$; (f) to replace a wf part $(\forall\mathbf{a})[\mathbf{C} \supset \mathbf{D}]$

by $[\mathbf{C} \supset (\forall \mathbf{a})\mathbf{D}]$, if $\mathbf{a}$ is not free in $\mathbf{C}$; (g) to replace a wf part $(\forall \mathbf{a})[[\mathbf{C}_1 \supset \mathbf{C}_2] \supset \mathbf{D}]$ by $\sim[(\forall \mathbf{a})[\sim\mathbf{C}_1 \supset \mathbf{D}] \supset \sim(\forall \mathbf{a})[\mathbf{C}_2 \supset \mathbf{D}]]$. By a series of applications of these reduction steps, together with steps which consist either in a transformation of a wf part by means of propositional calculus (and *342) or in an alphabetic change of bound variable, show that every wff $\mathbf{A}$ of the singulary functional calculus of first order, $F^{1,1}$, can be reduced to a wff $\mathbf{B}$ such that: (1) the only occurrences of quantifiers in $\mathbf{B}$ are in wf parts of the form $(\forall x) \,.\, \mathbf{D}_1 \supset \,.\, \mathbf{D}_2 \supset \,.\, \ldots \, \mathbf{D}_{n-1} \supset \mathbf{D}_n$, where n may be 1 or greater,[360] and where each $\mathbf{D}_i$ separately is either $\mathbf{f}_i(x)$ or $\sim\mathbf{f}_i(x)$, the functional variables $\mathbf{f}_1, \mathbf{f}_2, \ldots, \mathbf{f}_n$ being all different in the case of any one particular such wf part of $\mathbf{B}$;[361] and (2) $\vdash \mathbf{A} \equiv \mathbf{B}$.[362]

39.7. Apply the reduction process of the preceding exercises to the following wffs of $F^{1,1}$:

(1) $$(\exists x)(y) \,.\, F(x) \supset G(x) \supset \,.\, G(x) \supset H(x) \supset \,.\, F(y) \supset H(y)$$

(2) $$(\exists x)(y)(z) \,.\, F(x) \supset G(y) \supset H(x) \supset \,.\, F(z) \supset G(x) \supset H(z)$$

(3) $$F(x) \supset_x \,.\, F(y) \supset_y [G(x) \supset G(y)] \vee (z)F(z).$$

39.8. For a formulation of the singulary functional calculus of first order with primitive basis as in 39.2, supply the analogue of the reduction process and the metatheorem of 39.6. (In order to simplify the statement of the reduction process, make use of the full disjunctive normal form—cf. 39.3—and its dual, the *full conjunctive normal form.*)

39.9. For a formulation of the functional calculus of first order, let the primitive symbols be as described in exercises 30.6. And let there be two rules of inference, as follows ($\mathbf{a}$ and $\mathbf{b}$ being individual variables): *from* $\mathbf{A} \,|_\mathbf{a} \,.\, \mathbf{B} \,|_\mathbf{b}\, \mathbf{C}$ *and* $\mathbf{A}$ *to infer* $\mathbf{C}$, *if* $\mathbf{b}$ *is not free in* $\mathbf{C}$; *from* $\mathbf{A} \,|_\mathbf{a}\, \mathbf{B}$ *to infer* $\check{S}^{\mathbf{a}}_{\mathbf{b}}\mathbf{A}| \,|_\mathbf{b}\, \check{S}^{\mathbf{a}}_{\mathbf{b}}\mathbf{B}|$, *if no free occurrence of* $\mathbf{a}$ *in* $\mathbf{A}$ *or* $\mathbf{B}$ *is in a well-formed part of the form* $\mathbf{C} \,|_\mathbf{b}\, \mathbf{D}$. Find axiom schemata (seek to make them as few and as simple as possible) such that the system becomes equivalent to F^1 in an appropriate sense, and carry the development far enough to establish this equivalence. (Use may be made of results previously obtained regarding the propositional calculus, including those of exercises 25.)

39.10. Establish the equivalence to F^1, in an appropriate sense, of the

---

[360]In case n is 1, the wf part of $\mathbf{B}$ in question is simply $(\forall x)\mathbf{D}_1$, i.e., either $(\forall x)\mathbf{f}_1(x)$ or $(\forall x)\sim\mathbf{f}_1(x)$.

[361]But the same functional variable or some of the same functional variables may occur in two or more different such wf parts of $\mathbf{B}$.

[362]The result of this exercise is due to Heinrich Behmann in a paper in the *Mathematische Annalen*, vol. 86 (1922), pp. 163–229 (see especially pp. 190–191).

following system F^1_{gb}. The primitive symbols are the same as those of F^1 with the two additional primitive symbols, $\exists$ and $\rightarrow$. The wffs are of two kinds, *terms* and *sentences*. Namely, the terms are given by the same formation rules as those of F^1, with "wff" replaced by "term" (and "wf" by "a term") throughout, together with one additional rule: if $\mathbf{\Gamma}$ is a term and $\mathbf{a}$ is an individual variable, then $(\exists\mathbf{a})\mathbf{\Gamma}$ is a term. The sentences are all formulas $\mathbf{\Gamma}_1, \mathbf{\Gamma}_2, \ldots, \mathbf{\Gamma}_n \rightarrow \mathbf{\Delta}$; where n is any natural number (not excluding 0), and $\mathbf{\Gamma}_1$ and $\mathbf{\Gamma}_2$ . . . and $\mathbf{\Gamma}_n$ and $\mathbf{\Delta}$ are terms. Thus no wff contains more than one occurrence of $\rightarrow$, and a wff is a sentence or a term according as it does or does not contain an occurrence of $\rightarrow$. In a term, an occurrence of a variable $\mathbf{a}$ is bound if it is an occurrence in a wf part of either of the forms $(\forall\mathbf{a})\mathbf{B}$ or $(\exists\mathbf{a})\mathbf{B}$; otherwise free. In a sentence all occurrences of variables are bound. There is one axiom schema, namely $\mathbf{A} \rightarrow \mathbf{A}$, where $\mathbf{A}$ is a term. And the rules of inference are the eleven following, where $\mathbf{A}_0, \mathbf{A}_1, \mathbf{A}_2, \ldots, \mathbf{A}_n, \mathbf{A}, \mathbf{B}, \mathbf{C}$ are terms, and $\mathbf{a}$ is an individual variable, and $\mathbf{b}$ is an individual variable or an individual constant: (I) from $\mathbf{A}_1, \mathbf{A}_2, \ldots, \mathbf{A}_n \rightarrow \mathbf{B}$ to infer $\mathbf{A}_0, \mathbf{A}_1, \mathbf{A}_2, \ldots, \mathbf{A}_n \rightarrow \mathbf{B}$;[363] (II) from $\mathbf{A}_1, \mathbf{A}_2, \ldots, \mathbf{A}_{k-1}, \mathbf{A}_k, \mathbf{A}_{k+1}, \mathbf{A}_{k+2}, \ldots, \mathbf{A}_n \rightarrow \mathbf{B}$ to infer $\mathbf{A}_1, \mathbf{A}_2, \ldots, \mathbf{A}_{k-1}, \mathbf{A}_{k+1}, \mathbf{A}_k, \mathbf{A}_{k+2}, \ldots, \mathbf{A}_n \rightarrow \mathbf{B}$; (III) from $\mathbf{A}_1, \mathbf{A}_1, \mathbf{A}_2, \ldots, \mathbf{A}_n \rightarrow \mathbf{B}$ to infer $\mathbf{A}_1, \mathbf{A}_2, \ldots, \mathbf{A}_n \rightarrow \mathbf{B}$; (IV) from $\mathbf{A}_1, \mathbf{A}_2, \ldots, \mathbf{A}_n \rightarrow \mathbf{B}$ to infer $\mathbf{A}_1, \mathbf{A}_2, \ldots, \mathbf{A}_{n-1} \rightarrow \mathbf{A}_n \supset \mathbf{B}$; (V) from $\mathbf{A}_1, \mathbf{A}_2, \ldots, \mathbf{A}_n \rightarrow \mathbf{A} \supset \mathbf{B}$ and $\mathbf{A}_1, \mathbf{A}_2, \ldots, \mathbf{A}_n \rightarrow \mathbf{A}$ to infer $\mathbf{A}_1, \mathbf{A}_2, \ldots, \mathbf{A}_n \rightarrow \mathbf{B}$;[364] (VI) from $\mathbf{A}_1, \mathbf{A}_2, \ldots, \mathbf{A}_n \rightarrow \mathbf{B}$ and $\mathbf{A}_1, \mathbf{A}_2, \ldots, \mathbf{A}_n \rightarrow \sim\mathbf{B}$ to infer $\mathbf{A}_1, \mathbf{A}_2, \ldots, \mathbf{A}_{n-1} \rightarrow \sim\mathbf{A}_n$; (VII) from $\mathbf{A}_1, \mathbf{A}_2, \ldots, \mathbf{A}_n \rightarrow \sim\sim\mathbf{B}$ to infer $\mathbf{A}_1, \mathbf{A}_2, \ldots, \mathbf{A}_n \rightarrow \mathbf{B}$;[364] (VIII) from $\mathbf{A}_1, \mathbf{A}_2, \ldots, \mathbf{A}_n \rightarrow \mathbf{A}$ to infer $\mathbf{A}_1, \mathbf{A}_2, \ldots, \mathbf{A}_n \rightarrow (\forall\mathbf{a})\mathbf{A}$, if $\mathbf{a}$ is not free in $\mathbf{A}_1, \mathbf{A}_2, \ldots, \mathbf{A}_n$;[364] (IX) from $\mathbf{A}_1, \mathbf{A}_2, \ldots, \mathbf{A}_n \rightarrow (\forall\mathbf{a})\mathbf{A}$ to infer $\mathbf{A}_1, \mathbf{A}_2, \ldots, \mathbf{A}_n \rightarrow \check{S}^{\mathbf{a}}_{\mathbf{b}}\mathbf{A}|$, if no free occurrence of $\mathbf{a}$ in $\mathbf{A}$ is in a wf part of $\mathbf{A}$ of either of the forms $(\forall\mathbf{b})\mathbf{C}$ or $(\exists\mathbf{b})\mathbf{C}$;[364] (X) from $\mathbf{A}_1, \mathbf{A}_2, \ldots, \mathbf{A}_n \rightarrow \check{S}^{\mathbf{a}}_{\mathbf{b}}\mathbf{A}|$ to infer $\mathbf{A}_1, \mathbf{A}_2, \ldots, \mathbf{A}_n \rightarrow (\exists\mathbf{a})\mathbf{A}$, if no free occurrence of $\mathbf{a}$ in $\mathbf{A}$ is in a wf part of $\mathbf{A}$ of either of the forms $(\forall\mathbf{b})\mathbf{C}$ or $(\exists\mathbf{b})\mathbf{C}$;[364] (XI) from $\mathbf{A}_1, \mathbf{A}_2, \ldots, \mathbf{A}_n \rightarrow \mathbf{B}$ to infer $\mathbf{A}_1, \mathbf{A}_2, \ldots, \mathbf{A}_{n-1}, (\exists\mathbf{a})\mathbf{A}_n \rightarrow \mathbf{B}$, if $\mathbf{a}$ is not free in $\mathbf{A}_1, \mathbf{A}_2, \ldots, \mathbf{A}_{n-1}, \mathbf{B}$.[365] (As a first step toward establishing the desired equivalence show that, if $\mathbf{B}$ is any term of F^1_{gb} and $\mathbf{B}'$ is the corresponding wff of F^1, obtained from $\mathbf{B}$ by replacing $(\exists\ )$ everywhere by $\sim(\forall\ )\sim$, then $\rightarrow \mathbf{B}$ is a theorem of F^1_{gb} if and only if $\mathbf{B}'$ is a theorem of F^1.)

39.11. Also establish the equivalence to F^1, in an appropriate sense, of

[363] The special case that $n = 0$ is not excluded, i.e., from $\rightarrow \mathbf{B}$ to infer $\mathbf{A} \rightarrow \mathbf{B}$.

[364] Again, as a special case, n may be 0.

[365] The system of this exercise is intermediate between Gentzen's calculi *NK* and *LK* (see his paper cited in footnotes 294–295), with a modification due to Bernays (*Logical Calculus*, 1935–1936).

the following system F^1_g.[366] The primitive symbols and the terms are the same as those of F^1_{gb} (see the preceding exercise). The sentences ("Sequenzen") are all formulas $\mathbf{\Gamma}_1, \mathbf{\Gamma}_2, \ldots, \mathbf{\Gamma}_n \rightarrow \mathbf{\Delta}_1, \mathbf{\Delta}_2, \ldots, \mathbf{\Delta}_m$; where m and n are natural numbers (not excluding 0), and $\mathbf{\Gamma}_1$ and $\mathbf{\Gamma}_2 \ldots$ and $\mathbf{\Gamma}_n$ and $\mathbf{\Delta}_1$ and $\mathbf{\Delta}_2 \ldots$ and $\mathbf{\Delta}_m$ are terms.[367] Thus no wff contains more than one occurrence of $\rightarrow$, and a wff is a sentence or a term according as it does or does not contain an occurrence of $\rightarrow$. Free and bound occurrences of variables are defined as in the preceding exercise. Again there is one axiom schema, $\mathbf{A} \rightarrow \mathbf{A}$, where $\mathbf{A}$ is a term. And the rules of inference are the following, where $\mathbf{A}_0, \mathbf{A}_1, \mathbf{A}_2, \ldots, \mathbf{A}_n, \mathbf{B}_0, \mathbf{B}_1, \mathbf{B}_2, \ldots, \mathbf{B}_m, \mathbf{A}, \mathbf{B}, \mathbf{C}$ are terms, and $\mathbf{a}$ is an individual variable, and $\mathbf{b}$ is an individual variable or an individual constant:

Ia. From $\mathbf{A}_1, \mathbf{A}_2, \ldots, \mathbf{A}_n \rightarrow \mathbf{B}_1, \mathbf{B}_2, \ldots, \mathbf{B}_m$ to infer $\mathbf{A}_0, \mathbf{A}_1, \mathbf{A}_2, \ldots, \mathbf{A}_n \rightarrow \mathbf{B}_1, \mathbf{B}_2, \ldots, \mathbf{B}_m$.[368]

Ib. From $\mathbf{A}_1, \mathbf{A}_2, \ldots, \mathbf{A}_n \rightarrow \mathbf{B}_1, \mathbf{B}_2, \ldots, \mathbf{B}_m$ to infer $\mathbf{A}_1, \mathbf{A}_2, \ldots, \mathbf{A}_n \rightarrow \mathbf{B}_0, \mathbf{B}_1, \mathbf{B}_2, \ldots, \mathbf{B}_m$.[368]

IIa. From $\mathbf{A}_1, \mathbf{A}_2, \ldots, \mathbf{A}_{k-1}, \mathbf{A}_k, \mathbf{A}_{k+1}, \mathbf{A}_{k+2}, \ldots, \mathbf{A}_n \rightarrow \mathbf{B}_1, \mathbf{B}_2, \ldots, \mathbf{B}_m$ to infer $\mathbf{A}_1, \mathbf{A}_2, \ldots, \mathbf{A}_{k-1}, \mathbf{A}_{k+1}, \mathbf{A}_k, \mathbf{A}_{k+2}, \ldots, \mathbf{A}_n \rightarrow \mathbf{B}_1, \mathbf{B}_2, \ldots, \mathbf{B}_m$.[369]

IIb. From $\mathbf{A}_1, \mathbf{A}_2, \ldots, \mathbf{A}_n \rightarrow \mathbf{B}_1, \mathbf{B}_2, \ldots, \mathbf{B}_{l-1}, \mathbf{B}_l, \mathbf{B}_{l+1}, \mathbf{B}_{l+2}, \ldots, \mathbf{B}_m$ to infer $\mathbf{A}_1, \mathbf{A}_2, \ldots, \mathbf{A}_n \rightarrow \mathbf{B}_1, \mathbf{B}_2, \ldots, \mathbf{B}_{l-1}, \mathbf{B}_{l+1}, \mathbf{B}_l, \mathbf{B}_{l+2}, \ldots, \mathbf{B}_m$.[370]

IIIa. From $\mathbf{A}_1, \mathbf{A}_1, \mathbf{A}_2, \ldots, \mathbf{A}_n \rightarrow \mathbf{B}_1, \mathbf{B}_2, \ldots, \mathbf{B}_m$ to infer $\mathbf{A}_1, \mathbf{A}_2, \ldots, \mathbf{A}_n \rightarrow \mathbf{B}_1, \mathbf{B}_2, \ldots, \mathbf{B}_m$.[369]

IIIb. From $\mathbf{A}_1, \mathbf{A}_2, \ldots, \mathbf{A}_n \rightarrow \mathbf{B}_1, \mathbf{B}_1, \mathbf{B}_2, \ldots, \mathbf{B}_m$ to infer $\mathbf{A}_1, \mathbf{A}_2, \ldots, \mathbf{A}_n \rightarrow \mathbf{B}_1, \mathbf{B}_2, \ldots, \mathbf{B}_m$.[370]

IVa. From $\mathbf{A}_1, \mathbf{A}_2, \ldots, \mathbf{A}_n \rightarrow \mathbf{B}$ to infer $\mathbf{A}_1, \mathbf{A}_2, \ldots, \mathbf{A}_{n-1} \rightarrow \mathbf{A}_n \supset \mathbf{B}$.

IVb. From $\mathbf{A}_1, \mathbf{A}_2, \ldots, \mathbf{A}_n \rightarrow \mathbf{A}$ and $\mathbf{B} \rightarrow \mathbf{B}_1, \mathbf{B}_2, \ldots, \mathbf{B}_m$ to infer $\mathbf{A}_1, \mathbf{A}_2, \ldots, \mathbf{A}_n, \mathbf{A} \supset \mathbf{B} \rightarrow \mathbf{B}_1, \mathbf{B}_2, \ldots, \mathbf{B}_m$.[368]

Va. From $\mathbf{A}_1, \mathbf{A}_2, \ldots, \mathbf{A}_n \rightarrow \mathbf{B}_1, \mathbf{B}_2, \ldots, \mathbf{B}_m$ to infer $\mathbf{A}_1, \mathbf{A}_2, \ldots, \mathbf{A}_{n-1} \rightarrow {\sim}\mathbf{A}_n, \mathbf{B}_1, \mathbf{B}_2, \ldots, \mathbf{B}_m$.[369]

Vb. From $\mathbf{A}_1, \mathbf{A}_2, \ldots, \mathbf{A}_n \rightarrow \mathbf{B}_1, \mathbf{B}_2, \ldots, \mathbf{B}_m$ to infer $\mathbf{A}_1, \mathbf{A}_2, \ldots, \mathbf{A}_n, {\sim}\mathbf{B}_1 \rightarrow \mathbf{B}_2, \mathbf{B}_3, \ldots, \mathbf{B}_m$.[370]

VIa. From $\mathbf{A}_1, \mathbf{A}_2, \ldots, \mathbf{A}_n \rightarrow \mathbf{A}$ to infer $\mathbf{A}_1, \mathbf{A}_2, \ldots, \mathbf{A}_n \rightarrow (\forall \mathbf{a})\mathbf{A}$, if $\mathbf{a}$ is not free in $\mathbf{A}_1, \mathbf{A}_2, \ldots, \mathbf{A}_n$.[370]

---

[366]This is Gentzens' calculus *LK*, with some obvious minor simplifications (which were not adopted by Gentzen because he wished to maintain as close a similarity as possible between *LK* and the intuitionistic calculus *LJ*). Compare the discussion of Gentzen's methods in §29.

[367]In particular, the arrow standing alone constitutes a sentence.

[368]Here m or n or both may be 0.

[369]Here m may be 0.

[370]Here n may be 0.

VIb. From $\mathbf{B} \rightarrow \mathbf{B}_1, \mathbf{B}_2, \ldots, \mathbf{B}_m$ to infer $(\exists \mathbf{a})\mathbf{B} \rightarrow \mathbf{B}_1, \mathbf{B}_2, \ldots, \mathbf{B}_m$, if $\mathbf{a}$ is not free in $\mathbf{B}_1, \mathbf{B}_2, \ldots, \mathbf{B}_m$.[369]

VIIa. From $\check{S}^{\mathbf{a}}_{\mathbf{b}}\mathbf{A}| \rightarrow \mathbf{B}_1, \mathbf{B}_2, \ldots, \mathbf{B}_m$ to infer $(\forall \mathbf{a})\mathbf{A} \rightarrow \mathbf{B}_1, \mathbf{B}_2, \ldots, \mathbf{B}_m$, if no free occurrence of $\mathbf{a}$ in $\mathbf{A}$ is in a wf part of $\mathbf{A}$ of either of the forms $(\forall \mathbf{b})\mathbf{C}$ or $(\exists \mathbf{b})\mathbf{C}$.[369]

VIIb. From $\mathbf{A}_1, \mathbf{A}_2, \ldots, \mathbf{A}_n \rightarrow \check{S}^{\mathbf{a}}_{\mathbf{b}}\mathbf{A}|$ to infer $\mathbf{A}_1, \mathbf{A}_2, \ldots, \mathbf{A}_n \rightarrow (\exists \mathbf{a})\mathbf{A}$, if no free occurrence of $\mathbf{a}$ in $\mathbf{A}$ is in a wf part of $\mathbf{A}$ of either of the forms $(\forall \mathbf{b})\mathbf{C}$ or $(\exists \mathbf{b})\mathbf{C}$.[370]

VIII. From $\mathbf{A}_1, \mathbf{A}_2, \ldots, \mathbf{A}_k \rightarrow \mathbf{C}, \mathbf{B}_1, \mathbf{B}_2, \ldots, \mathbf{B}_l$ and $\mathbf{A}_{k+1}, \mathbf{A}_{k+2}, \ldots, \mathbf{A}_n, \mathbf{C} \rightarrow \mathbf{B}_l, \mathbf{B}_{l+1}, \ldots, \mathbf{B}_m$ to infer $\mathbf{A}_1, \mathbf{A}_2, \ldots, \mathbf{A}_n \rightarrow \mathbf{B}_1, \mathbf{B}_2, \ldots, \mathbf{B}_m$.[371]

39.12. Establish the equivalence to F^1_g of the system F^1_{gh} obtained by replacing the rule of inference VIII by the inverses of the two rules Va and Vb, and adding a rule of alphabetic change of bound variable—in the sense that the theorems of the two systems are identical.[372]

---

[371]Here we may have as special cases any or all of $k = 0$, $l = 0$, $k = n$, $l = m$.

[372]This is Gentzen's "Hauptsatz" for *LK*, as modified to conform to changes which we have made in the system. Gentzen's method is to show first how to eliminate an application of VIII from a proof, if such an application occurs only once and as the last step of the proof, and if $\mathbf{C}$ is not identical with any of the terms $\mathbf{B}_1, \mathbf{B}_2, \ldots \mathbf{B}_l, \mathbf{A}_{k+1}, \mathbf{A}_{k+2}, \ldots, \mathbf{A}_n$. This is not done in one step, but rather the application of VIII in the given proof is replaced by one or two applications of VIII which either come earlier in the proof or have a term $\mathbf{C}$ that contains a smaller total number of occurrences of the symbols $\supset$, $\sim$, $\forall$, $\exists$. Details of this, which involve a mathematical induction of rather complex form, may be found in Gentzen's original paper. But the reader who is interested in following the matter up is urged first to work this out for himself and then afterwards to look up Gentzen's treatment.

IV. The Pure Functional Calculus of First Order

40. An alternative formulation. In the case of a functional calculus of first order having a sufficient apparatus of variables, a formulation is possible, as already remarked, in which rules of substitution are used (in addition to the rules of *modus ponens* and generalization) and the axiom schemata of §30 are replaced by basic instances of them—so that the number of axioms is then finite. In this section we give such a formulation, F^{1p}_2, of the pure functional calculus of first order.

The primitive symbols are the eight improper symbols listed in §30, the individual variables, the propositional variables, and for each positive integer n the n-ary functional variables. The formation rules, 40i–v are the same as 30i–v except that the references to functional constants and individual constants in 30ii are deleted. The same abbreviations of wffs are used as described in §30, including the definition schemata D3–17. The rules of inference are the following:

***400.** From $\mathbf{A} \supset \mathbf{B}$ and $\mathbf{A}$ to infer $\mathbf{B}$. (*Rule of modus ponens.*)

***401.** From $\mathbf{A}$, if $\mathbf{a}$ is an individual variable, to infer $(\mathbf{a})\mathbf{A}$. (*Rule of generalization.*)

***402.** From $\mathbf{A}$, if $\mathbf{a}$ is an individual variable which is not free in $\mathbf{N}$ and $\mathbf{b}$ is an individual variable which does not occur in $\mathbf{N}$, if $\mathbf{B}$ results from $\mathbf{A}$ by substituting $\mathrm{S}^{\mathbf{a}}_{\mathbf{b}}\mathbf{N}|$ for a particular occurrence of $\mathbf{N}$ in $\mathbf{A}$, to infer $\mathbf{B}$. (*Rule of alphabetic change of bound variable.*)

***403.** From $\mathbf{A}$, if $\mathbf{a}$ and $\mathbf{b}$ are individual variables, if no free occurrence of $\mathbf{a}$ in $\mathbf{A}$ is in a wf part of $\mathbf{A}$ of the form $(\mathbf{b})\mathbf{C}$, to infer $\underset{\cdot}{\mathrm{S}}^{\mathbf{a}}_{\mathbf{b}}\mathbf{A}|$. (*Rule of substitution for individual variables.*)

***404$_0$.** From $\mathbf{A}$, if $\mathbf{p}$ is a propositional variable, to infer $\check{\mathrm{S}}^{\mathbf{p}}_{\mathbf{B}}\mathbf{A}|$. (*Rule of substitution for propositional variables.*)

***404$_n$.** From $\mathbf{A}$, if $\mathbf{f}$ is an n-ary functional variable and $\mathbf{x}_1, \mathbf{x}_2, \ldots, \mathbf{x}_n$ are distinct individual variables, to infer

$\check{\mathrm{S}}^{\mathbf{f}(\mathbf{x}_1, \mathbf{x}_2, \ldots, \mathbf{x}_n)}_{\mathbf{B}}\mathbf{A}|$. (*Rule of substitution for functional variables.*)

The axioms are the five following:

†405. $p \supset . q \supset p$

†406. $s \supset [p \supset q] \supset . s \supset p \supset . s \supset q$

†407. $\sim p \supset \sim q \supset_{\shortmid} . q \supset p$

†408. $p \supset_x F(x) \supset . p \supset (x)F(x)$

†409. $(x)F(x) \supset F(y)$

From results obtained in the preceding chapter (especially §35), there follows the equivalence of the systems F_2^{1p} and F^{1p} in the sense that every theorem of either system is a theorem also of the other. Hence also the derived rules of F^{1p} which were obtained in the preceding chapter may be extended at once to the system F_2^{1p}.

The developments of the following sections (§§41–47) belong to the theoretical syntax of the pure functional calculus of first order, and—except the results of §41, which concern the particular formulation F_2^{1p}—they apply to the pure functional calculus of first order indifferently in either of the formulations F^{1p} or F_2^{1p}. Many of the results can be extended to other functional calculi of first order, some even to an arbitrary functional calculus of first order in the formulation F^1. But we shall confine attention to the pure functional calculus of first order, leaving it to the reader to make such extensions of the results where obvious.

We remark that the method of the present section, for obtaining a formulation of the functional calculus of first order in which the axiom schemata of F^1 are replaced by a finite number of axioms, can be extended to any case in which functional variables of at least one type are present. Namely, the appropriate changes are made in the list of primitive symbols and in the formation rules. The rules of inference remain the same except that: (1) if individual constants are present, the appropriate changes are to be made in *403 and *404_n to allow for them (as in *351 and *352_n); and (2) if any of the rules *404_n become vacuous, they may be omitted. The five axioms remain the same as in F_2^{1p} if the required variables are present, and otherwise they receive an obvious modification.

In particular, for any positive integer m, a formulation $F_2^{1,m}$ of the m-ary functional calculus of first order may be obtained from F_2^{1p} by merely omitting from the list of primitive symbols all functional variables which are more than m-ary and omitting all the rules *404_n for which $n > m$. This formulation of the m-ary functional calculus of first order is easily seen to be equivalent (in the sense that the theorems are the same) to the m-ary functional calculus of first order in the formulation $F^{1,m}$ of Chapter III.

EXERCISES 40

40.0. Show that the theorems of the singulary functional calculus of first order are identical with those theorems of the pure functional calculus of first order in which all the functional variables are singulary.

40.1. In the system F_2^{1p} show that, without changing the class of theorems, the rule of generalization and the axioms †408 and †409 could be replaced by the two following rules of inference: from $\mathbf{A} \supset \mathbf{B}$, if $\mathbf{a}$ is an individual variable which is not free in $\mathbf{A}$, to infer $\mathbf{A} \supset (\mathbf{a})\mathbf{B}$; from $\mathbf{A} \supset (\mathbf{a})\mathbf{B}$, if $\mathbf{a}$ is not free in $\mathbf{A}$, to infer $\mathbf{A} \supset \mathbf{B}$.

40.2. For a formulation of the extended propositional calculus with primitive symbols as indicated in exercise 30.7, let the rules of inference be *modus ponens*, the rule of substitution (for propositional variables), and the two rules introduced in 40.1 as these are modified by taking $\mathbf{a}$ to be a propositional variable rather than an individual variable. And let the axioms be the same as the three axioms of P_B^I. Carry the development of the system far enough to establish a solution of its decision problem along the lines suggested in §28.[400] (Make use of the result of 18.3.)

40.3. In a partial system of extended propositional calculus, with primitive symbols as in 30.7, rules of inference *modus ponens*, generalization, and substitution, the two axioms of P^+, and axiom schemata $(\mathbf{a})\mathbf{A} \supset \mathbf{A}$, and $(\mathbf{a})[\mathbf{b} \supset \mathbf{A}] \supset . \mathbf{b} \supset (\mathbf{a})\mathbf{A}$ where $\mathbf{b}$ is not $\mathbf{a}$, show that under suitable definitions of conjunction, disjunction, equivalence, and negation the entire intuitionistic propositional calculus is contained. (See 19.6.)

41. Independence. From the equivalence of F_2^{1p} and F^{1p} it is easily seen that the rules $*404_2$, $*404_3$, . . . of F_2^{1p} are non-independent. For by means of the rules *402, *403, $*404_0$, $*404_1$ it is possible to infer an arbitrary instance of one of the five schemata *302–*306 from the corresponding one of the five axioms †405–†409.

Though not independent, the rules $*404_n$ ($n > 1$) are nevertheless in a certain sense not superfluous, since they restrict the class of sound interpretations of F_2^{1p}. Indeed an interpretation which is like the principal interpretation of F^{1p} except that functional variables with superscript greater than 1 are interpreted as functional constants (each one corresponding to a particular propositional function of individuals) is a sound interpretation of the system F^{1p}, and of the

[400]These axioms and rules of inference for the extended calculus are given in the paper of footnote 243, where they are credited to Tarski.

system obtained from F_2^{1p} by deleting the rules *404_n ($n > 1$), but is not a sound interpretation of F_2^{1p} itself.

The need for the rules *404_n ($n > 1$) may be seen from a syntactical standpoint if F_2^{1p} is thought of not as a self-sufficient system but as a system to which undefined terms and postulates are to be added in order to develop some special branch of mathematics by the formal axiomatic method (as described in the concluding paragraphs of §07). For such an added postulate may well contain, e.g., a binary functional variable in such a way that *404_2 must be used in making required inferences from it.

Except *404_n ($n > 1$), the rules and axioms of F_2^{1p} are independent. We go on to indicate briefly how this may be established.

Consider a formulation of the propositional calculus in which the rules of inference are substitution and *modus ponens*, and the five axioms are †405, †406, †407, $p \supset r \supset . p \supset r$, and $r \supset r$. Here the last two axioms are afps of †408 and †409 respectively, in the sense of §32. Hence from a given proof of any theorem **A** of F_2^{1p}, upon replacing each wff in the proof by a suitably chosen afp of it, we obtain a proof of an afp $\mathbf{A}_0$ of **A** as a theorem of this formulation of the propositional calculus. By the methods of §19 we may show, for this formulation of the propositional calculus, the independence of the rule of *modus ponens* and of each of the axioms †405, †406, †407. There follows, for F_2^{1p}, the independence of each of *400, †405, †406, †407. (Details are left to the reader.)

Consider the transformation upon the wffs of F_2^{1p} which consists in replacing all occurrences of $(\forall \mathbf{a})$ by $\sim(\forall \mathbf{a})\sim$, simultaneously for all individual variables **a**. I.e., briefly, consider the transformation which consists in replacing the universal quantifier everywhere by the existential quantifier. It may be verified that this transforms every axiom of F_2^{1p} except †409 into a theorem of F_2^{1p}; and, in an obvious sense, it transforms every primitive rule of inference of F_2^{1p} into a primitive or derived rule of F_2^{1p}. But †409 is transformed into $(\exists x)F(x) \supset F(y)$, which is not a theorem. (If $\vdash (\exists x)F(x) \supset F(y)$, then, by *330 and P, $\vdash F(x) \supset F(y)$, contrary to **324.) The independence of †409 follows.

Consider the transformation upon the wffs of F_2^{1p} which consists in replacing simultaneously every wf part of the form $(\forall \mathbf{a})\mathbf{C}$ by $(\forall \mathbf{a})\sim[\mathbf{C} \supset \mathbf{C}]$.[401] This transforms every axiom into a theorem, and every primitive rule of inference except *401 (the rule of generalization) into a primitive or derived

[401]More explicitly, by this transformation every quantifier-free formula is transformed into itself, and if **C** and **D** are transformed into **C**′ and **D**′ respectively, then $[\mathbf{C} \supset \mathbf{D}]$ is transformed into $[\mathbf{C}' \supset \mathbf{D}']$, $\sim\mathbf{C}$ is transformed into $\sim\mathbf{C}'$, and $(\forall \mathbf{a})\mathbf{C}$ is transformed into $(\forall \mathbf{a})\sim[\mathbf{C}' \supset \mathbf{C}']$.

rule. On the other hand it transforms the theorem $F(x) \supset_x F(x)$ into the non-theorem $(x){\sim} \centerdot F(x) \supset F(x) \supset \centerdot F(x) \supset F(x)$. There follows the independence of *401.

Consider the transformation upon the wffs of F_2^{1p} which consists in replacing $(\forall \mathbf{a})$ by ${\sim}(\forall \mathbf{a}){\sim}$ whenever **a** is a different variable than x. This transforms every axiom into a theorem and every primitive rule of inference except *402 (the rule of alphabetic change of bound variable) into a primitive or derived rule. It transforms the theorem $(y)F(y) \supset F(z)$ into the non-theorem $(\exists y)F(y) \supset F(z)$. There follows the independence of *402.

Consider the transformation upon the wffs of F_2^{1p} which consists in replacing every wf part of the form $(\forall \mathbf{a})\mathbf{C}$ by $\mathrm{S}_y^{\mathbf{a}}\mathbf{C}|$. (Or, as the transformation may also be described, every bound occurrence of an individual variable in the wff is replaced by the particular individual variable y, and then $(\forall y)$ is omitted wherever it occurs.) This transforms every axiom into a theorem and every primitive rule of inference except *403 (the rule of substitution for individual variables) into a primitive or derived rule. It transforms the theorem $(x)F(x) \supset F(z)$ into the non-theorem $F(y) \supset F(z)$ (cf. §32). There follows the independence of *403.

Consider the transformation upon the wffs of F_2^{1p} which consists in omitting ${\sim}$ wherever it occurs and at the same time replacing p everywhere by $[p \supset p]$. This transforms every axiom into a theorem and every primitive rule of inference except *404_0 (the rule of substitution for propositional variables) into a primitive or derived rule. It transforms the theorem ${\sim}q \supset q \supset q$ into the non-theorem $q \supset q \supset q$ (**320). There follows the independence of *404_0.

Consider the transformation upon the wffs of F_2^{1p} which consists in replacing $F(\mathbf{a})$ throughout by $[F(\mathbf{a}) \supset F(\mathbf{a})]$ (for every individual variable **a**, but only for the one functional variable F^1) and at the same time replacing $(\forall \mathbf{a})$ throughout by ${\sim}(\forall \mathbf{a}){\sim}$ (for every individual variable **a**). This transforms every axiom into a theorem and every primitive rule of inference except *404_1 (the rule of substitution for singulary functional variables) into a primitive or derived rule. It transforms the theorem $(x)G(x) \supset G(y)$ into the non-theorem $(\exists x)G(x) \supset G(y)$. There follows the independence of *404_1.

Finally, to establish the independence of †408 we use a more elaborate transformation upon the wffs of F_2^{1p}, which is described in steps as follows. First replace every individual variable by the individual variable next following it in alphabetic order, i.e., replace simultaneously x by y, y by z, z by x_1, and so on. Then change simultaneously every propositional variable to a singulary functional variable and every n-ary functional variable to an $(n + 1)$-ary functional variable in the following way: a propositional variable $\mathbf{a}$ is to be replaced by $\mathbf{b}(x)$, where $\mathbf{b}$ is the singulary functional variable having the same alphabetic position as $\mathbf{a}$ (i.e., if $\mathbf{a}$ is the ith propositional variable in alphabetic order, then $\mathbf{b}$ is the ith singulary functional variable in alphabetic order); and, $\mathbf{a}$ being an n-ary functional variable, each wf part $\mathbf{a}(\mathbf{c}_1, \mathbf{c}_2, \ldots, \mathbf{c}_n)$ is to be replaced by $\mathbf{b}(\mathbf{c}_1, \mathbf{c}_2, \ldots, \mathbf{c}_n, x)$ where $\mathbf{b}$ is the $(n + 1)$-ary functional variable having the same alphabetic position as $\mathbf{a}$. Then in every wf part having the form of an implication $[\mathbf{A} \supset \mathbf{B}]$ prefix an existential quantifier $\sim(\forall x)\sim$ to the antecedent $\mathbf{A}$ and to the consequent $\mathbf{B}$ (i.e., change $[\mathbf{A} \supset \mathbf{B}]$ to $[\sim(\forall x)\sim\mathbf{A} \supset \sim(\forall x)\sim\mathbf{B}]$), and at the same time change every universal quantifier $(\forall\mathbf{a})$ to $(\forall\mathbf{a})(\forall x)$, and change $\sim$ everywhere to $(\forall x)\sim$.

The result of applying this transformation to †408 is

$$(1)\quad (\exists x)[(\exists x)F(x) \supset_{yx} (\exists x)F(y, x)] \supset (\exists x) \,\mathbf{.}\, (\exists x)F(x) \supset (\exists x)\,(y)\,(x)F(y, x),$$

which is not a theorem of F_2^{1p}. On the other hand, application of this transformation to the remaining axioms of F_2^{1p} yields, in order,

$$(2)\quad (\exists x)F(x) \supset (\exists x) \,\mathbf{.}\, (\exists x)G(x) \supset (\exists x)F(x),$$

$$(3)\quad (\exists x)[(\exists x)F_1(x) \supset (\exists x) \,\mathbf{.}\, (\exists x)F(x) \supset (\exists x)G(x)] \supset (\exists x) \,\mathbf{.}\, (\exists x)[(\exists x)F_1(x) \supset (\exists x)F(x)] \supset (\exists x) \,\mathbf{.}\, (\exists x)F_1(x) \supset (\exists x)G(x),$$

$$(4)\quad (\exists x)[(\exists x)(x)\sim F(x) \supset (\exists x)(x)\sim G(x)] \supset (\exists x) \,\mathbf{.}\, (\exists x)G(x) \supset (\exists x)F(x),$$

$$(5)\quad (\exists x)(y)(x)F(y, x) \supset (\exists x)F(z, x),$$

which are theorems of F_2^{1p}.

Moreover, by this transformation every primitive rule of inference of F_2^{1p} is transformed into a primitive or derived rule. (In order to show this in the case of *400, it is necessary to make use of the fact, which is a corollary of **320, that no theorem of F_2^{1p} fails to contain an implication sign, and hence that no theorem of F_2^{1p} is transformed into a wff containing x as a free variable.) The independence of †408 follows.

EXERCISES 41

41.0. In order to complete the proof of independence of †408 as described above, supply details of the demonstration that (1) is a non-theorem of F_2^{1p} and that (2)–(5) are theorems. (For the first part, using rules of substitution for functional variables, proceed by showing that, if (1) is a theorem, then $G(x) \supset G(y)$ is a theorem, contrary to **324.)

41.1. Following Gödel, prove the independence of †408 by means of the transformation upon wffs of F_2^{1p} which consists in replacing every wf part of the form $(\forall \mathbf{a})\mathbf{f}(\mathbf{a})$, where **a** is an individual variable and **f** is a singulary functional variable, and also every wf part of the form $(\forall \mathbf{a})\mathbf{b}$, where **a** is an individual variable and **b** is a propositional variable, by $[p \not\subset p]$.

41.2. Following the analogy of §40, and using the same rules of inference *400–*404$_n$ as in §40, reformulate the system F^{1i} of exercise 38.6 as a pure intuitionistic functional calculus of first order F_2^{1ip} with a finite number of axioms. Discuss the independence of the axioms and rules of F_2^{1ip}. (Use may be made of the results of 26.18, 36.6–38.8, and §41.)

41.3. Investigate the independence of the axioms and rules of the formulation of the extended propositional calculus which was introduced in exercise 40.2.

42. Skolem normal form. A wff[402] is said to be in *Skolem normal form* if it is in prenex normal form without free individual variables and has a prefix of the form

$$(\exists \mathbf{a}_1)(\exists \mathbf{a}_2) \ldots (\exists \mathbf{a}_m)(\mathbf{b}_1)(\mathbf{b}_2) \ldots (\mathbf{b}_n),$$

where $m \geqq 1$ and $n \geqq 0$. In other words, a wff in prenex normal form is in Skolem normal form if it has no free individual variables and its prefix contains at least one existential quantifier and every existential quantifier in the prefix precedes every universal quantifier.[403]

In order to obtain what we shall call *the Skolem normal form of* a wff **A**, we apply the following reduction procedure:

i. First reduce **A** to its prenex normal form **B** by the method of §39.

ii. If $\mathbf{c}_1$ is the first in alphabetic order of the free individual variables of **B**, prefix the universal quantifier $(\forall \mathbf{c}_1)$ to **B**. Repeat this step until **B** has been reduced to a wff $\mathbf{C}_1$ which is in prenex normal form without free individual variables. (Thus $\mathbf{C}_1$ is $(\mathbf{c}_u)(\mathbf{c}_{u-1}) \ldots (\mathbf{c}_1)\mathbf{B}$, where $\mathbf{c}_1, \mathbf{c}_2, \ldots, \mathbf{c}_u$ are the free individual variables of **B** in alphabetic order. Of course u may be 0.)

iii. If $\mathbf{C}_1$ is in Skolem normal form, let **C** be the same as $\mathbf{C}_1$.

iv. If $\mathbf{C}_1$ has a null prefix (this will be the case if $\mathbf{C}_1$ is a wff of P), let **C** be $(\exists x) \centerdot F(x) \supset F(x) \supset \mathbf{C}_1$. Then **C** is in Skolem normal form.

[402] In §§42–47, "wff" shall mean "well-formed formula of the pure functional calculus of first order in either of the formulations F^{1p} or F_2^{1p}," except where the contrary is indicated by using the explicit wording "wff of" such and such a system.

[403] When we refer to the universal quantifiers in the prefix, we mean only the universal quantifiers which are without $\sim$ before and after, i.e., we exclude those universal quantifiers which occur as parts of the existential quantifiers (D14). This remark applies here and at various places below.

v. Except in the cases iii and iv, $\mathbf{C}_1$ must have the form

$$(\exists \mathbf{a}_1)(\exists \mathbf{a}_2) \ldots (\exists \mathbf{a}_k)(\mathbf{a}_{k+1})\mathbf{N}_1,$$

where $k \geqq 0$, and $\mathbf{N}_1$ is in prenex normal form and has $\mathbf{a}_1, \mathbf{a}_2, \ldots, \mathbf{a}_{k+1}$ as its only free individual variables. Let $\mathbf{f}_1$ be the first $(k+1)$-ary functional variable in alphabetic order which does not occur in $\mathbf{C}_1$. Let $\mathbf{C}_2$ be the prenex normal form of

$$(\exists \mathbf{a}_1)(\exists \mathbf{a}_2) \ldots (\exists \mathbf{a}_k) \centerdot (\mathbf{a}_{k+1})[\mathbf{N}_1 \supset \mathbf{f}_1(\mathbf{a}_1, \mathbf{a}_2, \ldots, \mathbf{a}_{k+1})] \supset (\mathbf{a}_{k+1})\mathbf{f}_1(\mathbf{a}_1, \mathbf{a}_2, \ldots, \mathbf{a}_{k+1}).$$

Then, if $\mathbf{C}_2$ is in Skolem normal form, let $\mathbf{C}$ be the same as $\mathbf{C}_2$. Otherwise repeat the reduction. I.e., $\mathbf{C}_2$ has the form

$$(\exists \mathbf{a}_1)(\exists \mathbf{a}_2) \ldots (\exists \mathbf{a}_{k'})(\mathbf{a}_{k'+1})\mathbf{N}_2,$$

where $k' > k$, and $\mathbf{N}_2$ is in prenex normal form and has $\mathbf{a}_1, \mathbf{a}_2, \ldots, \mathbf{a}_{k'+1}$ as its only free individual variables. Let $\mathbf{f}_2$ be the first $(k'+1)$-ary functional variable in alphabetic order which does not occur in $\mathbf{C}_2$. Let $\mathbf{C}_3$ be the prenex normal form of

$$(\exists \mathbf{a}_1)(\exists \mathbf{a}_2) \ldots (\exists \mathbf{a}_{k'}) \centerdot (\mathbf{a}_{k'+1})[\mathbf{N}_2 \supset \mathbf{f}_2(\mathbf{a}_1, \mathbf{a}_2, \ldots, \mathbf{a}_{k'+1})] \supset (\mathbf{a}_{k'+1})\mathbf{f}_2(\mathbf{a}_1, \mathbf{a}_2, \ldots, \mathbf{a}_{k'+1}).$$

Then, if $\mathbf{C}_3$ is in Skolem normal form, let $\mathbf{C}$ be the same as $\mathbf{C}_3$. Otherwise repeat the reduction again, reducing $\mathbf{C}_3$ to $\mathbf{C}_4$; and so on until a wff $\mathbf{C}_{n-l+1}$ in Skolem normal form is obtained, which is then $\mathbf{C}$. We shall see that $\mathbf{C}$ is $\mathbf{C}_{n-l+1}$, where l is the number of universal quantifiers which occur at the end of the prefix of $\mathbf{C}_1$, after the last existential quantifier, and where n is the total number of universal quantifiers in the prefix of $\mathbf{C}_1$ (which is in fact the same as that in the prefix of $\mathbf{C}$).

**420. Any wff $\mathbf{A}$ can be reduced to a wff $\mathbf{C}$ in Skolem normal form, by the procedure just described in i–v. This process of reduction is effective, and the resulting wff $\mathbf{C}$ in Skolem normal form is uniquely determined when $\mathbf{A}$ is given.

Proof. In the series of reductions described in v, by which $\mathbf{C}_1$ is reduced to $\mathbf{C}_2$, $\mathbf{C}_2$ to $\mathbf{C}_3$, and so on, the effect on the prefix at each step is to change the first universal quantifier to an existential quantifier and at the same time to

---

[404]To see this it is necessary to take into account the nature of the process of reduction to prenex normal form, as defined in §39. Thus, if the prefix of $\mathbf{N}_1$ is $\mathbf{\Pi}_1\mathbf{\Pi}_2 \ldots \mathbf{\Pi}_e$, the prefix of $\mathbf{C}_2$ is

$$(\exists \mathbf{a}_1)(\exists \mathbf{a}_2) \ldots (\exists \mathbf{a}_k)(\exists \mathbf{a}_{k+1})\mathbf{\Pi}_1\mathbf{\Pi}_2 \ldots \mathbf{\Pi}_e(\mathbf{b}_{l+1})$$

by the reduction steps (iii)–(v) of §39. Here $e = m + l - k - 1$, and $\mathbf{b}_{l+1}$ is the first individual variable in alphabetic order after $\mathbf{a}_{k+1}$, which does not occur in $\mathbf{C}_1$.)

add a universal quantifier at the end of the prefix (without other change in the prefix).[404] Thus at each step the number of universal quantifiers is reduced by 1 which occur in a position preceding any existential quantifier. The series of reductions must therefore terminate in a wff **C** in Skolem normal form. The remaining part of the theorem is then obvious.

Definition. The wff **C** which is obtained from the wff **A** by the procedure described in i–v is called the Skolem normal form of **A**.

*421. If **C** is the Skolem normal form of **A**, then $\vdash$ **A** if and only if $\vdash$ **C**.

Proof. We continue to use the same notations as in the statement of i–v above.

By *392, $\vdash \mathbf{A} \equiv \mathbf{B}$. Hence by P, $\vdash$ **A** if and only if $\vdash$ **B**.

By *301 and *306—else by *401, †409, and *404_1—$\vdash$ **B** if and only if $\vdash \mathbf{C}_1$.[405]

If **C** is obtained by iii, then **C** is the same as $\mathbf{C}_1$. Hence $\vdash$ **A** if and only if $\vdash$ **C**.

In case **C** is obtained by iv, we have by P:

$$\vdash \mathbf{C}_1 \equiv \centerdot F(x) \supset F(x) \supset \mathbf{C}_1.$$

Hence by generalizing upon x and then using *388, since $\mathbf{C}_1$ is without free individual variables, we have:

$$\vdash \mathbf{C}_1 \equiv (\exists x) \centerdot F(x) \supset F(x) \supset \mathbf{C}_1.$$

I.e., $\vdash \mathbf{C}_1 \equiv \mathbf{C}$. Hence by P, $\vdash \mathbf{C}_1$ if and only if $\vdash$ **C**. Hence $\vdash$ **A** if and only if $\vdash$ **C**.

Finally we consider the case that **C** is obtained by v. We must show that $\vdash \mathbf{C}_1$ if and only if $\vdash \mathbf{C}_2$; $\vdash \mathbf{C}_2$ if and only if $\vdash \mathbf{C}_3$; . . .; $\vdash \mathbf{C}_{n-l}$ if and only if $\vdash \mathbf{C}_{n-l+1}$. Since **C** is the same as $\mathbf{C}_{n-l+1}$, it will then follow that $\vdash$ **A** if and only if $\vdash$ **C**.

We state in detail the proof that $\vdash \mathbf{C}_1$ if and only if $\vdash \mathbf{C}_2$. The proofs that $\vdash \mathbf{C}_2$ if and only if $\vdash \mathbf{C}_3$, and so on, are precisely similar—the argument may therefore be completed by mathematical induction, a step that is left to the reader.

By *306 (cf. footnote 405) and P,

$$(\mathbf{a}_{k+1})\mathbf{N}_1 \vdash \mathbf{N}_1 \supset \mathbf{f}_1(\mathbf{a}_1, \mathbf{a}_2, \ldots, \mathbf{a}_{k+1}) \supset \mathbf{f}_1(\mathbf{a}_1, \mathbf{a}_2, \ldots, \mathbf{a}_{k+1}).$$

Hence by generalization upon $\mathbf{a}_{k+1}$, and *333,

[405] The axiom schemata of F^{1p} are, with obvious modification in wording, also theorem schemata of F^{1p}_2. And we shall hereafter use them by number in this dual role (just as we use the numbered theorem schemata of F^{1p} as being at the same time theorem schemata of F^{1p}_2).

$$(\mathbf{a}_{k+1})\mathbf{N}_1 \vdash (\mathbf{a}_{k+1})[\mathbf{N}_1 \supset \mathbf{f}_1(\mathbf{a}_1, \mathbf{a}_2, \ldots, \mathbf{a}_{k+1})] \supset (\mathbf{a}_{k+1})\mathbf{f}_1(\mathbf{a}_1, \mathbf{a}_2, \ldots, \mathbf{a}_{k+1}).$$

Hence by *367,

$$(\exists \mathbf{a}_1)(\exists \mathbf{a}_2) \ldots (\exists \mathbf{a}_k)(\mathbf{a}_{k+1})\mathbf{N}_1 \vdash (\exists \mathbf{a}_1)(\exists \mathbf{a}_2) \ldots (\exists \mathbf{a}_k) \centerdot (\mathbf{a}_{k+1})[\mathbf{N}_1 \supset \mathbf{f}_1(\mathbf{a}_1, \mathbf{a}_2, \ldots, \mathbf{a}_{k+1})] \supset (\mathbf{a}_{k+1})\mathbf{f}_1(\mathbf{a}_1, \mathbf{a}_2, \ldots, \mathbf{a}_{k+1}).$$

Hence by *392 and P, $\mathbf{C}_1 \vdash \mathbf{C}_2$. Hence if $\vdash \mathbf{C}_1$ then $\vdash \mathbf{C}_2$.

Now suppose that $\vdash \mathbf{C}_2$. Then by *392 and P,

$$\vdash (\exists \mathbf{a}_1)(\exists \mathbf{a}_2) \ldots (\exists \mathbf{a}_k) \centerdot (\mathbf{a}_{k+1})[\mathbf{N}_1 \supset \mathbf{f}_1(\mathbf{a}_1, \mathbf{a}_2, \ldots, \mathbf{a}_{k+1})] \supset (\mathbf{a}_{k+1})\mathbf{f}_1(\mathbf{a}_1, \mathbf{a}_2, \ldots, \mathbf{a}_{k+1}).$$

Hence by the rule of substitution for functional variables, substituting $\mathbf{N}_1$ for $\mathbf{f}_1(\mathbf{a}_1, \mathbf{a}_2, \ldots, \mathbf{a}_{k+1})$, we have:

$$\vdash (\exists \mathbf{a}_1)(\exists \mathbf{a}_2) \ldots (\exists \mathbf{a}_k) \centerdot (\mathbf{a}_{k+1})[\mathbf{N}_1 \supset \mathbf{N}_1] \supset (\mathbf{a}_{k+1})\mathbf{N}_1.$$

Now by P and generalization, $\vdash (\mathbf{a}_{k+1})[\mathbf{N}_1 \supset \mathbf{N}_1]$. Hence by *modus ponens*,

$$(\mathbf{a}_{k+1})[\mathbf{N}_1 \supset \mathbf{N}_1] \supset (\mathbf{a}_{k+1})\mathbf{N}_1 \vdash (\mathbf{a}_{k+1})\mathbf{N}_1.$$

Hence by *367,

$$(\exists \mathbf{a}_1)(\exists \mathbf{a}_2) \ldots (\exists \mathbf{a}_k) \centerdot (\mathbf{a}_{k+1})[\mathbf{N}_1 \supset \mathbf{N}_1] \supset (\mathbf{a}_{k+1})\mathbf{N}_1 \vdash (\exists \mathbf{a}_1)(\exists \mathbf{a}_2) \ldots (\exists \mathbf{a}_k)(\mathbf{a}_{k+1})\mathbf{N}_1.$$

Therefore $\vdash (\exists \mathbf{a}_1)(\exists \mathbf{a}_2) \ldots (\exists \mathbf{a}_k)(\mathbf{a}_{k+1})\mathbf{N}_1$. I.e., $\vdash \mathbf{C}_1$.

43. Validity and satisfiability. The rules a–f, given in small type in §30 as semantical rules determining a principal interpretation of F^{1p}, may be modified or reinterpreted in such a way as to give them a purely syntactical character.

Namely, in the statement of these rules in §30 we understood the words "range" and "value" each in a (presupposed) semantical sense—so that the rules are thereby relevant to the question what we take to be intended by a person who, using F^{1p} as an actual language for purposes of communication, asserts a particular one of its wffs—or, more exactly, so that the rules constitute a proposal of a norm, an ideal demand as to what shall be intended by such a person.

But we now reintroduce the rules a–f with a new meaning, according to which we do not take the words "range" and "value" in any semantical sense, but rather, after selecting a particular non-empty class as domain of individuals, we regard the rules as constituting a definition of the words "range" and "value" (in the case of the word "value," an inductive definition). On this basis, the "range" of a variable comes to be merely a certain class which is abstractly associated with the variable by the definition; and the "value" of a wff for a given system of "values" of the variables $\mathbf{a}_1, \mathbf{a}_2, \ldots, \mathbf{a}_n$ (all of the free variables of the wff being included among $\mathbf{a}_1, \mathbf{a}_2, \ldots, \mathbf{a}_n$)

comes to be merely a certain truth-value[406] which is abstractly associated with the wff and with n ordered pairs $\langle \mathbf{a}_1, a_1\rangle, \langle \mathbf{a}_2, a_2\rangle, \ldots, \langle \mathbf{a}_n, a_n\rangle$ in which each a_i is a member of the range of $\mathbf{a}_i$. Hereafter, when the words "range" and "value" occur in a syntactical discussion, they are to be understood in this syntactical sense, the fact that a passage is not in small type being sufficient indication that the words do not have their semantical sense. (Where needed for clarity, however, we may use such more explicit phrases as "value in the syntactical sense.")

These syntactical notions of *range* and *value* may of course be used independently of any interpretation of the system F^{1p} or F^{1p}_2—thus even if the system is used purely as a formal calculus, without interpretation—or even if it is used with some interpretation quite different from the principal interpretations as given in §30.

A non-empty domain of individuals having been selected, a wff is said to be *valid in* that domain if it has the value t for all possible values of its free variables, *satisfiable in* that domain if it has the value t for at least one system of possible values of its free variables. (Here, by a "possible" value of a variable is meant merely a value that belongs to the range of the variable according to rules a, b_n.)

A wff is said to be *valid* if it is valid in every non-empty domain, *satisfiable* if it is satisfiable in some non-empty domain.[407]

By the *universal closure* of a wff $\mathbf{B}$ we shall mean the wff $(\mathbf{c}_u)(\mathbf{c}_{u-1}) \ldots (\mathbf{c}_1)\mathbf{B}$, where $\mathbf{c}_1, \mathbf{c}_2, \ldots, \mathbf{c}_u$ are the free individual variables of $\mathbf{B}$ in alphabetic order. Similarly, the *existential closure* of $\mathbf{B}$ is the wff $(\exists\mathbf{c}_u)(\exists\mathbf{c}_{u-1}) \ldots$

---

[406]Observe that this reference in the definition to truth-values does not of itself render the definition semantical. Nevertheless, if preferred, any two other things may be used here instead of the two truth-values. For example, in the syntactical definition of "value" we might use the numbers 0 and 1 in place of the truth-values, truth and falsehood respectively, and then define a wff to be "valid" in a given domain if it has the value 0 for all possible values of its free variables.

[407]At this point §§07–09 of the introduction should be reread, especially the discussion in §09 of Tarski's syntactical definition of truth, and footnotes 142, 143. That we have given here a syntactical definition of validity rather than of truth is just because the pure functional calculus of first order has no wffs without free variables.

The notions of validity and of satisfiability may also be regarded as analogues, for the functional calculus of first order, of the notions of being a tautology and of not being a contradiction in the propositional calculus. Indeed, in the special case of a wff of the pure functional calculus of first order which is at the same time a wff of the propositional calculus, the former notions immediately reduce to the latter. And it is obvious that the discussion in the present section, regarding the distinction between the syntactical and the semantical notions of value, and the corresponding discussion at the beginning of §15 are closely parallel. But there is the important difference that an effective test was given for recognizing a wff of the propositional calculus (say of P_1, or of P_2) as being or not being a tautology, whereas no effective test is possible for recognizing a wff of the pure functional calculus of first order as being or not being valid.

$(\exists \mathbf{c}_1)\mathbf{B}$, where $\mathbf{c}_1, \mathbf{c}_2, \ldots, \mathbf{c}_u$ are the free individual variables of **B** in alphabetic order.

****430.** A wff **A** is valid in a given non-empty domain if and only if its negation $\sim\mathbf{A}$ is not satisfiable in that domain. A wff **A** is valid if and only if its negation $\sim\mathbf{A}$ is not satisfiable.

Proof. This follows at once by rule d, or, more correctly, by the clause corresponding to rule d in the definition of "value" in the syntactical sense.

****431.** A wff is satisfiable in a given non-empty domain if and only if its negation is not valid in that domain. A wff is satisfiable if and only if its negation is not valid.

Proof. Again this follows at once by rule d.

****432.** A wff is valid in a given non-empty domain if and only if its universal closure is valid in that domain. A wff is valid if and only if its universal closure is valid.

Proof. By rule f.

****433.** A wff is satisfiable in a given non-empty domain if and only if its existential closure is satisfiable in that domain. A wff is satisfiable if and only if its existential closure is satisfiable.

Proof. By rules d and f. For from these two rules together it follows that, for a given system of values of the free variables of $(\exists\mathbf{a})\mathbf{A}$, the value of $(\exists\mathbf{a})\mathbf{A}$ is t if the value of **A** is t for at least one value of **a**, and the value of $(\exists\mathbf{a})\mathbf{A}$ is f if the value of **A** is f for every value of **a**.

****434.** Every theorem is valid.[408]

Proof. Using either of the formulations F^{1p} and F^{1p}_2, we may show that all the axioms are valid and all the rules of inference preserve validity. Details are left to the reader.

****435.** A wff is valid in a given non-empty domain if and only if its prenex normal form is valid in that domain. A wff is valid if and only if its prenex normal form is valid.

---

[408]This metatheorem may be regarded as the analogue (for the pure functional calculus of first order) of ****150** or ****235** (for the propositional calculus). And in fact the method of proof is the same. Notice that ****324**, and hence ****323**, could now be proved as corollaries of ****434**. But by the proofs in §32 these two metatheorems were established on a much weaker basis than that required for ****434**.

Proof. If **B** is the prenex normal form of **A**, we have by *392 that ⊢ **A** ≡ **B**. Hence by **434, **A** ≡ **B** is valid. Hence by rules d and e, **A** and **B** have the same value for every system of values of their free variables. From this the metatheorem follows by the definition of validity.

**436. A wff is satisfiable in a given non-empty domain if and only if its prenex normal form is satisfiable in that domain. A wff is satisfiable if and only if its prenex normal form is satisfiable.

Proof. As in the previous proof, if **B** is the prenex normal form of **A**, then **A** and **B** have the same value for every system of values of their free variables. From this the metatheorem follows by the definition of satisfiability.

**437. A wff is valid in a given non-empty domain if and only if its Skolem normal form is valid in that domain. A wff is valid if and only if its Skolem normal form is valid.

Proof. The proof of this parallels exactly the proof of *421, except that wherever that proof makes use of a theorem the present proof must instead make use of the fact that that theorem is valid, and wherever that proof makes use of a rule of inference the present proof must instead make use of the fact that that rule of inference preserves validity (in an arbitrary non-empty domain).

A wff is said to be in the *Skolem normal form for satisfiability* if it is in prenex normal form without free individual variables and has a prefix of the form

$$(\mathbf{a}_1)(\mathbf{a}_2)\ldots(\mathbf{a}_m)(\exists\mathbf{b}_1)(\exists\mathbf{b}_2)\ldots(\exists\mathbf{b}_n),$$

where $m \geqq 1$, $n \geqq 0$.

Given any wff **A**, we may find the Skolem normal form of ~**A**. This will be a wff

$$(\exists\mathbf{a}_1)(\exists\mathbf{a}_2)\ldots(\exists\mathbf{a}_m)(\mathbf{b}_1)(\mathbf{b}_2)\ldots(\mathbf{b}_n)\mathbf{M}$$

in which $m \geqq 1$, $n \geqq 0$, and **M** is quantifier-free. Then the prenex normal form of the negation of this wff will be a wff

$$(\mathbf{a}_1)(\mathbf{a}_2)\ldots(\mathbf{a}_m)(\exists\mathbf{b}_1)(\exists\mathbf{b}_2)\ldots(\exists\mathbf{b}_n)\mathbf{M}',$$

where **M′** is obtained from **M** by either deleting or inserting an initial negation sign ~. This last wff, the prenex normal form of the negation of the Skolem normal form of the negation of **A**, is in Skolem normal form for satisfiability; we shall call it the *Skolem normal form of* **A** *for satisfiability*.

****438.** A wff is satisfiable in a given non-empty domain if and only if its Skolem normal form for satisfiability is satisfiable in that domain. A wff is satisfiable if and only if its Skolem normal form for satisfiability is satisfiable.

Proof. By **431, **437, **430, **436.

****439.** If a wff is valid in a given non-empty domain, it is valid in any non-empty domain having the same or a smaller number of individuals. If a wff is satisfiable in a given non-empty domain, it is satisfiable in any domain having the same or a larger number of individuals.

Proof. Suppose that **A** is satisfiable in the non-empty domain $\mathfrak{J}$, and let $\mathfrak{K}$ be a domain having the same or a larger number of individuals. Then a one-to-one correspondence can be found between $\mathfrak{J}$ and some part $\mathfrak{K}^\circ$ of the domain $\mathfrak{K}$ (where $\mathfrak{K}^\circ$ may coincide with $\mathfrak{K}$ or may be a proper part of $\mathfrak{K}$). If i is any individual in $\mathfrak{J}$, let i' be the corresponding individual in $\mathfrak{K}^\circ$ under this one-to-one correspondence. Also select a particular individual i_0 in $\mathfrak{J}$. If k is any individual in $\mathfrak{K}^\circ$, let $k^{\backprime}$ be the corresponding individual in $\mathfrak{J}$ under the foregoing one-to-one correspondence; and if k is any individual which is in $\mathfrak{K}$ but not in $\mathfrak{K}^\circ$, let $k^{\backprime}$ be i_0. If Φ is an m-ary propositional function *over* $\mathfrak{J}$, i.e., a propositional function whose range is the ordered m-tuples of individuals of $\mathfrak{J}$, let Φ' be an m-ary propositional function over $\mathfrak{K}$, determined by the rule that $\Phi'(k_1, k_2, \ldots, k_m)$ is $\Phi(k^{\backprime}_1, k^{\backprime}_2, \ldots, k^{\backprime}_m)$.

Let $\mathbf{a}_1, \mathbf{a}_2, \ldots, \mathbf{a}_n$ be the complete list of free variables of **A**. Since **A** is satisfiable in $\mathfrak{J}$, there is a system of values $a_1, a_2, \ldots, a_n$ of $\mathbf{a}_1, \mathbf{a}_2, \ldots, \mathbf{a}_n$ for which the value of **A** is truth, each a_i being an individual in $\mathfrak{J}$ or a propositional function over $\mathfrak{J}$. Then $a'_1, a'_2, \ldots, a'_n$ is a system of values of $\mathbf{a}_1, \mathbf{a}_2, \ldots, \mathbf{a}_n$ for which the value of **A** is truth, each a'_i being an individual in $\mathfrak{K}$ or a propositional function over $\mathfrak{K}$. Thus **A** is satisfiable in $\mathfrak{K}$.

This completes the proof of the second part of the metatheorem. From this the first part then follows by **430.

EXERCISES 43

43.0. Let **A** have no free individual variables and let **C** be the Skolem normal form of **A**. Show by an example that it is not in general true that $\vdash \mathbf{A} \equiv \mathbf{C}$. Is it always true that $\vdash \mathbf{A} \supset \mathbf{C}$? That $\vdash \mathbf{C} \supset \mathbf{A}$?

43.1. Find the Skolem normal form of each of the following:

(1) $$F(x) \supset F(x)$$

(2) $$(x)F(x) \supset (\exists x)F(x)$$

(3) $$p \supset (\exists x)p$$

(4) $$G(x) \supset_x H(x) \supset . (\exists x)G(x) \supset (\exists x)H(x)$$

(5) $$(\exists x)(\exists y)F(x, y, z) \supset (\exists y)(\exists x)F(x, y, z)$$

43.2. As was done in the text for F^{1p} and rules a–f, let the rules α–ζ of §30 be reinterpreted as syntactical definitions of "range" and "value" for the case of the system F^{1h}. Let a wff of F^{1h} be called *valid* if it has the value truth for all possible values of its free variables; and let a wff of F^{1h} without free variables be called *true* if it has the value truth. Also, given a wff **A** of F^{1h} in which the distinct individual variables (bound and free) that occur are $\mathbf{a}_1, \mathbf{a}_2, \ldots, \mathbf{a}_n$, let the *characteristic function* of any wf part **B** of **A** be defined by induction as follows: The characteristic function of $\Sigma(\mathbf{a}_i, \mathbf{a}_j, \mathbf{a}_k)$ is the n-ary function of natural numbers whose value for arguments $a_1, a_2, \ldots, a_n$ is 0 if $a_i + a_j = a_k$, and 1 in the contrary case; the characteristic function of $\Pi(\mathbf{a}_i, \mathbf{a}_j, \mathbf{a}_k)$ is the n-ary function of natural numbers whose value for arguments $a_1, a_2, \ldots, a_n$ is 0 if $a_i a_j = a_k$, and 1 in the contrary case; if the characteristic function of $\mathbf{B}_1$ is f_1, the characteristic function of $\sim\mathbf{B}_1$ is the n-ary function of natural numbers whose value for arguments $a_1, a_2, \ldots, a_n$ is $1 - f_1(a_1, a_2, \ldots, a_n)$; if the characteristic functions of $\mathbf{B}_1$ and $\mathbf{B}_2$ are f_1 and f_2 respectively, the characteristic function of $[\mathbf{B}_1 \supset \mathbf{B}_2]$ is the n-ary function of natural numbers whose value for arguments $a_1, a_2, \ldots, a_n$ is $(1 - f_1(a_1, a_2, \ldots, a_n))f_2(a_1, a_2, \ldots, a_n)$; if the characteristic function of $\mathbf{B}_1$ is f_1, the characteristic function of $(\forall \mathbf{a}_i)\mathbf{B}_1$ is the n-ary function of natural numbers whose value for arguments $a_1, a_2, \ldots, a_n$ is

$$1 - \prod_{a_i} (1 - f_1(a_1, a_2, \ldots, a_n)),$$

the product being taken over all natural numbers a_i, with the convention that the product is 1 if all the factors are 1, 0 if any factor is 0. Prove that a wff **A** of F^{1h} is *valid* (or in case **A** has no free variables, *true*) if and only if the characteristic function of **A** as a wf part of **A** vanishes identically, i.e., has the value 0 for all possible arguments.

43.3. Show that the wff

$$(\exists x)(y) . F(x, y) \supset . \sim F(y, x) \supset . F(x, x) \equiv F(y, y)$$

is valid in any domain consisting of not more than three individuals but is not valid in a domain of four individuals.

43.4. Prove **324 by showing that, if a quantifier-free formula is not a substitution instance of a tautology, there is a finite non-empty domain in which it is non-valid.

43.5. Show that the following wffs are valid in every non-empty finite domain but not valid in an infinite domain:[409]

$$(1)\qquad (\exists x)(y)(\exists z) \,.\, F(y, z) \supset F(x, z) \supset \,.\, F(x, x) \supset F(y, x)$$

$$(2)\quad (x_1)(x_2)(x_3)[F(x_1, x_1) \,.\, F(x_1, x_3) \supset F(x_1, x_2) \vee F(x_2, x_3)] \supset (\exists y)(z)F(y,z)$$

43.6. (1) Without relying on or presupposing the reduction of a wff to Skolem normal form by the method of §42, make a direct statement of a process for reducing a wff to its *Skolem normal form for satisfiability*. (2) Apply this process to the negation of the wff of exercise 43.5(1).

43.7. Prove directly that a wff is satisfiable if and only if its dual is not valid.

44. Gödel's completeness theorem.

**440. Every valid wff is a theorem. (*Gödel's completeness theorem.*)

Proof. By *421 and **437, it is sufficient to consider a wff **A** in Skolem normal form,

$$(\exists \mathbf{a}_1)(\exists \mathbf{a}_2) \ldots (\exists \mathbf{a}_m)(\mathbf{b}_1)(\mathbf{b}_2) \ldots (\mathbf{b}_n)\mathbf{M},$$

where **M** is quantifier-free and contains the individual variables $\mathbf{a}_1, \mathbf{a}_2, \ldots, \mathbf{a}_m, \mathbf{b}_1, \mathbf{b}_2, \ldots, \mathbf{b}_n$ (and no other individual variables). We shall show that either (1) **A** is a theorem, or (2) **A** is not valid.

Let us enumerate the (ordered) m-tuples of positive integers according to the following rule: If $i_1 + i_2 + \ldots + i_m < j_1 + j_2 + \ldots + j_m$, the m-tuple $\langle i_1, i_2, \ldots, i_m\rangle$ comes before (i.e., comes earlier in the enumeration than) the m-tuple $\langle j_1, j_2, \ldots, j_m\rangle$; if $i_1 + i_2 + \ldots + i_m = j_1 + j_2 + \ldots + j_m$, $i_1 = j_1$, $i_2 = j_2, \ldots, i_k = j_k$, $i_{k+1} < j_{k+1}$, the m-tuple $\langle i_1, i_2, \ldots, i_m\rangle$ again comes before the m-tuple $\langle j_1, j_2, \ldots, j_m\rangle$.[410]

Thus the m-tuples of positive integers are enumerated, or arranged in an infinite sequence. The first m-tuple in this enumeration is $\langle 1, 1, 1, \ldots, 1, 1, 1\rangle$ the second m-tuple is $\langle 1, 1, 1, \ldots, 1, 1, 2\rangle$: the third one is $\langle 1, 1, 1, \ldots,$

[409] These are modified forms of examples due, one to Kurt Schütte, and the other to Paul Bernays and Moses Schönfinkel.

[410] That is, the m-tuples $\langle i_1, i_2, \ldots, i_m\rangle$ are arranged in order of increasing sums $i_1 + i_2 \ldots + i_m$. And m-tuples having the same sum are arranged among themselves in lexicographic order.

1, 2, 1⟩; the fourth is ⟨1, 1, 1, . . ., 2, 1, 1⟩; and so on.[411] Evidently, no positive integer occurring in the kth m-tuple is greater than k (if $m \geqq 1$).

We let the kth m-tuple in this enumeration be $\langle [k1], [k2], \ldots, [km] \rangle$. I.e., we use $[kl]$ as a notation for the lth positive integer in the kth m-tuple.

Now let $\mathbf{B}_k$ be

$$\mathrm{S}^{\mathbf{a}_1 \quad \mathbf{a}_2 \quad \ldots \mathbf{a}_m \quad \mathbf{b}_1 \qquad \mathbf{b}_2 \qquad \ldots \mathbf{b}_n}_{x_{[k1]} x_{[k2]} \ldots x_{[km]} x_{(k-1)n+2} x_{(k-1)n+3} \ldots x_{kn+1}} \mathbf{M}|,$$

let $\mathbf{C}_k$ be

$$\mathbf{B}_1 \vee \mathbf{B}_2 \vee \ldots \vee \mathbf{B}_k,$$

and let $\mathbf{D}_k$ be

$$(x_1)(x_2) \ldots (x_{kn+1})\mathbf{C}_k.$$

We notice that the variables $x_{(k-1)n+2}, x_{(k-1)n+3}, \ldots, x_{kn+1}$, which are here substituted for $\mathbf{b}_1, \mathbf{b}_2, \ldots, \mathbf{b}_n$ are none of them the same as any of the variables $x_{[k1]}, x_{[k2]}, \ldots, x_{[km]}$, which are substituted for $\mathbf{a}_1, \mathbf{a}_2, \ldots, \mathbf{a}_m$. Moreover the variables $x_{(k-1)n+2}, x_{(k-1)n+3}, \ldots, x_{kn+1}$ are all different among themselves, and different from all the variables occurring in $\mathbf{B}_1, \mathbf{B}_2, \ldots, \mathbf{B}_{k-1}$. But all the variables $x_1, x_2, \ldots, x_{kn+1}$ occur in $\mathbf{C}_k$.

(It is possible that n may be 0, and the reader should observe that this special case makes no difficulty; but m is never less than 1.)

Since $\mathbf{M}$ is quantifier-free, it follows that $\mathbf{B}_k$ and $\mathbf{C}_k$ are also quantifier-free. And, except in the case $n = 0$, the complete list of free individual variables in $\mathbf{C}_k$ is $x_1, x_2, \ldots, x_{kn+1}$.

Lemma: For every k, $\mathbf{D}_k \vdash \mathbf{A}$.

We prove the lemma by mathematical induction with respect to k. In doing so, we assume that none of the variables $\mathbf{a}_1, \mathbf{a}_2 \ldots, \mathbf{a}_m, \mathbf{b}_1, \mathbf{b}_2, \ldots, \mathbf{b}_n$ are the same as any of the variables $x_1, x_2, x_3, \ldots$ (as may be brought about by alphabetic changes of bound variable in $\mathbf{A}$ if necessary).

By *330 and *modus ponens*, repeated m times,

$$(\mathbf{b}_1)(\mathbf{b}_2) \ldots (\mathbf{b}_n)\ \mathrm{S}^{\mathbf{a}_1 \mathbf{a}_2 \ldots \mathbf{a}_m}_{x_1 x_1 \ldots x_1} \mathbf{M}| \vdash \mathbf{A}.$$

Hence by *306 and *modus ponens*,

$$(x_1)(\mathbf{b}_1)(\mathbf{b}_2) \ldots (\mathbf{b}_n)\ \mathrm{S}^{\mathbf{a}_1 \mathbf{a}_2 \ldots \mathbf{a}_m}_{x_1 x_1 \ldots x_1} \mathbf{M}| \vdash \mathbf{A}.$$

Hence by alphabetic change of bound variable, repeated n times, $\mathbf{D}_1 \vdash \mathbf{A}$.

This is the case $k = 1$ of the lemma.

---

[411]Here, for purposes of illustration, we have taken $m > 6$; and the dots in each case represent a number of 1's. As further illustration the mth m-tuple is ⟨1, 2, 1, . . . , 1, 1, 1⟩; the $(m + 1)$th is ⟨2, 1, 1, . . . , 1, 1, 1⟩; the $(m + 2)$th is ⟨1, 1, 1, . . , 1, 1, 3⟩; the $(m + 3)$th is ⟨1, 1, 1, . . . , 1, 2, 2⟩; the $(m + 4)$th is ⟨1, 1, 1, . . . , 1, 3, 1⟩; the $(m + 5)$th is ⟨1, 1, 1, . . . , 2, 1, 2⟩.

Now suppose, for some particular k greater than 1, that $\mathbf{D}_{k-1} \vdash \mathbf{A}$. By *385 and *341 and P, repeating n times, we have that

$$(x_{(k-1)n+2})(x_{(k-1)n+3}) \ldots (x_{kn+1})[\mathbf{C}_{k-1} \vee \mathbf{B}_k] \vdash \mathbf{C}_{k-1} \vee (x_{(k-1)n+2})(x_{(k-1)n+3}) \ldots (x_{kn+1})\mathbf{B}_k.$$

From this, since $\mathbf{C}_{k-1} \vee \mathbf{B}_k$ is the wff $\mathbf{C}_k$, we have by *306 and *modus ponens*, repeated $(k-1)n+1$ times, that

$$\mathbf{D}_k \vdash \mathbf{C}_{k-1} \vee (x_{(k-1)n+2})(x_{(k-1)n+3}) \ldots (x_{kn+1})\mathbf{B}_k.$$

Hence by alphabetic change of bound variable, repeated n times,

$$\mathbf{D}_k \vdash \mathbf{C}_{k-1} \vee (\mathbf{b}_1)(\mathbf{b}_2) \ldots (\mathbf{b}_n)\ \mathrm{S}^{\mathbf{a}_1\ \mathbf{a}_2\ \ldots \mathbf{a}_m}_{x_{[k1]}x_{[k2]}\ldots x_{[km]}}\mathbf{M}|.$$

Also by *330 and the transitive law of implication, m times,

$$\vdash (\mathbf{b}_1)(\mathbf{b}_2) \ldots (\mathbf{b}_n)\ \mathrm{S}^{\mathbf{a}_1\ \mathbf{a}_2\ \ldots \mathbf{a}_m}_{x_{[k1]}x_{[k2]}\ldots x_{[km]}}\mathbf{M}| \supset \mathbf{A}.$$

Hence by P,

$$\mathbf{D}_k \vdash \mathbf{C}_{k-1} \vee \mathbf{A}.$$

Hence by generalizing upon $x_{(k-1)n+1}$ and using *386 and P, then generalizing upon $x_{(k-1)n}$ and using *386 and P, and so on, repeating $(k-1)n+1$ times, we have that

$$\mathbf{D}_k \vdash \mathbf{D}_{k-1} \vee \mathbf{A}.$$

Since $\mathbf{D}_{k-1} \vdash \mathbf{A}$, we have that $\vdash \mathbf{D}_{k-1} \supset \mathbf{A}$, and hence by P that $\mathbf{D}_k \vdash \mathbf{A}$.

This completes the proof of the lemma. Continuing the proof of **440, we distinguish two cases.[412]

Case 1: for some k, $\mathbf{C}_k$ is a theorem. Then by generalization ($kn+1$ times), $\mathbf{D}_k$ is a theorem. Hence by the lemma, $\mathbf{A}$ is a theorem.

Case 2: for every k, $\mathbf{C}_k$ is a non-theorem. Then for every k, by *311, it is possible to find such an assignment of truth-values to the *elementary parts* of $\mathbf{C}_k$ (i.e., the wf parts which have either the form of a propositional variable alone or the form $\mathbf{f}(\mathbf{c}_1, \mathbf{c}_2, \ldots, \mathbf{c}_i)$) that the value of $\mathbf{C}_k$, as obtained by the truth-tables of $\supset$ and $\sim$, is f. Or, as we shall say, it is possible to find a *falsifying assignment* of truth-values to the elementary parts of $\mathbf{C}_k$. (The same elementary part may occur more than once in $\mathbf{C}_k$, in which case of course the same truth-value is to be given to it at every occurrence; but different elementary parts may receive different truth-values, even if they differ only as to individual variables.)

[412]At this point of the proof we make use of the law of excluded middle (in the syntax language). But since no effective means is at hand to determine which of the two cases holds for a given $\mathbf{A}$, the method of the proof yields no solution of the decision problem of the pure functional calculus of first order.

Now let $\mathbf{E}_1$, $\mathbf{E}_2$, $\mathbf{E}_3$, . . ., be an enumeration of the different elementary parts occurring in $\mathbf{C}_1$, $\mathbf{C}_2$, $\mathbf{C}_3$, . . ., according to the following order: first the different elementary parts of $\mathbf{C}_1$, in the order of their first occurrence in $\mathbf{C}_1$; then the different elementary parts of $\mathbf{C}_2$ that do not occur in $\mathbf{C}_1$, in the order of their first occurrence in $\mathbf{C}_2$; then the different elementary parts of $\mathbf{C}_3$ that do not occur in $\mathbf{C}_1$, $\mathbf{C}_2$, in the order of their first occurrence in $\mathbf{C}_3$; and so on.

We proceed to make a "master assignment" of truth-values to $\mathbf{E}_1$, $\mathbf{E}_2$, $\mathbf{E}_3$, . . ., as follows. If $\mathbf{E}_1$ receives the value t in infinitely many of the falsifying assignments[413] of truth-values to the elementary parts of $\mathbf{C}_1$, $\mathbf{C}_2$, $\mathbf{C}_3$, . . ., we give $\mathbf{E}_1$ the value t in the master assignment; in the contrary case, $\mathbf{E}_1$ must receive the value f in infinitely many of the falsifying assignments of truth-values to the elementary parts of $\mathbf{C}_1$, $\mathbf{C}_2$, $\mathbf{C}_3$, . . ., and we give $\mathbf{E}_1$ the value f in the master assignment. Next we consider those infinitely many falsifying assignments of truth-values to the elementary parts of $\mathbf{C}_1$, $\mathbf{C}_2$, $\mathbf{C}_3$, . . . in which $\mathbf{E}_1$ receives the same truth-value as in the master assignment; if $\mathbf{E}_2$ receives the value t in infinitely many of these, we give $\mathbf{E}_2$ the value t in the master assignment; and in the contrary case we give $\mathbf{E}_2$ the value f in the master assignment. Next we consider those infinitely many falsifying assignments of truth-values to the elementary parts of $\mathbf{C}_1$, $\mathbf{C}_2$, $\mathbf{C}_3$, . . . in which $\mathbf{E}_1$ and $\mathbf{E}_2$ receive each the same truth-value as in the master assignment; if $\mathbf{E}_3$ receives the value t in infinitely many of these, we give $\mathbf{E}_3$ the value t in the master assignment; and in the contrary case we give $\mathbf{E}_3$ the value f in the master assignment. And so on.

Now suppose that the master assignment should result in the value t for one of $\mathbf{C}_1$, $\mathbf{C}_2$, $\mathbf{C}_3$, . . ., say for $\mathbf{C}_k$. The different elementary parts of $\mathbf{C}_k$ are contained in a finite initial segment of $\mathbf{E}_1$, $\mathbf{E}_2$, $\mathbf{E}_3$, . . ., say in $\mathbf{E}_1$, $\mathbf{E}_2$, . . ., $\mathbf{E}_l$. Let e_1, e_2, . . ., e_l be the truth-values assigned to $\mathbf{E}_1$, $\mathbf{E}_2$, . . ., $\mathbf{E}_l$ respectively, by the master assignment. Then in view of the form of $\mathbf{C}_1$, $\mathbf{C}_2$, $\mathbf{C}_3$, . . . as disjunctions, and in view of the truth-table of $\vee$, we have that no assignment of truth-values to the elementary parts of $\mathbf{C}_j$, $j > k$, can be a falsifying assignment if it includes the assignment of e_1, e_2, . . ., e_l to $\mathbf{E}_1$, $\mathbf{E}_2$, . . ., $\mathbf{E}_l$

[413]It may happen that there is more than one falsifying assignment of truth-values to the elementary parts of $\mathbf{C}_k$ (though the number is always finite for a fixed $\mathbf{C}_k$). In speaking of "the falsifying assignments of truth-values to the elementary parts of $\mathbf{C}_1$, $\mathbf{C}_2$, $\mathbf{C}_3$, . . . ," we mean to include, for each $\mathbf{C}_k$, all the various falsifying assignments of truth-values to the elementary parts of $\mathbf{C}_k$.

In the special case $n = 0$, it may also happen that falsifying assignments of truth-values to the elementary parts of $\mathbf{C}_k$ and of $\mathbf{C}_l$ coincide in the sense that the list of elementary parts involved is the same and the truth-values assigned are the same; nevertheless we count the two assignments as different if k and l are different.

respectively. But this contradicts the rule which was used in assigning the truth-value e_l to $\mathbf{E}_l$ in the construction of the master assignment.

It follows that the master assignment results in the value f for every $\mathbf{C}_k$. I.e., $\mathbf{C}_1, \mathbf{C}_2, \mathbf{C}_3, \ldots$ are simultaneously falsified by the master assignment.

Now we take the positive integers as domain of individuals, and proceed to assign values to the propositional and functional variables of **A** as follows. To a propositional variable p is given the same value as given to p in the master assignment. To an i-ary functional variable **f** is assigned as value an i-ary propositional function Φ of individuals, as determined by the following rule: $\Phi(u_1, u_2, \ldots, u_i)$ has the same truth-value which is assigned to $\mathbf{f}(x_{u_1}, x_{u_2}, \ldots, x_{u_i})$ in the master assignment; or if no truth-value is assigned to $\mathbf{f}(x_{u_1}, x_{u_2}, \ldots, x_{u_i})$ in the master assignment, the truth-value of $\Phi(u_1, u_2, \ldots, u_i)$ is t.

This assignment of values to the propositional and functional variables of **A** is at the same time an assignment of values to the propositional and functional variables of each $\mathbf{B}_k$ and of each $\mathbf{C}_k$. If we also assign to each individual variable x_u the positive integer u as value, we have an assignment of values to all the variables of $\mathbf{B}_k$ and of $\mathbf{C}_k$. This assignment gives to $\mathbf{C}_k$ the value f (since we have proved that the master assignment falsifies $\mathbf{C}_k$). Hence it also gives to $\mathbf{B}_k$ the value f (since $\mathbf{C}_k$ is $\mathbf{C}_{k-1} \vee \mathbf{B}_k$, and in view of the truth-table of $\vee$). Hence by rule f of §30 (or by the clause corresponding to rule f in the definition of "value" in the syntactical sense), it gives the value f also to

$$(\mathbf{b}_1)(\mathbf{b}_2) \ldots (\mathbf{b}_n) \; \mathrm{S}^{x_{(k-1)n+2}\, x_{(k-1)n+3} \ldots x_{kn+1}}_{\mathbf{b}_1 \quad \mathbf{b}_2 \quad \ldots \mathbf{b}_n} \mathbf{B}_k|,$$

i.e., to

$$(\mathbf{b}_1)(\mathbf{b}_2) \ldots (\mathbf{b}_n) \; \mathrm{S}^{\mathbf{a}_1 \quad \mathbf{a}_2 \quad \ldots \mathbf{a}_m}_{x_{[k1]}\, x_{[k2]} \ldots x_{[km]}} \mathbf{M}|.$$

This holds for all k; and as k runs through all values, we know that $\langle x_{[k1]}, x_{[k2]}, \ldots, x_{[km]} \rangle$ runs through all possible m-tuples of the variables x_u, and hence that all possible substitutions are made of variables x_u for $\mathbf{a}_1, \mathbf{a}_2, \ldots, \mathbf{a}_m$. Hence the value f is given to

$$(\exists \mathbf{a}_1)(\exists \mathbf{a}_2) \ldots (\exists \mathbf{a}_m)(\mathbf{b}_1)(\mathbf{b}_2) \ldots (\mathbf{b}_n)\mathbf{M}.$$

Thus we have found an assignment of values to the propositional and functional variables of **A** such that the value of **A** is f. Therefore **A** is not valid.

This completes the proof of **440. We notice as a corollary the following metatheorem:

441. If **A is a wff in Skolem normal form,

$$(\exists \mathbf{a}_1)(\exists \mathbf{a}_2) \ldots (\exists \mathbf{a}_m)(\mathbf{b}_1)(\mathbf{b}_2) \ldots (\mathbf{b}_n)\mathbf{M},$$

if $\mathbf{B}_k$ is the quantifier-free wff,

$$\mathrm{S}^{\mathbf{a}_1\ \mathbf{a}_2\ \ldots \mathbf{a}_m\ \mathbf{b}_1\ \mathbf{b}_2\ \ldots \mathbf{b}_n}_{x_{[k1]} x_{[k2]} \ldots x_{[km]} x_{(k-1)n+2} x_{(k-1)n+3} \ldots x_{kn+1}} \mathbf{M}|,$$

and if $\mathbf{C}_k$ is $\mathbf{B}_1 \vee \mathbf{B}_2 \vee \ldots \vee \mathbf{B}_k$, then **A** is a theorem (and is valid) if and only if there is some positive integer k such that $\mathbf{C}_k$ is a substitution instance of a tautology of P.

45. Löwenheim's theorem and Skolem's generalization. In case 2 of the foregoing proof of Gödel's completeness theorem we have shown about an arbitrary wff **A** which is in Skolem normal form and is not a theorem that it is not valid in the domain of positive integers. Hence, if a wff in Skolem normal form is valid in the domain of positive integers, it is a theorem, and therefore by **434 it is valid in every non-empty domain. Hence by **437, if any wff is valid in the domain of positive integers, it is valid in every non-empty domain.

Moreover, by **439, any enumerably infinite domain (in particular, e.g., the domain of natural numbers) may take the place here of the domain of positive integers. Thus follows:

**450. If a wff is valid in an enumerably infinite domain, it is valid in every non-empty domain.[414] (*Löwenheim's theorem.*)

As a corollary by **431, we have also:

**451. If a wff is satisfiable in any non-empty domain, it is satisfiable in an enumerably infinite domain.

Skolem's generalization of this is the metatheorem that, if a class of wffs (it may be an infinite class) is simultaneously satisfiable in any non-empty domain, then it is simultaneously satisfiable in an enumerably infinite domain.

Here the definition of simultaneous satisfiability is the obvious generalization of the definition of satisfiability of a single wff. Namely, a class Γ of wffs is said to be *simultaneously satisfiable in* a given non-empty domain of individuals if, of all the free variables of all the wffs of Γ taken together, there exists at least one system of possible values for which every wff of Γ

[414]Thus, if a wff is valid in the domain of natural numbers, it is valid also in non-enumerably infinite domains. If a wff is satisfiable in a non-enumerably infinite domain, it is satisfiable also in the domain of natural numbers.

has the value t. (In the case of a finite non-empty class Γ of wffs, it is clear that simultaneous satisfiability is equivalent to the satisfiability of a single wff, the conjunction of all the wffs of Γ.)

A class of wffs is said to be *simultaneously satisfiable* if it is simultaneously satisfiable in some non-empty domain of individuals.

We shall need also the following definitions (which we adopt not only for the case of the pure functional calculus of first order but also for other functional calculi of first and higher orders[415]).

Where Γ is any class of wffs and **B** is any wff, we say that $\Gamma \vdash \mathbf{B}$ if there are a finite number of wffs $\mathbf{A}_1, \mathbf{A}_2, \ldots, \mathbf{A}_m$ of Γ such that $\mathbf{A}_1, \mathbf{A}_2, \ldots, \mathbf{A}_m \vdash \mathbf{B}$.

A class Γ of wffs is called *inconsistent* if there exists a wff **B** such that $\Gamma \vdash \mathbf{B}$ and $\Gamma \vdash \sim\mathbf{B}$. If no such wff **B** exists, we say that Γ is *consistent*.

Where Γ is any class of wffs and **C** is any wff, we say that **C** is *consistent with* Γ if the class is consistent whose members are **C** and the members (wffs) of Γ; otherwise we say that **C** is *inconsistent with* Γ.

A class Γ of wffs is called a *maximal consistent* class if Γ is consistent and no wff **C** is consistent with Γ which is not a member of Γ.

We establish the following as a lemma:

**452. Every consistent class Γ of wffs can be extended to a maximal consistent class $\bar{\Gamma}$, i.e., there exists a maximal consistent class $\bar{\Gamma}$ among whose members are all the members of Γ.

Proof. We shall give a rule by which the maximal consistent class $\bar{\Gamma}$ is uniquely determined when Γ is given. (However, this rule will not be such as to provide in any sense an effective construction of the members of $\bar{\Gamma}$.)

First the wffs must be enumerated, as is possible by well-known methods (since the primitive symbols are enumerable, and the wffs are certain finite sequences of primitive symbols).[416] Then we shall speak of "the first wff," "the second wff," and so on, referring to this enumeration of the wffs.

---

[415]At one place below, we use also an obvious extension of the definition of simultaneous satisfiability to an applied functional calculus of first order.

[416]For example, we may use the following enumeration of the wffs.

First take the enumeration of the ordered m-tuples of positive integers which was introduced in §44. And let us speak of the kth m-tuple of positive integers to mean the kth m-tuple of positive integers in this enumeration. (In particular, we shall speak in this way of the kth ordered pair of positive integers.)

Then let the primitive symbols of the pure functional calculus of first order, as introduced in §30, be enumerated as follows. The first eight primitive symbols in the enumeration are the eight improper symbols, in the order in which they are listed in §30. The $(k + 8)$th primitive symbol in the enumeration is: the mth individual variable in alphabetic order, if the kth ordered pair of positive integers is $\langle 1, m \rangle$; the mth propositional variable in alphabetic order, if the kth ordered pair of positive integers is

And for every wff there is a positive integer n such that it is "the nth wff" (i.e., the nth wff of the enumeration).

Given any class Γ of wffs, we define the infinite sequence of classes $\Gamma^0, \Gamma^1, \Gamma^2, \ldots$, (by recursion) as follows: Γ^0 is the same as Γ. If the $(n+1)$th wff is consistent with Γ^n, then Γ^{n+1} is the class whose members are the $(n+1)$th wff and the members of Γ^n. Otherwise Γ^{n+1} is the same as Γ_n.

It follows by mathematical induction that $\Gamma^0, \Gamma^1, \Gamma^2, \ldots$ are consistent classes of wffs if Γ is consistent. For Γ^0 is the same as Γ. And the consistency of Γ^{n+1} follows at once from that of Γ^n.

We let $\bar{\Gamma}$ be the union of the classes $\Gamma^0, \Gamma^1, \Gamma^2, \ldots$. I.e., a wff **C** is a member of $\bar{\Gamma}$ if and only if there is some n such that **C** is a member of Γ^n.

Now if Γ is a consistent class, it follows that $\bar{\Gamma}$ is a consistent class.

For suppose that $\bar{\Gamma}$ is inconsistent. Then there are a finite number of wffs $\mathbf{A}_1, \mathbf{A}_2, \ldots, \mathbf{A}_m$ of $\bar{\Gamma}$, and a wff **B**, such that $\mathbf{A}_1, \mathbf{A}_2, \ldots, \mathbf{A}_m \vdash \mathbf{B}$ and $\mathbf{A}_1, \mathbf{A}_2, \ldots, \mathbf{A}_m \vdash \sim\mathbf{B}$. Say that $\mathbf{A}_1$ is the a_1th wff, $\mathbf{A}_2$ is the a_2th wff, ..., $\mathbf{A}_m$ is the a_mth wff; and let a be the greatest of the positive integers $a_1, a_2, \ldots, a_m$. Then all the wffs $\mathbf{A}_1, \mathbf{A}_2, \ldots, \mathbf{A}_m$ are members of Γ^a, and consequently Γ^a is inconsistent. But this contradicts our proof above that all the classes $\Gamma^0, \Gamma^1, \Gamma^2, \ldots$ are consistent (if Γ is consistent).

Moreover $\bar{\Gamma}$ is a maximal class, if Γ is consistent.

For let **C** be any wff which is consistent with $\bar{\Gamma}$. Say that **C** is the $(n+1)$th wff. Being consistent with $\bar{\Gamma}$, **C** must be consistent with Γ^n. Therefore, by the definition of Γ^{n+1}, **C** is a member of Γ^{n+1}. Therefore **C** is a member of $\bar{\Gamma}$.

Thus the proof of the lemma, **452, is completed. We observe that a corresponding lemma holds not only for the pure functional calculus of first order but also for any applied functional calculus of first order if the primitive symbols are enumerable. And indeed **452 can be extended to a wide variety of logistic systems, since the proof requires only a suitable notion of consistency of a class of wffs, and the enumerability of the primitive symbols of the system.[417]

$\langle 2, m\rangle$; and the mth i-ary functional variable in alphabetic order, if the kth ordered pair of positive integers is $\langle i+2, m\rangle$.

And let us speak of the μth primitive symbol to mean the μth primitive symbol in the foregoing enumeration.

Then let the formulas of the pure functional calculus of first order be enumerated by the rule that, if the kth order pair of positive integers is $\langle \mu, m\rangle$ and the μth m-tuple of positive integers is $\langle \mu_1, \mu_2, \ldots, \mu_m\rangle$, then the kth formula in the enumeration is $\boldsymbol{\mu}_1\boldsymbol{\mu}_2 \ldots \boldsymbol{\mu}_m$, where $\boldsymbol{\mu}_1$ is the μ_1th primitive symbol, $\boldsymbol{\mu}_2$ is the μ_2th primitive symbol, ..., $\boldsymbol{\mu}_m$ is the μ_mth primitive symbol.

Finally, from the enumeration of the formulas, delete all those which are not well-formed, so obtaining an enumeration of the wffs.

[417]This last is presumably a consequence of requirement (I) of §07.

We now consider an infinite sequence of applied functional calculi of first order $S_0, S_1, S_2, \ldots$, having as primitive symbols all the primitive symbols of the pure functional calculus of first order and in addition certain individual constants. Namely, the primitive symbols of S_0 are those of the pure functional calculus of first order and the individual constants $w_{0,0}, w_{1,0}, w_{2,0}, \ldots$; the primitive symbols of S_{n+1} are those of S_n and the additional individual constants $w_{0,n+1}, w_{1,n+1}, w_{2,n+1}, \ldots$.

Also we let S_ω be the applied functional calculus of first order which has as its primitive symbols the primitive symbols of all the systems $S_0, S_1, S_2, \ldots$. (Thus the individual constants of S_ω are all the constants $w_{m,n}$, for $m = 0, 1, 2, \ldots$ and $n = 0, 1, 2, \ldots$.)

In the same way that we have already remarked for the pure functional calculus of first order, it is possible to enumerate the wffs of S_ω. Then, in the case of each S_n, an enumeration of its wffs is obtained by deleting from the enumeration of the wffs of S_ω those which are not wffs of S_n. And referring to this enumeration we shall speak of "the first wff of S_n," "the second wff of S_n," and so on.

Using these enumerations, we can extend any consistent class Δ_n of wffs of S_n to a maximal consistent class $\bar{\Delta}_n$, of wffs of S_n, by the method stated above in the proof of **452.

Now let a consistent class Γ_0 be given of wffs of S_0 which have no free individual variables. We define the classes Γ_n^m ($m = 0, 1, 2, 3, \ldots$ and $n = 1, 2, 3, \ldots$) as follows: Γ_1^0 is $\bar{\Gamma}_0$. If the $(m + 1)$th wff of S_n, $n > 0$, has the form $\sim(\mathbf{a})\mathbf{A}$ and is a member of Γ_n^0, then Γ_n^{m+1} is the class whose members are

$$\sim \mathrm{S}^{\mathbf{a}}_{w_{m,n}} \mathbf{A}|$$

and the members of Γ_n^m; otherwise Γ_n^{m+1} is the same as Γ_n^m. And Γ_{n+1}^0 is $\bar{\Delta}_n$, where Δ_n is the union of the classes $\Gamma_n^0, \Gamma_n^1, \Gamma_n^2, \ldots$.

Evidently the members of Γ_n^m are wffs of S_n. And Γ_{n+1}^0 is a maximal consistent class of wffs of S_n.

Assume that, for some particular m, Γ_n^m is consistent but Γ_n^{m+1} is inconsistent. Then we must have the case that Γ_n^{m+1} is not the same as Γ_n^m but has the additional member

$$\sim \mathrm{S}^{\mathbf{a}}_{w_{m,n}} \mathbf{A}|.$$

By the inconsistency of Γ_n^{m+1}, and the deduction theorem,

$$\Gamma_n^m \vdash \sim \mathrm{S}^{\mathbf{a}}_{w_{m,n}} \mathbf{A}| \supset \mathbf{B}$$

and

$$\Gamma^m_n \vdash \sim \underset{\cdot}{\mathrm{S}}^{\mathbf{a}}_{w_{m,n}} \mathbf{A}| \supset \sim \mathbf{B}.$$

Hence by P,

$$\Gamma^m_n \vdash \underset{\cdot}{\mathrm{S}}^{\mathbf{a}}_{w_{m,n}} \mathbf{A}|$$

Let **x** be an individual variable which does not occur in this proof from hypotheses, and in it replace the constant $w_{m,n}$ everywhere by **x**; since $w_{m,n}$ does not occur in any of the members of Γ^m_n, we thus have:

$$\Gamma^m_n \vdash \underset{\cdot}{\mathrm{S}}^{\mathbf{a}}_{\mathbf{x}} \mathbf{A}|.$$

By generalizing upon **x**,[418] and then making one or more alphabetic changes of bound variable, we have that $\Gamma^m_n \vdash (\mathbf{a})\mathbf{A}$; but since $\sim(\mathbf{a})\mathbf{A}$ is a member of Γ^0_n and therefore of Γ^m_n, this contradicts the assumption that Γ^m_n is consistent.

Thus we have proved that, if Γ^m_n is consistent, then Γ^{m+1}_n is consistent. By mathematical induction, it follows that, if Γ^0_n is consistent, then Γ^m_n is consistent for every m, and therefore Γ^0_{n+1} is consistent. Since Γ^0_1 is the same as $\bar{\Gamma}^0$, and is therefore consistent, it follows by a second mathematical induction that Γ^0_n is consistent for every n.

Let Γ_ω be the union of the classes $\Gamma^0_1, \Gamma^0_2, \Gamma^0_3, \ldots$. Then Γ_ω is a maximal consistent class of wffs of S_ω. (For Γ_ω could be inconsistent only if, for some n, Γ^0_n were inconsistent. Further, if **C** is a wff of S_ω consistent with Γ_ω, then, for some n, **C** is a wff of S_n and is consistent with Γ^0_{n+1}; since Γ^0_{n+1} is a maximal consistent class of wffs of S_n, it follows that **C** is a member of Γ^0_{n+1} and therefore a member of Γ_ω.)

We need the following properties of Γ_ω:

d1. If **A** is a member of Γ_ω, then $\sim$**A** is a non-member of Γ_ω. (For otherwise Γ_ω would be inconsistent.)

d2. If **A** is a non-member of Γ_ω, then $\sim$**A** is a member of Γ_ω. (For, if **A** is a non-member of Γ_ω, then **A** must be inconsistent with Γ_ω; therefore, by the deduction theorem and P, $\Gamma_\omega \vdash \sim\mathbf{A}$; therefore $\sim$**A** is consistent with Γ_ω; therefore $\sim$**A** is a member of Γ_ω.)

e1. If **B** is a member of Γ_ω, then $\mathbf{A} \supset \mathbf{B}$ is a member of Γ_ω. (For by P, $\Gamma_\omega \vdash \mathbf{A} \supset \mathbf{B}$; thus $\mathbf{A} \supset \mathbf{B}$ is consistent with Γ_ω and therefore a member of Γ_ω.)

e2. If **A** is a non-member of Γ_ω, then $\mathbf{A} \supset \mathbf{B}$ is a member of Γ_ω. (For by d2, $\sim$**A** is a member of Γ_ω, and therefore by P, $\Gamma_\omega \vdash \mathbf{A} \supset \mathbf{B}$.)

e3. If **A** is a member of Γ_ω and **B** is a non-member of Γ_ω, then $\mathbf{A} \supset \mathbf{B}$ is a

[418]This is permissible because **x** does not occur in any of the members of Γ^m_n which are here actually used as hypotheses in the proof of

$$\underset{\cdot}{\mathrm{S}}^{\mathbf{a}}_{\mathbf{x}} \mathbf{A}|$$

non-member of Γ_ω. (For by d2, $\sim\mathbf{B}$ is a member of Γ_ω; hence, if $\mathbf{A} \supset \mathbf{B}$ were a member of Γ_ω, Γ_ω would be inconsistent, by an application of *modus ponens*.)

f1. If, for every individual constant $w_{m,n}$,

$$\overset{\mathbf{a}}{\underset{w_{m,n}}{\mathrm{S}}}\ \mathbf{A}|$$

is a member of Γ_ω, then $(\mathbf{a})\mathbf{A}$ is a member of Γ_ω. (For, if $(\mathbf{a})\mathbf{A}$ is not a member of Γ_ω, it follows by d2 that $\sim(\mathbf{a})\mathbf{A}$ is a member of Γ_ω; hence, for some n, $\sim(\mathbf{a})\mathbf{A}$ is a member of Γ_n^0; hence, by the way in which the classes Γ_n^{m+1} were defined, we have for some m that

$$\sim\overset{\mathbf{a}}{\underset{w_{m,n}}{\mathrm{S}}}\ \mathbf{A}|$$

is a member of Γ_n^{m+1} and therefore a member of Γ_ω; thus by d1 we have for some m and n that

$$\overset{\mathbf{a}}{\underset{w_{m,n}}{\mathrm{S}}}\ \mathbf{A}|$$

is a non-member of Γ_ω.)

f2. If, for at least one individual constant $w_{m,n}$,

$$\overset{\mathbf{a}}{\underset{w_{m,n}}{\mathrm{S}}}\ \mathbf{A}|$$

is a non-member of Γ_ω, then $(\mathbf{a})\mathbf{A}$ is a non-member of Γ_ω. (For, if $(\mathbf{a})\mathbf{A}$ is a member of Γ_ω, we have by *306 and *modus ponens*, for an arbitrary individual constant $w_{m,n}$, that

$$\Gamma_\omega \vdash \overset{\mathbf{a}}{\underset{w_{m,n}}{\mathrm{S}}}\ \mathbf{A}|,$$

and hence that

$$\overset{\mathbf{a}}{\underset{w_{m,n}}{\mathrm{S}}}\ \mathbf{A}|,$$

being consistent with Γ_ω, must be a member of Γ_ω.)

Taking the natural numbers as domain of individuals, we now assign values to all the propositional and functional variables of S_ω—or, what is the same thing, to all the propositional and functional variables of the pure functional of first order—as follows:

To a propositional variable $\mathbf{p}$ is assigned the value t if the wff $\mathbf{p}$ is a member of Γ_ω, the value f if the wff $\mathbf{p}$ is a non-member of Γ_ω. Letting $u_{m,n}$ be the natural number $\frac{1}{2}(m^2 + 2mn + n^2 + 3m + n)$,[419] we have for each

[419]This corresponds in an obvious way to the following enumeration of the ordered pairs of natural numbers,

$$\langle 0, 0\rangle, \langle 0, 1\rangle, \langle 1, 0\rangle, \langle 0, 2\rangle, \langle 1, 1\rangle, \langle 2, 0\rangle, \langle 0, 3\rangle, \ldots,$$

analogous to the enumeration of the ordered m-tuples of positive integers which was introduced in §44.

individual constant $w_{m,n}$ a unique corresponding natural number $u_{m,n}$, and for each natural number $u_{m,n}$ a unique corresponding individual constant $w_{m,n}$. Then to an i-ary functional variable **f** is assigned as value the i-ary propositional function Φ of individuals such that $\Phi(u_{m_1,n_1}, u_{m_2,n_2}, \ldots, u_{m_i,n_i})$ is t or f according as $\mathbf{f}(w_{m_1,n_1}, w_{m_2,n_2}, \ldots, w_{m_i,n_i})$ is a member or a non-member of Γ_ω.

Now notice the way in which d1 and d2 above are related to rule d of §30, and e1, e2, e3 are related to rule e of §30, and f1 and f2 are related to rule f of §30. From this relationship (or, more correctly, we should speak of the relationship to the clauses which correspond to rules d–f in the definition of "value" in the syntactical sense) it follows that every wff of S_ω without free individual variables has, for the system of values that we have just assigned to the propositional and functional variables of S_ω, the value t or the value f according as it is a member or a non-member of Γ_ω. (It is left to the reader to state the proof of this explicitly by mathematical induction.)

Since the members of Γ_0 are without free individual variables and are included among the members of Γ_ω, we have thus shown that Γ_0 is simultaneously satisfiable in the domain of natural numbers, hence, by an obvious extension of **439, simultaneously satisfiable in any enumerably infinite domain.

But Γ_0 was chosen as an arbitrary consistent class of wffs of S_0 without free individual variables. Hence *every consistent class of well-formed formulas of* S_0 *without free individual variables is simultaneously satisfiable in an enumerably infinite domain* (with such an assignment of values to the individual constants as to give a different value to each).

To extend this result to any consistent class of wffs of the pure functional calculus of first order, we have only to substitute, for the free occurrences of individual variables, individual constants $w_{n,0}$, in such a way that a different individual constant is substituted for the free occurrences of each different individual variable.

Thus we have the following metatheorem of the pure functional calculus of first order:

**453. Every consistent class of wffs is simultaneously satisfiable in an enumerably infinite domain.

We need also the following:

**454. Every simultaneously satisfiable class of wffs is consistent.

Proof. Let Γ be an inconsistent class of wffs. Then there are a finite number of wffs $\mathbf{A}_1, \mathbf{A}_2, \ldots, \mathbf{A}_m$ of Γ, and a wff **B**, such that $\mathbf{A}_1, \mathbf{A}_2, \ldots, \mathbf{A}_m \vdash \mathbf{B}$ and $\mathbf{A}_1, \mathbf{A}_2, \ldots, \mathbf{A}_m \vdash \sim\mathbf{B}$. By **321, $m > 0$. Therefore by the deduction theorem and P, $\vdash \sim . \mathbf{A}_1\mathbf{A}_2 \ldots \mathbf{A}_m$. Therefore by **434, $\sim . \mathbf{A}_1\mathbf{A}_2 \ldots \mathbf{A}_m$ is valid. In view of the truth-tables of negation and conjunction, it follows that $\mathbf{A}_1, \mathbf{A}_2, \ldots, \mathbf{A}_m$ cannot have the value t simultaneously, for any system of values of their free variables. Thus Γ cannot be simultaneously satisfiable.

From **453 and **454 we obtain as a corollary the result of Skolem which was mentioned at the beginning of this section:

**455. If a class of wffs is simultaneously satisfiable in any non-empty domain, it is simultaneously satisfiable in an enumerably infinite domain.

It is worth noticing that Gödel's completeness theorem and Löwenheim's theorem now follow as corollaries, so that we obtain alternative proofs of these theorems, in which there is no use of the Skolem normal form or of §44.

In the case of Gödel's theorem, this is seen as follows. Let **A** be any valid wff. Then by **430, the class whose single member is $\sim\mathbf{A}$ is not simultaneously satisfiable. Therefore by **453, this class is not consistent. Therefore, for some wff **B**, both $\sim\mathbf{A} \vdash \mathbf{B}$ and $\sim\mathbf{A} \vdash \sim\mathbf{B}$. Therefore by the deduction theorem and P, $\vdash \mathbf{A}$.

EXERCISES 45

45.0. Carry through the proof of **440 (as given in §44) explicitly for the case that $n = 0$, making such simplifications as are possible in this special case, and verifying that the proof is sound also for this case.

45.1. Establish **454, without use of **321 and **434, by showing directly that the process of proof from hypotheses, as defined in §36, preserves the property of having the value t for a given system of values of the free variables of the hypotheses (and for all values of the other free variables occurring).

45.2. On the basis of the definition of validity (§43) and of the two proofs of **440 that are given in the text (§§44 and 45), discuss the questions, (1) whether **440 may be used as a derived rule of inference of the pure functional calculus of first order, and (2) whether **440 may be used in the role of an axiom schema in a formulation of the pure functional calculus of first order. (See the discussion of the logistic method and the definition of a logistic system in §07, the distinction in §08 between elementary and theoretical syntax, the introduction of the idea of derived rules of inference in §12, and the remarks of footnotes 183, 221.)

45.3. Prove the completeness of the propositional calculus, in the formulation P_1, by applying the ideas used in the text in the proof of **452 and **453. Compare this completeness proof for the propositional calculus with the completeness proof of Chapter I, especially as regards the question of a stronger or weaker basis on which results are obtained (cf. the initial paragraphs of §08).

45.4. Let a class Γ of wffs be given, and a particular *valuation* of Γ in the domain of natural numbers, i.e., with the natural numbers as the individuals, a particular system of possible values of the free variables of the wffs of Γ. And suppose that, for this valuation, every wff of Γ has the value t. Show that the method which is employed in §45 (in the proofs of **452 and **453), to obtain a valuation of Γ for which every wff of Γ has the value t, can be made to yield the given particular valuation of Γ by a suitable choice (a) of the enumeration of the wffs that is used, and (b) of the correspondence that is used, not necessarily a one-to-one correspondence, between the constants $w_{m,n}$ and the natural numbers $u_{m,n}$.

45.5. Let a class Γ of wffs be called *disjunctively valid in* a given non-empty domain of individuals if, for each valuation of Γ in that domain, there exists at least one wff of Γ which has the value t. (1) If a class of wffs is disjunctively valid in an infinite domain, then the disjunction of some finite subclass of them is valid. (2) If a class of wffs is disjunctively valid in a finite domain, then the disjunction of some finite subclass of them is valid in that domain.

46. The decision problem, solution in special cases. Though the decision problem of the pure functional calculus of first order is known to be unsolvable—in the sense that no effective decision procedure exists which suffices to determine of an arbitrary wff whether or not it is a theorem[420]—there nevertheless exist solutions in a number of special cases[421] which have

[420]Alonzo Church in *The Journal of Symbolic Logic*, vol. 1 (1936), pp. 40–41, 101–102. Hilbert and Bernays, *Grundlagen der Mathematik*, vol. 2, Supplement II.

[421]By a solution of the decision problem in a special case we mean that there shall be given a special class of wffs, an effective procedure to determine of an arbitrary wff whether it belongs to this class (this will be obvious in most cases discussed below), and an effective procedure to determine of any wff of this class whether it is a theorem. To this we seek always to add an effective procedure by which to find a proof of any wff which has thus been ascertained to be a theorem—but this last requirement is subject to the reservations that are indicated in footnote 183.

(*Added in proof.*) A comprehensive treatment of solutions of the decision problem in special cases is in Wilhelm Ackermann's monograph, *Solvable Cases of the Decision Problem* (1954).

some substantial interest. Some of the simpler of these will be treated in this section.

We begin with a solution of the decision problem (due to Bernays and Schönfinkel) for the special case of:

I *Well-formed formulas having a prenex normal form such that, in the prefix, no existential quantifier precedes any universal quantifier.*

It will be sufficient, by §39, and *301, *306, to find a decision procedure for the universal closure of the prenex normal form of the wff. Hence the solution of this case of the decision problem is contained in the four following metatheorems:

*460. Let **M** be a quantifier-free formula, and let $\mathbf{b}_1, \mathbf{b}_2, \ldots, \mathbf{b}_n$ ($n \geqq 0$) be the complete list of individual variables in **M**. If any afp of **M** is a tautology,

$$\vdash (\exists \mathbf{b}_1)(\exists \mathbf{b}_2) \ldots (\exists \mathbf{b}_n)\mathbf{M}.$$

Proof. $\mathrm{S}^{\mathbf{b}_1\mathbf{b}_2\ldots\mathbf{b}_n}_{x\,x\,\ldots x}\mathbf{M}|$ is a substitution instance of the afp of **M** and is therefore a theorem by *311. Hence use *330 and *modus ponens.*

461. Let **M be a quantifier-free formula, and let $\mathbf{b}_1, \mathbf{b}_2, \ldots, \mathbf{b}_n$ ($n \geqq 0$) be the complete list of individual variables in **M**. If

$$\vdash (\exists \mathbf{b}_1)(\exists \mathbf{b}_2) \ldots (\exists \mathbf{b}_n)\mathbf{M},$$

every afp of **M** is a tautology.

Proof. Every afp of **M** becomes an afp of

$$(\exists \mathbf{b}_1)(\exists \mathbf{b}_2) \ldots (\exists \mathbf{b}_n)\mathbf{M}$$

upon prefixing $2n$ negation signs, $\sim$. Hence use **320 and the truth-table of negation.

*462. Let **M** be a quantifier-free formula, and let $\mathbf{a}_1, \mathbf{a}_2, \ldots, \mathbf{a}_m, \mathbf{b}_1, \mathbf{b}_2 \ldots, \mathbf{b}_n$ ($m \geqq 1$, $n \geqq 0$) be the complete list of individual variables in **M**. If the disjunction **D** of all the wffs[422]

$$\mathrm{S}^{\mathbf{b}_1\mathbf{b}_2\ldots\mathbf{b}_n}_{\mathbf{d}_1\mathbf{d}_2\ldots\mathbf{d}_n}\mathbf{M}|$$

is a substitution instance of a tautology of P, where $\mathbf{d}_1, \mathbf{d}_2, \ldots, \mathbf{d}_n$

[422]The order in which these m^n different wffs are combined into a disjunction is evidently immaterial, in view of the commutative and associative laws of disjunction. In order to make the decision procedure definite, it may be fixed in some arbitrary (effective) way.

are any among the variables $\mathbf{a}_1, \mathbf{a}_2, \ldots, \mathbf{a}_m$, taken in any order and not necessarily all different, then

$$\vdash (\mathbf{a}_1)(\mathbf{a}_2) \ldots (\mathbf{a}_m)(\exists \mathbf{b}_1)(\exists \mathbf{b}_2) \ldots (\exists \mathbf{b}_n)\mathbf{M}.$$

Proof. By *330 and *modus ponens*,

$$\mathrm{S}^{\mathbf{b}_1\mathbf{b}_2\ldots\mathbf{b}_n}_{\mathbf{d}_1\mathbf{d}_2\ldots\mathbf{d}_n}\mathbf{M}| \vdash (\exists \mathbf{b}_1)(\exists \mathbf{b}_2) \ldots (\exists \mathbf{b}_n)\mathbf{M}.$$

Hence by the deduction theorem and P,

$$\mathbf{D} \vdash (\exists \mathbf{b}_1)(\exists \mathbf{b}_2) \ldots (\exists \mathbf{b}_n)\mathbf{M}.$$

Therefore, since **D** is a substitution instance of a tautology, we have by *311 that

$$\vdash (\exists \mathbf{b}_1)(\exists \mathbf{b}_2) \ldots (\exists \mathbf{b}_n)\mathbf{M}.$$

Hence by generalization,

$$\vdash (\mathbf{a}_1)(\mathbf{a}_2) \ldots (\mathbf{a}_m)(\exists \mathbf{b}_1)(\exists \mathbf{b}_2) \ldots (\exists \mathbf{b}_n)\mathbf{M}.$$

****463.** Let **M** be a quantifier-free formula, and let $\mathbf{a}_1, \mathbf{a}_2, \ldots, \mathbf{a}_m, \mathbf{b}_1, \mathbf{b}_2, \ldots, \mathbf{b}_n$ ($m \geqq 1$, $n \geqq 0$) be the complete list of individual variables in **M**. If

$$\vdash (\mathbf{a}_1)(\mathbf{a}_2) \ldots (\mathbf{a}_m)(\exists \mathbf{b}_1)(\exists \mathbf{b}_2) \ldots (\exists \mathbf{b}_n)\mathbf{M},$$

then the disjunction **D** of all the wffs[422]

$$\mathrm{S}^{\mathbf{b}_1\mathbf{b}_2\ldots\mathbf{b}_n}_{\mathbf{d}_1\mathbf{d}_2\ldots\mathbf{d}_n}\mathbf{M}|$$

is a substitution instance of a tautology of P, where $\mathbf{d}_1, \mathbf{d}_2, \ldots, \mathbf{d}_n$ are any among the variables $\mathbf{a}_1, \mathbf{a}_2, \ldots, \mathbf{a}_m$, taken in any order and not necessarily all different.

Proof. By *306 and *modus ponens*, $(\exists \mathbf{b}_1)(\exists \mathbf{b}_2) \ldots (\exists \mathbf{b}_n)\mathbf{M}$ is a theorem. Therefore by **434 it is valid, hence, in particular, valid in a domain of m individuals $u_1, u_2, \ldots, u_m$.

Taking this finite domain of individuals, consider any system of possible values of the free variables of $(\exists \mathbf{b}_1)(\exists \mathbf{b}_2) \ldots (\exists \mathbf{b}_n)\mathbf{M}$ such that the values of $\mathbf{a}_1, \mathbf{a}_2, \ldots, \mathbf{a}_m$ are $u_1, u_2, \ldots, u_m$ respectively. For this system of values of its free variables $(\exists \mathbf{b}_1)(\exists \mathbf{b}_2) \ldots (\exists \mathbf{b}_n)\mathbf{M}$ has the value t. Hence by the definition of value (rules d and f of §30), for this same system of values of the variables and for certain values $u_{i_1}, u_{i_2}, \ldots, u_{i_n}$ of $\mathbf{b}_1, \mathbf{b}_2, \ldots, \mathbf{b}_n$ respectively, **M** has the value t. If $\mathbf{d}_1, \mathbf{d}_2, \ldots, \mathbf{d}_n$ are chosen as $\mathbf{a}_{i_1}, \mathbf{a}_{i_2}, \ldots, \mathbf{a}_{i_n}$ respectively, then

$$\mathrm{S}^{\mathbf{b}_1\mathbf{b}_2\ldots\mathbf{b}_n}_{\mathbf{d}_1\mathbf{d}_2\ldots\mathbf{d}_n}\mathbf{M}|$$

has the value t. Therefore—still for the same system of values of the free variables of $(\exists \mathbf{b}_1)(\exists \mathbf{b}_2) \ldots (\exists \mathbf{b}_n)\mathbf{M}$—the disjunction **D** has the value t.

Since the free variables of **D** are the same as those of $(\exists \mathbf{b}_1)(\exists \mathbf{b}_2) \ldots (\exists \mathbf{b}_n)\mathbf{M}$ we have thus shown that—for this finite domain of *m* individuals, and for any system of possible values of the free variables of **D** such that $\mathbf{a}_1, \mathbf{a}_2, \ldots, \mathbf{a}_m$ have the values $u_1, u_2, \ldots, u_m$ respectively—the value of **D** is t. Now given any assignment of truth-values to the elementary parts of **D**,[423] it is clear that (because the values of $\mathbf{a}_1, \mathbf{a}_2, \ldots, \mathbf{a}_m$ are all different) it will always be possible to choose the values of the propositional and functional variables of **D** in such a way as to reproduce the given assignment of truth-values to the elementary parts of **D**. Therefore **D** has the value t for every assignment of truth-values to its elementary parts. Therefore **D** is a substitution instance of a tautology of P.

In *462 and **463, it is now clear that the condition that **D** is a substitution instance of a tautology is equivalent to the condition that

$$(\mathbf{a}_1)(\mathbf{a}_2) \ldots (\mathbf{a}_m)(\exists \mathbf{b}_1)(\exists \mathbf{b}_2) \ldots (\exists \mathbf{b}_n)\mathbf{M}$$

is valid in a domain of *m* individuals. Also in *460 and **461, the condition that an afp of **M** is a tautology is equivalent to the condition that

$$(\exists \mathbf{b}_1)(\exists \mathbf{b}_2) \ldots (\exists \mathbf{b}_n)\mathbf{M}$$

is valid in a domain of a single individual. Hence we have the following corollary of *460–**463:

*464. Let **M** be a quantifier-free formula, and let $\mathbf{a}_1, \mathbf{a}_2, \ldots, \mathbf{a}_m, \mathbf{b}_1, \mathbf{b}_2, \ldots, \mathbf{b}_n$ $(m \geqq 0,\ n \geqq 0)$ be the complete list of individual variables in **M**. Then

$$(\mathbf{a}_1)(\mathbf{a}_2) \ldots (\mathbf{a}_m)(\exists \mathbf{b}_1)(\exists \mathbf{b}_2) \ldots (\exists \mathbf{b}_n)\mathbf{M}$$

is a theorem if it is valid in a domain of *m* individuals, or, in the case $m = 0$, of a single individual.

Since the definition of validity in a domain leads immediately to an effective test for validity in any particular finite domain—and since, by **434, a theorem must be valid in all non-empty domains, including finite domains—we may regard *464 as stating an alternative form of our solution of the special case I of the decision problem. Indeed it is in this latter form that

[423]Where the same elementary part occurs more than once in **D**, of course it is meant that the same truth-value is assigned to it at all of its occurrences.

(As defined in §30, the *elementary parts* of a wff are those wf parts which have either the form of a propositional variable alone or the form $\mathbf{f}(\mathbf{a}_1, \mathbf{a}_2, \ldots, \mathbf{a}_n)$.)

the solution is more usually stated. And we shall introduce a corresponding form of statement of the solution (referring to validity in a specified finite domain) also in some other cases below.

We turn now to consideration of another decision procedure, which is applicable in a variety of cases. It will be convenient first to state the decision procedure itself, before considering the question of characterizing a class of wffs to which it is applicable.

A particular occurrence of a wff **P** as a wf part of a wff **A** is called an *occurrence as a truth-functional constituent*, or, as we shall also say, an *occurrence as a* P-*constituent*, in **A** if it is not within the scope of a quantifier and does not have either of the forms $\sim\mathbf{B}$ or $[\mathbf{B}_1 \supset \mathbf{B}_2]$. And the *truth-functional constituents*, or the P-*constituents*, of **A** are those wffs which have occurrences as P-constituents in **A**.

It is clear that each of the P-constituents of a wff either is an elementary part or else is of the form $(\mathbf{a})\mathbf{B}$. Moreover, any wff can be thought of as obtained from a wff of P by substituting its P-constituents in an appropriate way for the propositional variables; and, for a particular wff, the wff of P and the required substitution are determined uniquely to within an alphabetic change of propositional variables.

As a first step in the decision procedure we are about to describe, we reduce (separately) each P-constituent of the given wff to prenex normal form, and if any of the P-constituents are then found to differ only by alphabetic changes of bound variables, we make the appropriate alphabetic changes of bound variables to render them identical.[424] Let **A** be the wff so obtained. (Evidently the given wff is a theorem if and only if **A** is a theorem.)

If **A** has just *m* different P-constituents, $\mathbf{P}_1, \mathbf{P}_2, \ldots, \mathbf{P}_m$, then for each of the 2^m different systems of truth-values of these P-constituents we may ascertain the corresponding truth-values of **A**. (The work of doing this may be arranged in the same way as described in §15 for the case of wffs of the propositional calculus.) If the value of **A** is found to be t for all systems of truth-values of its P-constituents, then $\vdash \mathbf{A}$ by *311. Otherwise we list all the *falsifying* systems of truth-values of the P-constituents of **A**, i.e., the systems of truth-values for which the corresponding value of **A** is f. And

[424]In practice it will be desirable also to make at this stage any preliminary simplifications by propositional calculus that are seen to be possible, both to the wff as a whole (the P-constituents being treated as units) and within the matrix of each separate P-constituent. Especially, if any of the P-constituents can be rendered identical by such simplifications of their matrices, together with alphabetic changes of bound variables, this should be done; and it may be desirable to test systematically for the possibility of this, by the truth-table decision procedure (§15).

then we make use of the following metatheorem, which has an obvious connection with the conjunctive normal form:

*465. Where **A** is any wff, let the complete list of the P-constituents of **A** be $\mathbf{P}_1, \mathbf{P}_2, \ldots, \mathbf{P}_m$, and let the complete list of falsifying systems of truth-values of the P-constituents of **A** consists in the systems of values $\tau_1^i, \tau_2^i, \ldots, \tau_m^i$ of $\mathbf{P}_1, \mathbf{P}_2, \ldots, \mathbf{P}_m$ respectively $(i = 1, 2, \ldots, n)$. Let $\mathbf{P}_j^i$ be $\mathbf{P}_j$ or $\sim\mathbf{P}_j$ according as τ_j^i is t or f. Then $\vdash$ **A** if and only if all of the wffs

$$\mathbf{P}_1^i \supset . \mathbf{P}_2^i \supset . \ldots \mathbf{P}_{m-1}^i \supset \sim\mathbf{P}_m^i$$

are theorems.

Proof. For every system of truth-values of $\mathbf{P}_1, \mathbf{P}_2, \ldots, \mathbf{P}_m$, the value of **A** is f if and only if the value of one of the wffs

$$\mathbf{P}_1^i \supset . \mathbf{P}_2^i \supset . \ldots \mathbf{P}_{m-1}^i \supset \sim\mathbf{P}_m^i$$

is f. Consequently

$$\mathbf{A} \equiv [\mathbf{P}_1^1 \supset . \mathbf{P}_2^1 \supset . \ldots \mathbf{P}_{m-1}^1 \supset \sim\mathbf{P}_m^1][\mathbf{P}_1^2 \supset . \mathbf{P}_2^2 \supset . \ldots \mathbf{P}_{m-1}^2 \supset \sim\mathbf{P}_m^2] \\ \ldots [\mathbf{P}_1^n \supset . \mathbf{P}_2^n \supset . \ldots \mathbf{P}_{m-1}^n \supset \sim\mathbf{P}_m^n]$$

is a substitution instance of a tautology, and therefore a theorem by *311. Hence, by P, if $\vdash$ **A**, all of the wffs

$$\mathbf{P}_1^i \supset . \mathbf{P}_2^i \supset . \ldots \mathbf{P}_{m-1}^i \supset \sim\mathbf{P}_m^i$$

are theorems, and conversely, if all of these wffs are theorems, then $\vdash$ **A**.

Thus the decision problem for **A** is reduced to the decision problem for the wffs

$$\mathbf{P}_1^i \supset . \mathbf{P}_2^i \supset . \ldots \mathbf{P}_{m-1}^i \supset \sim\mathbf{P}_m^i.$$

We deal with these latter wffs by reducing them to a prenex normal form, since decision procedure are known for wffs in prenex normal form with prefixes of various special kinds. According to the fixed procedure of §39 for reduction to *the* prenex normal form, the quantifiers are taken one by one in left-to-right order and brought forward into the prefix. Here, however, we vary this fixed procedure by allowing the quantifiers to be taken (and brought forward into the prefix) also in any other order that is feasible. And we endeavor in this way to obtain a prefix of one of the kinds for which a decision procedure is known. E.g., if we succeed in obtaining, in all of the *n* cases, a prefix in which no existential quantifier precedes any universal quantifier, we are then able to decide whether **A** is a theorem by using our previous solution of the decision problem for the special case I.

In particular, such a reduction to the special case I can always be obtained when the foregoing procedure is applied to any one of the:[425]

II *Well-formed formulas in which every truth-functional constituent either is quantifier-free or has a prenex normal form that has only universal quantifiers in the prefix.*

We mention also the following subcase of II as of especial importance:

II′ *Well-formed formulas in which there are no free individual variables and in which every truth-functional constituent either is quantifier-free (therefore a propositional variable) or has the form* (**a**)**M**, *where* **M** *is quantifier-free and contains no propositional variables.*

In the subcase II′ the two following simplifications of the decision procedure are possible, as it is left to the reader to verify:

(1) Suppose that the P-constituents of **A** are numbered in such an order that $\mathbf{P}_1, \mathbf{P}_2, \ldots, \mathbf{P}_k$ are the ones which are propositional variables, and the remaining P-constituents are $\mathbf{P}_{k+1}, \mathbf{P}_{k+2}, \ldots, \mathbf{P}_m$. Then in applying *465 we may simplify the conclusion of the metatheorem as follows: ⊢ **A** if and only if all of the wffs

$$\mathbf{P}^i_{k+1} \supset \centerdot\, \mathbf{P}^i_{k+2} \supset \centerdot \ldots \mathbf{P}^i_{m-1} \supset {\sim}\mathbf{P}^i_m$$

are theorems.

(2) By alphabetic changes of bound variables we may suppose that **A** has been brought into such a form that only the one individual variable x occurs. Then $\mathbf{P}_{k+1}, \mathbf{P}_{k+2}, \ldots, \mathbf{P}_m$ have the forms $(x)\mathbf{M}_{k+1}, (x)\mathbf{M}_{k+2}, \ldots, (x)\mathbf{M}_m$ respectively, where $\mathbf{M}_{k+1}, \mathbf{M}_{k+2}, \ldots, \mathbf{M}_m$ are quantifier free and contain no propositional variables and no individual variables other than x. From a particular falsifying system of truth-values $\tau^i_1, \tau^i_2, \ldots, \tau^i_m$ of $\mathbf{P}_1, \mathbf{P}_2, \ldots, \mathbf{P}_m$ select the last $m - k$ truth-values $\tau^i_{k+1}, \tau^i_{k+2}, \ldots, \tau^i_m$, and among these suppose that $\tau^i_{t_1}, \tau^i_{t_2}, \ldots, \tau^i_{t_l}$ are t and $\tau^i_{f_1}, \tau^i_{f_2}, \ldots, \tau^i_{f_{m-k-l}}$ are f. Then in order that

---

[425]As a matter of fact, whenever such a reduction to the special case I is possible, it will always be possible also to reduce more directly to case I. Namely, among the various prenex normal forms to which **A** can be reduced, there will always be one in whose prefix no existential quantifier precedes any universal quantifier. But by making use of *465 in the way described, it is possible more easily to control the kind of prefix which is obtained and often also, especially in case II,, to shorten the work otherwise.

When the procedure described in the text, by making use of *465, results in a reduction to one of the special cases V–IX that are listed at the end of this section, it is not necessarily true that one of the prenex normal forms of the given wff **A** will also fall under one of these cases. And in this way solutions may be obtained of additional special cases of the decision problem.

$$\mathbf{P}^i_{k+1} \supset \centerdot\ \mathbf{P}^i_{k+2} \supset \centerdot \ldots \mathbf{P}^i_{m-1} \supset \sim\mathbf{P}^i_m$$

shall be a theorem, it is necessary and sufficient that at least one of the quantifier-free formulas

$$\mathbf{M}_{t_1} \supset \centerdot\ \mathbf{M}_{t_2} \supset \centerdot \ldots \mathbf{M}_{t_l} \supset \mathbf{M}_{f_j}$$

shall be a substitution instance of a tautology ($j = 1, 2, \ldots, m - k - l$).

Returning to the general case of the above-described decision procedure (based on *465), we notice that, roughly speaking, the finer the division of **A** obtained by dividing **A** into its P-constituents, the greater is the chance of success in determining by this procedure whether **A** is a theorem. Therefore before applying the decision procedure to a given wff **A** it may be desirable first to reduce **A** as far as possible by means of the reduction steps (a)–(g) of exercise 39.6.[426]

In particular, as proved in exercise 39.6, if **A** contains none but singulary functional variables, the reduction process of that exercise suffices to reduce the universal closure of **A** to the case II′ which we have just treated. This is Quine's solution of the decision problem for the special case of:

III *Well-formed formulas of the singulary functional calculus of first order.*

The history of this case of the decision problem is described in §49. Besides Quine's solution (which goes back to Behmann), another approach may be based on the following metatheorem of Bernays and Schönfinkel:[427]

**466. If a wff of the singulary functional calculus of first order is valid in a domain of 2^N individuals, where N is the number of different functional variables appearing, then it is valid in all domains of individuals.

Proof. We may suppose, by **432, that the given wff has no free individual variables. Let the propositional variables appearing be $\mathbf{p}_1, \mathbf{p}_2, \ldots, \mathbf{p}_l$ and

---

[426]The reduction process may be shortened by adding corresponding reduction steps for connectives other than implication and negation, so as to be able to deal directly with a wff abbreviated by means of D3–11 (as well as D14) rather than first to rewrite it in unabbreviated form. The full disjunctive and full conjunctive normal forms may also be found useful, as in 39.8. Or the implicative normal form (15.4) may replace the full disjunctive normal form, and the full conjunctive normal form may be used in the modified version that appears in the proof of *465. Details of organizing the reduction process in the most efficient manner are left to the reader.

[427]In practice the decision procedure of **466 is generally longer and therefore less advantageous than Quine's. But even Quine's procedure may become forbiddingly long in comparatively simple cases, as the reader may see by applying it, for instance, to 46.12(3).

the functional variables, $\mathbf{f}_1, \mathbf{f}_2, \ldots, \mathbf{f}_N$; and consider a system of values $\tau_1, \tau_2, \ldots, \tau_l, \Phi_1, \Phi_2, \ldots, \Phi_N$ of these variables in order, the domain of individuals being some arbitrary non-empty domain $\mathfrak{U}$. Let the individuals of $\mathfrak{U}$ be divided into classes by the rule that u_1 and u_2 belong to the same class if and only if the truth-values $\Phi_1(u_1), \Phi_2(u_1), \ldots, \Phi_N(u_1)$ are identical with the truth-values $\Phi_1(u_2), \Phi_2(u_2), \ldots, \Phi_N(u_2)$ respectively. Thus are obtained at most 2^N non-empty classes of individuals, call them $v_1, v_2, \ldots, v_n$ $(1 \leqq n \leqq 2^N)$. And let singulary propositional functions $\Psi_1, \Psi_2, \ldots, \Psi_N$ of these classes be defined by the rule that $\Psi_i(v_j)$ is the same truth-value as $\Phi_i(u)$, where u is any member of the class v_j.

Now we may take also the finite domain $\mathfrak{V}$, consisting of the individuals $v_1, v_2, \ldots, v_n$ and consider the values $\tau_1, \tau_2, \ldots, \tau_l, \Psi_1, \Psi_2, \ldots, \Psi_N$ of $\mathbf{p}_1, \mathbf{p}_2, \ldots, \mathbf{p}_l, \mathbf{f}_1, \mathbf{f}_2, \ldots, \mathbf{f}_N$ respectively. For this system of values of its free variables the given wff has the value t (because, being valid in a domain of 2^N individuals, it is by **439 valid in the domain $\mathfrak{V}$). But from the way in which the propositional functions Ψ_i were defined it follows that the given wff has the same value for the domain $\mathfrak{V}$ and for the system of values $\tau_1, \tau_2, \ldots, \tau_l, \Psi_1, \Psi_2, \ldots, \Psi_N$ of its free variables that it does for the domain $\mathfrak{U}$ and for the system of values $\tau_1, \tau_2, \ldots, \tau_l, \Phi_1, \Phi_2, \ldots, \Phi_N$ of its free variables. Therefore the value of the given wff is t also for the domain $\mathfrak{U}$ and for the latter system of values of its free variables.

Thus we have shown about the given wff that its value is t for an arbitrary system of values of its free variables and for an arbitrarily chosen domain $\mathfrak{U}$. I.e., we have shown that it is valid.

It will be observed that **466 is stated not as a solution of a special case of the decision problem (i.e., of the decision problem for provability) but rather as a solution of a special case of what we shall call the *decision problem for validity*, i.e., the problem of finding an effective procedure to determine validity.

By Gödel's completeness theorem (as proved in §44, and by another method in §45) it is true in one sense that a solution of a special case of the decision problem for validity is also a solution, in the same special case, of the decision problem. But in another sense—which we have not attempted to make precise—this is not true, as may be seen from the fact that the proof of **466 provides no effective method of finding a proof of a wff **A** which passes the test of containing none but singulary functional variables and being valid in a domain of 2^N individuals.[428]

---

[428]On the other hand, our demonstration of Quine's solution of the special case III

Closely related to the decision problem for validity is the *decision problem for satisfiability*, i.e., the problem of finding an effective procedure to determine satisfiability. By **430 and **431, every solution of a special case of either of these problems leads to a solution of a corresponding special case of the other, so that the two problems need not be considered separately. In much of the existing literature on the subject, it is the decision problem for satisfiability to which attention is primarily given. And apropos of the importance of this problem it should be observed that (by Gödel's completeness theorem) the consistency of a logistic system obtained by adding postulates[429] to a simple applied functional calculus of first order is always equivalent, in an obvious way, to the satisfiability of a corresponding wff of the pure functional calculus of first order.

The pure functional calculus of first order becomes a formalized language upon adopting one of the principal interpretations (§30). The domain of individuals on which the interpretation is based may be either infinite or finite. In the former case the *semantical decision problem* of the language (as defined in §15) is equivalent to the decision problem for validity, in the sense that any solution of a special case of either problem is also a solution, in the same special case, of the other. In the latter case the semantical decision problem of the language is equivalent to the decision problem for validity in the same finite domain and is therefore completely solved (on the assumption that the finite domain is given in such a way that the number of individuals is known).

We shall not here treat further the question of special cases of any of these decision problems, but we conlcude merely by recording the existence of solutions of the decision problem or of the decision problem for validity in each of the following cases (either explicitly in the literature or easily obtained by methods existing in the literature):[430]

---

of the decision problem does (implicitly) provide such an effective method of finding a proof of a wff which passes the test. In order to accomplish this also in connection with the Bernays-Schönfinkel solution of case III, the method may be followed which is suggested below in exercise 46.1.

[429]Compare §55, as well as the discussion of the *axiomatic method* at the end of §07.

[430]See Wilhelm Ackermann in the *Mathematische Annalen*, vol. 100 (1928), pp. 638–649; Thoralf Skolem in the *Norsk Matematisk Tidsskrift*, vol. 10 (1928), pp. 125–142; Jacques Herbrand in the *Comptes Rendus des Séances de la Société des Sciences et des Lettres de Varsovie*, Classe III, vol. 24 (1931), pp. 12–56; Kurt Gödel in Menger's *Ergebnisse eines Mathematischen Kolloquiums*, no. 2 (for 1929–1930, published 1932), pp. 27–28; László Kalmár in the *Mathematische Annalen*, vol. 108 (1933), pp. 466–484; Gödel in the *Monatshefte für Mathematik und Physik*, vol. 40 (1933), pp. 433–443; Kurt Schütte in the *Mathematische Annalen*, vol. 109 (1934), pp. 572–603, and vol. 110 (1934). pp. 161–194; Ackermann in the *Mathematische Annalen*, vol. 112 (1936), pp. 419–432.

Whenever in these papers the results are given in the form of solutions of special cases of the decision problem for satisfiability, they may be restated as solutions of

IV *Well-formed formulas* **A** *such that in each elementary part at most one variable has occurrences at which it is a bound variable of* **A**.

V′ *Well-formed formulas having a prenex normal form in which the matrix satisfies the condition of being a disjunction of elementary parts and negations of elementary parts or equivalent by laws of the propositional calculus to such a disjunction.*[431]

VI *Well-formed formulas having a prenex normal form with only one existential quantifier in the prefix, i.e., with a prefix of the form* $(\mathbf{a}_1)(\mathbf{a}_2)\ldots(\mathbf{a}_m)(\exists\mathbf{b})(\mathbf{c}_1)(\mathbf{c}_2)\ldots(\mathbf{c}_l)$.[432]

VII *Well-formed formulas having a prenex normal form with a prefix of the form* $(\mathbf{a}_1)(\mathbf{a}_2)\ldots(\mathbf{a}_m)(\exists\mathbf{b}_1)(\exists\mathbf{b}_2)(\mathbf{c}_1)(\mathbf{c}_2)\ldots(\mathbf{c}_l)$.[433]

VIII *Well-formed formulas having a prenex normal form with a prefix of the form* $(\mathbf{a}_1)(\mathbf{a}_2)\ldots(\mathbf{a}_m)(\exists\mathbf{b}_1)(\exists\mathbf{b}_2)\ldots(\exists\mathbf{b}_n)(\mathbf{c}_1)(\mathbf{c}_2)\ldots(\mathbf{c}_l)$ *and a matrix in which every elementary part that contains any of the variables* $\mathbf{b}_1, \mathbf{b}_2, \ldots, \mathbf{b}_n$ *contains either all of the variables* $\mathbf{b}_1, \mathbf{b}_2, \ldots, \mathbf{b}_n$ *or at least one of the variables* $\mathbf{c}_1, \mathbf{c}_2, \ldots, \mathbf{c}_l$.[434]

---

corresponding special cases of the decision problem for validity, and for the sake of uniformity in summarizing the results we have done this systematically. The decision problem in the sense of footnote 421 is dealt with explicitly only by Herbrand.

[431]This case is solved by Herbrand, *loc.cit.* An equivalent condition on the matrix is that it shall have the form $\mathbf{A}_1 \supset . \mathbf{A}_2 \supset . \ldots \mathbf{A}_{n-1} \supset \mathbf{A}_n$, where $n \geqq 1$ and each $\mathbf{A}_i$ ($i = 1, 2, \ldots, n$) is either an elementary part or the negation of an elementary part, or shall be equivalent by laws of the propositional calculus to a matrix of this form. Still another equivalent condition is that the value of the matrix shall be f for at most one assignment of truth-values to the elementary parts.

[432]Ackermann, Skolem, and Herbrand, *loc.cit.* According to Ackermann (1928), a wff of class VI which contains no free individual variables and no functional variables that are more than binary is valid if it is valid in a domain $m + ((ml + l)^\nu - 1)/(ml + l - 1)$ individuals, where N is the number of different functional variables appearing and

$$\nu = 3 \times 2^{N(2m+1)(m+1)^2l^2} + 1.$$

Or in case $m = 0$, $l = 1$, the wff is valid if valid in a domain of 3×2^N individuals. If ternary or higher functional variables appear, then a similar result may be found by Ackermann's methods.

This provides a strictly theoretical solution of case VI of the decision problem for validity, and is hardly available for use in practice. A more practicable decision procedure, however, may be obtained from any one of the three papers, and is indicated in exercises at the end of this section.

[433]Gödel, Kalmár, and Schütte, *loc.cit.* According to Schütte, a wff of class VII that contains no free individual variables is valid if it is valid in a domain of $m + 2^{20\nu}$ individuals, where N is the number of different functional variables appearing, none of the functional variables is more than h-ary, and

$$\nu = Nl^2 2^h (m + 1)^{h+4}.$$

Again there is a more practicable decision procedure which may be obtained from the papers of Gödel or that of Kalmár.

[434]Skolem, *loc.cit.*

IX *Well-formed formulas having a prenex normal form with a prefix terminating in* $(\mathbf{c}_1)(\mathbf{c}_2)\ldots(\mathbf{c}_l)$ *and a matrix in which every elementary part that contains any of the variables occurring in the prefix contains at least one of the variables* $\mathbf{c}_1, \mathbf{c}_2, \ldots, \mathbf{c}_l$.[434]

X *Well-formed formulas of the form* $(\mathbf{a}_1)(\mathbf{a}_2)\ldots(\mathbf{a}_n)\mathbf{M} \supset (\exists\mathbf{b})(\mathbf{c})\mathbf{f}(\mathbf{b}, \mathbf{c})$, *where* $n \leqq 4$, *and* $\mathbf{M}$ *is quantifier-free and contains no variables other than* $\mathbf{f}, \mathbf{a}_1, \mathbf{a}_2, \ldots, \mathbf{a}_n$.[435]

Treatment or partial treatment of all of these cases except VII and X will be indicated briefly in exercises which follow at the end of this section, as well as of some other cases of lesser importance.

In most of the cases it is possible to put the solution of the decision problem for validity in the form that, if a wff of the class in question is valid in a domain of a specified finite number of individuals, then it is valid (though this is seldom the most efficient form of the solution for use in practice, i.e., in applying the decision procedure to particular wffs). Case X is of some interest as an exception to this. For it includes wffs that are valid in every finite domain without being valid in an infinite domain, as may be shown by the example 43.5(2).

EXERCISES 46

46.0. In order to establish the simplified decision procedure for case II′ of the decision problem, prove the rules (1) and (2) which are given above in connection with this case.

46.1. (1) Consider a wff $\mathbf{A}$ of the singulary functional calculus of first order (case III) and let the different functional variables appearing in $\mathbf{A}$ be $\mathbf{f}_1, \mathbf{f}_2, \ldots, \mathbf{f}_N$. We may suppose, by **432, that there are no free individual variables in $\mathbf{A}$. According to Quine's solution of case III of the decision problem, as described above, the reduction process of exercise 39.6 is first to be applied to $\mathbf{A}$. By a modification of this reduction process, show that $\mathbf{A}$ may be reduced to a wff $\mathbf{B}$ such that $\vdash \mathbf{A} \equiv \mathbf{B}$, and all the P-constituents of $\mathbf{B}$ other than propositional variables are of the form $(x)\,.\,\mathbf{D}_1 \supset .\,\mathbf{D}_2 \supset .\ldots\mathbf{D}_{N-1} \supset \mathbf{D}_N$ where each $\mathbf{D}_i$ separately is either $\mathbf{f}_i(x)$ or $\sim\mathbf{f}_i(x)$ $(i = 1, 2, \ldots, N)$. But only 2^N different P-constituents of this form are possible. Hence prove **466 by applying to $\mathbf{B}$ the decision procedure of case II′.

[435]The solution of this case is in Ackermann's paper of 1936, cited in footnote 430. A modification of one part of Ackermann's decision procedure, reducing its length for application in practice, is given by J. J. Gégalkine in the *Recueil Mathématique*, new series vol. 6 (1939), pp. 185–198.

The solution of the decision problem in case X should be compared with the reduction of the decision problem which is stated in footnote 447.

(2) Show also that $\vdash \mathbf{A}$ if and only if in every falsifying system of truth-values of the P-constituents of **B** all the P-constituents other than propositional variables have the value t.

46.2. Apply Quine's solution of case III to each of the following wffs:

(1) $$(\exists x)(y) \,.\, F(x) \equiv p \supset .\, F(y) \equiv p$$

(2) $$(\exists x)(y) \,.\, F(x) \equiv F(y)$$

(3) $$(\exists x)[F(x) \supset G(x)] \equiv (\exists x)(\exists y)[F(x) \supset G(y)]$$

(4)[436] $$F(x) \supset_x [F(y) \supset G(x)] \supset .\, p \supset .\, (x)F(x) \supset G(y)$$

(5) $$(\exists x)(\exists y)(z_1)(z_2) \,.\, F(y) \supset G(z_1) \supset G(x) \sim F(z_1) \supset . \\ F(x) \vee G(x) \supset H(x) \supset H(z_2) \,.\, H(y) \supset .\, F(z_2) \vee G(z_2) \supset H(z_2)$$

46.3. Solve case IV of the decision problem by employing the same reduction process (cf. exercise 39.6) as in Quine's solution of case III. Illustrate by using this method to determine which of the following wffs are theorems:

(1) $$(\exists x)(y_1)(y_2) \,.\, {\sim}F(x, z) \supset F(z, y_1) \supset .\, F(y_2, z) \supset F(z, y_2)$$

(2) $$(\exists z)F(x, z) \supset (z)G(x, z) \supset .\, (z)[G(z, z) \supset F(z, y)] \supset .\, F(x,y) \equiv (z)G(x,z)$$

46.4. Solve case IV of the decision problem by reducing it to case III, finding for every wff **A** of class IV a corresponding wff of class III which is a theorem if and only if **A** is a theorem. (*Suggestion:* Make use of the idea of replacing each elementary part of **A** by an elementary part involving only a singulary functional variable.) Check your solution by applying it to the two following wffs and verifying that the same results are obtained as when the decision procedure for case I is applied to them:

$$F(x, y) \supset_x F(y, x) \supset \sim .\, F(x, y) \supset_x {\sim}F(y, x)$$

$$F(x, y) \supset F(y, x) \supset_x G(x, y) \supset \sim .\, F(x, y) \supset F(y, x) \supset_x {\sim}G(x, y)$$

46.5. As explained above, every solution of a special case of the decision problem for validity leads to a solution of a corresponding special case of the decision problem for satisfiability. State special cases of the decision problem for satisfiability which thus correspond to cases I–IV of the decision problem for validity; and state a decision procedure for each of them, directly (i.e., without referring to decision procedures for cases I–IV of the decision problem for validity).

46.6. Let case VI_0^1 be the subcase of case VI in which there are no free individual variables, and $m = 0$, $l = 1$. I.e., in case VI_0^1 the given wff **A**

[436]This is a modified form of an example used by Quine.

has as prenex normal form $(\exists \mathbf{b})(\mathbf{c})\mathbf{M}$, where $\mathbf{M}$ is the matrix and contains no individual variables except $\mathbf{b}$ and $\mathbf{c}$.

Suppose that no propositional variables appear and the only functional variable appearing is a binary functional variable $\mathbf{f}$. Taking the positive integers as the domain of individuals, consider the following attempt to find a value of the functional variable $\mathbf{f}$ for which the value of $\mathbf{A}$ is f (falsehood). For the value 1 of $\mathbf{b}$ we must find a corresponding value of $\mathbf{c}$ for which $\mathbf{M}$ has the value f, and we may suppose without loss of generality that this corresponding value of $\mathbf{c}$ is 2. The (distinct) elementary parts of $\mathbf{M}$ are some or all of $\mathbf{f}(\mathbf{b}, \mathbf{c})$, $\mathbf{f}(\mathbf{c}, \mathbf{b})$, $\mathbf{f}(\mathbf{b}, \mathbf{b})$, $\mathbf{f}(\mathbf{c}, \mathbf{c})$; by assigning appropriate truth-values to these we can, in 0 or more ways, give to $\mathbf{M}$ the value f. Thus, if Φ is the propositional function which is to be the value of $\mathbf{f}$, we determine the possibilities as to what $\Phi(1, 2)$, $\Phi(2, 1)$, $\Phi(1, 1)$, $\Phi(2, 2)$ may be. Then we must consider also the value 2 of $\mathbf{b}$ and find corresponding to it a value of $\mathbf{c}$ for which $\mathbf{M}$ has the value f. Without loss of generality we may suppose that this new value of $\mathbf{c}$ is 3. Again we consider the truth-values to be assigned to $\mathbf{f}(\mathbf{b}, \mathbf{c})$, $\mathbf{f}(\mathbf{c}, \mathbf{b})$, $\mathbf{f}(\mathbf{b}, \mathbf{b})$, $\mathbf{f}(\mathbf{c}, \mathbf{c})$ so as to give to $\mathbf{M}$ the value f; and thus we determine the possibilities as to what $\Phi(2, 3)$, $\Phi(3, 2)$, $\Phi(2, 2)$, $\Phi(3, 3)$ may be. This gives us two separate determinations of what $\Phi(2, 2)$ is to be, and it is seen that there are the following alternatives. (i) It may happen that the two determinations of the value $\Phi(2, 2)$ of Φ cannot be reconciled with each other by using any of the possible assignments of truth-values to $\mathbf{f}(\mathbf{b}, \mathbf{c})$, $\mathbf{f}(\mathbf{c}, \mathbf{b})$, $\mathbf{f}(\mathbf{b}, \mathbf{b})$, $\mathbf{f}(\mathbf{c}, \mathbf{c})$ that give to $\mathbf{M}$ the value f (either by using the same assignment of truth-values to $\mathbf{f}(\mathbf{b}, \mathbf{c})$, $\mathbf{f}(\mathbf{c}, \mathbf{b})$, $\mathbf{f}(\mathbf{b}, \mathbf{b})$, $\mathbf{f}(\mathbf{c}, \mathbf{c})$ both times or by using two different assignments); then $\mathbf{A}$ is valid. (ii) It may happen that the two determinations of the value $\Phi(2, 2)$ of Φ can be reconciled with each other; then we may go on to find corresponding to the value 3 of $\mathbf{b}$ a value 4 of $\mathbf{c}$ for which $\mathbf{M}$ has the value f, and corresponding to the value 4 of $\mathbf{b}$ a value 5 of $\mathbf{c}$ for which $\mathbf{M}$ has the value f, and so forth; because no further hindrance can be encountered, it follows that $\mathbf{A}$ is not valid. Thus the issue depends on whether or not it is possible to find a value Φ of $\mathbf{f}$ such that $\mathbf{M}$ has the value falsehood both for the values 1, 2 of $\mathbf{b}$, $\mathbf{c}$ and for the values 2, 3 of $\mathbf{b}$, $\mathbf{c}$.

(1) Supply details of the argument which is outlined in the preceding paragraph, and complete it so as to show that $\mathbf{A}$ is valid if and only if the disjunction

$$\mathrm{S}^{\mathbf{b}\,\mathbf{c}}_{x_1 x_2}\mathbf{M}| \vee \mathrm{S}^{\mathbf{b}\,\mathbf{c}}_{x_2 x_3}\mathbf{M}|$$

is a substitution instance of a tautology, or, as we shall say, if and only if this disjunction is *tautologous*.

(2) Extend this result to the more general subase of case VI_0^1 in which any number of propositional variables appear and a single functional variable (not necessarily binary). I.e., show in this case also that **A** is valid if and only if the disjunction

$$\mathrm{S}_{x_1 x_2}^{\mathbf{b}\,\mathbf{c}}\,\mathbf{M}| \vee \mathrm{S}_{x_2 x_3}^{\mathbf{b}\,\mathbf{c}}\,\mathbf{M}|$$

is tautologous.

(3) Complete the solution of this special case of the decision problem by stating explicitly a proof of **A** if the foregoing disjunction is tautologous.

46.7. By the same method solve the further subcase of case VI_0^1 of the decision problem in which there appear any number of propositional variables and just two functional variables. Show in this case that **A** is a theorem if and only if the disjunction

$$\mathrm{S}_{x_1 x_2}^{\mathbf{b}\,\mathbf{c}}\,\mathbf{M}| \vee \mathrm{S}_{x_2 x_3}^{\mathbf{b}\,\mathbf{c}}\,\mathbf{M}| \vee \mathrm{S}_{x_3 x_4}^{\mathbf{b}\,\mathbf{c}}\,\mathbf{M}| \vee \mathrm{S}_{x_4 x_5}^{\mathbf{b}\,\mathbf{c}}\,\mathbf{M}|$$

is tautologous.

46.8. By the same method solve case VI_0^1 of the decision problem. Namely, show that **A** is a theorem if and only if the disjunction

$$\mathrm{S}_{x_1 x_2}^{\mathbf{b}\,\mathbf{c}}\,\mathbf{M}| \vee \mathrm{S}_{x_2 x_3}^{\mathbf{b}\,\mathbf{c}}\,\mathbf{M}| \vee \ldots \vee \mathrm{S}_{x_{2N}\, x_{2N+1}}^{\mathbf{b}\quad \mathbf{c}}\,\mathbf{M}|$$

is tautologous, where N is the number of different functional variables appearing.

46.9. Apply the decision procedure of 46.6–46.8 to determine which of the following wffs are theorems:

(1) $(\exists x)(y) \,.\, F(x, y) \equiv F(x, x) \supset .\, F(x, y) \equiv F(y, y)$

(2)[437] $(\exists x)(y) \,.\, F(x, x) \supset F(y, y) \supset F(x, y)G(x) \supset G(y)$

(Notice that it is not asked to write out explicitly the proof of a wff which is found to be a theorem. Therefore instead of making use of the disjunction which, according to 46.6–46.8, is tautologous if and only if the given wff is a theorem, it may often be found more convenient just to follow through the same procedure by which this disjunction was obtained, i.e., the procedure described in the second paragraph of 46.6, or a suitable generalization of this procedure.)

46.10. As a corollary of 46.8 establish the result of Bernays and Schönfinkel that, in case VI_0^1, **A** is valid if it is valid in a domain of $2^{N'}$ individuals, where N' is the number of different functional variables appearing or the number 2, whichever is greater.

---

[437]This is essentially the same as one of Skolem's examples.

46.11. The method used in 46.6–46.8 to solve case VI^1_0 of the decision problem can as a matter of fact be extended to solve case VI in general.[438]

(1) Use this method to solve the case VI^2_0, in which the given wff $\mathbf{A}$ has a prenex normal form $(\exists\mathbf{b})(\mathbf{c}_1)(\mathbf{c}_2)\mathbf{M}$, where $\mathbf{M}$ is the matrix and contains no individual variables except $\mathbf{b}$, $\mathbf{c}_1$, $\mathbf{c}_2$. Show in this case that $\mathbf{A}$ is a theorem if and only if the disjunction

$$\mathrm{S}^{\mathbf{b}\,\mathbf{c}_1\mathbf{c}_2}_{x_1x_2x_3}\mathbf{M}| \vee \mathrm{S}^{\mathbf{b}\,\mathbf{c}_1\mathbf{c}_2}_{x_2x_4x_5}\mathbf{M}| \vee \mathrm{S}^{\mathbf{b}\,\mathbf{c}_1\mathbf{c}_2}_{x_3x_6x_7}\mathbf{M}| \vee \mathrm{S}^{\mathbf{b}\,\mathbf{c}_1\mathbf{c}_2}_{x_4x_8x_9}\mathbf{M}| \vee \ldots \vee \mathrm{S}^{\mathbf{b}\;\;\mathbf{c}_1\;\;\mathbf{c}_2}_{x_\mu x_{2\mu} x_{2\mu+1}}\mathbf{M}|$$

is tautologous, where N is the number of different functional variables appearing and

$$\mu = 2^{2^N} - 1.$$

(2) Use this method to solve the case VI^1_1, in which the given wff $\mathbf{A}$ has a prenex normal form $(\mathbf{a})(\exists\mathbf{b})(\mathbf{c})\mathbf{M}$, where $\mathbf{M}$ is the matrix and contains no individual variables except $\mathbf{a}$, $\mathbf{b}$, $\mathbf{c}$. Show in this case that $\mathbf{A}$ is a theorem if and only if the disjunction

$$\mathrm{S}^{\mathbf{abc}}_{xxx_1}\mathbf{M}| \vee \mathrm{S}^{\mathbf{ab}\,\mathbf{c}}_{xx_1x_2}\mathbf{M}| \vee \mathrm{S}^{\mathbf{ab}\,\mathbf{c}}_{xx_2x_3}\mathbf{M}| \vee \mathrm{S}^{\mathbf{ab}\,\mathbf{c}}_{xx_3x_4}\mathbf{M}| \vee \ldots \vee \mathrm{S}^{\mathbf{ab}\;\;\;\mathbf{c}}_{xx_{2^\nu-1}x_{2^\nu}}\mathbf{M}|$$

is tautologous, where ν is the sum of the *weights* of the different functional variables that appear, the weight of an h-ary functional variable $\mathbf{f}$ being the number of different wffs of the form $\mathbf{f}(\mathbf{d}_1, \mathbf{d}_2, \ldots, \mathbf{d}_h)$ which occur as elementary parts in $\mathrm{S}^{\mathbf{c}}_{\mathbf{b}}\mathbf{M}|$, with the exception of the one wff $\mathbf{f}(\mathbf{a}, \mathbf{a}, \ldots, \mathbf{a})$ (which is not to be counted). (Taking the natural numbers as the domain of individuals, attempt to give to $\mathbf{A}$ the value f; for this it is sufficient to find one value of $\mathbf{a}$ for which the value of $(\exists\mathbf{b})(\mathbf{c})\mathbf{M}$ is f, and it may be supposed

---

[438]The close relationship should be noticed between this method and the method which was later used by Gödel in his proof of completeness of the functional calculus of first order. Indeed the disjunctions which are used in 46.6–46.8 and in 46.11(1) are the same as the disjunctions $\mathbf{C}_k$ of §44, each for a certain particular value of k.

In working with part (1) of exercise 46.11, it is recommended that the reader replace the notations,

$$\mathrm{S}^{\mathbf{b}\,\mathbf{c}_1\mathbf{c}_2}_{x_1x_2x_3}\mathbf{M}|, \qquad \mathrm{S}^{\mathbf{b}\,\mathbf{c}_1\mathbf{c}_2}_{x_2x_4x_5}\mathbf{M}|, \qquad \text{and so on,}$$

by the simpler notations $\mathbf{M}x_1x_2x_3$, $\mathbf{M}x_2x_4x_5$, and so on. Similarly, in part (2) the notations,

$$\mathrm{S}^{\mathbf{abc}}_{xxx_1}\mathbf{M}|, \qquad \mathrm{S}^{\mathbf{ab}\,\mathbf{c}}_{xx_1x_2}\mathbf{M}|, \qquad \text{and so on,}$$

may be replaced by $\mathbf{M}xxx_1$, $\mathbf{M}xx_1x_2$, and so on respectively. This simplified notation for substitution, essentially that of the Hilbert school, may conveniently be used in a context in which all substitutions are for the same list of variables, and especially when it is always a variable (or other single symbol) that is substituted for each variable. It will be useful also in connection with the exercises immediately following, and at many other places. However, in the text we shall retain the more explicit notation for substitution.

without loss of generality that this value of **a** is 0; then proceed as in 46.6, or as in 46.8.)

46.12. Apply the decision procedures of 46.11 to determine which of the following wffs are theorems:

(1) $(\exists x)(y)(z) \mathbin{.} F(y, z) \supset [G(y) \supset H(x)] \supset F(x, x) \supset \mathbin{.}$
$F(z, x) \supset G(x) \supset H(z) \supset \mathbin{.} F(x, y) \supset F(z, z)$

(2) $(x)(\exists y)(z) \mathbin{.} F(y, x) \supset [F(x, z) \supset F(x, y)] \mathbin{.}$
$F(x, y) \supset \mathbin{.} {\sim}F(x, z) \supset F(y, x)F(z, y)$

(3) $(\exists x)(y)(z) \mathbin{.} F(y) \supset G(y) \equiv F(x) \supset \mathbin{.} F(y) \supset H(y) \equiv G(x) \supset \mathbin{.}$
$F(y) \supset G(y) \supset H(y) \equiv H(x) \supset F(z)G(z)H(z)$

46.13. As remarked by Skolem in 1928, the same method may also be extended to cases in which there is more than one existential quantifier.

Take as an example the case of the prefix $(\exists \mathbf{b}_1)(\exists \mathbf{b}_2)(\mathbf{c})$. In connection with the prefix $(\exists \mathbf{b})(\mathbf{c})$, we used the following successive pairs of values of the variables **b**, **c**: $\langle 1, 2\rangle, \langle 2, 3\rangle, \langle 3, 4\rangle, \langle 4, 5\rangle$, and so on indefinitely Similarly, in connection with the prefix $(\exists \mathbf{b}_1)(\exists \mathbf{b}_2)(\mathbf{c})$, we may parallel the method of 46.6–46.8 as closely as possible, using the following successive triples of positive integers as values of the variables $\mathbf{b}_1$, $\mathbf{b}_2$, **c**: $\langle 1, 1, 2\rangle$; $\langle 1, 2, 3\rangle, \langle 2, 1, 4\rangle, \langle 2, 2, 5\rangle$; $\langle 1, 3, 6\rangle, \langle 3, 1, 7\rangle, \langle 2, 3, 8\rangle, \langle 3, 2, 9\rangle, \langle 3, 3, 10\rangle$; $\langle 1, 4, 11\rangle$, $\langle 4, 1, 12\rangle$, $\langle 2, 4, 13\rangle$, $\langle 4, 2, 14\rangle$, $\langle 3, 4, 15\rangle$, $\langle 4, 3, 16\rangle$, $\langle 4, 4, 17\rangle$; $\langle 1, 5, 18\rangle$, $\langle 5, 1, 19\rangle$, $\langle 2, 5, 20\rangle$, $\langle 5, 2, 21\rangle$, $\langle 3, 5, 22\rangle$, $\langle 5, 3, 23\rangle$, $\langle 4, 5, 24\rangle$, $\langle 5, 4, 25\rangle$, $\langle 5, 5, 26\rangle$; $\langle 1, 6, 27\rangle$, $\langle 6, 1, 28\rangle$, ... The enumeration of the ordered pairs of positive integers which is here employed has been modified, as compared to that used in §44. But this modification is non-essential from the point of view of §44, and we may therefore take the wffs $\mathbf{C}_k$ of §44 as modified correspondingly, i.e., by using the modified enumeration of the ordered pairs of positive integers. If in a special case we can find a particular value K of k about which we can prove that either $\mathbf{C}_K$ is tautologous or none of the wffs $\mathbf{C}_k$ is tautologous, then a solution of this special case of the decision problem follows by direct application of the methods of §44.[439]

(1) Apply this method to solve case V″ of the decision problem, in which the given wff **A** is in Skolem normal form and at the same time satisfies the conditions of case V′. Show in this case that $K = (l + 1)^n$, where n is the number of existential quantifiers in the prefix, and l the number of universal quantifiers. (Make use of the fact that **M** has the value f for at most one system of truth-values of its elementary parts.)

---

[439]For the assistance of the reader we add the following table, the significance of

(2) Apply this method to solve the subcase IX′ of case IX in which there are no free individual variables and the prefix is $(\exists \mathbf{b}_1)(\exists \mathbf{b}_2) \ldots (\exists \mathbf{b}_n)(\mathbf{c}_1)(\mathbf{c}_2) \ldots (\mathbf{c}_l)$, showing in this case that the given wff $\mathbf{A}$ is a theorem if and only if

$$\mathrm{S}^{\mathbf{b}_1\mathbf{b}_2\ldots\mathbf{b}_n}_{\mathbf{b}_1\mathbf{b}_1\ldots\mathbf{b}_1}\mathbf{M}|$$

is tautologous.

---

which will be clear by analogy with the explanation given in 46.6:

$\mathbf{b}_1$ $\mathbf{b}_2$ $\mathbf{c}$	$\mathbf{f}(\mathbf{b}_1, \mathbf{b}_1)$	$\mathbf{f}(\mathbf{b}_2, \mathbf{b}_2)$	$\mathbf{f}(\mathbf{b}_1, \mathbf{b}_2)$	$\mathbf{f}(\mathbf{b}_2, \mathbf{b}_1)$	$\mathbf{f}(\mathbf{b}_1, \mathbf{c})$	$\mathbf{f}(\mathbf{c}, \mathbf{b}_1)$	$\mathbf{f}(\mathbf{b}_2, \mathbf{c})$	$\mathbf{f}(\mathbf{c}, \mathbf{b}_2)$	$\mathbf{f}(\mathbf{c}, \mathbf{c})$
1 1 2	$\Phi(1,1)$	$\Phi(1,1)$	$\Phi(1,1)$	$\Phi(1,1)$	$\Phi(1,2)$	$\Phi(2,1)$	$\Phi(1,2)$	$\Phi(2,1)$	$\Phi(2,2)$
1 2 3	$\Phi(1,1)$	$\Phi(2,2)$	$\Phi(1,2)$	$\Phi(2,1)$	$\Phi(1,3)$	$\Phi(3,1)$	$\Phi(2,3)$	$\Phi(3,2)$	$\Phi(3,3)$
2 1 4	$\Phi(2,2)$	$\Phi(1,1)$	$\Phi(2,1)$	$\Phi(1,2)$	$\Phi(2,4)$	$\Phi(4,2)$	$\Phi(1,4)$	$\Phi(4,1)$	$\Phi(4,4)$
2 2 5	$\Phi(2,2)$	$\Phi(2,2)$	$\Phi(2,2)$	$\Phi(2,2)$	$\Phi(2,5)$	$\Phi(5,2)$	$\Phi(2,5)$	$\Phi(5,2)$	$\Phi(5,5)$
1 3 6	$\Phi(1,1)$	$\Phi(3,3)$	$\Phi(1,3)$	$\Phi(3,1)$	$\Phi(1,6)$	$\Phi(6,1)$	$\Phi(3,6)$	$\Phi(6,3)$	$\Phi(6,6)$
3 1 7	$\Phi(3,3)$	$\Phi(1,1)$	$\Phi(3,1)$	$\Phi(1,3)$	$\Phi(3,7)$	$\Phi(7,3)$	$\Phi(1,7)$	$\Phi(7,1)$	$\Phi(7,7)$
2 3 8	$\Phi(2,2)$	$\Phi(3,3)$	$\Phi(2,3)$	$\Phi(3,2)$	$\Phi(2,8)$	$\Phi(8,2)$	$\Phi(3,8)$	$\Phi(8,3)$	$\Phi(8,8)$
3 2 9	$\Phi(3,3)$	$\Phi(2,2)$	$\Phi(3,2)$	$\Phi(2,3)$	$\Phi(3,9)$	$\Phi(9,3)$	$\Phi(2,9)$	$\Phi(9,2)$	$\Phi(9,9)$
3 3 10	$\Phi(3,3)$	$\Phi(3,3)$	$\Phi(3,3)$	$\Phi(3,3)$	$\Phi(3,10)$	$\Phi(10,3)$	$\Phi(3,10)$	$\Phi(10,3)$	$\Phi(10,10)$
1 4 11	$\Phi(1,1)$	$\Phi(4,4)$	$\Phi(1,4)$	$\Phi(4,1)$	$\Phi(1,11)$	$\Phi(11,1)$	$\Phi(4,11)$	$\Phi(11,4)$	$\Phi(11,11)$
4 1 12	$\Phi(4,4)$	$\Phi(1,1)$	$\Phi(4,1)$	$\Phi(1,4)$	$\Phi(4,12)$	$\Phi(12,4)$	$\Phi(1,12)$	$\Phi(12,1)$	$\Phi(12,12)$
2 4 13	$\Phi(2,2)$	$\Phi(4,4)$	$\Phi(2,4)$	$\Phi(4,2)$	$\Phi(2,13)$	$\Phi(13,2)$	$\Phi(4,13)$	$\Phi(13,4)$	$\Phi(13,13)$
4 2 14	$\Phi(4,4)$	$\Phi(2,2)$	$\Phi(4,2)$	$\Phi(2,4)$	$\Phi(4,14)$	$\Phi(14,4)$	$\Phi(2,14)$	$\Phi(14,2)$	$\Phi(14,14)$
3 4 15	$\Phi(3,3)$	$\Phi(4,4)$	$\Phi(3,4)$	$\Phi(4,3)$	$\Phi(3,15)$	$\Phi(15,3)$	$\Phi(4,15)$	$\Phi(15,4)$	$\Phi(15,15)$
4 3 16	$\Phi(4,4)$	$\Phi(3,3)$	$\Phi(4,3)$	$\Phi(3,4)$	$\Phi(4,16)$	$\Phi(16,4)$	$\Phi(3,16)$	$\Phi(16,3)$	$\Phi(16,16)$
4 4 17	$\Phi(4,4)$	$\Phi(4,4)$	$\Phi(4,4)$	$\Phi(4,4)$	$\Phi(4,17)$	$\Phi(17,4)$	$\Phi(4,17)$	$\Phi(17,4)$	$\Phi(17,17)$
1 5 18	$\Phi(1,1)$	$\Phi(5,5)$	$\Phi(1,5)$	$\Phi(5,1)$	$\Phi(1,18)$	$\Phi(18,1)$	$\Phi(5,18)$	$\Phi(18,5)$	$\Phi(18,18)$
5 1 19	$\Phi(5,5)$	$\Phi(1,1)$	$\Phi(5,1)$	$\Phi(1,5)$	$\Phi(5,19)$	$\Phi(19,5)$	$\Phi(1,19)$	$\Phi(19,1)$	$\Phi(19,19)$
2 5 20	$\Phi(2,2)$	$\Phi(5,5)$	$\Phi(2,5)$	$\Phi(5,2)$	$\Phi(2,20)$	$\Phi(20,2)$	$\Phi(5,20)$	$\Phi(20,5)$	$\Phi(20,20)$
5 2 21	$\Phi(5,5)$	$\Phi(2,2)$	$\Phi(5,2)$	$\Phi(2,5)$	$\Phi(5,21)$	$\Phi(21,5)$	$\Phi(2,21)$	$\Phi(21,2)$	$\Phi(21,21)$
3 5 22	$\Phi(3,3)$	$\Phi(5,5)$	$\Phi(3,5)$	$\Phi(5,3)$	$\Phi(3,22)$	$\Phi(22,3)$	$\Phi(5,22)$	$\Phi(22,5)$	$\Phi(22,22)$
5 3 23	$\Phi(5,5)$	$\Phi(3,3)$	$\Phi(5,3)$	$\Phi(3,5)$	$\Phi(5,23)$	$\Phi(23,5)$	$\Phi(3,23)$	$\Phi(23,3)$	$\Phi(23,23)$
4 5 24	$\Phi(4,4)$	$\Phi(5,5)$	$\Phi(4,5)$	$\Phi(5,4)$	$\Phi(4,24)$	$\Phi(24,4)$	$\Phi(5,24)$	$\Phi(24,5)$	$\Phi(24,24)$
5 4 25	$\Phi(5,5)$	$\Phi(4,4)$	$\Phi(5,4)$	$\Phi(4,5)$	$\Phi(5,25)$	$\Phi(25,5)$	$\Phi(4,25)$	$\Phi(25,4)$	$\Phi(25,25)$
5 5 26	$\Phi(5,5)$	$\Phi(5,5)$	$\Phi(5,5)$	$\Phi(5,5)$	$\Phi(5,26)$	$\Phi(26,5)$	$\Phi(5,26)$	$\Phi(26,5)$	$\Phi(26,26)$
1 6 27	$\Phi(1,1)$	$\Phi(6,6)$	$\Phi(1,6)$	$\Phi(6,1)$	$\Phi(1,27)$	$\Phi(27,1)$	$\Phi(6,27)$	$\Phi(27,6)$	$\Phi(27,27)$
6 1 28	$\Phi(6,6)$	$\Phi(1,1)$	$\Phi(6,1)$	$\Phi(1,6)$	$\Phi(6,28)$	$\Phi(28,6)$	$\Phi(1,28)$	$\Phi(28,1)$	$\Phi(28,28)$
2 6 29	$\Phi(2,2)$	$\Phi(6,6)$	$\Phi(2,6)$	$\Phi(6,2)$	$\Phi(2,29)$	$\Phi(29,2)$	$\Phi(6,29)$	$\Phi(29,6)$	$\Phi(29,29)$
. . .	. . .	. . .	. . .	. . .	. . .	. . .	. . .	. . .	. . .

This table has been constructed for the case that only a single binary functional variable $\mathbf{f}$ appears (and Φ is the propositional function which is to be the value of $\mathbf{f}$). The reader may find it helpful to construct similar tables for one or two other cases, say the case of two binary functional variables and the case of one ternary functional variable.

For part (3) of this exercise the first two columns of the above table, headed $\mathbf{f}(\mathbf{b}_1\ \mathbf{b}_1)$ and $\mathbf{f}(\mathbf{b}_2\ \mathbf{b}_2)$, are to be deleted. And for part (2), the first four columns of the table are

(3) Apply this method to solve the subcase $\mathrm{VIII}_0^{2,1}$ of case VIII, in which there are no free individual variables and the prefix is $(\exists \mathbf{b}_1)(\exists \mathbf{b}_2)(\mathbf{c})$. (i) On the hypothesis that there is only a single binary functional variable appearing, supply a quantifier-free disjunction, as short as possible and not necessarily one of the wffs $\mathbf{C}_k$, such that the given wff **A** is a theorem if and only if this disjunction is tautologous. Do the same thing also on the hypothesis: (ii) that there is one binary functional variable appearing and any number of singulary functional variables; (iii) that there are just two binary functional variables $\mathbf{f}$ and $\mathbf{g}$ appearing and that the only elementary parts which occur are $\mathbf{f}(\mathbf{b}_1, \mathbf{b}_2)$, $\mathbf{g}(\mathbf{b}_1, \mathbf{b}_2)$, $\mathbf{f}(\mathbf{b}_1, \mathbf{c})$, $\mathbf{g}(\mathbf{b}_2, \mathbf{c})$, $\mathbf{f}(\mathbf{c}, \mathbf{c})$, $\mathbf{g}(\mathbf{c}, \mathbf{c})$. Then (iv) show how to solve case $\mathrm{VIII}_0^{2,1}$ generally, not necessarily seeking the shortest decision procedure or the smallest number K, but establishing the success of the method by finding an upper bound of K.[440]

(4) Apply this method to solve the subcase VIII_0 of case VIII, in which there are no free individual variables and $m = 0$. (Again find an upper bound of K.)[440]

---

to be deleted. For part (4) the reader should construct a new table, similar to that for part (3) but involving a greater number of individual variables (the case $n = 3$, $l = 2$ may be taken as illustrative).

In any of these tables, let two rows be called *related* if there is at least one entry, consisting of Φ (or Ψ, etc.) with particular numbers as arguments, that appears in both rows. For example, in the table used for part (1), the first and fifth rows are related, and the fifth and ninth rows are related, but the first and ninth rows are not related. In the table used for part (3), the first, second, and third rows are mutually related; also the first and fourth rows are related, but not the second and fourth rows or the first and fifth rows.

In the tables for parts (3) and (4), it will be seen that each row is related to the rows obtained by a permutation of the values assigned to $\mathbf{b}_1, \mathbf{b}_2, \ldots, \mathbf{b}_n$, and that *otherwise each row is related to at most one earlier row of the table.* Use may be made of this in proving the existence of the number K, or in finding an upper bound of K.

[440]For the solution of case $\mathrm{VIII}_0^{2,1}$, one approach is the following. (We state the matter for this particular subcase, but it will be seen that the same idea is applicable to case VIII_0, and indeed to case VIII generally.)

In the table constructed as described in the preceding footnote, let a row be called a *single-row* if the values assigned to $\mathbf{b}_1$ and $\mathbf{b}_2$ are the same (e.g., the first row in the table, the fourth row, and the ninth row are single-rows). And excepting the single-rows, let each row of the table be associated with the row obtained from it by interchanging the values assigned to $\mathbf{b}_1$ and $\mathbf{b}_2$, and let the resulting pair of associated rows be called a *row-pair* (e.g., the second and third rows in the table are a row-pair, likewise the fifth and sixth rows, and so on). For any single-row, taken in isolation, there is a finite class C_1 of possible assignments of truth-values that falsify the matrix **M** (i.e., give to **M** the value f) and at the same time satisfy the condition that the same truth-value must be assigned to two elementary parts which are so related that they become identical when $\mathbf{b}_1$ and $\mathbf{b}_2$ are replaced by the same variable **b**. Similarly, for any row-pair there is a finite class C_2 of pairs of possible assignments of truth-values (one assignment for each row) that falsify **M** and at the same time satisfy the condition that the same

46.14. Apply the decision procedures of 46.13 to determine which of the following wffs are theorems:

(1) $(\exists x)(\exists y)(z) \mathbin{.} F(x, x) \supset \mathbin{.} F(y, y) \supset \mathbin{.} F(x, z) \supset F(z, y)$

(2) $(\exists x)(\exists y)(z) \mathbin{.} F(x, z) \equiv F(z, y) \supset \mathbin{.} F(z, y) \equiv F(z, z) \supset \mathbin{.}$
$F(x, y) \equiv F(y, x) \supset \mathbin{.} F(x, y) \equiv F(x, z)$

(3) $\exists(x)(\exists y)(z) \mathbin{.} F(x, z) \supset \mathbin{.} F(y, z) \supset \mathbin{.} F(x, y) \equiv F(z, z) \supset \mathbin{.}$
$F(y, x) \lor F(z, z) \supset \mathbin{.} F(z, x) \lor F(z, y)$

(4) $(\exists x)(\exists y)(z) \mathbin{.} F(x, y)F(y, x) \not\equiv F(x, z) \supset \mathbin{.} F(x, z) \equiv F(z, x) \supset \mathbin{.}$
$F(x, z) \equiv F(y, z) \supset \mathbin{.} F(y, x) \supset F(x, y) \equiv F(z, z) \supset \mathbin{.}$
$F(x, y) \equiv F(y, x) \equiv F(z, y)$
(Answer: (4) is a theorem.)

(5) $(\exists x)(\exists y)(z) \mathbin{.} F(x, y) \supset F(y, z)F(z, z) \mathbin{.} F(x, y)G(x, y) \supset G(x, z)G(z, z)$

---

truth-value must be assigned to any two elementary parts, one in each row, which can be obtained one from the other by interchanging $\mathbf{b}_1$ and $\mathbf{b}_2$. Let S_1 be a non-empty subclass of C_1, and let S_2 be a non-empty subclass of C_2.

It is necessary to consider five different patterns of correspondence that occur between a single-row or row-pair in the table and an earlier (related) single-row or row-pair. These are, namely, the patterns which appear: (i) in the correspondence between the fourth row and the first row in the table (or between the twenty-fifth row and the fourth row, or between the hundredth row and the ninth row, and so on); (ii) in the correspondence between the row-pair consisting of the second and third rows, on the one hand, and the first row, on the other hand; (iii) between the row-pair consisting of the fifth and sixth rows, on the one hand, and that consisting of the second and third rows, on the other hand; (iv) between the row-pair consisting of the seventh and eigth rows, on the one hand, and that consisting of the second and third rows, on the other hand; (v) between the ninth row and the row-pair consisting of the second and third rows.

In case (i) we must ascertain that, for an arbitrary member of S_1 used as the assignment of truth-values in the earlier single-row, there is a corresponding member of S_1 which may be used simultaneously as the assignment of truth-values in the later single-row. In case (ii) we must ascertain that, for an arbitrary member of S_1 used as the assignment of truth-values in the single-row, there is a corresponding member of S_2 which may be used simultaneously as the assignment of truth-values in the row-pair. In each of cases (iii), (iv) we must ascertain that, for an arbitrary member of S_2 used as the assignment of truth-values in the earlier row-pair, there is a corresponding member of S_2 which may be used simultaneously as the assignment of truth-values in the later row-pair. In case (v) we must ascertain that, for an arbitrary member of S_2 used as the assignment of truth-values in the row-pair, there is a corresponding member of S_2 which may be used simultaneously as the assignment of truth-values in the single-row.

Since C_1 and C_2 are finite, the number of different pairs S_1, S_2 of non-empty subclasses of C_1, C_2 is finite. In the case of any particular wff **A** the complete list of pairs S_1, S_2 may be written down, and for each such pair S_1, S_2 it may be determined whether the conditions just stated are satisfied. If one pair S_1, S_2 is found for which these conditions are satisfied, the wff **A** is non-valid; in the contrary case, **A** is valid.

(6) $(\exists x)(\exists y)(z) \,.\, F(x, y) \supset [F(x, z) \equiv G(y, z)] \supset .$
$F(x, y) \equiv [F(z, z) \supset G(z, z)] \supset . G(x, y) \equiv G(z, z)$

(7) $(\exists x)(\exists y)(z) \,.\, F(x, z) \supset . F(z, z) \supset G(z, z) \equiv F(x, y) \supset .$
$G(z, z) \supset F(z, z) \equiv G(x, y) \supset . G(x, y) \supset F(y, x) \equiv G(y, z) \supset .$
$F(z, y) \equiv F(y, x)$

46.15. Extend the method of 46.13 (1) (compare also 46.11 (2)) to solve case V′ of the decision problem. First work out the method in application to the following particular examples:

(1) $(x)(\exists y_1)(\exists y_2)(z) \,.\, F(x, z) \supset . F(y_1, z) \supset . F(y_2, z) \supset . F(y_1, x) \supset F(z, y_2)$

(2) $(x_1)(x_2)(\exists y)(z) \,.\, F(x_1, y) \supset . F(z, x_1) \supset . F(z, y) \supset F(x_2, y) \vee F(x_2, z)$

(3) $(\exists x)(y)(\exists z) \,.\, F(x, y) \supset . F(z, x) \supset F(y, y)$

(4) $(\exists x)(y)(\exists z) \,.\, F(x, y, z) \supset F(y, z, z)$

(5) $(\exists x)(y)(\exists z_1)(\exists z_2) \,.\, F(x, y, z_1, z_2, z_1) \supset F(z_1, x, y, z_1, z_2)$

(6) $(\exists x_1)(x_2)(\exists x_3)(x_4) \,.\, F(x_1, x_2, x_3) \supset F(x_2, x_3, x_4)$

(7) $(x_1)(\exists x_2)(x_3)(\exists x_4) \,.\, F(x_1, x_2, x_3) \supset F(x_4, x_4, x_1)$

Then (8) state the method in general and show that it provides a solution of case V′ of the decision problem for validity. Finally (9) show how to obtain a proof of a wff which has been found by this method to be valid.

46.16. (1) By the same method solve also the following case V of the decision problem:

V *Well-formed formulas with a prenex normal form in which the matrix satisfies the condition of not having the value falsehood for two different assignments of truth-values to its elementary parts unless the two assignments differ in the truth-value for at least one elementary part that contains none of the variables occurring in the prefix.*

And illustrate by applying the solution to the following particular examples (after dropping universal quantifiers from the beginning of the prefix if necessary):

(2) $(\exists y_1)(\exists y_2)(z) \,.\, F(x, y_1) \supset F(z, x) \supset F(x, x) \supset F(x, x) F(y_1, y_2)$

(3) $(x)(\exists y_1)(\exists y_2)(z) \,.\, F(x, z) \supset [F(y_1, z) \supset F(y_2, x)] \supset F(x, x) \supset .$
$F(x, x) \,.\, F(y_1, y_2) \supset F(z, z)$

(4) $(x_1)(x_2)(\exists y_1)(\exists y_2)(z) \,.$
$F(x_1, x_2, y_1) \supset F(y_2, y_1, z) \supset [F(x_1, x_1, x_2) \supset F(x_1, x_2, x_2)] \supset .$
$F(x_2, y_1, y_2) \supset F(y_1, z, z) \supset [F(x_1, x_2, x_2) \supset F(x_1, x_1, x_2)] \supset .$
$F(y_1, y_2, z) \supset F(x_2, x_2, y_1) \,.\, F(x_1, x_1, x_2) \equiv F(x_1, x_2, x_2)$

46.17. (1) In the same way, extend the method of 46.13(2) to solve case IX of the decision problem. And illustrate by applying the solution to the following particular examples:

(2) $(x_1)(x_2)(\exists y_1)(\exists y_2)(z_1)(z_2) \centerdot F(x_2, z_1) \supset \centerdot$
$F(y_1, z_2) \supset F(y_1, z_1)F(y_2, z_1) \vee F(x_2, z_2)F(y_2, z_2)$

(3) $(\exists x_1)(\exists x_2)(y) \centerdot [F(x_1, y) \equiv F(x_2, y) \equiv F(z_1, z_2)] \vee [F(z_1, y) \equiv F(z_2, y)]$

(4) $(\exists x_1)(x_2)(\exists x_3)(x_4) \centerdot [F(x_1, x_4) \equiv F(x_4, x_3) \equiv F(x_3, x_4) \equiv F(x_4, x_1)] \centerdot$
$F(x_2, x_4) \equiv F(x_4, x_3) \equiv F(x_3, x_4) \equiv F(x_4, x_2)$

(5) $(\exists x_1)(x_2)(\exists x_3)(x_4) \centerdot F(x_1, x_4) \equiv F(x_2, x_4) \supset \centerdot$
$F(x_1, x_4) \equiv F(x_4, x_3) \equiv F(x_3, x_4) \equiv F(x_4, x_2)$

46.18. (1) In the same way, extend the method of 46.13(3), (4) to solve case VIII of the decision problem.[440] Illustrate by applying the decision procedure to the following particular examples:

(2) $(x)(\exists y_1)(\exists y_2)(z) \centerdot F(x, z) \equiv F(z, x) \supset \centerdot$
$F(x, z) \equiv F(y_2, z) \centerdot F(y_1, z) \supset F(y_1, y_2)$

(3) $(x_1)(x_2)(\exists y_1)(\exists y_2)(z) \centerdot F(x_1, y_2, x_1, z) \supset \centerdot$
$F(x_1, y_1, x_1, y_2) \equiv F(y_1, x_2, y_1, y_2) \supset \centerdot$
$F(x_1, y_1, x_1, y_2) \supset [F(x_1, y_2, y_1, y_2) \supset F(x_1, z, y_1, z)] \centerdot$
$F(x_1, z, y_1, z) \supset \centerdot F(x_1, y_1, x_1, y_2) \equiv F(x_1, y_2, y_1, y_2)$

(4) $(x_1)(x_2)(\exists y_1)(\exists y_2)(z) \centerdot F(x_1, x_2) \supset \centerdot F(y_1, y_2) \supset F(x_2, z) \vee F(y_2, z) \supset \centerdot$
$F(y_1, y_2) \supset [F(x_2, z) \equiv F(y_1, z)] \supset F(z, z) \supset \centerdot$
$F(y_1, y_2) \centerdot F(y_1, z) \equiv F(y_2, z)$

(5) $(x)(\exists y_1)(\exists y_2)(\exists y_3)(z) \centerdot$
$F(y_1, y_2, y_3) \supset [F(x, x, z) \supset F(y_2, y_3, y_1) \vee F(y_3, y_1, y_2)] \supset \centerdot$
$F(y_3, y_1, y_2) \supset F(y_1, y_2, y_3)F(y_2, y_3, y_1) \equiv F(y_2, y_1, z) \supset \centerdot$
$F(y_2, y_3, y_1) \supset F(y_1, y_2, y_3)F(y_3, y_1, y_2) \equiv F(y_1, z, y_2) \supset \centerdot$
$F(y_3, y_1, y_2) \supset \sim F(y_2, y_3, y_1) \supset F(y_1, y_2, y_3) \equiv F(z, y_2, y_1) \supset \centerdot$
$F(y_1, y_2, y_3)F(y_2, y_3, y_1)F(y_3, y_1, y_2) \equiv F(z, z, z)$

46.19. Apply the remark of footnote 425 in order to reduce the solution of each of the following additional special cases of the decision problem to that of cases I, V–IX:

XI′ *Well-formed formulas in which every truth-functional constituent is in prenex normal form with an elementary part or the negation of an elementary part as its matrix.*

XI *Well-formed formulas in which every truth-functional constituent is in prenex normal form with a matrix that has at most one elementary part containing any of the variables that occur in its prefix.*

XII *Well-formed formulas in which the prenex normal forms of the truth-functional constituents have prefixes of the following forms only:* $(\mathbf{a}_1)$, $(\mathbf{a}_1)(\mathbf{a}_2)$, $(\mathbf{a}_1)(\exists\mathbf{b}_1)$, $(\mathbf{a}_1)(\mathbf{a}_2)(\exists\mathbf{b}_1)$, $(\mathbf{a}_1)(\exists\mathbf{b}_1)(\exists\mathbf{b}_2)$, $(\mathbf{a}_1)(\mathbf{a}_2)(\exists\mathbf{b}_1)(\exists\mathbf{b}_2)$.

XIII *Well-formed formulas in which the prenex normal form of each truth-functional constituent* $\mathbf{P}_i$ *has a prefix of one of the forms* $(\mathbf{b}_1)(\mathbf{b}_2)\ldots(\mathbf{b}_n)$ *or* $(\mathbf{a}_1)(\mathbf{a}_2)\ldots(\mathbf{a}_{m_i})(\exists\mathbf{b}_1)(\exists\mathbf{b}_2)\ldots(\exists\mathbf{b}_n)$ *and a matrix in which every elementary part other than a propositional variable contains all of the variables* $\mathbf{b}_1, \mathbf{b}_2, \ldots, \mathbf{b}_n$*—where the number n is the same for all the constituents* $\mathbf{P}_i$*, and the numbers* m_i *are each of them less than or equal to n.*

46.20. Consider the following additional case of the decision problem:

XIV *Well-formed formulas having a prenex normal form with a prefix of the form* $(\mathbf{a}_1)(\mathbf{a}_2)\ldots(\mathbf{a}_m)(\exists\mathbf{b}_1)(\exists\mathbf{b}_2)\ldots(\exists\mathbf{b}_n(\mathbf{c}_1)(\mathbf{c}_2)\ldots(\mathbf{c}_l)$ *and a matrix in which the complete list of functional variables occurring is* $\mathbf{f}_1, \mathbf{f}_2, \ldots, \mathbf{f}_M, \mathfrak{g}_1, \mathfrak{g}_2, \ldots, \mathfrak{g}_N$*, such that no elementary part with one of the functional variables* $\mathbf{f}_i$ *contains any of the variables* $\mathbf{c}_1, \mathbf{c}_2, \ldots, \mathbf{c}_l$*, and each elementary part with one of the functional variables* $\mathfrak{g}_i$ *contains either none of the variables* $\mathbf{b}_1, \mathbf{b}_2, \ldots, \mathbf{b}_n$ *or at least one of the variables* $\mathbf{c}_1, \mathbf{c}_2, \ldots, \mathbf{c}_l$.

(1) By the method described in 46.13, solve the subcase XIV_0 in which there are no free individual variables and $m = 0$, showing in this case that $K = 1$. (2) Extend this method to solve case XIV in general. (3) By taking $l = 0$, find a solution of case I as a corollary of the solution of case XIV. (4) Illustrate the solution of case XIV by applying it to the following example:

$$(x_1)(x_2)(\exists y_1)(\exists y_2)(z) \centerdot F(x_1, x_2) \supset \centerdot G(x_1, x_2) \supset \centerdot$$
$$G(x_2, z) \equiv G(y_2, z) \supset [F(y_1, y_2) \supset F(x_2, y_2)] \supset \centerdot$$
$$G(x_2, z) \equiv G(y_1, z) \supset F(x_1, y_1)F(x_2, y_1)F(y_1, y_2).$$

46.21. Consider the general method for the solution of the decision problem which is outlined in footnotes 439 and 440, and study the question of extending it to cases in which a row in the table (or row-pair, etc.) may be related to more than one earlier row (row-pair, etc.). Explain why the method cannot be extended to an arbitrary such case; and seek for any special cases of this sort to which the extension may be possible. Consider in particular the case VII_0^1 of any wff **A** having a prenex normal form in which there are no free individual variables and the prefix is $(\exists \mathbf{b}_1)(\exists \mathbf{b}_2)(\mathbf{c})$.

46.22. The problem traditionally treated under the head of the *categorical syllogism* may be represented as follows in connection with an applied functional calculus of first order having singulary functional constants among its primitive symbols. Let a sentence be said to express a *categorical proposition* if it has one of the four forms $\mathbf{f}(x) \supset_x \mathbf{g}(x)$, $\mathbf{f}(x) \supset_x \sim\mathbf{g}(x)$, $(\exists x) \centerdot \mathbf{f}(x)\mathbf{g}(x)$, $(\exists x) \centerdot \mathbf{f}(x) \sim\mathbf{g}(x)$, where (in each case) **f** and **g** are singulary functional constants.[441] It is required to find all valid forms of inference in which there are two premisses, and the premisses and conclusion each of them have one of the four categorical forms. But cases are to be excluded in which there is essentially only one premiss, i.e., in which there is a simpler valid inference according to which the conclusion in question would follow from one of the two premisses alone.

For example, among the required forms of inference are the following which correspond to the traditional syllogisms in *Darii*, *Ferio*, and *Feriso* respectively, and which are to be distinguished as all three different: from $\mathbf{g}(x) \supset_x \mathbf{h}(x)$ and $(\exists x) \centerdot \mathbf{f}(x)\mathbf{g}(x)$ to infer $(\exists x) \centerdot \mathbf{f}(x)\mathbf{h}(x)$; from $\mathbf{g}(x) \supset_x \sim\mathbf{h}(x)$ and $(\exists x) \centerdot \mathbf{f}(x)\mathbf{g}(x)$ to infer $(\exists x) \centerdot \mathbf{f}(x) \sim\mathbf{h}(x)$; from $\mathbf{g}(x) \supset_x \sim\mathbf{h}(x)$ and $(\exists x) \centerdot \mathbf{g}(x)\mathbf{f}(x)$ to infer $(\exists x) \centerdot \mathbf{f}(x) \sim\mathbf{h}(x)$.

Evidently such forms of inference can be tested by writing for each one a corresponding *leading principle*, expressed as a wff of the pure functional calculus of first order (compare exercise 15.9). And the form of inference is to be considered valid if and only if its leading principle is valid. For example, the leading principle of Darii is $G(x) \supset_x H(x) \supset \centerdot (\exists x)[F(x)G(x)] \supset (\exists x) \centerdot F(x)H(x)$; and it may be verified by the decision procedure for case III, as given above, that this leading principle is valid, and that neither of the simpler leading principles $G(x) \supset_x H(x) \supset (\exists x) \centerdot F(x)H(x)$, $(\exists x)[F(x)G(x)] \supset (\exists x) \centerdot F(x)H(x)$ is valid.

---

[441]Traditionally the four forms are called A, E, I, O respectively and are rendered in words as: all F's are G's, no F's are G's, some F's are G's, some F's are not G's. Notice, however, that the version here suggested of the traditional doctrine of categorical propositions and the categorical syllogism is not put forward as *the correct* interpretation but rather only as one possible or plausible interpretation.

The fact is that the traditional doctrine is not sufficiently definite and coherent—and different writers are not sufficiently in agreement—to make it clear what is the best or most faithful representation of it in a logistic system. For there is, on the one hand, the difficulty about "existential import," as it is called (some aspects of the traditional doctrine would seem to be better represented if A and E were taken as $(\exists x)\mathbf{f}(x) \centerdot \mathbf{f}(x) \supset_x \mathbf{g}(x)$ and $(\exists x)\mathbf{f}(x) \centerdot \mathbf{f}(x) \supset_x \sim\mathbf{g}(x)$ respectively, instead of in the way suggested in the exercise). And there is, on the other hand, the question whether the traditional "terms" should not rather be construed as common names (see footnotes 4, 6) or as variables instead of class names or functional constants.

By the method indicated, solve the problem of finding all such valid forms of inference (valid categorical syllogisms).

46.23. Implicit in some of the foregoing exercises (see 46.11(2), 46.13, 46.15–46.18) is a metatheorem due to Herbrand,[442] namely a generalization of **441 to the case of a wff **A** in prenex normal form with an arbitrary prefix, the only further difference in the statement of the generalized metatheorem being in the substitution by which the quantifier-free wff $\mathbf{B}_k$ is obtained from **M**. State this generalization of **441 explicitly: (1) for the case that the prefix is (∃**b**)(**c**)(∃**d**)(**e**) and the free individual variables of **A** are $\mathbf{a}_1$ and $\mathbf{a}_2$; (2) for the case of an arbitrary prefix and an arbitrary number of free individual variables in **A**.

46.24. By means of the metatheorem of 46.23, prove the completeness (in the sense of §44) of the following described formulation, F^{1p}_{h}, of the pure functional calculus of first order, due to Herbrand:[442] The primitive sentence connectives are negation and disjunction. The primitive quantifiers are the universal and existential quantifiers. The axioms are all quantifier-free tautologous wffs. And the rules of inference, none requiring more than a single premiss, are as follows: the rule of alphabetic change of bound variable (*402); the rule of generalization (*401); from **A** to infer (∃**b**)**B**, where **b** is an individual variable which does not occur in **A**, and **B** is obtained from **A** by replacing zero or more free occurrences of the individual variable **a** in **A** (not necessarily all free occurrences of **a** in **A**) by **b**; to replace a wf part (**a**)[**C** ∨ **D**] by (**a**)**C** ∨ **D**, if **a** is not free in **D**; to replace a wf part (∃**a**)(**C** ∨ **D**] by (∃**a**)**C** ∨ **D**, if **a** is not free in **D**; to replace a wf part (**a**) ~**C** by ~(∃**a**)**C**; to replace a wf part (∃**a**) ~**C** by ~(**a**)**C**; to replace a wf part **P** ∨ **Q** by **Q** ∨ **P**; to replace a wf part **P** ∨ [**Q** ∨ **R**] by [**P** ∨ **Q**] ∨ **R**; to replace a wf part [**P** ∨ **Q**] ∨ **R** by **P** ∨ [**Q** ∨ **R**]; to replace a wf part **P** ∨ **P** by **P**.

47. Reductions of the decision problem.

A *reduction of the decision problem* (of the pure functional calculus of first order) consists in a special class Γ of wffs and an effective procedure by which, when an arbitrary wff **A** is given, a corresponding wff $\mathbf{A}_\Gamma$ of the class Γ can be found such that **A** is a theorem if and only if $\mathbf{A}_\Gamma$ is a theorem, and by which, further, a proof of **A** can be found if a proof of $\mathbf{A}_\Gamma$ is known. For example, **420 and *421 constitute a reduction of the decision problem, the class Γ being in this case the class of wffs in Skolem normal form.

---

[442] *Recherches sur la Théorie de la Démonstration*, Warsaw 1930. This is Herbrand's dissertation at the University of Paris. (*Added in proof.* See in this connection a paper of Burton Dreben in the *Proceeding of the National Academy of Sciences of the U.S.A.*, vol. 38 (1952), pp. 1047–1052.)

A *reduction of the decision problem for validity* consists in a special class Γ of wffs and an effective procedure by which, when an arbitrary wff **A** is given, a corresponding wff $\mathbf{A}_\Gamma$ of the class Γ can be found which is valid if and only if **A** is valid. Similarly, a *reduction of the decision problem for satisfiability* consists in a special class of wffs and an effective procedure by which, when an arbitrary wff **A** is given, a corresponding wff of the special class can be found which is satisfiable if and only if **A** is satisfiable.

Clearly, every reduction of the decision problem for satisfiability leads to a corresponding reduction of the decision problem for validity, and *vice versa.* (The correspondence between **437 and **438 may be taken as an illustration of this.) Thus it is necessary to treat only one of the two kinds of reduction. Wherever results in the literature are stated as reductions of the decision problem for satisfiability, we shall here reproduce them in the other form, i.e., we shall state the corresponding reduction of the decision problem for validity.

With the exception of the reduction to Skolem normal form, and reductions which (like that to prenex normal form) can be regarded as included in this or which (like those of 39.5, *465) follow by little more than propositional calculus, reductions of the decision problem in our present sense, i.e., of the decision problem for provability, have rarely received treatment in the literature, perhaps only in the work of Herbrand. Since for many purposes the weaker result is sufficient, we shall deal in the remainder of this section with reductions of the decision problem for validity; and it will be convenient to express such reductions by saying that the class Γ is a *reduction class.*

In view of the unsolvability of the general decision problem of the pure functional calculus of first order (whether for provability or validity), it is evident that, *if Γ is a reduction class, then the special case of the decision problem for wffs of the class Γ is unsolvable.* And this may be regarded as being a part of the significance of reductions of the decision problem for validity.

As a lemma for later proofs, we first establish the following metatheorem, the idea of which is due to Herbrand:[443]

470. Let **A be any wff, let $\mathbf{p}_1, \mathbf{p}_2, \ldots, \mathbf{p}_M$ be the complete list of distinct propositional variables occurring in **A**, and let $\mathbf{f}_1, \mathbf{f}_2, \ldots, \mathbf{f}_N$ be the complete list of distinct functional variables occurring in **A**. Suppose that $\mathbf{f}_i$ is an h_i-ary functional variable ($i = 1, 2, \ldots, N$), and let $h - 1$ be the greatest of the numbers $h_1, h_2, \ldots, h_N$. Choose

[443] In the paper cited in footnote 430.

distinct individual variables[444] $\mathbf{x}_1, \mathbf{x}_2, \ldots, \mathbf{x}_{M+N}, \mathbf{y}_1, \mathbf{y}_2, \ldots, \mathbf{y}_{h-1}$, of which $\mathbf{x}_1, \mathbf{x}_2, \ldots, \mathbf{x}_{M+N}$ do not occur in **A**, and choose an h-ary functional variable **f**.[444] Let $\mathbf{C}_j$ be $\mathbf{f}(\mathbf{x}_j, \mathbf{x}_j, \ldots, \mathbf{x}_j)$ $(j = 1, 2, \ldots, M)$ let $\mathbf{F}_i$ be $\mathbf{f}_i(\mathbf{y}_1, \mathbf{y}_2, \ldots, \mathbf{y}_{h_i})$ $(i = 1, 2, \ldots, N)$; let $\mathbf{D}_i$ be $\mathbf{f}(\mathbf{x}_{M+i}, \mathbf{x}_{M+i}, \ldots, \mathbf{x}_{M+i}, \mathbf{y}_1, \mathbf{y}_2, \ldots, \mathbf{y}_{h_i})$ $(i = 1, 2, \ldots, N)$; and let **B** be

$$\check{S}^{\mathbf{p}_1}_{\mathbf{C}_1}\ \check{S}^{\mathbf{p}_2}_{\mathbf{C}_2} \ldots \check{S}^{\mathbf{p}_M}_{\mathbf{C}_M}\ \check{S}^{\mathbf{F}_1}_{\mathbf{D}_1}\ \check{S}^{\mathbf{F}_2}_{\mathbf{D}_2} \ldots \check{S}^{\mathbf{F}_N}_{\mathbf{D}_N}\ \mathbf{A}| \ldots ||| \ldots ||.$$

Then **B** is valid if and only if **A** is valid.

Proof. If $\vdash$ **A**, then $\vdash$ **B** by *352. By **440 and **434 it follows that, if **A** is valid, then **B** is valid.

By *301 and *306, we may suppose that **A** contains no free individual variables. Taking the positive integers as domain of individuals, assume that **B** is valid, and consider any system of values $\tau_1, \tau_2, \ldots, \tau_M, \Phi_1, \Phi_2, \ldots, \Phi_N$ of the free variables $\mathbf{p}_1, \mathbf{p}_2, \ldots, \mathbf{p}_M, \mathbf{f}_1, \mathbf{f}_2, \ldots, \mathbf{f}_N$ of **A**. Let the values $1, 2, \ldots, M + N$ be assigned to the free variables $\mathbf{x}_1, \mathbf{x}_2, \ldots, \mathbf{x}_{M+N}$ of **B** respectively, and let a value Φ of the free variable **f** of **B** be determined as follows: $\Phi(j, j, \ldots, j) = \tau_j$ $(j = 1, 2, \ldots, M)$; $\Phi(M+i, M+i, \ldots, M+i, u_1, u_2, \ldots, u_{h_i}) = \Phi_i(u_1, u_2, \ldots, u_{h_i})$ $(i = 1, 2, \ldots, N)$; $\Phi(u_1, u_2, \ldots, u_h) = \mathrm{t}$ in all other cases. The value of **B** for this system of values of its free variables is evidently the same as the value of **A** for the system of values $\tau_1, \tau_2, \ldots, \tau_M, \Phi_1, \Phi_2, \ldots, \Phi_N$ of the free variables of **A**. And since the value of **B** is t, it follows that the value of **A** is t.

Thus we have shown that, if **B** is valid in the domain of positive integers, then **A** is valid in the domain of positive integers. Hence by **450, if **B** is valid, then **A** is valid.

It follows that the class of wffs containing only one functional variable, and no propositional variables, is a reduction class. However, we go on at once to obtain stronger reductions than this.

According to a result of Löwenheim, the class of wffs containing only binary functional variables is a reduction class. By a refinement of Löwenheim's method it is possible to obtain the result that the class of wffs containing only a single binary functional variable (no other functional variables, and no propositional variables) is a reduction class.[445] We proceed to show how this may be done.

Given an arbitrary wff **A**, we first reduce it by **470 to a wff **B** which con-

---

[444] To render the reduction process effective, the choice must be made explicit in some manner, say by taking in each case the first available variable or variables, according to the alphabetic order of the variables.

[445] First proved by Kalmár in *Compositio Mathematica*, vol. 4 (1936), pp. 137–144.

tains only a single h-ary functional variable $\mathbf{f}$. If $h \neq 2$, we choose a binary functional variable $\mathfrak{g}$, and $2h + 1$ distinct individual variables $\mathbf{c}_1, \mathbf{c}_2, \ldots, \mathbf{c}_h, \mathbf{d}_1, \mathbf{d}_2, \ldots, \mathbf{d}_{h+1}$, of which $\mathbf{d}_1, \mathbf{d}_2, \ldots, \mathbf{d}_{h+1}$ do not occur in $\mathbf{B}$.[444] And we let $\mathbf{G}$ be the conjunction

$$\mathfrak{g}(\mathbf{d}_1, \mathbf{d}_2)\mathfrak{g}(\mathbf{d}_2, \mathbf{d}_3) \ldots \mathfrak{g}(\mathbf{d}_h, \mathbf{d}_{h+1})\mathfrak{g}(\mathbf{d}_{h+1}, \mathbf{d}_1)\mathfrak{g}(\mathbf{d}_1, \mathbf{c}_1)\mathfrak{g}(\mathbf{d}_2, \mathbf{c}_2) \ldots \\ \mathfrak{g}(\mathbf{d}_h, \mathbf{c}_h) \sim\mathfrak{g}(\mathbf{d}_1, \mathbf{d}_1) \sim\mathfrak{g}(\mathbf{c}_1, \mathbf{d}_2) \sim\mathfrak{g}(\mathbf{c}_2, \mathbf{d}_3) \ldots \sim\mathfrak{g}(\mathbf{c}_h, \mathbf{d}_{h+1})\mathfrak{g}(\mathbf{d}_{h+1}, \mathbf{c}_1).$$

Then letting $\mathbf{C}$ be

$$\check{S}^{\mathbf{f}(\mathbf{c}_1, \mathbf{c}_2, \ldots, \mathbf{c}_h)}_{(\exists \mathbf{d}_1)(\exists \mathbf{d}_2)\ldots(\exists \mathbf{d}_{h+1})\mathbf{G}} \mathbf{B}|,$$

we show that $\mathbf{C}$ is valid if and only if $\mathbf{B}$ is valid, therefore if and only if $\mathbf{A}$ is valid.

If $\vdash \mathbf{B}$, then $\vdash \mathbf{C}$ by *352. By **440 and **434 it follows that, if $\mathbf{B}$ is valid, then $\mathbf{C}$ is valid.

By *301 and *306 we may consider, instead of $\mathbf{B}$ and $\mathbf{C}$, their universal closures $\mathbf{B}'$ and $\mathbf{C}'$. Taking the natural numbers as domain of individuals, assume that $\mathbf{C}'$ is valid, and consider an arbitrary value Φ of the single free variable $\mathbf{f}$ of $\mathbf{B}'$. Then let a value Ψ of the single free variable $\mathfrak{g}$ of $\mathbf{C}'$ be detemined as follows.

The (ordered) h-tuples of natural numbers are enumerated in such a way that the natural numbers occurring in the kth h-tuple are all less than k. (Analogously to the enumeration used in §44, this may be done by arranging the h-tuples $\langle v_1, v_2, \ldots, v_h \rangle$ in order of increasing sums $v_1 + v_2 + \ldots + v_h$, h-tuples having the same sum being arranged among themselves in lexicographic order.) Then $\Psi(u, u) = \mathrm{t}$ except when $u = k(h + 1) + 1$ (i.e., except when u is congruent to 1 mod $h + 1$). If there is a natural number k such that $u_1 = k(h + 1) + 1$, $u_2 = k(h+1) + 2, \ldots, u_{h+1} = (k + 1)(h+1)$, then $\Psi(u_1, u_2) = \Psi(u_2, u_3) = \ldots = \Psi(u_h, u_{h+1}) = \Psi(u_{h+1}, u_1) = \mathrm{t}$. If $\langle v_1, v_2, \ldots, v_h \rangle$ is the $(k + 1)$th h-tuple and $u_l = k(h + 1) + l$ $(l = 1, 2, \ldots, h + 1)$, then $\Psi(u_1, v_1) = \Psi(u_2, v_2) = \ldots = \Psi(u_h, v_h) = \mathrm{t}$, and $\Psi(u_{h+1}, v_1) = \Phi(v_1, v_2, \ldots, v_h)$. And in all remaining cases $\Psi(u, v) = \mathrm{f}$.

In view of the special properties of the propositional function Ψ, the value of $\mathbf{C}'$ for the value Ψ of $\mathfrak{g}$ is the same as the value of $\mathbf{B}'$ for the value Φ of $\mathbf{f}$. And since the value of $\mathbf{C}'$ is t, it follows that the value of $\mathbf{B}'$ is t.

Thus we have shown that, if $\mathbf{C}'$ is valid in the domain of natural numbers, then $\mathbf{B}'$ is valid in that domain. It follows that, if $\mathbf{C}$ is valid, then $\mathbf{B}$ is valid.

This completes the proof, since $\mathbf{C}$ contains only the single binary functional variable $\mathfrak{g}$, and we have shown altogether that $\mathbf{C}$ is valid if and only if $\mathbf{A}$ is valid.

Notice that in place of $(\exists \mathbf{d}_1)(\exists \mathbf{d}_2) \ldots (\exists \mathbf{d}_{h+1})\mathbf{G}$ in the foregoing proof we might equally well have used $(\mathbf{d}_1)(\mathbf{d}_2) \ldots (\mathbf{d}_{h+1})\mathbf{H}$, where $\mathbf{H}$ is

$$\mathfrak{g}(\mathbf{d}_1, \mathbf{d}_2) \supset \centerdot \mathfrak{g}(\mathbf{d}_2, \mathbf{d}_3) \supset \centerdot \ldots \mathfrak{g}(\mathbf{d}_h, \mathbf{d}_{h+1}) \supset \centerdot \mathfrak{g}(\mathbf{d}_{h+1}, \mathbf{d}_1) \supset \centerdot$$
$$\mathfrak{g}(\mathbf{d}_1, \mathbf{c}_1) \supset \centerdot \mathfrak{g}(\mathbf{d}_2, \mathbf{c}_2) \supset \centerdot \ldots \mathfrak{g}(\mathbf{d}_h, \mathbf{c}_h) \supset \centerdot \sim\mathfrak{g}(\mathbf{d}_1, \mathbf{d}_1) \supset \centerdot$$
$$\sim\mathfrak{g}(\mathbf{c}_1, \mathbf{d}_2) \supset \centerdot \sim\mathfrak{g}(\mathbf{c}_2, \mathbf{d}_3) \supset \centerdot \ldots \sim\mathfrak{g}(\mathbf{c}_h, \mathbf{d}_{h+1}) \supset \mathfrak{g}(\mathbf{d}_{h+1}, \mathbf{c}_1).$$

For both of the wffs $(\exists \mathbf{d}_1)(\exists \mathbf{d}_2) \ldots (\exists \mathbf{d}_{h+1})\mathbf{G}$ and $(\mathbf{d}_1)(\mathbf{d}_2) \ldots (\mathbf{d}_{h+1})\mathbf{H}$ alike have the value $\Phi(v_1, v_2, \ldots, v_h)$ for the system of values $\Psi, v_1, v_2, \ldots, v_h$ of their free variables $\mathfrak{g}, \mathbf{c}_1, \mathbf{c}_2, \ldots, \mathbf{c}_h$ (where Ψ is the propositional function which was introduced above). By taking advantage of this observation, we now go on to establish the following still stronger reduction of the decision problem for validity:

****471.** The class of wffs in Skolem normal form which contain only a single binary functional variable (no other functional variables, and no propositional variables) is a reduction class.

Proof. By **437 we may suppose that the given wff $\mathbf{A}$ is already in Skolem normal form.

We consider first the case that $\mathbf{A}$ contains only a single h-ary functional variable $\mathbf{f}$, making use in this case of the wffs $\mathbf{G}$ and $\mathbf{H}$ that were introduced above. Therefore let $\mathbf{A}$ be

$$(\exists \mathbf{a}_1)(\exists \mathbf{a}_2) \ldots (\exists \mathbf{a}_m)(\mathbf{b}_1)(\mathbf{b}_2) \ldots (\mathbf{b}_n)\mathbf{M},$$

and let the distinct elementary parts of the matrix $\mathbf{M}$ of $\mathbf{A}$ be $\mathbf{E}_1, \mathbf{E}_2, \ldots, \mathbf{E}_\mu$, where $\mathbf{E}_i$ is $\mathbf{f}(\mathbf{c}_{i1}, \mathbf{c}_{i2}, \ldots, \mathbf{c}_{ih})$ $(i = 1, 2, \ldots, \mu)$. Since the P-constituents of $\mathbf{M}$ are the same as its elementary parts, we have by *465 that a certain equivalence $\mathbf{M} \equiv \mathbf{M}_1$ is valid, where $\mathbf{M}_1$ has the form of a conjunction

$$[\mathbf{E}_1^1 \supset \centerdot \mathbf{E}_2^1 \supset \centerdot \ldots \mathbf{E}_{\mu-1}^1 \supset \sim\mathbf{E}_\mu^1][\mathbf{E}_1^2 \supset \centerdot \mathbf{E}_2^2 \supset \centerdot \ldots \mathbf{E}_{\mu-1}^2 \supset \sim\mathbf{E}_\mu^2] \ldots$$
$$[\mathbf{E}_1^\nu \supset \centerdot \mathbf{E}_2^\nu \supset \centerdot \ldots \mathbf{E}_{\mu-1}^\nu \supset \sim\mathbf{E}_\mu^\nu],$$

each $\mathbf{E}_i^j$ being either $\mathbf{E}_i$ or $\sim\mathbf{E}_i$ $(i = 1, 2, \ldots, \mu,$ and $j = 1, 2, \ldots, \nu)$. Therefore $\mathbf{A}$ is valid if and only if $\mathbf{A}_1$ is valid, where $\mathbf{A}_1$ is

$$(\exists \mathbf{a}_1)(\exists \mathbf{a}_2) \ldots (\exists \mathbf{a}_m)(\mathbf{b}_1)(\mathbf{b}_2) \ldots (\mathbf{b}_n)\mathbf{M}_1.$$

Choose distinct individual variables $\mathbf{c}_1, \mathbf{c}_2, \ldots, \mathbf{c}_h, \mathbf{d}_1, \mathbf{d}_2, \ldots, \mathbf{d}_{h+1}$, of which $\mathbf{d}_1, \mathbf{d}_2, \ldots, \mathbf{d}_{h+1}$ do not occur in $\mathbf{A}_1$,[444] let $\mathbf{M}_2$ be obtained from $\mathbf{M}_1$ by replacing each part $\mathbf{E}_i^j$ by

$$\check{S}^{\mathbf{f}(\mathbf{c}_1, \mathbf{c}_2, \ldots, \mathbf{c}_h)}_{(\mathbf{d}_1)(\mathbf{d}_2)\ldots(\mathbf{d}_{h+1})\mathbf{H}}\, \mathbf{E}_i^j|$$

and let $\mathbf{A}_2$ be

$$(\exists\mathbf{a}_1)(\exists\mathbf{a}_2)\ldots(\exists\mathbf{a}_m)(\mathbf{b}_1)(\mathbf{b}_2)\ldots(\mathbf{b}_n)\mathbf{M}_2.$$

Then $\mathbf{A}_2$ is

$$\check{S}^{\mathbf{f}(\mathbf{c}_1,\mathbf{c}_2,\ldots,\mathbf{c}_h)}_{(\mathbf{d}_1)(\mathbf{d}_2)\ldots(\mathbf{d}_{h+1})\mathbf{H}}\mathbf{A}_1|,$$

and by *352, **440, and **434 it follows that if $\mathbf{A}_1$ is valid, then $\mathbf{A}_2$ is valid.

Let $\mathbf{M}_3$ be obtained from $\mathbf{M}_1$ by replacing each part $\mathbf{E}_i^j$ by

$$\check{S}^{\mathbf{f}(\mathbf{c}_1,\mathbf{c}_2,\ldots,\mathbf{c}_h)}_{(\exists\mathbf{d}_1)(\exists\mathbf{d}_2)\ldots(\exists\mathbf{d}_{h+1})\mathbf{G}}\mathbf{E}_i^j|$$

or

$$\check{S}^{\mathbf{f}(\mathbf{c}_1,\mathbf{c}_2,\ldots,\mathbf{c}_h)}_{(\mathbf{d}_1)(\mathbf{d}_2)\ldots(\mathbf{d}_{h+1})\mathbf{H}}\mathbf{E}_i^j|$$

according as $\mathbf{E}_i^j$ is $\mathbf{E}_i$ or $\sim\mathbf{E}_i$, and let $\mathbf{A}_3$ be

$$(\exists\mathbf{a}_1)(\exists\mathbf{a}_2)\ldots(\exists\mathbf{a}_m)(\mathbf{b}_1)(\mathbf{b}_2)\ldots(\mathbf{b}_n)\mathbf{M}_3.$$

Let $\mathbf{A}_4$ be

$$(\exists\mathbf{d}_1)(\exists\mathbf{d}_2)\ldots(\exists\mathbf{d}_{h+1})\mathbf{G}\supset_{\mathbf{c}_1\mathbf{c}_2\ldots\mathbf{c}_h}(\mathbf{d}_1)(\mathbf{d}_2)\ldots(\mathbf{d}_{h+1})\mathbf{H}\supset\mathbf{A}_3.$$

By the same argument that was used above (employing again the domain of natural numbers and the same binary propositional function Ψ) we may show that, if $\mathbf{A}_4$ is valid, then $\mathbf{A}_1$ is valid.

Letting $\mathbf{K}$ be the conjunction of all the wffs

$$S^{\mathbf{c}_1\mathbf{c}_2\ldots\mathbf{c}_h}_{\mathbf{c}_{i1}\mathbf{c}_{i2}\ldots\mathbf{c}_{ih}}(\exists\mathbf{d}_1)(\exists\mathbf{d}_2)\ldots(\exists\mathbf{d}_{h+1})\mathbf{G}\supset(\mathbf{d}_1)(\mathbf{d}_2)\ldots(\mathbf{d}_{h+1})\mathbf{H}|,$$

we have by P that $\vdash\mathbf{M}_2\supset.\,\mathbf{K}\supset\mathbf{M}_3$. Hence by *306 and P,

$$\vdash\mathbf{M}_2\supset.\,(\exists\mathbf{d}_1)(\exists\mathbf{d}_2)\ldots(\exists\mathbf{d}_{h+1})\mathbf{G}\supset_{\mathbf{c}_1\mathbf{c}_2\ldots\mathbf{c}_h}(\mathbf{d}_1)(\mathbf{d}_2)\ldots(\mathbf{d}_{h+1})\mathbf{H}\supset\mathbf{M}_3.$$

Hence by *301, *333, *335, *365, *382, and P, $\vdash\mathbf{A}_2\supset\mathbf{A}_4$. By *modus ponens*, and since we know that the valid wffs are the same as the theorems (by **440 and **434), it follows that, if $\mathbf{A}_2$ is valid, then $\mathbf{A}_4$ is valid. Consequently, if $\mathbf{A}_1$ is valid, then $\mathbf{A}_4$ is valid.

Then the prenex normal form of $\mathbf{A}_4$ is in Skolem normal form, contains only the single binary functional variable $\mathfrak{g}$, and is valid if and only if $\mathbf{A}$ is valid.

This completes the proof of the case that $\mathbf{A}$ contains only one functional variable. Turning now to the general case, we suppose that $\mathbf{A}$ is in Skolem normal form

$$(\exists\mathbf{a}_1)(\exists\mathbf{a}_2)\ldots(\exists\mathbf{a}_m)(\mathbf{b}_1)(\mathbf{b}_2)\ldots(\mathbf{b}_n)\mathbf{M},$$

$\mathbf{M}$ being the matrix, and contains M different propositional variables $\mathbf{p}_1, \mathbf{p}_2, \ldots, \mathbf{p}_M$ and N different functional variables $\mathbf{f}_1, \mathbf{f}_2, \ldots, \mathbf{f}_N$. Let $\mathbf{f}_i$ be an h_i-ary functional variable ($i = 1, 2, \ldots, N$), and let h be the greatest of the numbers $h_1, h_2, \ldots, h_N$.

In place of the two wffs **G** and **H** which were used in the first part of the proof, we now use $2(M+N)$ wffs $\mathbf{G}_1, \mathbf{G}_2, \ldots, \mathbf{G}_{M+N}, \mathbf{H}_1, \mathbf{H}_2, \ldots, \mathbf{H}_{M+N}$. Namely, where $\mathbf{g}$ is a binary functional variable and $\mathbf{c}_1, \mathbf{c}_2, \ldots, \mathbf{c}_h, \mathbf{d}_1, \mathbf{d}_2, \ldots, \mathbf{d}_{h+M+N}$ are distinct individual variables of which $\mathbf{d}_1, \mathbf{d}_2, \ldots, \mathbf{d}_{h+M+N}$ do not occur in **A**,[444] we take $\mathbf{G}_\alpha$ to be the conjunction

$$\mathbf{g}(\mathbf{d}_1, \mathbf{d}_2)\mathbf{g}(\mathbf{d}_2, \mathbf{d}_3) \ldots \mathbf{g}(\mathbf{d}_{h+M+N-1}, \mathbf{d}_{h+M+N})\mathbf{g}(\mathbf{d}_{h+M+N}, \mathbf{d}_1)$$
$$\mathbf{g}(\mathbf{d}_1, \mathbf{c}_1)\mathbf{g}(\mathbf{d}_2, \mathbf{c}_2) \ldots \mathbf{g}(\mathbf{d}_h, \mathbf{c}_h) \sim\mathbf{g}(\mathbf{d}_1, \mathbf{d}_1) \sim\mathbf{g}(\mathbf{c}_1, \mathbf{d}_2)$$
$$\sim\mathbf{g}(\mathbf{c}_2, \mathbf{d}_3) \ldots \sim\mathbf{g}(\mathbf{c}_h, \mathbf{d}_{h+1})\mathbf{g}(\mathbf{d}_{h+\alpha}, \mathbf{c}_1)$$

($\alpha = 1, 2, \ldots, M+N$), and we take $\mathbf{H}_\alpha$ to be

$$\mathbf{g}(\mathbf{d}_1, \mathbf{d}_2) \supset \centerdot\, \mathbf{g}(\mathbf{d}_2, \mathbf{d}_3) \supset \centerdot \ldots \mathbf{g}(\mathbf{d}_{h+M+N-1}, \mathbf{d}_{h+M+N}) \supset \centerdot\, \mathbf{g}(\mathbf{d}_{h+M+N}, \mathbf{d}_1) \supset \centerdot$$
$$\mathbf{g}(\mathbf{d}_1, \mathbf{c}_1) \supset \centerdot\, \mathbf{g}(\mathbf{d}_2, \mathbf{c}_2) \supset \centerdot \ldots \mathbf{g}(\mathbf{d}_h, \mathbf{c}_h) \supset \centerdot \sim\mathbf{g}(\mathbf{d}_1, \mathbf{d}_1) \supset \centerdot \sim\mathbf{g}(\mathbf{c}_1, \mathbf{d}_2) \supset \centerdot$$
$$\sim\mathbf{g}(\mathbf{c}_2, \mathbf{d}_3) \supset \centerdot \ldots \sim\mathbf{g}(\mathbf{c}_h, \mathbf{d}_{h+1}) \supset \mathbf{g}(\mathbf{d}_{h+\alpha}, \mathbf{c}_1)$$

($\alpha = 1, 2, \ldots, M+N$).

The same use is made as before of the natural numbers as domain of individuals. And given a system of values $\tau_1, \tau_2, \ldots, \tau_M, \Phi_1, \Phi_2, \ldots, \Phi_N$ of the propositional and functional variables $\mathbf{p}_1, \mathbf{p}_2, \ldots, \mathbf{p}_M, \mathbf{f}_1, \mathbf{f}_2, \ldots, \mathbf{f}_N$, the propositional function Ψ, used as a value of the variable $\mathbf{g}$, is now determined as follows. The same enumeration is used of the ordered h-tuples of natural numbers. $\Psi(u, u) =$ t except when $u = k(h+M+N)+1$ (i.e., except when u is congruent to 1 modulo $h+M+N$). If there is a natural number k such that $u_1 = k(h+M+N)+1$, $u_2 = k(h+M+N)+2$, $\ldots, u_{h+M+N} = (k+1)(h+M+N)$, then $\Psi(u_1, u_2) = \Psi(u_2, u_3) = \ldots = \Psi(u_{h+M+N-1}, u_{h+M+N}) = \Psi(u_{h+M+N}, u_1) =$ t. If $\langle v_1, v_2, \ldots, v_h\rangle$ is the $(k+1)$th h-tuple of natural numbers and $u_l = k(h+M+N)+l$ ($l = 1, 2, \ldots, h+M+N$), then $\Psi(u_1, v_1) = \Psi(u_2, v_2) = \ldots = \Psi(u_h, v_h) =$ t, and $\Psi(u_{h+j}, v_1) = \tau_j$ ($j = 1, 2, \ldots, M$), and $\Psi(u_{h+M+i}, v_1) = \Phi_i(v_1, v_2, \ldots, v_{h_i})$ ($i = 1, 2, \ldots, N$). And in all remaining cases $\Psi(u, v) =$ f.

In place of the substitutions

$$\check{S}^{\mathbf{f}(\mathbf{c}_1, \mathbf{c}_2, \ldots, \mathbf{c}_h)}_{(\exists \mathbf{d}_1)(\exists \mathbf{d}_2)\ldots(\exists \mathbf{d}_{h+1})\mathbf{G}} \quad \text{and} \quad \check{S}^{\mathbf{f}(\mathbf{c}_1, \mathbf{c}_2, \ldots, \mathbf{c}_h)}_{(\mathbf{d}_1)(\mathbf{d}_2)\ldots(\mathbf{d}_{h+1})\mathbf{H}}$$

which were used in the first part of the proof, we now use the substitutions

$$\check{S}^{\mathbf{p}_j}_{(\exists \mathbf{d}_1)(\exists \mathbf{d}_2)\ldots(\exists \mathbf{d}_{h+M+N})\mathbf{G}_j}, \qquad \check{S}^{\mathbf{p}_j}_{(\mathbf{d}_1)(\mathbf{d}_2)\ldots(\mathbf{d}_{h+M+N})\mathbf{H}_j},$$

$$\check{S}^{\mathbf{f}_i(\mathbf{c}_1, \mathbf{c}_2, \ldots, \mathbf{c}_h)}_{(\exists \mathbf{d}_1)(\exists \mathbf{d}_2)\ldots(\exists \mathbf{d}_{h+M+N})\mathbf{G}_{M+i}}, \qquad \check{S}^{\mathbf{f}_i(\mathbf{c}_1, \mathbf{c}_2, \ldots, \mathbf{c}_h)}_{(\mathbf{d}_1)(\mathbf{d}_2)\ldots(\mathbf{d}_{h+M+N})\mathbf{H}_{M+i}},$$

where $j = 1, 2, \ldots, M$, and $i = 1, 2, \ldots, N$.

With these indications, we leave it to the reader to supply the remainder of the proof, following the same plan used in the first part of the proof.

By similar methods, involving the use of an enumerably infinite domain of individuals (such as the positive integers or the natural numbers) and of an enumeration of the ordered pairs or of the ordered h-tuples of individuals, many other reductions of the decision problem for validity can be obtained. We shall indicate briefly the proof of one more such result, and then conclude this section by stating without proof some of the other results which can be found in the literature.

**472. The class of wffs which are in Skolem normal form with just three existential quantifiers in the prefix and which contain just four binary functional variables (no other functional variables, and no propositional variables) is a reduction class.

Proof. By **471 we may suppose that the given wff **A** is in Skolem normal form

$$(\exists \mathbf{a}_1)(\exists \mathbf{a}_2) \ldots (\exists \mathbf{a}_m)(\mathbf{b}_1)(\mathbf{b}_2) \ldots (\mathbf{b}_n)\mathbf{M},$$

M being the matrix, and contains only a single binary functional variable $\mathfrak{g}$. We may suppose also that $m > 3$, the required reduction being obvious in the contrary case. Let $\mathfrak{g}_1, \mathfrak{g}_2, \mathfrak{g}_3$ be binary functional variables which are distinct from one another and from $\mathfrak{g}$, let $\mathbf{x}_1, \mathbf{x}_2, \ldots, \mathbf{x}_{m-1}, \mathbf{y}, \mathbf{z}, \mathbf{c}_1, \mathbf{c}_2, \mathbf{c}_3$ be distinct individual variables which do not occur in **A**, and let **B** be:

$$(\mathbf{x}_1)(\mathbf{y})(\mathbf{z})(\exists \mathbf{c}_1)(\exists \mathbf{c}_2)(\exists \mathbf{c}_3)[\mathfrak{g}_1(\mathbf{x}_1, \mathbf{c}_1)\mathfrak{g}_2(\mathbf{x}_1, \mathbf{c}_2)\mathfrak{g}_1(\mathbf{c}_3, \mathbf{y})\mathfrak{g}_2(\mathbf{c}_3, \mathbf{z}) \;\blacksquare$$
$$\mathfrak{g}_1(\mathbf{x}_1, \mathbf{y})\mathfrak{g}_1(\mathbf{x}_1, \mathbf{z}) \supset \mathfrak{g}_3(\mathbf{y}, \mathbf{z}) \;\blacksquare\; \mathfrak{g}_2(\mathbf{x}_1, \mathbf{y})\mathfrak{g}_2(\mathbf{x}_1, \mathbf{z}) \supset \mathfrak{g}_3(\mathbf{y}, \mathbf{z}) \;\blacksquare\; \mathfrak{g}_3(\mathbf{y}, \mathbf{z}) \supset \blacksquare$$
$$\mathfrak{g}_1(\mathbf{y}, \mathbf{x}_1) \equiv \mathfrak{g}_1(\mathbf{z}, \mathbf{x}_1) \;\blacksquare\; \mathfrak{g}_2(\mathbf{y}, \mathbf{x}_1) \equiv \mathfrak{g}_2(\mathbf{z}, \mathbf{x}_1) \;\blacksquare\; \mathfrak{g}(\mathbf{y}, \mathbf{x}_1) \equiv \mathfrak{g}(\mathbf{z}, \mathbf{x}_1) \;\blacksquare$$
$$\mathfrak{g}(\mathbf{x}_1, \mathbf{y}) \equiv \mathfrak{g}(\mathbf{x}_1, \mathbf{z})] \supset$$
$$(\exists \mathbf{x}_1)(\mathbf{x}_2)(\mathbf{x}_3) \ldots (\mathbf{x}_{m-1})(\mathbf{a}_1)(\mathbf{a}_2) \ldots (\mathbf{a}_m)(\mathbf{b}_1)(\mathbf{b}_2) \ldots (\mathbf{b}_n) \;\blacksquare\; \mathfrak{g}_1(\mathbf{x}_1, \mathbf{a}_1) \supset \blacksquare$$
$$\mathfrak{g}_2(\mathbf{x}_1, \mathbf{x}_2) \supset \blacksquare\; \mathfrak{g}_1(\mathbf{x}_2, \mathbf{a}_2) \supset \blacksquare\; \mathfrak{g}_2(\mathbf{x}_2, \mathbf{x}_3) \supset \blacksquare \ldots \mathfrak{g}_1(\mathbf{x}_{m-1}, \mathbf{a}_{m-1}) \supset \blacksquare$$
$$\mathfrak{g}_2(\mathbf{x}_{m-1}, \mathbf{a}_m) \supset \mathbf{M}$$

By *381, **B** can be reduced to an equivalent wff **C** in which the quantifiers $(\mathbf{x}_1)$ and $(\exists \mathbf{x}_1)$ have been deleted from the antecedent and the consequent of **B** respectively and have been replaced by an initially placed quantifier $(\exists \mathbf{x}_1)$. The prenex normal form of **C** then satisfies the required conditions, that it is in Skolem normal form with just three existential quantifiers in the prefix and contains just four binary functional variables $\mathfrak{g}_1, \mathfrak{g}_2, \mathfrak{g}_3, \mathfrak{g}$.

We have at once that the prenex normal form of **C** is valid if and only if **B** is valid. That **B** is valid if and only if **A** is valid we leave to the reader to prove, with the aid of the following remark.

Take the natural numbers as domain of individuals, and choose any enumeration of the ordered pairs of natural numbers. Given an arbitrary value Ψ of $\mathbf{g}$, a system of values Ψ_1, Ψ_2, Ψ_3 of $\mathbf{g}_1$, $\mathbf{g}_2$, $\mathbf{g}_3$, may be determined as follows so as to give to the antecedent of **B** the value t: $\Psi_1(u, v) =$ t if and only if v is the first number in the $(u + 1)$th ordered pair; $\Psi_2(u, v) =$ t if and only if v is the second number in the $(u + 1)$th ordered pair; $\Psi_3(u,v) =$ t if and only if $u = v$. For this particular system of values of $\mathbf{g}_1$, $\mathbf{g}_2$, $\mathbf{g}_3$ it is clear that the consequent of **B** has the value t if and only if **A** has the value t. It is true that, for a given value of $\mathbf{g}$, other systems of values of $\mathbf{g}_1$, $\mathbf{g}_2$, $\mathbf{g}_3$ may in general be found so as to give to the antecedent of **B** the value t; but (as may be read from the antecedent of **B** itself) these other systems of values of $\mathbf{g}_1$, $\mathbf{g}_2$, $\mathbf{g}_3$ must always have certain properties in common with the system Ψ_1, Ψ_2, Ψ_3 which are sufficient to ensure that the consequent of **B** has the value t if and only if **A** has the value t.

As will be indicated in exercises below, the reduction process of **472 may readily be modified so as to obtain only three binary functional variables in the wffs of the reduction class instead of four, or, alternatively, so as to obtain one binary and one ternary functional variable (the other conditions remaining in either case unchanged). By more elaborate methods of the same kind it is even possible to reduce this to a single binary functional variable.

According to known results, including that just mentioned, each of the following classes of wffs is a reduction class (where it shall be understood in each case, without separate mention, that the wffs are to contain no free individual variables and no propositional variables, and that either the wff itself or its indicated antecedent and consequent are to be in prenex normal form):

Wffs with prefix $(\exists\mathbf{a}_1)(\exists\mathbf{a}_2)(\exists\mathbf{a}_3)(\mathbf{b}_1)(\mathbf{b}_2)\ldots(\mathbf{b}_n)$ which contain a single binary functional variable.[446]

Wffs with prefix $(\mathbf{a})(\exists\mathbf{b})(\mathbf{c})(\exists\mathbf{d}_1)(\exists\mathbf{d}_2)\ldots(\exists\mathbf{d}_n)$ which contain a single binary functional variable.[447]

---

[446]The reduction to wffs of this prefix containing none but binary functional variables is due to Gödel in *Monatshefte für Mathematik und Physik*, vol. 40 (1933), pp. 433–443 (another proof by Skolem in *Acta Scientiarum Mathematicarum*, vol. 7 (1935), pp. 193–199). The further reduction to a single binary functional variable is due to Lászlo Kalmár and János Surányi in *The Journal of Symbolic Logic*, vol. 12 (1947), pp. 65–73. The proof of **472 which is given in outline above is by Gödel's method.

[447]The reduction to wffs of this prefix is due to Ackermann in the paper cited in footnotes 430, 435, and the further reduction to a single binary functional variable is due to Kalmár in *The Journal of Symbolic Logic*, vol. 4 (1939), pp. 1–9. The result proved by Ackermann can be stated in the somewhat stronger form that the class of wffs of the form

$$(\exists\mathbf{a})(\mathbf{a}_1)(\mathbf{a}_2)\ldots(\mathbf{a}_n)\mathbf{M} \supset (\exists\mathbf{b})(\mathbf{c})\mathbf{f}(\mathbf{b}, \mathbf{c}),$$

where **M** is quantifier-free and contains no individual variables other than $\mathbf{a}$, $\mathbf{a}_1$, $\mathbf{a}_2$, $\ldots$, $\mathbf{a}_n$ and **f** is a binary functional variable occurring in **M**, is a reduction class.

Wffs with prefix $(\exists \mathbf{b}_1)(\exists \mathbf{b}_2)(\mathbf{c})(\exists \mathbf{d}_1)(\exists \mathbf{d}_2) \ldots (\exists \mathbf{d}_n)$ which contain a single binary functional variable.[448]

Wffs with prefix $(\exists \mathbf{a}_1)(\exists \mathbf{a}_2) \ldots (\exists \mathbf{a}_n)(\mathbf{b})$ which contain a single binary functional variable.[449]

Wffs of the form $(\mathbf{a})(\mathbf{b})(\mathbf{c})\mathbf{M}_1 \supset (\exists \mathbf{a})(\exists \mathbf{b})(\mathbf{c})\mathbf{M}_2$, where $\mathbf{M}_1$ and $\mathbf{M}_2$ are quantifier-free and contain none but binary functional variables.[450]

Hence also wffs with prefix $(\exists \mathbf{a})(\exists \mathbf{b})(\exists \mathbf{c})(\mathbf{d})$ which contain none but binary functional variables.[450]

And also wffs with prefix $(\exists \mathbf{a})(\exists \mathbf{b})(\mathbf{c})(\exists \mathbf{d})$ which contain none but binary functional variables.[450]

---

[448]The reduction to wffs of this prefix is due to Józef Pepis in *Fundamenta Mathematicae*, vol. 30 (1938), pp. 257–348. More fully, Pepis's result in this paper is that the class of wffs of the form

$$(\mathbf{a}_1)(\mathbf{a}_2) \ldots (\mathbf{a}_n)\mathbf{M} \supset (\exists \mathbf{b}_1)(\exists \mathbf{b}_2)(\mathbf{c})\mathbf{f}(\mathbf{b}_1, \mathbf{b}_2, \mathbf{c}),$$

where $\mathbf{M}$ is quantifier-free, and contains no individual variables other than $\mathbf{a}_1$, $\mathbf{a}_2$, $\ldots$, $\mathbf{a}_n$, and contains besides the ternary functional variable $\mathbf{f}$ only one singulary functional variable, is a reduction class. Or $\mathbf{f}(\mathbf{b}_1, \mathbf{b}_2, \mathbf{c})$ may be replaced by the disjunction $\mathbf{f}_1(\mathbf{b}_1, \mathbf{c}) \vee \mathbf{f}_2(\mathbf{b}_2, \mathbf{c})$, in which case $\mathbf{M}$ contains the two binary functional variables $\mathbf{f}_1$ and $\mathbf{f}_2$ and one singulary functional variable. The reduction to the prefix $(\exists \mathbf{b}_1)(\exists \mathbf{b}_2)(\mathbf{c})(\exists \mathbf{d}_1)(\exists \mathbf{d}_2) \ldots (\exists \mathbf{d}_n)$ and a single binary functional variable is due to Kalmár and Surányi in *The Journal of Symbolic Logic*, vol. 15 (1950), pp. 161–173.

[449]The reduction to wffs of this prefix is again due to Pepis, being a corollary of the fuller result quoted in the preceding footnote. The further reduction to a single binary functional variable is due to Surányi in *Matematikai és Fizikai Lapok*, vol. 50 (1943), pp. 51–74 (see also the paper of Kalmár and Surányi which is cited in the preceding footnote).

The same paper of Pepis contains also a number of other results, in the direction of reducing the number of functional variables required in connection with various prefixes. Some of these have since been superseded by stronger results, but the following seems to be worth quoting: in the Ackermann normal form as given in footnote 447, $\mathbf{M}$ may be restricted to contain, besides the binary functional variable $\mathbf{f}$, only one singulary and one ternary functional variable, or else only one singulary and two binary functional variables (as preferred).

[450]These reductions are due to Surányi in the paper cited in the preceding footnote. (*Added in proof.*) The same reductions are also obtained by Surányi in a paper in *Acta Mathematica Academiae Scientiarum Hungaricae*, vol. 1 (1950), pp. 261–271. Another paper by Surányi in the same periodical, vol. 2 (1951), pp. 325–335, adds to the list of reduction classes the two following: wffs with prefix $(\mathbf{a})(\exists \mathbf{b})(\mathbf{c})(\exists \mathbf{d})(\exists \mathbf{e})$ which contain none but singulary and binary functional variables, including at most seven binary functional variables; and wffs with prefix $(\exists \mathbf{a})(\mathbf{b})(\mathbf{c})(\exists \mathbf{d})(\exists \mathbf{e})$ which contain none but singulary and binary functional variables, including at most seven binary functional variables. Two papers by Kalmár, *ibid.*, vol. 1 (1950), pp. 64–73, and vol. 2 (1951), pp. 19–38, add the three following reduction classes: wffs with prefix $(\exists \mathbf{a}_1)(\exists \mathbf{a}_2)(\mathbf{b}_1)(\mathbf{b}_2) \ldots (\mathbf{b}_n)(\exists \mathbf{c})$ which contain a single binary functional variable; wffs with prefix $(\exists \mathbf{a})(\mathbf{b}_1)(\mathbf{b}_2) \ldots (\mathbf{b}_n)(\exists \mathbf{c}_1)(\exists \mathbf{c}_2)$ which contain a single binary functional variable; and wffs with prefix $(\mathbf{a}_1)(\mathbf{a}_2) \ldots)(\mathbf{a}_n)(\exists \mathbf{b})(\mathbf{c}_1)(\mathbf{c}_2) \ldots (\mathbf{c}_{48})(\exists \mathbf{d}_1)(\exists \mathbf{d}_2)$ which contain a single binary functional variable. A reduction of the decision problem for satisfiability to that concerning validity in every finite domain, and hence a reduction of the decision problem for validity to that concerning satisfiability in some finite domain, is contained in another paper of Kalmár, *ibid.*, vol. 2 (1951), pp. 125–141 (the unsolvability of the decision problem concerning validity in every finite domain had already been proved by Trachtenbrot in the paper cited in footnote 567).

EXERCISES 47

47.0. Extend the result of **470 to obtain a reduction of the decision problem for provability, by showing how to find a proof of **A** if a proof of **B** is known.

47.1. Supply in detail the last part of the proof of **471, which was omitted in the text.

47.2. For the proof of **472, supply in detail the omitted demonstration that **B** is valid if and only if **A** is valid.

47.3. Show that, in **472, the number of binary functional variables may be reduced from four to three by introducing a binary functional variable **h**, replacing $\mathfrak{g}_2(\mathbf{d}, \mathbf{e})$ everywhere by $\mathbf{h}(\mathbf{d}, \mathbf{e}) \sim\mathbf{h}(\mathbf{e}, \mathbf{d})$, and $\mathfrak{g}_3(\mathbf{d}, \mathbf{e})$ everywhere by $\mathbf{h}(\mathbf{d}, \mathbf{e})\mathbf{h}(\mathbf{e}, \mathbf{d})$.

47.4. Show that, in **472, the reduction process may be modified so as to obtain one binary and one ternary functional variable, replacing $\mathfrak{g}_1(\mathbf{d}, \mathbf{e})$ everywhere by $\mathbf{h}(\mathbf{d}, \mathbf{e}, \mathbf{e})$, $\mathfrak{g}_2(\mathbf{d}, \mathbf{e})$ by $\mathbf{h}(\mathbf{d}, \mathbf{e}, \mathbf{d})$, and $\mathfrak{g}_3(\mathbf{d}, \mathbf{e})$ by $\sim\mathbf{h}(\mathbf{d}, \mathbf{d}, \mathbf{e})$

47.5. The leading idea of the reduction process of **472 is to make use of an enumeration of the ordered pairs of natural numbers in order to replace a sequence of existential quantifiers $(\exists\mathbf{a}_1)(\exists\mathbf{a}_2) \ldots (\exists\mathbf{a}_m)$ by a single existential quantifier $(\exists\mathbf{x}_1)$, at the expense of increasing the number of universal quantifiers. With appropriate modifications, the same idea may be used to replace a sequence of universal quantifiers by a single universal quantifier, at the expense of increasing the number of existential quantifiers. For example, $(\mathbf{b}_1)(\mathbf{b}_2)\mathbf{M}$ might be replaced by $(\mathbf{x}_1)(\exists\mathbf{b}_1)(\exists\mathbf{b}_2) \,\blacksquare\, \mathfrak{g}_1(\mathbf{x}_1, \mathbf{b}_1)\mathfrak{g}_2(\mathbf{x}_1, \mathbf{b}_2)\mathbf{M}$, which will have the same value as $(\mathbf{b}_1)(\mathbf{b}_2)\mathbf{M}$ if $\mathfrak{g}_1$ and $\mathfrak{g}_2$ have the values Ψ_1 and Ψ_2 that are given in the proof of **472. Investigate the question what additional reductions of the decision problem for validity can be obtained (beyond those of **470–**472) by using this method, together with the methods and results of **420–**421, **470–**472.

48. Functional calculus of first order with equality. *The functional calculus of first order with equality* is a logistic system obtained from the functional calculus of first order by adding a binary functional constant I and certain axioms (or postulates, according to the point of view) that contain I. Or, alternatively, it may be described as obtained by adjoining additional axioms to an applied functional calculus of first order among whose primitive symbols is the binary functional constant I. The wffs of the system are the same as the wffs of this applied functional calculus of first order, but of course there are additional theorems in consequence of the added axioms.

We shall speak of the *pure functional calculus of first order with equality* if the primitive symbols include all propositional and functional variables (as listed in §30) and no functional constants except I; an *applied functional calculus of first order with equality* if there are other functional constants in addition to I; a *simple applied functional calculus of first order with equality* if there are other functional constants in addition to I and no functional variables. Besides these there is the *simple calculus of equality*, obtained by adding appropriate axioms to the simple applied functional calculus of first order which has I as its only functional constant.

If the formulation of §30 is used for the functional calculus of first order, the axioms to be added are the single axiom

$$I(x, x)$$

and the infinite list of axioms given by the axiom schema

$$I(\mathbf{a}, \mathbf{b}) \supset . \mathbf{A} \supset \mathbf{B}.$$

where **a** is an individual variable or an individual constant, **b** is an individual variable or an individual constant, and **B** is obtained from **A** by replacing one particular occurrence of **a** by **b**, this particular occurrence of **a** being within the scope neither of a quantifier (**a**) nor of a quantifier (**b**). The formulation of the functional calculus of first order with equality that is obtained by adding the functional constant I and the foregoing axioms to F^1 we shall call F^I. And in particular the formulation $\mathrm{F}^{I\mathrm{p}}$ of the pure functional calculus of first order with equality is obtained by adding the functional constant I and these axioms to $\mathrm{F}^{1\mathrm{p}}$.

For the simple calculus of equality we may begin with the formulation F^1 of a simple applied functional calculus of first order having I as its only functional constant. To this we may add the axiom $I(x, x)$ and all the axioms given by the above axiom schema, so obtaining the formulation $\ddot{\mathrm{E}}$ of the simple calculus of equality. It is sufficient, however, to add only the three following axioms:

$$I(x, x) \qquad (\textit{Reflexive law of equality.})$$

$$I(x, y) \supset I(y, x) \qquad (\textit{Commutative law of equality.})$$

$$I(x, y) \supset . I(y, z) \supset I(x, z) \qquad (\textit{Transitive law of equality.})$$

And the formulation of the simple calculus of equality that is obtained in this way we call E.

For the pure functional calculus of first order we may use also the formulation $\mathrm{F}^{1\mathrm{p}}_2$ of §40. By adding to this the functional constant I and two axioms,

$$I(x, x)$$
$$I(x, y) \supset . F(x) \supset F(y),$$

we obtain a formulation of the pure functional of first order with equality which we shall call F_2^{Ip}.

In all of these calculi the notations $=$ and $\neq$, more familiar than I, may be introduced by definition as follows:

D18. $[\mathbf{a} = \mathbf{b}] \rightarrow I(\mathbf{a}, \mathbf{b})$

D19. $[\mathbf{a} \neq \mathbf{b}] \rightarrow \sim I(\mathbf{a}, \mathbf{b})$

And of course all the definitions and methods of abbreviation of wffs continue in force which were introduced for the functional calculus of first order in §30.

For the principal interpretation of all of these systems it is intended that I shall denote the relation of equality, or identity, between individuals.

For example, in the case of the pure functional calculus of first order with equality, after choosing some non-empty class as the individuals, we fix the principal interpretation by the same semantical rules a–f as given in §30 (for the pure functional calculus of first order), together with two additional rules as follows:

g_1. If **a** is an individual variable, the value of $I(\mathbf{a}, \mathbf{a})$ is t for all values of **a**.

g_2. If **a** and **b** are distinct individual variables, the value of $I(\mathbf{a}, \mathbf{b})$ is t if the value of **a** is the same as the value of **b**, and the value of $I(\mathbf{a}, \mathbf{b})$ is f if the values of **a** and **b** are different.

The syntactical definitions of validity and satisfiability (§43), as well as the metatheorem that every theorem is valid (**434), can be extended in obvious fashion to the pure functional calculus of first order with equality, and, especially in some of the exercises following, we shall assume that this has been done.

EXERCISES 48

48.0. In the formulation E of the simple calculus of equality, the commutative and transitive laws of equality may be replaced by Euclid's axiom that "things equal to the same thing are also equal to each other," expressed as follows in the notation of the system:

$$I(x, z)I(y, z) \supset I(x, y)$$

Thus is obtained a formulation É of the simple calculus of equality which has only two added axioms instead of three. Show that E and É are equivalent in the sense that their theorems are the same.

48.1. For each of the systems E and É show that the added axioms, containing I, are independent.

48.2. Show that the two formulations Ë and E of the simple calculus of equality are equivalent in the sense that their theorems are the same. (Compare the proof of *340, which may here be paralleled in certain respects.)

48.3. Show that the two formulations F^{Ip} and F^{Ip}_2 of the pure functional calculus of first order with equality are equivalent in the sense that their theorems are the same. (The same method may be used by which the equivalence of F^{1p} and F^{1p}_2 was proved. But notice, in particular, that the added axioms here introduce some new questions in connection with the rules of substitution.)

48.4. For a formulation of a simple applied functional calculus of first order with equality, if the number of functional constants is finite, show that a finite number of added axioms is sufficient, as follows: the reflexive, commutative, and transitive laws of equality; for each singulary functional constant **f** an axiom,

$$I(x, y) \supset \mathbf{.}\, \mathbf{f}(x) \supset \mathbf{f}(y);$$

for each binary functional constant **f** other than I, two axioms,

$$I(x, y) \supset \mathbf{.}\, \mathbf{f}(x, z) \supset \mathbf{f}(y, z),$$

$$I(x, y) \supset \mathbf{.}\, \mathbf{f}(z, x) \supset \mathbf{f}(z, y);$$

for each ternary functional constant, three analogous axioms; and so on until axioms of this kind have been introduced for all the functional constants. (Again, compare the method of proof of *340.)

48.5. By using an idea similar to that of the preceding exercise, but applied to functional variables rather than functional constants, show how, for any wff **A** of F^{Ip}, to find a corresponding wff **A**′ of F^{1p} which is valid if and only if **A** is valid and which is a theorem of F^{1p} if and only if **A** is a theorem of F^{Ip}. Hence extend the Gödel completeness theorem, **440, to the pure functional calculus of first order with equality.

48.6. Use the same method to prove the following extension of **450 to the pure functional calculus of first order with equality: if a wff of F^{Ip} is valid in every non-empty finite domain and is also valid in an enumerably infinite domain, then it is valid in every non-empty domain.

48.7. Find and prove similar extensions of **453 and **455 to the pure functional calculus of first order with equality.

48.8. Prove the consistency of F^I by making use of the afp of a wff as in §32.

48.9. Extend the metatheorem **323 to F^I.

48.10. Extend the metatheorem **325 to F^I.

48.11. Extend the principles of duality *372—*374 to F^I. (Wffs of F^I are to be rewritten by means of D18 and D19 in such a way that the symbol I no longer appears explicitly. Then in dualizing. the notations $=$ and $\neq$ are to be interchanged, as well as $\supset$ and $\not\supset$, disjunction and conjunction, $\equiv$ and $\not\equiv$, $\subset$ and $\not\subset$, $\bar{\vee}$ and $|$, $\forall$ and $\exists$.)

48.12. By using the reduction found in 48.5, solve the decision problem for the case of wffs of F^{Ip} which have a prenex normal form such that, in the prefix, no existential quantifier precedes any universal quantifier. (Notice that this includes, in particular, the case of quantifier-free formulas of F^{Ip}.)

48.13. Solve the decision problem for quantifier-free formulas of F^{Ip} directly, by a method as closely similar as possible to the truth-table decision procedure by which it is determined whether a quantifier-free formula of F^{1p} is tautologous. Hence restate the solution of the decision problem for the special case of 48.12, in a form as similar as possible to that of *460–**463.

48.14. Solve the decision problem for the singulary functional calculus of first order with equality, i.e., for the class of wffs of F^{Ip} in which all the functional variables occurring are singulary.

Suggestion: Following Behmann, we may add the following reduction steps to the reduction steps (a)–(g) of exercise 39.6: (α) to replace a wf part $(\mathbf{a})[\mathbf{a} = \mathbf{b}]$ by $(\mathbf{a})(\mathbf{c})[\mathbf{a} = \mathbf{c}]$, if $\mathbf{a}$ and $\mathbf{b}$ are distinct individual variables and $\mathbf{c}$ is the first individual variable in alphabetic order other than $\mathbf{a}$ and $\mathbf{b}$; (β) to replace a wf part $(\mathbf{a})[\mathbf{a} \neq \mathbf{b}]$ by $(\mathbf{a})[\mathbf{a} \neq \mathbf{a}]$, if $\mathbf{a}$ and $\mathbf{b}$ are distinct individual variables; (γ) to replace a wf part $(\mathbf{a})[\mathbf{a} \neq \mathbf{b}_1 \supset . \mathbf{a} \neq \mathbf{b}_2 \supset . \ldots . \mathbf{a} \neq \mathbf{b}_n \supset \mathbf{a} = \mathbf{b}]$ by the conjunction $\mathbf{A}_n[\mathbf{b}_1 = \mathbf{b}_2 \supset \mathbf{A}_{n-1}][\mathbf{b}_1 = \mathbf{b}_3 \supset \mathbf{A}_{n-1}] \ldots [\mathbf{b}_n = \mathbf{b} \supset \mathbf{A}_{n-1}][\mathbf{b}_1 = \mathbf{b}_2 \supset . \mathbf{b}_1 = \mathbf{b}_3 \supset \mathbf{A}_{n-2}][\mathbf{b}_1 = \mathbf{b}_2 \supset . \mathbf{b}_1 = \mathbf{b}_4 \supset \mathbf{A}_{n-2}] \ldots [\mathbf{b}_{n-2} = \mathbf{b} \supset . \mathbf{b}_{n-1} = \mathbf{b}_n \supset \mathbf{A}_{n-2}][\mathbf{b}_{n-1} = \mathbf{b}_n \supset . \mathbf{b}_{n-1} = \mathbf{b} \supset \mathbf{A}_{n-2}] \ldots\ldots [\mathbf{b}_1 = \mathbf{b}_2 \supset . \mathbf{b}_1 = \mathbf{b}_3 \supset . \ldots \mathbf{b}_1 = \mathbf{b} \supset \mathbf{A}_0]$, if $\mathbf{a}, \mathbf{b}_1, \mathbf{b}_2, \ldots, \mathbf{b}_n, \mathbf{b}$ are distinct individual variables, $\mathbf{c}_1, \mathbf{c}_2, \ldots, \mathbf{c}_n, \mathbf{c}$ are the first $n + 1$ individual variables in alphabetic order distinct from each other and from $\mathbf{a}, \mathbf{b}_1, \mathbf{b}_2, \ldots, \mathbf{b}_n, \mathbf{b}$, and $\mathbf{A}_i$ is

$$(\mathbf{a})(\mathbf{c}_1)(\mathbf{c}_2)\ldots(\mathbf{c}_i)(\mathbf{c}) . \mathbf{a} \neq \mathbf{c}_1 \supset . \mathbf{a} \neq \mathbf{c}_2 \supset . \ldots . \mathbf{a} \neq \mathbf{c}_i \supset . \mathbf{c}_1 \neq \mathbf{c}_2 \supset . \mathbf{c}_1 \neq \mathbf{c}_3 \supset . \ldots . \mathbf{c}_1 \neq \mathbf{c}_i \supset . \mathbf{c}_1 \neq \mathbf{c} \supset . \mathbf{c}_2 \neq \mathbf{c}_3 \supset . \ldots \ldots . \mathbf{c}_i \neq \mathbf{c} \supset \mathbf{a} = \mathbf{c}$$

($i = 0, 1, 2, \ldots, n$); (δ) to replace a wf part $(\mathbf{a})[\mathbf{a} = \mathbf{a} \supset \mathbf{A}]$ by $(\mathbf{a})\mathbf{A}$; (ε) to replace a wf part $(\mathbf{a})[\mathbf{a} = \mathbf{b} \supset \mathbf{A}]$ by

$$S_{\mathbf{b}}^{\mathbf{a}} \mathbf{A}|,$$

if **a** and **b** are distinct individual variables and **A** is quantifier-free; (ζ) to replace a wf part $(\mathbf{a})[\mathbf{a} \neq \mathbf{a} \supset \mathbf{A}]$ by $(\mathbf{a})[\mathbf{a} = \mathbf{a}]$; ($\eta$) to replace a wf part $(\mathbf{a})[\mathbf{a} \neq \mathbf{b}_1 \supset . \mathbf{a} \neq \mathbf{b}_2 \supset . \ldots \mathbf{a} \neq \mathbf{b}_n \supset \mathbf{A}]$ by the conjunction $\mathbf{A}_n[\mathbf{B}_1 \supset \mathbf{A}_{n-1}][\mathbf{B}_2 \supset \mathbf{A}_{n-1}] \ldots [\mathbf{B}_n \supset \mathbf{A}_{n-1}][\mathbf{b}_1 = \mathbf{b}_2 \supset \mathbf{A}_{n-1}][\mathbf{b}_1 = \mathbf{b}_3 \supset \mathbf{A}_{n-1}] \ldots [\mathbf{b}_{n-1} = \mathbf{b}_n \supset \mathbf{A}_{n-1}][\mathbf{B}_1 \supset . \mathbf{B}_2 \supset \mathbf{A}_{n-2}][\mathbf{B}_1 \supset . \mathbf{B}_3 \supset \mathbf{A}_{n-2}] \ldots [\mathbf{B}_{n-1} \supset . \mathbf{B}_n \supset \mathbf{A}_{n-2}][\mathbf{b}_1 = \mathbf{b}_2 \supset . \mathbf{B}_3 \supset \mathbf{A}_{n-2}][\mathbf{b}_1 = \mathbf{b}_2 \supset . \mathbf{B}_4 \supset \mathbf{A}_{n-2}] \ldots [\mathbf{b}_{n-1} = \mathbf{b}_n \supset . \mathbf{B}_{n-2} \supset \mathbf{A}_{n-2}][\mathbf{b}_1 = \mathbf{b}_2 \supset . \mathbf{b}_1 = \mathbf{b}_3 \supset \mathbf{A}_{n-2}][\mathbf{b}_1 = \mathbf{b}_2 \supset . \mathbf{b}_1 = \mathbf{b}_4 \supset \mathbf{A}_{n-2}] \ldots [\mathbf{b}_{n-3} = \mathbf{b}_n \supset . \mathbf{b}_{n-2} = \mathbf{b}_{n-1} \supset \mathbf{A}_{n-2}][\mathbf{b}_{n-2} = \mathbf{b}_{n-1} \supset . \mathbf{b}_{n-2} = \mathbf{b}_n \supset \mathbf{A}_{n-2}] \ldots \ldots [\mathbf{B}_1 \supset . \mathbf{B}_2 \supset . \ldots \mathbf{B}_n \supset \mathbf{A}_0]$, if $\mathbf{a}, \mathbf{b}_1, \mathbf{b}_2, \ldots, \mathbf{b}_n$ are distinct individual variables and $\mathbf{c}_1, \mathbf{c}_2, \ldots, \mathbf{c}_n$ are the first n individual variables in alphabetic order distinct from each other and from $\mathbf{a}, \mathbf{b}_1, \mathbf{b}_2, \ldots, \mathbf{b}_n$, and if, further, **A** is quantifier-free and contains no individual variables except **a**, and $\mathbf{B}_i$ and $\mathbf{C}_i$ are respectively

$$\mathrm{S}^{\mathbf{a}}_{\mathbf{b}_i}\mathbf{A}| \quad \text{and} \quad \mathrm{S}^{\mathbf{a}}_{\mathbf{c}_i}\mathbf{A}|$$

$(i = 1, 2, \ldots, n)$, and $\mathbf{A}_i$ is

$$(\mathbf{a})(\mathbf{c}_1)(\mathbf{c}_2) \ldots (\mathbf{c}_i) . \mathbf{a} \neq \mathbf{c}_1 \supset . \mathbf{a} \neq \mathbf{c}_2 \supset . \ldots \mathbf{a} \neq \mathbf{c}_i \supset . \mathbf{c}_1 \neq \mathbf{c}_2 \supset . \mathbf{c}_1 \neq \mathbf{c}_3 \supset . \ldots \mathbf{c}_{i-1} \neq \mathbf{c}_i \supset . \sim\mathbf{C}_1 \supset . \sim\mathbf{C}_2 \supset . \ldots \sim\mathbf{C}_i \supset \mathbf{A}$$

$(i = 0, 1, 2, \ldots, n)$ (thus in particular $\mathbf{A}_0$ is $(\mathbf{a})\mathbf{A}$).

48.15. Solve the decision problem for wffs **A** of $\mathrm{F}^{I\mathrm{p}}$ such that in each elementary part not containing I at most one variable has occurrences at which it is a bound variable of **A**.

48.16. (1) Extend the method of 46.8 to solve the decision problem for wffs of $\mathrm{F}^{I\mathrm{p}}$ having a prenex normal form $(\exists\mathbf{b})(\mathbf{c})\mathbf{M}$ in which **M** is the matrix and contains no individual variables except **b** and **c**. And illustrate by applying the solution to the following particular examples:

(2) $(\exists x)(y) . F(x) \supset [F(y) \supset x = y] \supset . F(x) \equiv G(y) \supset . F(y) \equiv G(x)$

(3) $(\exists x)(y) . [F(x) \equiv G(x)] \vee [F(y) \equiv G(y) \equiv x = y]$

48.17. (1) Extend the method of 46.11(2) to solve the decision problem for wffs of $\mathrm{F}^{I\mathrm{p}}$ having a prenex normal form $(\mathbf{a})(\exists\mathbf{b})(\mathbf{c})\mathbf{M}$ in which **M** is the matrix and contains no individual variables except **a**, **b**, **c**. And illustrate by applying the solution to the following particular examples:

(2) $(x)(\exists y)(z) . F(x, x) \supset . F(x, z) \supset x = y \vee y = z \supset . F(x, z) \equiv F(x, y) \equiv F(y, y) \supset . F(y, y) \equiv F(z, z)$

(3) $(x)(\exists y)(z) . F(x) \supset . G(x) \supset . F(y) \supset [G(y) \supset x = y] \supset . x \neq z \supset . G(y) \equiv F(z) \supset . F(y) \equiv G(z)$

(4) $$(x)(\exists y)(z) \centerdot x \neq z \vee y \neq z$$

48.18. Apply the decision procedure of 48.14 to (2) the example 48.16(2), and (3) the example 48.17(3).

48.19. (1) Solve the decision problem for wffs of F^{Ip} having a prenex normal form $(\mathbf{a}_1)(\mathbf{a}_2)(\exists\mathbf{b})(\mathbf{c})\mathbf{M}$ in which $\mathbf{M}$ is the matrix and contains no individual variables except $\mathbf{a}_1$, $\mathbf{a}_2$, $\mathbf{b}$, $\mathbf{c}$. Illustrate by applying the decision procedure to show that the following wffs are theorems of F^{Ip}:

(2) $$(x_1)(x_2)(\exists y)(z) \centerdot F(x_1, x_2) \supset \centerdot F(x_1, z) \equiv F(y, y) \supset \centerdot x_1 \neq z \vee x_2 \neq z \supset [F(x_1, z) \supset F(x_1, y)] \supset \centerdot F(x_1, y) \vee F(x_1, z) \supset F(x_2, x_2)$$

(3) $$(x_1)(x_2)(\exists y)(z) \centerdot F(x_1) \supset \centerdot F(x_2) \supset \centerdot G(x_1) \supset \centerdot G(x_2) \supset \centerdot x_1 \neq y \supset [F(z) \supset G(z)] \supset \centerdot H(x_1) \supset H(x_2) \supset [G(y) \supset F(y)] \supset \centerdot G(y) \equiv F(z) \supset \centerdot F(y) \equiv G(z)$$

48.20. Apply the decision procedure of 48.15 to the example 48.19(2).

48.21. (1) State and solve a special case of the decision problem of F^{Ip} which is analogous to case V′ of the decision problem of F^{1p}. (Cf. 46.15.) (2) Illustrate by applying the solution to the following particular example:

$$(\exists x)(\exists y)(z) \centerdot \sim F(x, x) \vee F(x, z) \vee F(y, z) \vee [x = z \equiv \centerdot F(x, x) \equiv F(z, z)] \vee \centerdot y = z \equiv \centerdot F(y, y) \equiv F(z, z)$$

48.22. Prove the following metatheorem: Let Γ be a (finite or infinite) class of wffs of F^{Ip}. Let the complete list of free individual variables occurring in wffs of Γ be $\mathbf{a}_1, \mathbf{a}_2, \mathbf{a}_3, \ldots$. Let the complete list of propositional variables occurring be $\mathbf{p}_1, \mathbf{p}_2, \mathbf{p}_3, \ldots$, and let the complete list of functional variables occurring be $\mathbf{f}_1, \mathbf{f}_2, \mathbf{f}_3, \ldots$ (of course any or all of these lists may be infinite), and suppose that $\mathbf{f}_i$ is an h_i-ary functional variable ($i = 1, 2, 3, \ldots$) Suppose further that Γ is simultaneously satisfied in the domain of positive integers by the system of values $v_1, v_2, v_3, \ldots, \tau_1, \tau_2, \tau_3, \ldots, \Phi_1, \Phi_2, \Phi_3, \ldots$ of the variables $\mathbf{a}_1, \mathbf{a}_2, \mathbf{a}_3, \ldots, \mathbf{p}_1, \mathbf{p}_2, \mathbf{p}_3, \ldots, \mathbf{f}_1, \mathbf{f}_2, \mathbf{f}_3, \ldots$. Then in the domain of rational integers (i.e., positive integers, negative integers, and 0) there exist propositional functions $\Psi_1, \Psi_2, \Psi_3, \ldots$, such that Ψ_i is an h_i-ary propositional function of rational integers ($i = 1, 2, 3, \ldots$), and for arbitrary positive integers $u_1, u_2, \ldots u_{h_i}$ the truth-value $\Psi_i(u_i, u_2, \ldots, u_{h_i})$ is the same as $\Phi_i(u_1, u_2, \ldots, u_{h_i})$, and Γ is simultaneously satisfied in the domain of rational integers by the system of values $v_1, v_2, v_3, \ldots, \tau_1, \tau_2, \tau_3,$

. . ., $\Psi_1, \Psi_2, \Psi_3, \ldots$ of the variables $\mathbf{a}_1, \mathbf{a}_2, \mathbf{a}_3, \ldots, \mathbf{p}_1, \mathbf{p}_2, \mathbf{p}_3, \ldots, \mathbf{f}_1, \mathbf{f}_2, \mathbf{f}_3,$ . . . .[451]

Suggestion: We may suppose without loss of generality that the individual variables $\mathbf{a}_1, \mathbf{a}_2, \mathbf{a}_3, \ldots$, if any, are the particular variables $z_1, z_2, z_3, \ldots$. Adjoin to Γ all of the wffs $x_j \neq x_k$ for which the subscripts j and k are distinct positive integers, also all of the wffs

$$\mathbf{f}_i(x_{u_1}, x_{u_2}, \ldots, x_{u_{h_i}})$$

for which $\Phi_i(u_1, u_2, \ldots, u_{h_i})$ is truth, and also all of the wffs

$$\sim\mathbf{f}_i(x_{u_1}, x_{u_2}, \ldots, x_{u_{h_i}})$$

for which $\Phi_i(u_1, u_2, \ldots, u_{h_i})$ is falsehood. Let the class of wffs so obtained be Γ'; and let Γ'' be obtained from Γ' by adjoining further all of the wffs $y_j \neq x_k$ for which the subscripts j and k are arbitrary positive integers, and all of the wffs $y_j \neq y_k$ for which the subscripts j and k are distinct positive integers. Show that Γ' is consistent, hence that every finite subclass of Γ'' is consistent. Hence use the result of exercise 48.7 to show that Γ'' is simultaneously satisfiable in an enumerably infinite domain $\mathfrak{J}$. The individuals of the domain $\mathfrak{J}$ which serve as values of $x_1, x_2, x_3, \ldots$ may be identified with the positive integers 1, 2, 3, . . . respectively. Besides these $\mathfrak{J}$ necessarily includes infinitely many other individuals, which may then be identified in some arbitrary way with the non-positive integers 0, —1, —2,. .

48.23. As a corollary of the foregoing, prove the following metatheorem: Let Γ be a (finite or infinite) class of wffs of F^{1p}, and let one of the functional variables occurring in wffs of Γ be the binary functional variable **s**. Suppose

[451]The reference to the particular domains of positive integers and of rational integers is evidently non-essential, the substance of the metatheorem being that an enumerably infinite *model* of Γ (i.e., an enumerably infinite domain together with such a system of values of the free variables as to satisfy Γ simultaneously in that domain) is always capable of an enumerably infinite extension. The result is substantially due to A. Malcev in a paper in the *Recueil Mathématique*, vol. **43** (n.s. vol. 1) (**1936**), pp. 323–336, and the proof which is suggested above employs some of Malcev's ideas. Although Malcev's own proof is defective in regard to the use which he makes of the Skolem normal form for satisfiability, it appears that the defect is not difficult to remedy—by supplying an appropriate discussion of the relationship between a model of Γ and a model of the class Γ_S obtained from Γ by first making a suitably chosen alphabetic change of functional variables and then reducing every wff to Skolem normal form for satisfiability in such a way that the new functional variables introduced are all distinct from each other and from functional variables previously present. However, it seems to be preferable to avoid use of the Skolem normal form for satisfiability by substituting a proof like that suggested above.

It should be added that Malcev proves only that every infinite model of Γ has an extension (which might be a finite extension). But his methods can be made to yield the stronger result that there is an enumerably infinite extension.

On the other hand, Malcev deals with non-enumerably infinite models as well as enumerably infinite models, a matter into which we do not enter here.

that Γ is simultaneously satisfiable in the domain of positive integers in such a way that the value of **s** is the *successor relation,* i.e., the relation σ such that $\sigma(u, v)$ is truth if and only if $u + 1 = v$. Then Γ is also simultaneously satisfiable in an enumerably infinite domain $\mathfrak{J}$, with Φ as the value of **s**, in such a way that there is no one-to-one transformation of $\mathfrak{J}$ into the positive integers under which the relation Φ is transformed into the relation σ.[452]

48.24. Prove that, if a class of wffs of F^{Ip} is simultaneously satisfiable in some non-empty finite domain of individuals but is not simultaneously satisfiable in an enumerably infinite domain, then there is a greatest finite domain in which it is simultaneously satisfiable.

49. Historical notes. The chief features which distinguish the functional calculi of first order (and of higher orders) from the propositional calculus, namely, the notion of propositional function and the use of quantifiers, originated with Frege in his *Begriffsschrift* of 1879.

Somewhat later, and independently, quantifiers were introduced by C. S. Peirce,[453] who credits the idea to O. H. Mitchell. Still later, quantifiers appear in the work of Schröder, Peano, Russell, and others. The terms "quantifier" and "quantification" are Peirce's. The notation which we have been using for quantifiers is Russell's modification of the Peano notation.

The separation of the functional calculi of first order from those of higher order is implicit in Russell's theory of types,[454] or perhaps even earlier in Frege's hierarchy of "Stufen" or Schröder's hierarchy of "reine Mannigfaltigkeiten." The consideration by Löwenheim,[455] and afterwards by Skolem,[456] of "Zählausdrücke" and "Zählgleichungen" in connection with the Schröder calculus is in effect a treatment of the functional calculus of first order with equality. The singulary functional calculi of first and second order, with and without equality, were also treated by Behmann.[457] But the first explicit formulation of the functional calculus of first order as an in-

[452]From this there follows quickly the result of Skolem according to which no categorical system of postulates for the positive integers (whether the number of postulates is finite of infinite) can be expressed in the notation of a simple applied functional calculus of first order with equality. See exercise 55.18 and footnote 547.

[453]See *American Journal of Mathematics,* vol. 7 (1885), p. 194. Peirce's reference is probably to a paper by Mitchell in *Studies in Logic* (1883); but one essential point, the use of an operator variable in connection with the quantifier, was contributed by Peirce himself as a modification of Mitchell's notation.

[454]Bertrand Russell, "Mathematical Logic as Based on the Theory of Types," published in the *American Journal of Mathematics,* vol. 30 (1908), pp. 222–262.

[455]In the *Mathematische Annalen,* vol. 76 (1915), pp. 447–470.

[456]In papers published in *Skrifter Utgit av Videnskapsselskapet i Kristiania,* I. Matematisk-naturvidenskabelig Klasse, volumes for 1919 and 1920.

[457]In the *Mathematische Annalen,* vol. 86 (1922), pp. 163–229.

dependent logistic system is perhaps in the first edition of Hilbert and Ackermann's *Grundzüge der Theoretischen Logik* (1928).

For the functional calculus of first order and the functional calculus of second order (see Chapter V) Hilbert and Ackermann in their first edition employ the names "engerer Funktionenkalkül" and "erweiterter Funktionenkalkül" respectively. In their second edition (1938), partly following Hilbert and Bernays, they change these names to "engerer Prädikatenkalkül" and "Prädikatenkalkül der zweiten Stufe." This change is based on a usage of the word "Prädikat" (predicate)[458] which appears already in the first edition of Hilbert and Ackermann, but which we wish to avoid. In this book we have taken the term "functional calculus" from Hilbert and Ackermann's first edition, but have borrowed the numbering of orders from their second edition (where they use "Prädikatenkalkül der ersten Stufe" as synonymous with "engerer Prädikatenkalkül").

The axioms and rules of inference for the system F^1 are essentially those of Russell in his paper of 1908,[454] with some modifications, and with Russell's axioms for the propositional calculus replaced by those of Łukasiewicz. Russell, however, does not make it unmistakably clear whether he is stating single axioms or axiom schemata. It is possible to resolve this ambiguity in favor of axiom schemata, as in F^1. Later statements by Russell seem to favor on the whole the interpretation as single axioms, but then his rules of inference must be augmented by adding rules of substitution, as in F^{1p}_2.

Especially difficult is the matter of a correct statement of the rule of substitution for functional variables. An inadequate statement of this rule for the pure functional calculus of first order appears in the first edition of Hilbert and Ackermann (1928). There are better statements of the rule in Carnap's *Logische Syntax der Sprache* and in Quine's *A System of Logistic* (1934), but neither of these is fully correct. In the first volume of Hilbert and Bernays's *Grundlagen der Mathematik* (1934) the error of Hilbert and Ackermann is noted,[459] and a correct statement of a rule of substitution for

[458]By Hilbert and Ackermann, and by Hilbert and Bernays, the name "Prädikat" is applied to the same things which we call "propositional functions" and which Hilbert and Bernays call also "logische Funktionen" (see their *Grundlagen der Mathematik*, vol. 1 (1934), pp. 7, 126, 190). We prefer here the usage of Carnap (*Logische Syntax der Sprache*, 1934), who applies the name "Prädikat" to what is called by Hilbert and Bernays "Prädikatensymbol." Indeed Carnap's usage is nearer to the familiar use of "predicate" as a grammatical term in connection with the natural languages, and therefore seems to run less risk of leading in practice to confusion of use and mention (cf. §08).

[459]A revised statement of the rule is given also in the second edition of Hilbert and Ackermann's book (1938, see pp. 56–57), but this is still open to some objection. In the third edition the rule is correctly stated (1949, see pp. 60–61).

functional variables is given for the first time. However, Hilbert and Bernays's form of the rule could not be used in this book, because its correctness depends on a special feature of their formation rule corresponding to our 30v, according to which $(\forall \mathbf{a})\mathbf{B}$ is not wf if **B** contains **a** as a bound variable.[460] And our form of the rule is to be thought of rather as compiled by combining the versions of Carnap and of Quine.[461]

In §32 the proof of consistency of F^1 which depends on **320 is taken from the first edition of Hilbert and Ackermann (1928). It is given in a form to make its character unmistakable as being purely syntactical (rather than semantical). But it may also be described as depending on the remark, that the axioms are valid in a domain consisting of a single individual and the rules of inference preserve this property. And in this form it becomes obvious how the method may be extended to prove the consistency of the functional calculi of higher order, in particular of the functional calculus of order ω. This is Herbrand's proof[462] of the consistency of the functional cal-

[460]This contravenes the idea, which is implicit in the account given in §02, and which would seem to the writer natural on its own account, that a constant, as distinct from a form, may be used with the same meaning in any context (without regard to variables appearing). And, more serious, it imposes in connection with the use of abbreviative definition the practically intolerable burden of remembering for every definiendum the particular bound variables that occur in the definiens. However, this latter difficulty does not arise for Hilbert and Bernays, because, as already noticed, they do not make use of abbreviative definition.

[461]In the case of logistic systems which involve operators other than quantifiers, such as the abstraction operator λ or the description operator ι (see §06), correct statement of the rule of substitution for functional variables becomes still more troublesome and lengthy. For an example of a statement of the rule in such a case, reference may be made to Gödel's *On Undecidable Propositions of Formal Mathematical Systems* (mimeographed lecture notes of 1934), where the full statement was included at the suggestion of S. C. Kleene; also to a reproduction of this statement, with modifications to adapt it to another system, in the present writer's review of the above-mentioned book of Quine in the *Bulletin of the American Mathematical Society*, vol. 41 (1935), pp. 598–603. (The statement in the Gödel notes is, however, not quite correct, but requires to be amended by adding to 4b on page 10 the additional condition that no bound variable of G(x) is free in A.)

In the case of systems having the abstraction operator λ, it is possible to replace the rule of substitution for functional variables by a number of simpler primitive rules which may be thought of as constituting an analysis of it, as we shall see in Chapter X. Because of the complications which attend the rule of substitution for functional variables, even in the comparatively simple case of the functional calculus of first order, there therefore seems to be some ground for preferring systems (like that of Chapter X) which have the operator λ. However, the functional calculi, not having this operator, have been more extensively studied; and they do have an argument of economy in their favor, in view of Russell's discovery that description and abstraction operators can for many purposes be dispensed with.

[462]In his dissertation, cited in footnote 442—Warsaw 1930, see p. 51 and pp. 57–60. Independently of Herbrand, and of one another, this consistency proof was later found also by Tarski, then by Gentzen, and by E. W. Beth. The remark is added by Beth (*Nieuw Archief voor Wiskunde*, ser. 2 vol. 19 nos 1–2 (1936), pp. 59–62) that the same method can be used to prove consistency of the predicative and ramified functional

culi of first and higher orders; it remains applicable if axioms of choice or multiplicative axioms are added (as Herbrand remarks) and if axioms of extensionality are also added, but not of course upon addition of any sort of axiom of infinity.[463]

The remark is made in the first edition of Hilbert and Ackermann that the functional calculus of first order (in a formulation which is somewhat different from F^1 or F_2^{1p}, but easily seen to be equivalent) is not complete with respect to the transformation of **A** into **~A**, and the question of completeness in the weaker sense of **440 is put as an unsolved problem. The first proof of completeness in the latter sense is that of Gödel,[464] which is reproduced in §44. Another proof of completeness of the functional calculus of first order is due to Leon Henkin[465] and is reproduced in §45 (see further §54).

Independence of axioms for the pure functional calculus of first order was first treated by Gödel,[466] and for a formulation which is nearer to that of Russell[454] than our F^{1p} or F_2^{1p}. Indeed Gödel adds also the axioms $x = x$ and $x = y \supset . F(x) \supset F(y)$, and establishes the independence of the axioms of the resulting formulation of the pure functional calculus of first order with equality. He does not prove the independence of the rules of inference, but makes only the statement that this can easily be done.

For the Hilbert-Ackermann formulation of the pure functional calculus of first order, independence of both the axioms and the rules of inference was treated by McKinsey.[467] However, McKinsey understands the independence of a rule of inference in a weaker sense than that which we have adopted, and his proofs are not in all cases sufficient to show the independence of Hilbert and Ackermann's rules in the strong sense.[468] The second edition of Hilbert and Ackermann (1938) contains a demonstration of the

---

calculi—to which axioms of reducibility may be added, if desired, as well as axioms of choice and of extensionality, but not of course any axiom of infinity.

[463]The terminology will be explained in Chapters V and VI.

[464]*Monatshefte für Mathematik und Physik*, vol. 37 (1930), pp. 349–360. The essential points of a completeness proof by a method similar to that of Gödel are also in Herbrand's dissertation of 1930—compare exercises 46.23, 46.24. The germ of the method used by Herbrand and by Gödel is to be found already in Skolem's paper of 1928 (cited in footnote 430).

[465]In his dissertation (Princeton University, 1947) and in a paper in *The Journal of Symbolic Logic*, vol 14 (1949), pp. 159–166.

[466]In the paper cited in footnote 464. Compare exercise 41.1.

[467]*American Journal of Mathematics*, vol. 58 (1936), pp. 336–344.

[468]As indicated in the discussion in §41 of the rules $*404_n$ $(n > 1)$, the weaker sense of independence is not without its importance. But it seems desirable to prove independence in the strong sense when possible. (The question of the separate independence of the different rules $*404_n$ does not arise for McKinsey or for Hilbert and Ackermann because they take all of these rules together as a single rule, or, in the case of Hilbert and Ackermann, all of them but $*404_0$.)

independence of their axioms and rules, credited to Bernays, in which this defect is overcome.

The results of §34 are found in the first edition of Hilbert and Ackermann (1928) in a form which differs only in detail from ours. And see also the discussion of "truth-functions" and "formal equivalence" in the introduction to the first volume of *Principia Mathematica* (1910).

Use of the prenex normal form was introduced by C. S. Peirce, although in a different terminology and notation.[469] Peirce uses the term "Boolian" for what we here, following *Principia Mathematica*, call the matrix, and speaks of the prefix as "Quantifier" or "quantifiers." The process of reduction to prenex normal form which is explained in §39 is to be found in substance in the first volume of *Principia Mathematica*,[470] though it is there somewhat obscured by the peculiar doctrine that (in effect) only formulas in prenex normal form are to be considered wf, other formulas which we would treat as wf being construed, by abbreviative definition, simply as standing for their prenex normal forms. And from this source the reduction process appears in Behmann's paper of 1922, already referred to, and again in a paper of C. H. Langford.[471]

Origin of the functional calculus of first order with equality is difficult to fix. In a sense, it is implicit already in the work of Peirce and Schröder. Especially good is the treatment of this subject in the first volume of Hilbert and Bernays, which contains much that we have not here touched upon. From this source we have taken, in particular, the results which are indicated in exercises 48.0, 48.4. The idea of the reduction indicated in exercise 48.5 is due to Kalmár[472] and Gödel,[473] and the result of the exercise is due to Gödel.[473] The simple calculus of equality has been treated in detail by Heinrich Scholz.[474]

Developments of the last three or four decades in regard to questions of validity and satisfiability, the decision problem, and related matters may

---

[469]See a paper in *The Monist*, vol. 7 (1897), pp. 161–217; also his paper, already referred to, in the *American Journal of Mathematics*, vol. 7 (1885), pp. 180–202; and an otherwise unpublished addendum to the latter which appears in his *Collected Papers*, vol. 3 (1933), pp. 239–249.

[470]Of course it is not important in this connection that in *Principia* disjunction and negation are used as primitive connectives rather than our implication and negation, as the process of reduction to prenex normal form appropriate to one system of primitive sentence connectives is very easily modified to fit another. And in fact the Introduction to the second edition of *Principia* indicates the modification to be made for the case of Sheffer's stroke as sole primitive sentence connective.

[471]In the *Bulletin of the American Mathematical Society*, vol 32 (1926), see p. 701.

[472]*Acta Scientarum Mathematicarum*, vol. 4 no. 4 (1929), pp. 248–252.

[473]In the paper cited in footnote 464.

[474]In his *Metaphysik als Strenge Wissenschaft*, 1941.

perhaps be dated from Löwenheim's paper of 1915.[475] This contains the following results regarding the functional calculus of first order with equality: a solution of the decision problem for validity in the case that only singulary functional variables appear; a reduction of the general case of the decision problem for validity to that in which only binary functional variables appear; recognition of the existence of wffs that are valid in every finite domain but not valid in an infinite domain, and a demonstration that no wff containing only singulary functional variables can have this property; finally, a proof of the metatheorem now known as Löwenheim's theorem, i.e., **450 and the extension of **450 which is stated in exercise 48.6.

After the pioneering work of Löwenheim there followed the contributions of Skolem in his papers of 1919 and 1920.[476] The first paper contains, in effect, a solution of the decision problem for validity for the singulary functional calculus of second order, including at the same time an improved form of the solution for the singulary functional calculus of first order with equality. In the paper of 1920 the Skolem normal form for satisfiability is introduced and is used to obtain a simpler proof of Löwenheim's theorem. The point of view of satisfiability is adopted in this paper rather than that of validity (as by Löwenheim), and Löwenheim's theorem is therefore restated in the form of **451 and the extension of **451 to F^I. Also Skolem's generalization of Löwenheim's theorem, **455, is here proved for the first time.

Behmann's paper[477] of 1922 contains the result of exercise 39.6, and solutions of the decision problem for validity for the singulary functional calculus of first order, the singulary functional calculus of first order with equality, and the singulary functional calculus of second order. For the singulary functional calculus of first order Behmann's method, with some modifications due to Quine,[478] is reproduced in §46 above. And for the singulary functional calculus of first order with equality Behmann's method is sketched in exercise 48.14. The latter method is similar to that of Skolem in some important respects, but seems to have been found independently by Behmann.

The reduction of a wff **A** of the singulary functional calculus of first order to the form **B** which is described in exercise 46.1(1) is due in substance to

---

[475]Cited in footnote 455.
[476]Cited in footnote 456.
[477]Cited in footnote 457.
[478]See Quine's paper in *The Journal of Symbolic Logic*, vol. 10 (1945), pp. 1–12. Compare also the modified form of Behmann's method which is given by Hilbert and Bernays, *Grundlagen der Mathematik*, vol. 1 (1934), pp. 193–195.

Herbrand,[479] and the resulting form of the solution of the decision problem of the singulary functional calculus of first order which is given in 46.1(2) is due to Quine.[480]

The first treatment of cases of the decision problem in which functional variables other than singulary may appear is in a paper by Paul Bernays and Moses Schönfinkel in 1928.[481] This paper contains a solution of case I of the decision problem of the functional calculus of first order which (except that only the decision problem for validity is treated) is substantially the same as that given in §46 above. Also a solution of case VI_0^1 of the decision problem for validity, and the solution of case III (singulary functional calculus of first order) which is reproduced above in **466 and its proof.

The subsequent history of work on the decision problem has already been given in some detail in §§46 and 47, including exercises and footnotes to these sections. It remains only to mention the paper of F. P. Ramsey[482] dealing with the special case of the decision problem of the pure functional calculus of first order with equality for which a solution is indicated in exercises 48.12, 48.13. (The method of these two exercises is, however, much simpler than that of Ramsey.)

In a paper of 1929,[483] Skolem gives a new proof of his generalization of Löwenheim's theorem in which the result is freed of dependence on the axiom of choice,[484] and at the same time use of the Skolem normal form is avoided.

The metatheorem **453 is due to Gödel,[485] as well as the extension of **453 to the pure functional calculus of first order with equality (48.7). The proof of **453 which is given in §45 is due to Henkin, as well as the proof of Skolem's generalization of Löwenheim's theorem (**455) which is based on this,[486] and the remark of exercise 45.4.

---

[479]In his dissertation, cited in footnote 442, Chapter 2, §9.2.

[480]In his *O Sentido da Nova Lógica*, São Paulo, Brazil, 1944.

[481]*Mathematische Annalen*, vol. 99 (1928), pp. 342–372.

[482]*Proceedings of the London Mathematical Society*, ser. 2 vol. 30 (1930), pp. 264–286. Reprinted in Ramsey's *The Foundations of Mathematics and Other Logical Essays*, pp. 82–111.

[483]*Skrifter utgitt av det Norske Videnskaps-Akademi i Oslo*, I. Matematisk-naturvidenskapelig Klasse, volume for 1929.

[484]I.e., the axiom of choice is not used in the syntax language. (See the discussion of the axiom of choice in Chapter VI.)

[485]In the paper cited in footnote 464.

[486]In his dissertation, and in the paper cited in footnote 465.

V. Functional Calculi of Second Order

The *functional calculus of second order* or, as we shall also say (in order to distinguish from the ramified functional calculi of second order which are described in § 58 below), the *simple functional calculus of second order* has, in addition to notations of the functional calculus of first order, quantifiers with propositional or functional variables as operator variables. As in the case of the functional calculus of first order, there are various different systems, (simple) functional calculi of second order, which we shall treat simultaneously. The particular formulation selected for treatment in this chapter we call F^2_2 (the subscript referring to the particular formulation of the propositional calculus which is contained). Or, where necessary to distinguish the different functional calculi of second order, F^{2p}_2 is the formulation of the pure functional calculus of second order treated in this chapter, $F^{2,1}_2$ the singulary functional calculus of second order, $F^{2,2}_2$ the binary functional calculus of second order, and so on.

50. The primitive basis of F^2_2. The primitive symbols of F^2_2 are identical with those of F^1 or F^1_2 (see §30). The *pure functional calculus of second order* F^{2p}_2 includes among its primitive symbols all the individual, propositional, and functional variables, but no (individual or functional) constants. The *n-ary functional calculus of second order* $F^{2,n}_2$ includes all the individual and propositional variables, all the functional variables which are no more than n-ary, and no constants. An *applied functional calculus of second order* includes at least one constant, as well as all the individual variables and at least one kind of functional variables.

In order that the system be considered a functional calculus of second order at all, of course functional variables of one kind at least should be included among the primitive symbols. We shall confine our treatment to the case that both propositional variables and singulary functional variables (at least) are present, and in particular we use variables of both these kinds in the axioms. This is, however, not an essential point, and modification of the treatment to fit other cases may be left to the reader.

The formation rules of F^2_2 are the same as those of F^1, with removal of the restriction to individual variables in the fifth rule:

50i. A propositional variable standing alone is a wff.

50ii. If **f** is an n-ary functional variable or an n-ary functional constant, and if $\mathbf{a}_1, \mathbf{a}_2, \ldots, \mathbf{a}_n$ are individual variables or individual constants or both (not necessarily all different), then $\mathbf{f}(\mathbf{a}_1, \mathbf{a}_2, \ldots, \mathbf{a}_n)$ is a wff.

50iii. If $\mathbf{\Gamma}$ is wf, then $\sim\mathbf{\Gamma}$ is wf.

50iv. If $\mathbf{\Gamma}$ and $\mathbf{\Delta}$ are wf, then $[\mathbf{\Gamma} \supset \mathbf{\Delta}]$ is wf.

50v. If $\mathbf{\Gamma}$ is wf and **a** is any variable, then $(\forall\mathbf{a})\mathbf{\Gamma}$ is wf.

As in the case of F^1, an effective test of well-formedness follows, as well as uniqueness of the analysis of a wff into one of the forms $\sim\mathbf{A}$, $[\mathbf{A} \supset \mathbf{B}]$, $(\forall\mathbf{a})\mathbf{A}$, and analogues of the metatheorems **313—**316. The terms *antecedent, consequent, principal implication sign, converse, elementary part* are introduced with the same meaning as for F^1.

The distinction between *bound variables* and *free variables* is made in the same way as in §30. But in the functional calculus of second order, not only individual variables but also propositional and functional variables may have bound occurrences.

A wff will be called an *n-ary form* if it has exactly n different free variables, and it will be called a *constant*, or a *closed* wff, if it has no free variables. As in F^1, all forms are *propositional forms*, and all closed wffs are *sentences*.

The same methods of abbreviating wffs are used as for F^1, including the same conventions for omission of brackets, and the definition schemata D3–17. In D13–17 it is to be understood that the variables $\mathbf{a}, \mathbf{a}_1, \mathbf{a}_2, \ldots, \mathbf{a}_n$ may be of any kinds, propositional or functional as well as individual. Additional definitions and definition schemata may be introduced from time to time as required. And in particular we introduce at once the two following definitions:

D20. $f \rightarrow (s)s$

D21. $t \rightarrow (\exists s)s$

The rules of inference, axiom schemata, and axioms of F^2_2 are the following:

*500. From $\mathbf{A} \supset \mathbf{B}$ and **A** to infer **B**. (*Rule of modus ponens.*)

*501. From **A**, if **a** is any variable, to infer $(\mathbf{a})\mathbf{A}$. (*Rule of generalization.*)

*502. From **A**, if **a** is an individual variable which is not free in **N** and **b** is an individual variable which does not occur in **N**, if **B** results from **A** by substituting $\mathrm{S}^{\mathbf{a}}_{\mathbf{b}}\mathbf{N}|$ for a particular occurrence of **N** in **A**, to infer **B**. (*Rule of alphabetic change of bound individual variable.*)

*503. From **A**, if **a** is an individual variable, if **b** is an individual variable or an individual constant, if no free occurrence of **a** in **A** is in a wf part of **A** of the form (**b**)**C**, to infer $\text{Ş}^{\mathbf{a}}_{\mathbf{b}}\mathbf{A}|$.[500]

(Rule of substitution for individual variables.)

†505. $p \supset . q \supset p$

†506. $s \supset [p \supset q] \supset . s \supset p \supset . s \supset q$

†507. $\sim p \supset \sim q \supset . q \supset p$

†508. $p \supset_x F(x) \supset . p \supset (x)F(x)$

*508_0. $\mathbf{A} \supset_{\mathbf{p}} \mathbf{B} \supset . \mathbf{A} \supset (\mathbf{p})\mathbf{B}$, where **p** is any propositional variable which is not a free variable of **A**.

*508_n. $\mathbf{A} \supset_{\mathbf{f}} \mathbf{B} \supset . \mathbf{A} \supset (\mathbf{f})\mathbf{B}$, where **f** is an n-ary functional variable which is not a free variable of **A**.

†509. $(x)F(x) \supset F(y)$

*509_0. $(\mathbf{p})\mathbf{A} \supset \check{\text{S}}^{\mathbf{p}}_{\mathbf{B}}\mathbf{A}|$, where **p** is any propositional variable.[500]

*509_n. $(\mathbf{f})\mathbf{A} \supset \check{\text{S}}^{\mathbf{f}(\mathbf{x}_1, \mathbf{x}_2, \ldots, \mathbf{x}_n)}_{\mathbf{B}}\mathbf{A}|$, where **f** is an n-ary functional variable and $\mathbf{x}_1, \mathbf{x}_2, \ldots, \mathbf{x}_n$ are distinct individual variables.[500]

As in the case of F^{1p} (or F^{1p}_2), the principal interpretation of F^{2p}_2 depends on a domain of individuals, which must be non-empty. Once the domain of individuals is chosen, the principal interpretation is given by the same semantical rules a–f as in §30, with the single change that in rule f the restriction is removed that the variable **a** must be an individual variable. I.e., rule f is replaced by the following:

f^2. Let **a** be any variable and let **A** be any wff. For a given system of values of the free variables of $(\forall\mathbf{a})\mathbf{A}$, the value of $(\forall\mathbf{a})\mathbf{A}$ is t if the value of **A** is t for every value of **a**; and the value of $(\forall\mathbf{a})\mathbf{A}$ is f if the value of **A** is f for at least one value of **a**.

51. Propositional calculus and laws of quantifiers. Deduction theorem.

By *509 (with *501 and *500) the rule of substitution for propositional and functional variables follows as a derived rule of F^2_2:

*510_0. If **p** is a propositional variable, if $\vdash \mathbf{A}$, then

$$\vdash \check{\text{S}}^{\mathbf{p}}_{\mathbf{B}}\mathbf{A}|.$$

*510_n. If **f** is an n-ary functional variable and $\mathbf{x}_1, \mathbf{x}_2, \ldots, \mathbf{x}_n$ are distinct individual variables, if $\vdash \mathbf{A}$, then

$$\check{\text{S}}^{\mathbf{f}(\mathbf{x}_1, \mathbf{x}_2, \ldots, \mathbf{x}_n)}_{\mathbf{B}}\mathbf{A}|.$$

Now by *502, *503, and *510 we may obtain from †505–†508, †509 all of the axiom schemata (*302–*306) of F^1 as theorem schemata of F^2_2. Since

[500] The syntactical notations "Ş" and "Š" have the meanings which are explained in §§30, 35.

the two rules of inference (i.e., *300 and *301) of F^1 are included in *500 and *501, it follows that every theorem of F^1 is a theorem of F^2_2, with the understanding that we take calculi F^1 and F^2_2 that have the same list of primitive symbols. Analogously to the use of the term in §31 let us understand by a *substitution instance* of a wff **A** of F^1 any wff **B** of the logistic system under consideration (in this chapter, the system F^2_2) such that **B** is obtained from **A** by a finite succession of the substitution steps of *503, *510_0, and *510_n. Then follows:

*511. Every substitution instance of a theorem of F^1 is a theorem of F^2_2.

The role of *511 as a derived rule of F^2_2 is similar to that of *311 as a derived rule of F^1. In using *511 in this way, we may refer to it by the phrase "by F^1," or "by P" (in case propositional calculus only is involved); or we may simply refer to one of the theorem schemata of F^1 by number, treating it as a theorem schema of F^2_2.

It is also possible to establish, as theorem schemata of F^2_2, analogues of the theorem schemata of §33 in which **a** and **b** are allowed to be variables of arbitrary kind, instead of merely individual variables. (In the analogues of *330 and *339, but not in that of *336, **a** and **b** must be variables which are of the same kind.) The proofs follow closely those given in §33 and are left to the reader.

The following analogue of *340 may also be established as a theorem schema of F^2_2 (the proof follows closely that in §34, using in case 3 the analogue of *334 in place of *334 itself):

*512. If **B** results from **A** by substitution of **N** for **M** at zero or more places (not necessarily at all occurrences of **M** in **A**), and if the variables $\mathbf{a}_1, \mathbf{a}_2, \ldots, \mathbf{a}_n$ include at least those free variables of **M** and **N** which occur also as bound variables of **A**, then $\vdash \mathbf{M} \equiv_{\mathbf{a}_1\mathbf{a}_2\ldots\mathbf{a}_n} \mathbf{N} \supset . \mathbf{A} \equiv \mathbf{B}$.

Hence, as in §34, we obtain the rule of substitutivity of equivalence as a derived rule:

*513. If **B** results from **A** by substitution of **N** for **M** at zero or more places (not necessarily at all occurrences of **M** in **A**), if $\vdash \mathbf{M} \equiv \mathbf{N}$ and $\vdash \mathbf{A}$, then $\vdash \mathbf{B}$.

The following theorem schema is also a consequence of *512:

*514. $\vdash S^{\mathbf{p}}_{t}\mathbf{A}| \supset . S^{\mathbf{p}}_{f}\mathbf{A}| \supset (\mathbf{p})\mathbf{A}$, where **p** is a propositional variable which is not a bound variable of **A**.

Proof. By $*509_0$, $\vdash (s)\sim s \supset \sim . q \supset q$.
Hence by P, $\vdash \sim(s)\sim s$.
I.e. (cf. D21), $\vdash t$.
Also, by $*509_0$, $\vdash f \supset \mathbf{p}$.
By $*512$, $\vdash \mathbf{p} \equiv t \supset . \mathbf{A} \equiv S^{\mathbf{p}}_{t}\mathbf{A}|$.
And, by $*512$, $\vdash \mathbf{p} \equiv f \supset . \mathbf{A} \equiv S^{\mathbf{p}}_{f}\mathbf{A}|$.
Hence (using the four preceding lines) we have by P that

$$\vdash S^{\mathbf{p}}_{t}\mathbf{A}| \supset . S^{\mathbf{p}}_{f}\mathbf{A}| \supset \mathbf{A}.$$

Hence by $*501$, $\vdash (\mathbf{p}) . S^{\mathbf{p}}_{t}\mathbf{A}| \supset . S^{\mathbf{p}}_{f}\mathbf{A}| \supset \mathbf{A}$.
Then use $*508_0$ and P.

The rule of alphabetic change of bound propositional and functional variables may now be proved in exact analogy to the proof of *350 in §35, by using the analogue of *339 and using *513 in place of *342:

*515. If **a** is a propositional or functional variable which is not free in **N**, and **b** is a variable of the same kind as **a** not occurring in **N**, if **B** results from **A** by substituting $S^{\mathbf{a}}_{\mathbf{b}}\mathbf{N}|$ for a particular occurrence of **N** in **A**, and if $\vdash \mathbf{A}$, then $\vdash \mathbf{B}$.

The definition of *proof from hypotheses* for F^2_2 is closely analogous to that given in §36 for F^1. The changes are that the axioms of F^1 are replaced by those of F^2_2, and *300 is replaced by *500, *301 by *501, *350 by *502 and *515, *351 by *503, and *352 by *510. Then the deduction theorem may be proved in the same way as in §36:

*516. If $\mathbf{A}_1, \mathbf{A}_2, \ldots, \mathbf{A}_n \vdash \mathbf{B}$, then $\mathbf{A}_1, \mathbf{A}_2, \ldots, \mathbf{A}_{n-1} \vdash \mathbf{A}_n \supset \mathbf{B}$.

Also we may prove an analogue of *362:

*517. If every wff which occurs at least once in the list $\mathbf{A}_1, \mathbf{A}_2, \ldots, \mathbf{A}_n$ also occurs at least once in the list $\mathbf{C}_1, \mathbf{C}_2, \ldots, \mathbf{C}_r$ and if $\mathbf{A}_1, \mathbf{A}_2, \ldots, \mathbf{A}_n \vdash \mathbf{B}$, then $\mathbf{C}_1, \mathbf{C}_2, \ldots, \mathbf{C}_r \vdash \mathbf{B}$.

By first proving analogues of the theorem schemata *364 and *365, we may establish the following derived rules facilitating the use of the existential quantifier in connection with the deduction theorem:

*518. If $\mathbf{A}_1, \mathbf{A}_2, \ldots, \mathbf{A}_n \vdash \mathbf{B}$, and **a** is any variable which does not occur as a free variable in $\mathbf{A}_1, \mathbf{A}_2, \ldots, \mathbf{A}_{n-r}$, **B**, then $\mathbf{A}_1, \mathbf{A}_2, \ldots, \mathbf{A}_{n-r}, (\exists \mathbf{a}) . \mathbf{A}_{n-r+1}\mathbf{A}_{n-r+2} \ldots \mathbf{A}_n \vdash \mathbf{B}$. $(r = 1, 2, \ldots, n.)$

*519. If $\mathbf{A}_1, \mathbf{A}_2, \ldots, \mathbf{A}_n \vdash \mathbf{B}$, and **a** is any variable which does not occur as a free variable in $\mathbf{A}_1, \mathbf{A}_2, \ldots, \mathbf{A}_{n-r}$, then $\mathbf{A}_1, \mathbf{A}_2, \ldots, \mathbf{A}_{n-r}, (\exists \mathbf{a}) . \mathbf{A}_{n-r+1}\mathbf{A}_{n-r+2} \ldots \mathbf{A}_n \vdash (\exists \mathbf{a})\mathbf{B}$. $(r = 1, 2, \ldots, n.)$

The discussion of duality for F^2_2 follows closely that in §37 and may be left to the reader. The definition of the dual is word for word the same,[501] as well as the statement of the three principles of duality which correspond to *372–*374 and which may be shown to hold also for F^2_2.

Finally, analogues of the theorem schemata of §§37–38 may be proved in which **a** and **b** are allowed to be variables of arbitrary kind; and hence the reduction to prenex normal form (§39) may also be extended to F^2_2.

It should be noticed that all the derived rules of this section including *510 and *515, in contrast with the remark of footnote 340 about *352, have been established in such a way as to show that they will continue to hold for a system obtained from F^2_2 by the addition of any further axioms.

52. Equality.

In §48 we saw how the functional calculus of first order can be augmented by adding a functional constant I to denote the relation of equality, or identity, between individuals, together with appropriate axioms containing I. We could of course do the same thing in connection with the functional calculus of second order, so obtaining a functional calculus of second order with equality. But this is unnecesssary because it is in fact possible to introduce the relation of equality by definition in F^2_2. I.e., it is possible to find a wff of F^2_2 which has the individual variables **a** and **b** as its only free variables and which has the value t or f (in any principal interpretation of F^2_2) according as **a** and **b** do or do not have the same value; one such wff of F^2_2 is the definiens in D22 below, and the notation $=$ may thus be introduced by the abbreviative definition D22.[502]

In the functional calculi of fourth and higher orders we are able, by an exactly

---

[501]In addition to D3–11 and D14, abbreviation by D20–23 may be allowed in an expression to be dualized. In this case, *f* and *t* are to be interchanged in dualizing, and also $=$ and $\neq$.

[502]This definition is due to Leibniz in the form, "Eadem sunt quorum unum potest substitui alteri salva veritate." See Erdmann's *God. Guil. Leibnitii Opera Philosophica*, vol. 1 (1840), p. 94, and Gerhardt's *Die Philosophischen Schriften von Gottfried Wilhelm Leibniz*, vol. 7 (1890), pp. 228, 236 (also in English translation in the appendix of Lewis's *A Survey of Symbolic Logic*). In this form there is a certain confusion of use and mention: *things* are identical if the *name* of one can be substituted for that of the other without loss of truth. Nevertheless the important idea of the definition is to be credited to Leibniz.

Frege adopts Leibniz's definition unchanged in *Die Grundlagen der Arithmetik* (1884). In Frege's *Grundgesetze der Arithmetik*, vol. 1 (1893), the confusion of use and mention is corrected, but the principle appears in the form of an axiom rather than a definition: $\varphi(x = y) \supset \varphi((F)[F(x) \supset F(y)])$. The first statement of the principle in the form of a definition of identity and without the confusion of use and mention seems to have been by C. S. Peirce in 1885 (*American Journal of Mathematics*, vol. 7, see page 199). In Russell's *The Principles of Mathematics* (1903) the definition appears in the form: "x is identical with y if y belongs to every class to which x belongs, in other words, if 'x is a u' implies 'y is a u' for all values of u." In *Principia Mathematica*, vol. 1 (1910), we find the notation $=$ introduced by an abbreviative definition which is substantially the same as our D22.

analogous definition, to introduce the relation of equality also between things other than individuals, allowing **a** and **b** to be, e.g., propositional or functional variables (of the same type).

We add now the two following definition schemata, in which **a** is an individual variable or individual constant and **b** is an individual variable or individual constant, and then we go on to a number of theorems and derived rules in the statement of which we make use of the definitions:

D22. $[\mathbf{a} = \mathbf{b}] \to F(\mathbf{a}) \supset_F F(\mathbf{b})$

D23. $[\mathbf{a} \neq \mathbf{b}] \to (\exists F) \mathbin{.} F(\mathbf{a}) \not\subset F(\mathbf{b})$

†520. $x = x$ (*Reflexive law of equality.*)

Proof. By P, $\vdash F(x) \supset F(x)$.
Generalize upon F (*501).

†521. $x = y \supset \mathbin{.} y = x$ (*Commutative law of equality.*)

Proof. By $*509_1$, $\vdash x = y \supset \check{S}^{F(y)}_{y=x} F(x) \supset F(y)|$.
Hence by *modus ponens* (*500), $x = y \vdash x = x \supset \mathbin{.} y = x$.
Hence by †520 and *modus ponens*, $x = y \vdash y = x$.
Then use the deduction theorem.

†522. $x = y \supset \mathbin{.} y = z \supset \mathbin{.} x = z$ (*Transitive law of equality.*)

Proof. By †521, $x = y \vdash y = x$.
Hence by $*509_1$, $x = y \vdash \check{S}^{F(y)}_{y=z} F(y) \supset F(x)|$.
Then use the deduction theorem.

†523. $x = y \equiv \mathbin{.} y = x$ (*Complete commutative law of equality.*)

Proof. By †521, *503, and P.

†524. $x = y \supset \mathbin{.} F(x) \equiv F(y)$

Proof. By $*509_1$, $x = y \vdash F(x) \supset F(y)$.
By †521 and $*509_1$, $x = y \vdash F(y) \supset F(x)$.
Then use P and the deduction theorem.

†525. $x \neq y \equiv \sim \mathbin{.} x = y$

Proof. By †523 and P, $\vdash \sim(F)[F(y) \supset F(x)] \equiv \sim \mathbin{.} x = y$.
By P, $\vdash F(y) \supset F(x) \equiv \sim \mathbin{.} F(x) \not\subset F(y)$.
Then use *513.

†526. $F(x) \supset \mathbin{.} \sim F(y) \supset x \neq y$

Proof. By $*509_1$, $\vdash x = y \supset \mathbin{.} F(x) \supset F(y)$.
Hence by P, $\vdash F(x) \supset \mathbin{.} \sim F(y) \supset \sim \mathbin{.} x = y$.
Then use †525 and P.

*527. If **a** is an individual variable or an individual constant, if **b** is an individual variable or an individual constant, if **B** results from **A** by substitution of **b** for zero or more free occurrences of **a**, no one of which is within a wf part of **A** of the form (**b**)**C**—and not necessarily at all free occurrences of **a** in **A**—then $\vdash \mathbf{a} = \mathbf{b} \supset . \mathbf{A} \supset \mathbf{B}$

Proof. Let **x** be an individual variable which does not occur in **A**. Take all the occurrences of **a** in **A** at which **b** is substituted in obtaining **B**, and for every such occurrence of **a** substitute **x**. And let **X** be the wff which results from **A** by this substitution.

By $*509_1$:

$$\vdash \mathbf{a} = \mathbf{b} \supset . \check{S}_{\mathbf{X}}^{F(\mathbf{x})} F(\mathbf{a}) \supset F(\mathbf{b})|.$$

I.e., $\vdash \mathbf{a} = \mathbf{b} \supset . \mathbf{A} \supset \mathbf{B}$.

From this of course we have as corollaries, by *modus ponens*:

*528. If **a** is an individual variable or an individual constant, if **b** is an individual variable or an individual constant, if **B** results from **A** by substitution of **b** for zero or more free occurrences of **a**, no one of which is within a wf part of **A** of the form (**b**)**C**—and not necessarily at all free occurrences of **a** in **A**—then $\mathbf{a} = \mathbf{b} \vdash \mathbf{A} \supset \mathbf{B}$.

*529. If **a** is an individual variable or an individual constant, if **b** is an individual variable or an individual constant, if **B** results from **A** by substitution of **b** for zero or more free occurrences of **a**, no one of which is within a wf part of **A** of the form (**b**)**C**—and not necessarily at all free occurrences of **a** in **A**—then $\mathbf{a} = \mathbf{b}, \mathbf{A} \vdash \mathbf{B}$.

(*Rule of substitutivity of equality.*)

EXERCISES 52

52.0. Prove the following theorems of F^2_2:

(1) $$x = z \equiv_z y = z \equiv . x = y$$

(2) $$x = y \equiv . F(x) \equiv_F F(y)$$

(3) $$F(x) \equiv (\exists y) . F(y) . x = y$$

(4) $$F(x) \supset_x x \neq y \equiv \sim F(y)$$

(5) $$F(x_1, x_2) \supset_F [F(y_1, y_2) \supset F(z_1, z_2)] \equiv [x_1 = z_1][x_2 = z_2] \vee [y_1 = z_1][y_2 = z_2]$$

52.1. A formulation of the functional calculus of second order is to have the same primitive symbols as F_2^2 with the omission of $\sim$, the notation $\sim\mathbf{A}$ being introduced by definition in the way suggested in §28. Show how the rules and axioms of F_2^2 are to be modified, and in an appropriate sense establish the equivalence of the resulting system to F_2^2.

52.2. A formulation of the functional calculus of second order is to have the same primitive symbols as F_2^2 with omission of $\sim$ and of all propositional variables. Show how notation $\sim\mathbf{A}$ is to be defined and how the rules and axioms of F_2^2 are to be modified, and in an appropriate sense establish the equivalence of the resulting system to F_2^2. Then generalize this to the case that not only $\sim$ and the propositional variables are omitted but also all functional variables that are less than n-ary, the n-ary functional variables being retained.

52.3. Extend **434 to $\mathrm{F}_2^{2\mathrm{p}}$. Hence using the result of 48.5, show that every wff of $\mathrm{F}_2^{I\mathrm{p}}$ which is a theorem of $\mathrm{F}_2^{2\mathrm{p}}$ is also a theorem of $\mathrm{F}_2^{I\mathrm{p}}$.

52.4. Solve the decision problem for the singulary functional calculus of second order $\mathrm{F}^{2,1}$, by adding to the reduction steps (a)–(g) of exercise 39.6 and the reduction steps (α)–(η) of exercise 48.14, in the first place analogues of (a)–(g) in which $\mathbf{a}$ is a singulary functional variable instead of an individual variable, and secondly the following reduction steps (in which $\mathbf{p}$ is any propositional variable, and $\mathbf{f}$ is any singulary functional variable, and $\mathbf{a}, \mathbf{a}_1, \mathbf{a}_2, \ldots, \mathbf{a}_m, \mathbf{b}_1, \mathbf{b}_2, \ldots, \mathbf{b}_n, \mathbf{c}_1, \mathbf{c}_2, \ldots, \mathbf{c}_i$ are distinct individual variables): (A) to replace a wf part $(\mathbf{p})\mathbf{E}$ by the conjunction $\mathrm{S}^{\mathbf{p}}_{f}\mathbf{E}|\mathrm{S}^{\mathbf{p}}_{t}\mathbf{E}|$; (B) to replace a wf part $(\mathbf{f})[\mathbf{f}(\mathbf{a}_1) \supset . \mathbf{f}(\mathbf{a}_2) \supset . \ldots \mathbf{f}(\mathbf{a}_m) \supset \sim\mathbf{f}(\mathbf{a})]$ by f; (C) to replace a wf part $(\mathbf{f})[\sim\mathbf{f}(\mathbf{a}_1) \supset . \sim\mathbf{f}(\mathbf{a}_2) \supset . \ldots \sim\mathbf{f}(\mathbf{a}_m) \supset \mathbf{f}(\mathbf{a})]$ by f; (D) to replace a wf part $(\mathbf{f})[\sim\mathbf{f}(\mathbf{a}_1) \supset . \sim\mathbf{f}(\mathbf{a}_2) \supset . \ldots \sim\mathbf{f}(\mathbf{a}_m) \supset . \mathbf{f}(\mathbf{b}_1) \supset . \mathbf{f}(\mathbf{b}_2) \supset . \ldots \mathbf{f}(\mathbf{b}_n) \supset \mathbf{f}(\mathbf{a})]$ by $\mathbf{b}_1 \neq \mathbf{a}_1 \supset . \mathbf{b}_2 \neq \mathbf{a}_1 \supset . \ldots \mathbf{b}_n \neq \mathbf{a}_1 \supset . \mathbf{b}_1 \neq \mathbf{a}_2 \supset . \mathbf{b}_2 \neq \mathbf{a}_2 \supset . \ldots \mathbf{b}_n \neq \mathbf{a}_2 \supset . \ldots \ldots \mathbf{b}_1 \neq \mathbf{a}_m \supset . \mathbf{b}_2 \neq \mathbf{a}_m \supset . \ldots \mathbf{b}_n \neq \mathbf{a}_m \supset . \mathbf{b}_1 \neq \mathbf{a} \supset . \mathbf{b}_2 \neq \mathbf{a} \supset . \ldots \mathbf{b}_n = \mathbf{a}$; (E) to replace a wf part $(\mathbf{f})[\mathbf{E}_1 \supset . \mathbf{E}_2 \supset . \ldots \mathbf{E}_n \supset (\mathbf{a})\mathbf{D}]$ by $(\mathbf{a})(\mathbf{f})[\mathbf{E}_1 \supset . \mathbf{E}_2 \supset . \ldots \mathbf{E}_n \supset \mathbf{D}]$ if $\mathbf{a}$ is not a free variable of $\mathbf{E}_1, \mathbf{E}_2, \ldots, \mathbf{E}_n$; (F) to replace a wf part $(\mathbf{f})[\mathbf{E}_1 \supset . \mathbf{E}_2 \supset . \ldots \mathbf{E}_n \supset \sim(\mathbf{a})\mathbf{D}]$ by $(\mathbf{f})[\mathbf{E}'_1 \supset . \mathbf{E}'_2 \supset . \ldots \mathbf{E}'_n \supset \sim(\mathbf{a})\mathbf{D}']$, where $\mathbf{E}'_j$ is

$$\check{\mathrm{S}}^{\mathbf{f}(\mathbf{a})}_{\mathbf{f}(\mathbf{a})\equiv\mathbf{D}}\,\mathbf{E}_j|$$

($j = 1, 2, \ldots, n$), and $\mathbf{D}'$ is

$$\check{\mathrm{S}}^{\mathbf{f}(\mathbf{a})}_{\mathbf{f}(\mathbf{a})\equiv\mathbf{D}}\,\mathbf{D}|,$$

provided that $\mathbf{E}_1, \mathbf{E}_2, \ldots, \mathbf{E}_n, \mathbf{D}$ contain no bound propositional or functional variables other than the bound functional variable F in wf parts of the form $\mathbf{b} = \mathbf{c}$ or $\mathbf{b} \neq \mathbf{c}$, and $\mathbf{D}$ contains no bound individual variables; (G) to replace a wf part $(\mathbf{f})[\mathbf{E}_1 \supset . \mathbf{E}_2 \supset . \ldots \mathbf{E}_n \supset \sim(\mathbf{a})(\mathbf{c}_1)(\mathbf{c}_2) \ldots (\mathbf{c}_i) . \mathbf{a} \neq \mathbf{c}_1 \supset . \mathbf{a} \neq \mathbf{c}_2 \supset . \ldots \mathbf{a} \neq \mathbf{c}_i \supset . \mathbf{c}_1 \neq \mathbf{c}_2 \supset . \mathbf{c}_1 \neq \mathbf{c}_3 \supset . \ldots \mathbf{c}_{i-1} \neq \mathbf{c}_i \supset . \sim\mathbf{C}_1 \supset . \sim\mathbf{C}_2 \supset . \ldots \sim\mathbf{C}_i \supset \mathbf{A}]$, where $\mathbf{C}_j$ is

$$\mathrm{S}^{\mathbf{a}}_{\mathbf{c}_j} \mathbf{A}|$$

$(j = 1, 2, \ldots, i)$, by $(\mathbf{c}_1)(\mathbf{c}_2) \ldots (\mathbf{c}_i)(\mathbf{f})[\mathbf{E}_1 \supset . \mathbf{E}_2 \supset . \ldots \mathbf{E}_n \supset \sim(\mathbf{a}) . \mathbf{a} \neq \mathbf{c}_1 \supset . \mathbf{a} \neq \mathbf{c}_2 \supset . \ldots \mathbf{a} \neq \mathbf{c}_i \supset \mathbf{A}]$, provided that $\mathbf{E}_1, \mathbf{E}_2, \ldots, \mathbf{E}_n, \mathbf{A}$ do not contain $\mathbf{c}_1, \mathbf{c}_2, \ldots, \mathbf{c}_i$ as free variables, and contain no bound propositional or functional variables other than the bound functional variable F in wf parts of the form $\mathbf{b} = \mathbf{c}$ or $\mathbf{b} \neq \mathbf{c}$, and $\mathbf{A}$ contains no bound individual variables. (In all of these reduction steps of course m or n may as a special case be 0.)

52.5. With the aid of the foregoing results discuss the completeness, in various senses, (1) of the singulary functional calculus of second order $\mathrm{F}^{2,1}_2$, and (2) of the logistic system obtained from $\mathrm{F}^{2,1}_2$ by adding the following infinite list of axioms: $(\exists x_1)(\exists x_2) . x_1 \neq x_2$, $(\exists x_1)(\exists x_2)(\exists x_3) . x_1 \neq x_2 . x_1 \neq x_3 . x_2 \neq x_3$, $(\exists x_1)(\exists x_2)(\exists x_3)(\exists x_4) . x_1 \neq x_2 . x_1 \neq x_3 . x_1 \neq x_4 . x_2 \neq x_3 . x_2 \neq x_4 . x_3 \neq x_4$, ...

52.6. The *elimination problem* of the functional calculus of second order is the problem to find an effective procedure by which from a given wff **A** of the functional calculus of second order there is obtained a wff **B** of the functional calculus of first order with equality, such that $\mathbf{A} \equiv \mathbf{B}$ is a theorem of the functional calculus of second order.[503] We shall here require also an effective procedure by which to find a proof of $\mathbf{A} \equiv \mathbf{B}$. The wff **B** is then called the *resultant* of **A**.

(1) Solve the elimination problem of the singulary functional calculus of second order $\mathrm{F}^{2,1}_2$ by means of the reduction steps (a)–(g), (α)–(η), (A)–(G) of 39.6, 48.14, and 52.4.[504]

(2) Apply the elimination procedure found in (1) to get the resultant of the following wff **A**:[505]

$$(\exists F) . F(x) \supset_x G_1(x) . G_2(x) \supset_x F(x) . (\exists x)[F(x)H(x)] \supset . H(x) \supset_x F(x)$$

Show that the resultant can be simplified to:

$$G_2(x) \supset_x G_1(x) . (\exists x)[G_2(x)H(x)] \supset . H(x) \supset_x G_1(x)$$

52.7. (1) Solve the elimination problem for the special case of wffs of F^{2p}_2 of the form

[503]The elimination problem goes back to Schröder. See an account of the matter by Ackermann in a paper in the *Mathematische Annalen*, vol. 110 (1934), pp. 390–413.

[504]This solution is due to Skolem and Behmann in their papers of 1919 and 1922, referred to in §49. There is also a sketch of a solution in Löwenheim's paper of 1915.

[505]This example is taken from Ackermann's paper, cited in footnote 503.

$$(\exists \mathbf{f}) \centerdot \mathbf{C} \centerdot \mathbf{f}(\mathbf{a}, \mathbf{b}) \supset_{\mathbf{ab}} \mathbf{D}$$

where **f** is a binary functional variable and **a** and **b** are distinct individual variables, **C** and **D** contain no bound propositional or functional variables, **D** does not contain **f**, and the matrix **M** of the prenex normal form of **C** has the property that[506]

$$\vdash \mathbf{M} \supset \check{S}_{t}^{\mathbf{f}(\mathbf{a},\mathbf{b})} \mathbf{M}|.$$

Show that in this case a resultant is

$$\check{S}_{\mathbf{D}'}^{\mathbf{f}(\mathbf{a},\mathbf{b})} \mathbf{C}'|,$$

where **C**′ and **D**′ differ from **C** and **D** by (at most) certain alphabetic changes of bound variables.[507] Hence in particular find resultants of the following wffs of $\mathbf{F}_2^{2p}$:

(2) $(\exists F) \centerdot (x)(y)[F(x, y) \supset G(x, y)](x)(\exists y)[H(x, y) \supset F(x, y)]$

(3) $(\exists F)(x)(y) \centerdot F(x, y) \supset G(x, y) \centerdot H_1(x, y) \supset F(x, x) \centerdot H_2(x, y) \supset F(x, x)F(y, y)$

52.8. Similarly solve the elimination problem for the special case of wffs of F_2^{2p} of the form

$$(\exists \mathbf{f}) \centerdot \mathbf{C} \centerdot \mathbf{D} \supset_{\mathbf{ab}} \mathbf{f}(\mathbf{a}, \mathbf{b})$$

where **f** is a binary functional variable and **a** and **b** are distinct individual variables, **C** and **D** contain no bound propositional or functional variables, **D** does not contain **f**, and the matrix **M** of the prenex normal form of **C** has the property that[508]

$$\vdash \mathbf{M} \supset \check{S}_{f}^{\mathbf{f}(\mathbf{a},\mathbf{b})} \mathbf{M}|.$$

---

[506]In other words, **M** can be reduced to a disjunctive normal form (in the sense of footnote 299) in which **f** nowhere appears with a negation sign before it.

[507]The solution of this special case of the elimination problem, as well as of the special cases of 52.8–52.10 and the particular examples 52.7(2) and 52.7(3), are due to Ackermann in his paper cited in footnote 503.

Ackermann's paper contains also a proof of the unsolvability of the general elimination problem of the functional calculus of second order, in the sense that for some wffs there is no possible resultant; and a generalization of the elimination problem is treated in which the resultant of a wff of the functional calculus of second order is to be a class of wffs of the functional calculus of first order.

A note by Ackermann in the *Mathematische Annalen*, vol. 111 (1935), pp. 61–63, contains solutions of a few further special cases of the elimination problem—not quite for the functional calculus of second order, but for a system obtained from the functional calculus of second order by adding as axioms, summarized in an axiom schema, certain special cases of an axiom of choice which are expressible in the notation of the functional calculus of second order (see §56 and footnote 555).

[508]I.e., **M** has a disjunctive normal form in which **f** nowhere appears without a negation sign before it. Or alternatively we might define the *parity* (oddness or evenness)

52.9. Generalize the results of 52.7(1) and 52.8, (1) by replacing **f** by an n-ary functional variable, and (2) by replacing **f**(**a**, **b**) by **f**($\mathbf{a}_1$, $\mathbf{a}_2$, . . ., $\mathbf{a}_n$) ≡ **E**, where **E** is quantifier-free and does not contain **f**.

52.10. Parallel to the foregoing series of solutions of the elimination problem for wffs beginning with an existential quantifier (∃**f**), find a series of solutions of the elimination problem for wffs beginning with a universal quantifier (**f**) (by considering the negations of the latter wffs).

52.11. Following Behmann, show that the decision problem of the extended propositional calculus may, instead of the method described in §28, be solved by a reduction process like that of exercise 39.6, i.e., by a reversal of the process of reduction to prenex normal form.

52.12. Apply the decision procedure found in the foregoing exercise to the four examples of exercise 28.0.

52.13. How may the decision procedure of exercise 52.4 be modified in the light of exercise 52.11?

53. Consistency of F^2_2. As already suggested in §49, it is possible to prove the consistency of F^2_2 by a very elementary syntactical argument, closely similar to that used in §32 to prove the consistency of F^1.

For this purpose let us take a formulation of the extended propositional calculus in which the primitive connectives and operator are implication, negation, and the universal quantifier, and let us modify as follows the decision procedure described in §28. Take t and f not as abbreviations of wffs of F^2_2 or of the extended propositional calculus but as primitive constants of a formulation of the propositional calculus. Given a wff **A** of the extended propositional calculus, replace a wff part (**b**)**B** in it by the conjunction

$$\dot{S}^{\mathbf{b}}_{t}\mathbf{B}|\dot{S}^{\mathbf{b}}_{f}\mathbf{B}|,$$

and iterate this until all occurrences of the universal quantifier have been removed. If the quantifier-free formula $\mathbf{A}_0$ which is thus obtained is a tautology (of the appropriate formulation of the propositional calculus), we shall say that **A** is *valid*.

Thus we have an effective test for the validity of any wff of the extended propositional calculus.

From any wff of F^2_2 we obtain an *associated formula of the extended prop-*

---

of each occurrence of an elementary part in **M** as follows: when an elementary part stands alone, this is an even occurrence of the elementary part; in ∼**K** the parity of each occurrence of an elementary part is reversed as compared to **K**; in **K** ⊃ **L** the parity is reversed for each occurrence of an elementary part in **K** but remains the same in **L**. Then the requirement here is that **f** shall appear only at odd places in **M**; and in 52.7(1) the requirement is that **f** shall appear only at even places.

ositional calculus (abbreviated "afep") as follows. First we delete all those occurrences of the universal quantifer in which the operator variable is an individual variable. Then, if $\mathbf{f}_1, \mathbf{f}_2, \ldots, \mathbf{f}_m$ are the distinct functional variables and functional constants that appear, we select m distinct propositional variables $\mathbf{p}_1, \mathbf{p}_2, \ldots, \mathbf{p}_m$ not previously occurring, and we replace every wf part $\mathbf{f}_i(\mathbf{a}_1, \mathbf{a}_2, \ldots, \mathbf{a}_{n_i})$ by $\mathbf{p}_i$, and we replace every universal quantifier $(\forall \mathbf{f}_i)$ by $(\forall \mathbf{p}_i)$ $(i = 1, 2, \ldots, m)$.

We need not (for our present purpose) distinguish among the different afeps of a given wff of F^2_2, since they differ among themselves only by alphabetic changes of bound and free variables.

Now every axiom of F^2_2 has a valid afep, and the rules of inference preserve the property of having a valid afep, as we leave it to the reader to verify in detail. Hence follows:

**530. Every theorem of F^2_2 has a valid afep.

Since any afep of $\sim\mathbf{C}$ is the negation of an afep of $\mathbf{C}$, it follows that not both $\sim\mathbf{C}$ and $\mathbf{C}$ can have a valid afep, hence by **530 that not both $\sim\mathbf{C}$ and $\mathbf{C}$ can be theorems of F^2_2. Thus we have:

**531. F^2_2 is consistent with respect to the transformation of $\mathbf{C}$ into $\sim\mathbf{C}$.

**532. F^2_2 is absolutely consistent.

**533. F^2_2 is consistent in the sense of Post.

Similar syntactical consistency proofs are possible also for the functional calculi of higher order, employing, instead of the afep, an associated formula of protothetic or of higher protothetic.[509]

54. Henkin's completeness theorem.[510]

The principal interpretation of F^{2p}_2 for a given non-empty domain $\mathfrak{J}$ of individuals is given by rules a–e, f^2 of §50. We now introduce also what we call *the interpretation of* F^{2p}_2 for a given domain $\mathfrak{J}$ of individuals and given domains (classes) $\mathfrak{F}_1, \mathfrak{F}_2, \mathfrak{F}_3, \ldots$ of propositional functions of individuals, where all members of $\mathfrak{F}_1$ must be singulary propositional functions whose range is the

[509] See §49 and footnote 462. For the singulary functional calculus of order ω the proof is carried out in especial detail by Gentzen in the *Mathematische Zeitschrift*, vol. 41 (1936), pp. 357–366, though in a slightly different form from that indicated here.

[510] The method which is used in this section to prove a weak completeness theorem for the functional calculus of second order is due to Leon Henkin (in his dissertation, Princeton University, 1947). It is essentially the same as the method used in §45 (also due to Henkin, cf. footnote 465) to prove Gödel's completeness theorem for the functional calculus of first order. And it may be extended to functional calculi of higher order, though for systems containing a suitable form of the axiom of choice the modified method which is used by Henkin in a paper in *The Journal of Symbolic Logic*, vol. 15 (1950), pp. 81–91, may be preferable.

individuals, all members of $\mathfrak{F}_2$ must be binary propositional functions whose range is the ordered pairs of individuals, and so on. Namely, these new interpretations of F^{2p}_2 are given by the same rules which give the principal interpretations, except that rules b_1, b_2, b_3, . . . are replaced by the following:

$b\mathfrak{F}_1$. The singulary functional variables are variables having $\mathfrak{F}_1$ as their range.

$b\mathfrak{F}_2$. The binary functional variables are variables having $\mathfrak{F}_2$ as their range.

$b\mathfrak{F}_n$. The n-ary functional variables are variables having $\mathfrak{F}_n$ as their range.

Among these interpretations of F^{2p}_2 it is clear that not all are sound. But those among them which are sound, and are not principal interpretations, we call *secondary interpretations* of F^{2p}_2.

We must leave open temporarily the question of existence of such secondary interpretations. But an affirmative answer to this question will follow from results obtained below. In fact it will follow that there exist secondary interpretations of F^{2p}_2 in which all of the domains $\mathfrak{J}$, $\mathfrak{F}_1$, $\mathfrak{F}_2$, $\mathfrak{F}_3$, . . . are enumerably infinite.

Analogously to what was done in §43, the semantical rules, just described in small type, may be restated and reinterpreted in such a way as to give them a purely syntactical character. Namely, the words "range" and "value" are replaced everywhere by the phrases "range with respect to $\mathfrak{J}$, $\mathfrak{F}_1$, $\mathfrak{F}_2$, $\mathfrak{F}_3$, . . ." and "value with respect to $\mathfrak{J}$, $\mathfrak{F}_1$, $\mathfrak{F}_2$, $\mathfrak{F}_3$, . . ." respectively. And the rules are then regarded as constituting a (syntactical) definition of these latter phrases.

A wff of F^{2p}_2 is said to be *valid with respect to* the system of domains $\mathfrak{J}$, $\mathfrak{F}_1$, $\mathfrak{F}_2$, $\mathfrak{F}_3$, . . . if it has, with respect to $\mathfrak{J}$, $\mathfrak{F}_1$, $\mathfrak{F}_2$, $\mathfrak{F}_3$, . . ., the value t for all possible values of its free variables;[511] and *satisfiable with respect to* the system of domains $\mathfrak{J}$, $\mathfrak{F}_1$, $\mathfrak{F}_2$, $\mathfrak{F}_3$, . . ., if it has, with respect to $\mathfrak{J}$, $\mathfrak{F}_1$, $\mathfrak{F}_2$, $\mathfrak{F}_3$, . . ., the value t for at least one system of possible values of its free variables. (Here, by a "possible" value of a variable is meant a value that belongs to the range of the variable with respect to $\mathfrak{J}$, $\mathfrak{F}_1$, $\mathfrak{F}_2$, $\mathfrak{F}_3$, ...)

A system of domains $\mathfrak{J}$, $\mathfrak{F}_1$, $\mathfrak{F}_2$, $\mathfrak{F}_3$, . . . is said to be *normal* if all the axioms of F^{2p}_2 are valid with respect to it, and every rule of inference of F^{2p}_2 has the property of preserving validity with respect to it (i.e., the property that, whenever the premisses of the rule are valid with respect to $\mathfrak{J}$, $\mathfrak{F}_1$, $\mathfrak{F}_2$, $\mathfrak{F}_3$, . . ., the conclusion is also valid with respect to $\mathfrak{J}$, $\mathfrak{F}_1$, $\mathfrak{F}_2$, $\mathfrak{F}_3$, . . .). Evidently, in a normal system of domains, no domain is empty.

A wff of F^{2p}_2 is said to be *valid in* the non-empty domain $\mathfrak{J}$ of individuals if it is valid with respect to the system of domains $\mathfrak{J}$, $\mathfrak{F}_1$, $\mathfrak{F}_2$, $\mathfrak{F}_3$, . . ., where $\mathfrak{F}_1$ is the class of all propositional functions having the individuals (all members of $\mathfrak{J}$) as their range, $\mathfrak{F}_2$ is the class of all propositional functions

[511] As usual, if there are no free variables, then, by "having the value t for all possible values of its free variables," we understand simply, denoting t. (Compare footnote 312.)

which have all ordered pairs of individuals as their range, and so on.[512] And a wff is said to be *satisfiable in* the non-empty domain $\mathfrak{J}$ of individuals if it is satisfiable with respect to this same system of domains.

A wff is *valid* if it is valid in every non-empty domain $\mathfrak{J}$ of individuals; *satisfiable* if it is satisfiable in some non-empty domain $\mathfrak{J}$ of individuals. A wff is *secondarily valid* if it is valid with respect to every normal system of domains; *secondarily satisfiable* if it is satisfiable with respect to some normal system of domains. It can be shown that every secondarily valid wff is valid, and every satisfiable wff is secondarily satisfiable (compare the proof of **434).

The *universal closure* of a wff **B** in which no free functional variable is more than n-ary is the wff

$$(\mathbf{c}^n_{u_n})(\mathbf{c}^n_{u_n-1})\ldots(\mathbf{c}^n_1)(\mathbf{c}^{n-1}_{u_{n-1}})(\mathbf{c}^{n-1}_{u_{n-1}-1})\ldots(\mathbf{c}^{n-1}_1)\ldots$$
$$(\mathbf{c}^1_{u_1})(\mathbf{c}^1_{u_1-1})\ldots(\mathbf{c}^1_1)(\mathbf{c}^0_{u_0})(\mathbf{c}^0_{u_0-1})\ldots(\mathbf{c}^0_1)(\mathbf{c}_u)(\mathbf{c}_{u-1})\ldots(\mathbf{c}_1)\mathbf{B},$$

where $\mathbf{c}^k_1, \mathbf{c}^k_2, \ldots, \mathbf{c}^k_{u_k}$ are the free k-ary functional variables of **B** in alphabetic order ($k = 1, 2, \ldots, n$), $\mathbf{c}^0_1, \mathbf{c}^0_2, \ldots, \mathbf{c}^0_{u_0}$ are the free propositional variables of **B** in alphabetic order, and $\mathbf{c}_1, \mathbf{c}_2, \ldots, \mathbf{c}_u$ are the free individual variables of **B** in alphabetic order. The *existential closure* of **B** is similarly defined, with existential quantifiers replacing the universal quantifiers.

The following metatheorems about F^{2p}_2 are proved in the same way as their analogues in §43:

540. A wff **A is valid with respect to a given normal system of domains if and only if $\sim$**A** is not satisfiable with respect to that system of domains; valid in a given non-empty domain of individuals if and only if $\sim$**A** is not satisfiable in that domain; valid if and only if $\sim$**A** is not satisfiable; secondarily valid if and only if $\sim$**A** is not secondarily satisfiable.

541. A wff **A is satisfiable with respect to a given normal system of domains if and only if $\sim$**A** is not valid with respect to that system of domains; satisfiable in a given non-empty domain of individuals if and only if $\sim$**A** is not valid in that domain; satisfiable if and only if $\sim$**A** is not valid; secondarily satisfiable if and only if $\sim$**A** is not secondarily valid.

[512] The domain of individuals being fixed as some particular non-empty domain, a closed wff may be said to be *true* if it is valid in that domain. Since the sentences of the (pure) functional calculus of second order are the same as the closed wffs, this may be taken as the syntactical equivalent of the semantical property of being a true sentence, as described in §09.

**542. A wff is valid with respect to a given normal system of domains if and only if its universal closure is valid with respect to that system of domains; valid in a given non-empty domain of individuals if and only if its universal closure is valid in that domain; valid if and only if its universal closure is valid; secondarily valid if and only if its universal closure is secondarily valid.

**543. A wff is satisfiable with respect to a given normal system of domains if and only if its existential closure is satisfiable (or equivalently, valid) with respect to that system of domains; satisfiable in a given non-empty domain of individuals if and only if its existential closure is satisfiable in that domain; secondarily satisfiable if and only if its existential closure is secondarily satisfiable.

*544. Every theorem of F_2^{2p} is secondarily valid, and therefore also valid.

As in §45, if Γ is any class of wffs of any of the functional calculi of second order, we say that $\Gamma \vdash \mathbf{B}$ if there are a finite number of wffs $\mathbf{A}_1, \mathbf{A}_2, \ldots, \mathbf{A}_m$ of Γ such that $\mathbf{A}_1, \mathbf{A}_2, \ldots, \mathbf{A}_m \vdash \mathbf{B}$. Γ is *inconsistent* if $\Gamma \vdash f$, and in the contrary case Γ is *consistent*. $\mathbf{C}$ is *inconsistent with* Γ or *consistent with* Γ according as the class whose members are $\mathbf{C}$ and the members of Γ is inconsistent or consistent. Γ is a *maximal consistent class of closed well-formed formulas* if Γ is consistent and no closed wff $\mathbf{C}$ is consistent with Γ which is not a member of Γ.

A class Γ of wffs of F_2^{2p} is said to be *simultaneously satisfiable with respect to* a system of domains if, with respect to that system of domains, all the wffs of Γ have the value t simultaneously for at least one system of possible values of all their free variables taken together. And Γ is said to be *simultaneously satisfiable in* the non-empty domain $\mathfrak{J}$ of individuals if it is simultaneously satisfiable with respect to the system of domains $\mathfrak{J}, \mathfrak{F}_1, \mathfrak{F}_2, \mathfrak{F}_3, \ldots$, where $\mathfrak{F}_1$ is the class of all propositional functions having the class $\mathfrak{J}$ of individuals as their range, $\mathfrak{F}_2$ is the class of all propositional functions having the class of ordered pairs of individuals as their range, and so on. And Γ is said to be *simultaneously satisfiable* if it is simultaneously satisfiable in some non-empty domain $\mathfrak{J}$.

We consider an applied functional calculus of second order, S, having as primitive symbols all the primitive symbols of F_2^{2p}, together with the individual constants $w_0, w_1, w_2, \ldots$ and, for every positive integer k, the k-ary functional constants $w_0^k, w_1^k, w_2^k, \ldots$. By an adaptation of the method of footnote 416, we fix a particular enumeration of the closed wffs of S, and referring to this enumeration, we speak of "the first closed wff of S," "the second closed wff of S," and so on. Moreover, using this enumeration, we can extend an arbitrary consistent class Γ of closed wffs of S to a maximal consistent class $\bar{\Gamma}$ of closed wffs of S—by the same method which was used in §45 (compare **452 and its proof).

Now let **H** be a closed wff of F_2^{2p} which is not a theorem.

The class Γ_0 whose single member is $\sim$**H** is then a consistent class of wffs—of F_2^{2p}, and therefore also of S. We define the classes Γ_n by the following recursion rule:[513] If the $(n + 1)$th closed wff of S has the form $(\mathbf{a})\mathbf{A}$, where **a** is an individual variable, and if in the list $w_0, w_1, w_2, \ldots$ the first constant that does not occur either in **A** or in any member of Γ_n is w_m, then Γ_{n+1} is the class whose members are

$$\overset{\mathbf{a}}{\underset{w_m}{\text{Ṣ}}}\mathbf{A}| \supset (\mathbf{a})\mathbf{A}$$

and the members of Γ_n; if the $(n + 1)$th closed wff of S has the form $(\mathbf{a})\mathbf{A}$, where **a** is a k-ary functional variable ($k = 1, 2, 3, \ldots$), and if in the list $w_0^k, w_1^k, w_2^k, \ldots$ the first constant that does not occur either in **A** or in any member of Γ_n is w_m^k, then Γ_{n+1} is the class whose members are

$$\overset{\mathbf{a}}{\underset{w_m^k}{\text{Ṣ}}}\mathbf{A}| \supset (\mathbf{a})\mathbf{A}$$

and the members of Γ_n; and otherwise Γ_{n+1} is the same as Γ_n.

Then (for $n = 0, 1, 2, \ldots$) Γ_n is a finite class of closed wffs of S. We shall show by mathematical induction that every Γ_n is consistent.

Suppose that, for some particular n, Γ_n is consistent but Γ_{n+1} is inconsistent. Then we must have the case that Γ_{n+1} is not the same as Γ_n but has the additional member

$$\overset{\mathbf{a}}{\underset{\mathbf{w}}{\text{Ṣ}}}\mathbf{A}| \supset (\mathbf{a})\mathbf{A},$$

where **w** is a suitably chosen constant (as described above). By the inconsistency of Γ_{n+1}, and the deduction theorem (*516),

$$\Gamma_n \vdash \overset{\mathbf{a}}{\underset{\mathbf{w}}{\text{Ṣ}}}\mathbf{A}| \supset (\mathbf{a})\mathbf{A} \supset f.$$

In this proof from hypotheses, replace **w** everywhere by a new variable **x**, which is of the same type as **a** and which does not otherwise occur. Since **w** does not occur in **A** or in any of the members of Γ_n, we thus have

[513] The device which is used at this point was suggested to the writer by Henkin in July 1950. As compared to the procedure at the corresponding point in §45, or in Henkin's dissertation, it has the advantage of making it possible to replace the infinite sequence of applied functional calculi $S_1, S_2, S_3, \ldots$ by the single applied functional calculus S. A similar simplification could have been made in §45 in the proof of **453; but, because the difference in length is slight, we have there retained the older form of the proof (which it is thought may be instructive).

$$\Gamma_n \vdash \underset{\cdot}{\mathrm{S}}^{\mathbf{a}}_{\mathbf{x}}\mathbf{A}| \supset (\mathbf{a})\mathbf{A} \supset f.$$

Hence by generalizing upon **x** and then using the theorem schemata *380 and *383 (or analogues of them),

$$\Gamma_n \vdash (\mathbf{x})\underset{\cdot}{\mathrm{S}}^{\mathbf{a}}_{\mathbf{x}}\mathbf{A}| \supset (\mathbf{a})\mathbf{A} \supset f.$$

By alphabetic change of bound variable,[514]

$$\Gamma_n \vdash (\mathbf{a})\mathbf{A} \supset (\mathbf{a})\mathbf{A} \supset f.$$

But since $(\mathbf{a})\mathbf{A} \supset (\mathbf{a})\mathbf{A}$ is a theorem (by P), we then have $\Gamma_n \vdash f$, contrary to the supposed consistency of Γ_n.

Since Γ_0 is consistent, and the consistency of Γ_{n+1} follows from that of Γ_n (as just shown), therefore every Γ_n is consistent.

Let Γ be the union of the classes $\Gamma_0, \Gamma_1, \Gamma_2, \ldots$, and let $\bar{\Gamma}$ be the extension of Γ to a maximal consistent class of closed wffs of S.

We need the following properties of $\bar{\Gamma}$ (where in d1–e3, **A** and **B** are closed wffs of S):

d1. If **A** is a member of $\bar{\Gamma}$, then $\sim\mathbf{A}$ is a non-member of $\bar{\Gamma}$. (For otherwise $\bar{\Gamma}$ would be inconsistent, by P.)

d2. If **A** is a non-member of $\bar{\Gamma}$, then $\sim\mathbf{A}$ is a member of $\bar{\Gamma}$. (For if **A** is a non-member of $\bar{\Gamma}$, then **A** must be inconsistent with $\bar{\Gamma}$; therefore, by the deduction theorem and P, $\bar{\Gamma} \vdash \sim\mathbf{A}$; therefore $\sim\mathbf{A}$ is consistent with $\bar{\Gamma}$; therefore $\sim\mathbf{A}$ is a member of $\bar{\Gamma}$.)

e1. If **B** is a member of $\bar{\Gamma}$, then $\mathbf{A} \supset \mathbf{B}$ is a member of $\bar{\Gamma}$. (For by P, $\bar{\Gamma} \vdash \mathbf{A} \supset \mathbf{B}$; thus $\mathbf{A} \supset \mathbf{B}$ is consistent with $\bar{\Gamma}$ and therefore a member of $\bar{\Gamma}$.)

e2. If **A** is a non-member of $\bar{\Gamma}$, then $\mathbf{A} \supset \mathbf{B}$ is a member of $\bar{\Gamma}$. (For by d2, $\sim\mathbf{A}$ is a member of $\bar{\Gamma}$, and therefore by P, $\bar{\Gamma} \vdash \mathbf{A} \supset \mathbf{B}$.)

e3. If **A** is a member of $\bar{\Gamma}$ and **B** is a non-member of $\bar{\Gamma}$, then $\mathbf{A} \supset \mathbf{B}$ is a non-member of $\bar{\Gamma}$. (For by d2, $\sim\mathbf{B}$ is a member of $\bar{\Gamma}$; hence if $\mathbf{A} \supset \mathbf{B}$ were a member of $\bar{\Gamma}$, $\bar{\Gamma}$ would be inconsistent.)

f1. If **a** is an individual or functional variable, and if, for every constant **w** of the same type as the variable **a**,

$$\underset{\cdot}{\mathrm{S}}^{\mathbf{a}}_{\mathbf{w}}\mathbf{A}|$$

is a member of $\bar{\Gamma}$, then $(\mathbf{a})\mathbf{A}$ is a member of $\bar{\Gamma}$. (For in consequence of the way in which the classes Γ_n were defined, there is some constant **w** of the same type as the variable **a** such that

---

[514]If **a** does not occur as a bound variable in **A**, one application of *515 is sufficient. Otherwise the required result may be obtained by three or more successive applications of *515.

$$\underset{\cdot}{S}{}^{\mathbf{a}}_{\mathbf{w}}\mathbf{A}| \supset (\mathbf{a})\mathbf{A}$$

is a member of Γ, therefore a member of $\bar{\Gamma}$.)

f1′. If $\mathbf{a}$ is a propositional variable, and if both $\underset{\cdot}{S}{}^{\mathbf{a}}_{t}\mathbf{A}|$ and $\underset{\cdot}{S}{}^{\mathbf{a}}_{f}\mathbf{A}|$ are members of $\bar{\Gamma}$, then $(\mathbf{a})\mathbf{A}$ is a member of $\bar{\Gamma}$. (By *514, *515.)

f2. If $\mathbf{a}$ is an individual or functional variable, and if, for at least one constant $\mathbf{w}$ of the same type as the variable $\mathbf{a}$,

$$\underset{\cdot}{S}{}^{\mathbf{a}}_{\mathbf{w}}\mathbf{A}|$$

is a non-member of $\bar{\Gamma}$, then $(\mathbf{a})\mathbf{A}$ is a non-member of $\bar{\Gamma}$. (By †509 and *509_n.)

f2′. If $\mathbf{a}$ is a propositional variable, and if either

$$\underset{\cdot}{S}{}^{\mathbf{a}}_{t}\mathbf{A}| \quad \text{or} \quad \underset{\cdot}{S}{}^{\mathbf{a}}_{f}\mathbf{A}|$$

is a non-member of $\bar{\Gamma}$, then $(\mathbf{a})\mathbf{A}$ is a non-member of $\bar{\Gamma}$. (By *509_0.)

To each of the individual constants w_n we now assign as *associated natural number* the number n (i.e., 0 is the associated natural number of w_0, 1 is the associated natural number of w_1, and so on). And to each of the k-ary functional constants w^k_n we assign as *associated propositional function* the k-ary propositional function Φ^k_n of natural numbers determined by the rule that $\Phi^k_n(u_1, u_2, \ldots, u_k)$ is t or f according as $w^k_n(w_{u_1}, w_{u_2}, \ldots, w_{u_k})$ is a member or a non-member of $\bar{\Gamma}$.

We shall also speak of the natural number n as *associated to* the constant w_n, and of the propositional function Φ^k_n as *associated to* the constant w^k_n.

Let the domain $\mathfrak{J}$ consist of the natural numbers, let the domain $\mathfrak{F}_1$ consist of the associated propositional functions of all the singular functional constants w^1_n, let $\mathfrak{F}_2$ consist of the associated propositional functions of all the binary functional constants w^2_n, and so on. Then each of the domains $\mathfrak{J}, \mathfrak{F}_1, \mathfrak{F}_2, \mathfrak{F}_3, \ldots$ is finite or enumerably infinite.[515]

With respect to the system of domains $\mathfrak{J}, \mathfrak{F}_1, \mathfrak{F}_2, \mathfrak{F}_3, \ldots$, the value of a wff $\mathbf{X}$ of F^{2p}_2, for a given system of values of its free variables $\mathbf{x}_1, \mathbf{x}_2, \ldots, \mathbf{x}_m$, is t or f according as

$$\underset{\cdot}{S}{}^{\mathbf{x}_1 \mathbf{x}_2 \ldots \mathbf{x}_m}_{\mathbf{w}_1 \mathbf{w}_2 \ldots \mathbf{w}_m}\mathbf{X}|$$

is a member or a non-member of $\bar{\Gamma}$, where $\mathbf{w}_i$ is the constant, or one of the constants, to which the value of $\mathbf{x}_i$ is associated ($i = 1, 2, \ldots, m$), or in case $\mathbf{x}_i$ is a propositional variable, $\mathbf{w}_i$ is t or f according as the value of $\mathbf{x}_i$ is t or f.

[515]The possibility that the domains $\mathfrak{F}_1, \mathfrak{F}_2, \mathfrak{F}_3, \ldots$ may all be finite is realized if, for example, we take $\mathbf{H}$ to be the wff $(\exists x)(\exists y)(\exists F) \,.\, F(x) \sim F(y)$, which has a non-valid afep and is therefore not a theorem (**530). On the other hand, if we take $\mathbf{H}$ to be the negation of one of the axioms of infinity (see §57), then all the domains must be enumerably infinite.

This follows, in fact, from the definition of validity with respect to a system of domains and from the properties d1, d2, e1, e2, e3, f1, f1′, f2, f2′ listed above.

Taking **X** to be the particular wff $\sim$**H**, which has no free variables, and which is of course a member of $\bar{\Gamma}$, we have therefore that $\sim$**H** is valid with respect to the system of domains $\mathfrak{J}, \mathfrak{F}_1, \mathfrak{F}_2, \mathfrak{F}_3, \ldots$.

Taking **X** to be any axiom of F_2^{2p}, we have that

$$\mathrm{S}^{\mathbf{x}_1 \mathbf{x}_2 \ldots \mathbf{x}_m}_{\mathbf{w}_1 \mathbf{w}_2 \ldots \mathbf{w}_m} \mathbf{X}|$$

is always a theorem of S and therefore a member of $\bar{\Gamma}$. Therefore again **X** is valid with respect to the system of domains $\mathfrak{J}, \mathfrak{F}_1, \mathfrak{F}_2, \mathfrak{F}_3, \ldots$.

In order to establish that $\mathfrak{J}, \mathfrak{F}_1, \mathfrak{F}_2, \mathfrak{F}_3, \ldots$ are a normal system of domains, we have to show further that each of the four rules of inference of F_2^{2p} preserves validity with respect to this system of domains. In each case this may be done by considering the universal closure of the premiss or premisses of the rule and the universal closure of the conclusion, since it is obviously always a derived rule of inference that the universal closure of the conclusion may be inferred from the universal closure of the premisses. If the premisses are valid with respect to the system of domains in question, then by **542 their universal closures are valid with respect to that system of domains, and are therefore members of $\bar{\Gamma}$; therefore the universal closure of the conclusion is a member of $\bar{\Gamma}$; therefore the universal closure of the conclusion is valid with respect to the system of domains in question; therefore finally by **542 the conclusion itself is valid with respect to that system of domains.

Thus we have proved the metatheorem:

545. If a closed wff **H of F_2^{2p} is not a theorem, there exists a normal system of finite or enumerably infinite domains with respect to which $\sim$**H** is valid.

Now consider a secondarily valid wff **A** of F_2^{2p}, and let **H** be the universal closure of **A**. By **542, **H** is also secondarily valid. Therefore there can be no normal system of domains with respect to which $\sim$**H** is valid. Therefore by **545, **H** is a theorem of F_2^{2p}. Therefore **A** is a theorem of F_2^{2p}.

Thus we have as a corollary of **545 the following metatheorem (Henkin's completeness theorem for the pure functional calculus of second order):

**546. Every secondarily valid wff of F_2^{2p} is a theorem.

From one point of view, Henkin's completeness theorem for the functional calculus of second order is much like Gödel's completeness theorem for the functional calculus of first order, since the semantical significance in both cases is that all those wffs are theorems which, under all of a certain class of interpretations of the calculus, have the value t for all systems of values of their free variables. There is, however, the important difference that in the case of the pure functional calculus of first order the interpretations are the principal interpretations, whereas in the case of the pure functional calculus of second order they include also the secondary interpretations. At issue is the question of the intention in formulating the calculus; and properly speaking, Henkin's theorem has the meaning of a completeness theorem only if it was intended in formulating the pure functional calculus of second order to treat the secondary as well as the principal interpretations. In fact it is impossible to extend the pure functional calculus of second order by adding rules and axioms in such a way that the theorems come to coincide with the wffs which have the value t for all systems of values of their free variables, under all the principal interpretations. (This last will follow from the famous incompleteness theorems of Gödel, to be discussed in a later chapter.)

There is also another point of view from which Henkin's completeness theorem for the functional calculus of second order is weak compared to that of Gödel for the functional calculus of first order. For we have in connection with the latter theorem that there is one particular interpretation (the principal interpretation with the natural numbers as the individuals) under which the theorems coincide with the wffs that have the value t for all systems of values of their free variables. But in the case of the pure functional calculus of second order there appears no such one interpretation relative to which we have completeness; and if in particular we adopt the principal interpretation with the natural numbers as the individuals, there are various independent axioms (not containing any new primitive symbols) which we may be led to add, and some of the most immediate of which are given in §§56, 57.

As further corollaries we have:

**547. Every wff of F_2^{2p} which is valid with respect to all normal systems of finite and enumerably infinite domains is secondarily valid, and valid.

**548. Every wff of F_2^{2p} whose negation is not a theorem is satisfiable with respect to some normal system of finite and enumerably infinite domains.

EXERCISES 54

54.0. The wffs of the extended propositional calculus (in the formulation of it which is used in §53) are included among the wffs of F_2^{2p}. Hence the definition in §54 of validity of wffs of F_2^{2p} applies in particular to wffs of the extended propositional calculus. (1) Prove that this definition of validity of wffs of the extended propositional calculus is equivalent to that of §53 in the sense that the class of valid wffs is the same. (2) Prove that a wff of the extended propositional calculus is a theorem of F_2^{2p} if and only if it is valid.

54.1. In view of the solution of the elimination problem of the singulary functional calculus of second order in 52.6(1), and of results concerning F^{Ip} obtained in exercises 48, prove that, if a wff of the singulary functional calculus of second order $F_2^{2,1}$ is valid in every non-empty finite domain, then it is valid, and a theorem.

54.2. Determine which of the following things are true of every normal system of domains $\mathfrak{J}$, $\mathfrak{F}_1$, $\mathfrak{F}_2$, $\mathfrak{F}_3$, . . ., and supply a proof or disproof in each case: (1) $\mathfrak{F}_1$ contains the null class and the universal class of individuals.[516] (2) If $\mathfrak{F}_1$ contains any two classes of individuals, it always contains also the class which is the union of those two classes. (3) $\mathfrak{F}_2$ contains the relation of identity or equality between individuals, and the relation of non-identity or diversity between individuals.[516] (4) If $\mathfrak{F}_2$ contains any relation between individuals, then $\mathfrak{F}_1$ always contains the domain and the converse domain of that relation.[517] (5) If $\mathfrak{F}_2$ contains any two relations between individuals, it always contains also their relative product.[518]

54.3 Let $\mathfrak{J}$, $\mathfrak{F}_1$, $\mathfrak{F}_2$, $\mathfrak{F}_3$, . . ., be a system of domains of the kind described in the first paragraph of §54. Show that it is a normal system of domains if and only if $\mathfrak{J}$ is non-empty and, for every wff **A** of F_2^{2p}, for every list $\mathbf{a}_1, \mathbf{a}_2, \ldots, \mathbf{a}_n$ of distinct individual variables ($n \geqq 1$), and for every system $\mathfrak{S}$ of values (with respect to $\mathfrak{J}$, $\mathfrak{F}_1$, $\mathfrak{F}_2$, $\mathfrak{F}_3$ . . .) of the free variables of **A** other than $\mathbf{a}_1, \mathbf{a}_2, \ldots, \mathbf{a}_n$, there is a propositional function Φ in $\mathfrak{F}_n$ such that $\Phi(a_1, a_2, \ldots, a_n)$ is always the same as the value of **A** with respect to $\mathfrak{J}$, $\mathfrak{F}_1$, $\mathfrak{F}_2$, $\mathfrak{F}_3$, . . . for the values $a_1, a_2, \ldots, a_n$ of $\mathbf{a}_1, \mathbf{a}_2, \ldots, \mathbf{a}_n$ and the system of values $\mathfrak{S}$ of the remaining free variables of **A**. (Notice that this characterization of a normal system of domains, unlike that in the text, is independent of the axioms and rules of inference of F_2^{2p}.)

54.4. Show (as is assumed in the text) that every consistent class of wffs of F_2^{2p} is a consistent class of wffs of the applied functional calculus S.

---

[516]As explained in §04, we take classes to be singulary propositional functions and relations to be binary propositional functions.

[517]The *domain* of a relation is the class of things (individuals) which bear that relation to at least one thing; and the *converse domain* of a relation is the class of things to which at least one thing bears that relation. (The converse domain of a relation is thus the same as the domain of the converse of the relation, where "converse" has the meaning explained in §03.)

[518]The *relative product* of two relations Φ and Ψ is the relation which holds between two things (individuals) a and b if and only if there is at least one thing c such that a bears the relation Φ to c and c bears the relation Ψ to b.

For example, if we take the individuals to be human beings, the relative product of the relation *husband* and the relation *daughter* is the relation *son-in-law*; and the relative product of the relation *parent* and the relation *parent* is the relation *grandparent*.

54.5. By the methods of §54 prove: Every consistent class of wffs of F_2^{2p} is simultaneously satisfiable with respect to some normal system of finite and enumerably infinite domains.

54.6. Supply a proof of Gödel's completeness theorem for the functional calculus of first order which parallels as closely as possible the proof of the metatheorems **545 and **546 in §54.

55. Postulate theory.[519] From the viewpoint which is explained in §07, when a system of postulates is used as basis for the formal treatment of some mathematical theory or branch of mathematics (say, for example, arithmetic, or Euclidean plane geometry), the postulates have to be thought of as added to an underlying logic. And indeed for the precise syntactical definition of the particular branch of mathematics it is necessary to state not only the specific mathematical postulates but also a formalization of the underlying logic, since the class of theorems belonging to the branch of mathematics in question is determined by both the postulates and the underlying logic, and could be changed by a change in either.[520]

---

[519]For other discussions of postulate theory from the logistic standpoint see for example Carnap's *Abriss der Logistik* (Vienna, 1929), Carnap's *The Logical Syntax of Language* (New York and London, 1937), and Hilbert and Bernays's *Grundlagen der Mathematik* (Berlin, 1934, 1939). Though the reader must allow for some differences in approach and terminology, we believe that our account of the matter is in essential agreement with that of these authors.

[520]The point may be illustrated by the case of elementary number theory versus analysis, since these two branches of mathematics may be based if we like on the very same system of postulates, but with different underlying logics. Namely, elementary number theory may be defined by adding the postulates (A_1), given below, to an applied functional calculus of first order containing all propositional and functional variables as well as the functional constants which appear in the postulates. And analysis may be defined by adding the same postulates to an applied functional calculus of fourth order—since, in the resulting system, rational, real, and complex numbers may be introduced by any of various well-known methods for defining these numbers in terms of the natural numbers.

The foregoing statement is open to certain reservations because of uncertainty as to exactly what should be understood by "elementary number theory" and "analysis" as they appear in common (informal) mathematical usage. It is the feeling of the writer, however, that "elementary number theory" is preferably understood in such a way as not to exclude the expression of certain generalities about classes and functions of natural numbers—as, e.g., the proposition that in every class of natural numbers there is a least number.

On the other hand, in the system obtained by adding the postulates (A_1) to an applied functional calculus of second order there is a certain sense in which a large part of analysis can already be obtained, by means of appropriate artifices which are beyond the scope of our present discussion. The decision to go as high as the functional calculus of fourth order in specifying the underlying logic of mathematical analysis is thus open to some question, but again it would seem to the writer that this best represents the existing informal usage.

When the underlying logic is a functional calculus of second or higher order it is usual, as noted below, to replace the postulates (A_1) by the more economical system of postulates (A_2). However, this need not affect our present illustration, since the postulates (A_1) could always be retained if desired.

The present section is devoted to a digression for the further treatment of this matter, and the introduction of a number of examples. This seems to be an appropriate place for such a discussion, because in many cases it is sufficient to take the underlying logic to be a functional calculus of first or second order.

As a first example we take the following system of postulates for arithmetic, which we shall call (A_0), and for which the underlying logic is to be (syntactically) a simple applied functional calculus of first order. There are two undefined terms, the ternary functional constants Σ and Π; and the notations Z_0, Z_1, and $=$ are introduced by the following definition schemata,[521] in which **a** and **c** are any variables, **b** is the next individual variable in alphabetic order after **a**, and **d** and **e** are the first two individual variables in alphabetic order distinct from each other and from **a** and **c**:

$$Z_0(\mathbf{a}) \rightarrow (\mathbf{b})\Sigma(\mathbf{a}, \mathbf{b}, \mathbf{b})$$

$$Z_1(\mathbf{a}) \rightarrow (\mathbf{b})\Pi(\mathbf{a}, \mathbf{b}, \mathbf{b})$$

$$[\mathbf{a} = \mathbf{c}] \rightarrow (\mathbf{d})(\mathbf{e}) \,\blacksquare\, \Sigma(\mathbf{a}, \mathbf{d}, \mathbf{e}) \supset \Sigma(\mathbf{c}, \mathbf{d}, \mathbf{e})$$

The postulates include first of all the twelve following:

$$(\exists z)\Sigma(x, y, z)$$

$$\Sigma(x_1, x_2, y_1) \supset \blacksquare\, \Sigma(x_2, x_3, y_2) \supset \blacksquare\, \Sigma(y_1, x_3, z) \supset \Sigma(x_1, y_2, z)$$

$$\Sigma(x, y, z) \supset \Sigma(y, x, z)$$

$$\Sigma(x_1, y, z) \supset \blacksquare\, \Sigma(x_2, y, z) \supset \blacksquare\, x_1 = x_2$$

$$(\exists x)\Sigma(x, y, y)$$

$$(\exists z)\Pi(x, y, z)$$

$$\Pi(x_1, x_2, y_1) \supset \blacksquare\, \Pi(x_2, x_3, y_2) \supset \blacksquare\, \Pi(y_1, x_3, z) \supset \Pi(x_1, y_2, z)$$

---

[521]In connection with the informal statement of these postulates, additional undefined terms 0, 1, and = (some or all of them) would often be listed. However, the logistic method shows these additional undefined terms to be unnecessary. Individual constants 0 and 1 indeed are not provided for by the definitions which we give, but the notations Z_0 and Z_1 serve the essential purposes which would be served by the actual inclusion of individual constants 0 and 1 as undefined terms.

For the informal statement of the postulates, "natural number" would also ordinarily be listed as an undefined term. This additional undefined term has not so much been eliminated by our present method as incorporated into the undefined terms Σ and Π, because we regard the range of a function as determined by the function, and as being given as soon as the function itself is given. (To change the range would be to change the function itself to a different function.) Or alternatively, if the reader prefers, he may regard the undefined term "natural number" of the informal statement as represented by the individual variables—which have the individuals, i.e., the natural numbers, as their range.

$$\Pi(x, y, z) \supset \Pi(y, x, z)$$

$$\Pi(x_1, y, z) \supset . \Pi(x_2, y, z) \supset Z_0(y) \vee x_1 = x_2$$

$$(\exists x)\Pi(x, y, y)$$

$$\Pi(x_1, x_2, y_2) \supset . \Pi(x_1, x_3, y_3) \supset . \Sigma(x_2, x_3, y_1) \supset . \Sigma(y_2, y_3, z) \supset \Pi(x_1, y_1, z)$$

$$Z_1(y) \supset . \Sigma(x, y, z) \supset \sim Z_0(z)$$

Then in addition to these twelve, there is an infinite list of *postulates of mathematical induction*—as given by the following postulate schema, in which **A** is any wff not containing the variable z and **B** is $\text{Ṣ}_z^x\mathbf{A}|$:

$$Z_0(x) \supset . Z_1(y) \supset . \mathbf{A} \supset . \mathbf{A} \supset_x (\exists z)[\Sigma(x, y, z)\mathbf{B}] \supset (x)\mathbf{A}$$

This system of postulates (A_0) is to be added to the simple applied functional calculus of first order F^{1h} (see §30) as underlying logic. The resulting system is a formulation of what we shall call *elementary arithmetic*.[522]

Of the logistic system A^0, obtained by adding the postulates (A_0) to F^{1h}, the principal interpretation is the same as for F^{1h} itself, and is given by the semantical rules α–ζ of §30. The separation of the semantical rules into two categories, as spoken of in §07, is by assigning the rules α_0, β_0, γ, δ, ε, ζ to the underlying logic, and thus putting them in the first category, while the rules α_1, β_1, and β_2 are put in the second category.

As a result, for the value a of **a**, $Z_0(\mathbf{a})$ denotes t or f according as a is or is not 0, $Z_1(\mathbf{a})$ denotes t or f according as a is or is not 1. We may thus take these notations as meaning respectively that a is 0 and that a is 1. And in a similar way we may take the notation $=$ as meaning equality of natural numbers. For although this is a different sense of "$=$" or "equals" from that given by D22, there is no reason that we should not change the sense of "equals" in connection with A^0, since the propositional function is not changed in extension, and since the mathematical theory is not thereby altered formally or prevented from serving its purpose.

The first five postulates of (A_0) express, in order, the existence of the sum of two natural numbers, the associative law of addition, the commutative law of addition, the law of cancellation for addition, and (in a weak sense) the existence of an identity element for addition. The next five postulates express the five corresponding properties of multiplication of natural numbers. The eleventh postulate expresses the distributive law. And the twelfth postulate expresses that the result of adding 1 to a natural number is never 0.

Each of the postulates of mathematical induction—as given by the postulate schema—either expresses a certain particular case of the principle of mathematical induction, if **A** contains no free variables except x, or else, if **A** contains other free variables, expresses a principle which (though it has some generality)

---

[522]Introducing this term in a sense which we distinguish from that of "elementary number theory."

is still to be regarded as obtained from the general principle of mathematical induction by a specialization.

In view of the lack of functional variables, it is not possible in A^0 to express the principle of mathematical induction as a general law. However, such an expression of the general principle of mathematical induction appears below as the final postulate of the system of postulates (A_1).

In spite of the restriction imposed by the lack of functional variables, substantially all the propositions of elementary number theory, as usually understood, can be expressed and proved in A^0, excepting only those which directly require functional variables (such as, e.g., the principle of mathematical induction, or the principle that is stated in exercise 55.12). We shall not carry this out here, even in part, but refer the reader to the treatment of an equivalent system by Hilbert and Bernays in *Grundlagen der Mathematik*. A crucial point is the method, due to Gödel,[523] of introducing by definition other numerical functions than addition and multiplication—e.g., exponentiation, the factorial, the quotient and remainder upon division, the nth prime number as a function of n, and indeed recursive functions generally.

Another example is the following system of postulates (A_1). The undefined terms are Σ and Π, and the definition schemata for the notations Z_0, Z_1, $=$ are the same as in the case of (A_0). There are thirteen postulates, of which the first twelve are the same as the first twelve postulates of (A_0), and the thirteenth is the *postulate of mathematical induction*,

$$Z_0(x_1) \supset . Z_1(y) \supset . F(x_1) \supset . F(x) \supset_x (\exists z)[\Sigma(x, y, z)F(z)] \supset (x)F(x).$$

The underlying logic is (or is formalized as) the functional calculus F^{1a}_2, i.e., a functional calculus of first order which has the ternary functional constants Σ and Π, and in addition all propositional and functional variables, the same as the functional calculus F^{1a} of 30.4 except that the rules and axioms of §40 are used instead of those of §30. And the logistic system obtained by adding the postulates (A_1) to this underlying logic we shall take as a formulation of *elementary number theory*, or (as we shall also say) of *first-order arithmetic*.[524]

The postulates (A_1) might of course also be added to a functional calculus of second or higher order as underlying logic, so obtaining a stronger system,

---

[523] *Monatshefte für Mathematik und Physik*, vol. 38 (1931), see pp. 191–193.

[524] Cf. footnote 520. To logistic formulations of either elementary arithmetic or first-order arithmetic the name *Hilbert arithmetic* is often given because of the introduction of systems of this kind by Hilbert and his school. See a paper by Hilbert in *Abhandlungen aus dem Mathematischen Seminar der Hamburgischen Universität*, vol. 6 (1928), pp. 65–85 (reprinted in the seventh edition of Hilbert's *Grundlagen der Geometrie*); also a paper by Ackermann in the *Mathematische Annalen*, vol. 117 (1940), pp. 162–194; as well as the treatment of the systems Z, Z*, Z**, Z′, etc. by Hilbert and Bernays in *Grundlagen der Mathematik*.

but instead of this we prefer to employ the following different system of postulates (A_2), which are equivalent to (A_1) when so used.

The postulates (A_2) are essentially a form of Peano's postulates for the natural numbers,[525] as modified for use in the present context. There is a single undefined term, the binary functional constant S. The notations $=$ and $\neq$ are those introduced in D22 and D23. And the notations Z_0, Z_1 are introduced by the following definition schemata, in which **a** is any individual variable, and **b** is the next individual variable in alphabetic order after **a**:

$$Z_0(\mathbf{a}) \to (\mathbf{b}){\sim}S(\mathbf{b}, \mathbf{a})$$

$$Z_1(\mathbf{a}) \to (\exists \mathbf{b}) \,\centerdot\, Z_0(\mathbf{b})S(\mathbf{b}, \mathbf{a})$$

The postulates are the five following:

$$(\exists y)S(x, y)$$

$$S(x, y) \supset \centerdot\, S(x, z) \supset \centerdot\, y = z$$

$$S(y, x) \supset \centerdot\, S(z, x) \supset \centerdot\, y = z$$

$$(\exists x)Z_0(x)$$

$$Z_0(x) \supset \centerdot\, F(x) \supset \centerdot\, F(y) \supset_y [S(y, z) \supset_z F(z)] \supset (y)F(y)$$

In the interpretation, the individuals are again the natural numbers. And S denotes the relation of *having as successor*, so that, if a and b are the values of **a** and **b**, then $S(\mathbf{a}, \mathbf{b})$ denotes t or f according as b is or is not equal to $a + 1$. Detailed statement of the semantical rules, for the case of a functional calculus of second order as underlying logic, may be supplied by analogy and is left to the reader.

The notations $Z_0(\mathbf{a})$ and $Z_1(\mathbf{a})$, for a value a of **a**, again may be taken as meaning, respectively, that a is 0 and that a is 1. The sense is indeed changed as compared to the notations Z_0 and Z_1 used in connection with (A_0), or with (A_1). But the fact that the corresponding propositional functions are the same in extension is sufficient for the purpose of the mathematical theory.

A similar remark applies to the notations Σ and Π which are introduced below by definition to replace the notations Σ and Π that appeared as undefined terms in the postulates (A_0), or (A_1).

The logistic system obtained by adding the postulates (A_0) to F^{1h} we call A^0. That obtained by adding the postulates (A_1) to F^{1a}_2 we call A^1. Those obtained by adding the postulates (A_2) to a functional calculus of

[525] *Arithmetices Principia, Nova Methodo Exposita*, Turin, 1889; *Formulaire de Mathématiques*, vol. II §2, Turin, 1898. As Peano points out, his postulates are in the treatise of Richard Dedekind, *Was Sind und was Sollen die Zahlen?* (1888), though not quite as postulates. Some of the essentials, however, are already contained in a paper by C. S. Peirce in the *American Journal of Mathematics*, vol. 4 (1881), pp. 85–95.

nth order ($n = 2, 3, 4, \ldots$) which has all propositional and functional variables appropriate to its order, and has in addition the binary functional constant S, we call $A^2, A^3, A^4, \ldots$. And we call A^n ($n = 1, 2, 3, 4, \ldots$) a formulation of n*th-order arithmetic*.

The detailed development of the system A^2 will be the subject of a later chapter. At this place, we carry the matter no further, except to state the following definition schemata, introducing notations Σ and Π to replace the primitive notations Σ and Π of A^0 and A^1:[526]

$$\Sigma(\mathbf{a}, \mathbf{b}, \mathbf{c}) \rightarrow [Z_0(\mathbf{b}_0) \supset_{\mathbf{b}_0} . \, \mathbf{a} = \mathbf{c}_0 \supset_{\mathbf{c}_0} F(\mathbf{b}_0, \mathbf{c}_0)][F(\mathbf{b}_0, \mathbf{c}_0) \supset_{\mathbf{b}_0 \mathbf{c}_0} . \, S(\mathbf{b}_0, \mathbf{b}_1) \supset_{\mathbf{b}_1} . \, S(\mathbf{c}_0, \mathbf{c}_1) \supset_{\mathbf{c}_1} F(\mathbf{b}_1, \mathbf{c}_1)] \supset_F F(\mathbf{b}, \mathbf{c})$$

$$\Pi(\mathbf{a}, \mathbf{b}, \mathbf{c}) \rightarrow [Z_0(\mathbf{b}_0) \supset_{\mathbf{b}_0} . \, Z_0(\mathbf{c}_0) \supset_{\mathbf{c}_0} F(\mathbf{b}_0, \mathbf{c}_0)][F(\mathbf{b}_0, \mathbf{c}_0) \supset_{\mathbf{b}_0 \mathbf{c}_0} . \, S(\mathbf{b}_0, \mathbf{b}_1) \supset_{\mathbf{b}_1} . \, \Sigma(\mathbf{a}, \mathbf{c}_0, \mathbf{c}_1) \supset_{\mathbf{c}_1} F(\mathbf{b}_1, \mathbf{c}_1)] \supset_F F(\mathbf{b}, \mathbf{c})$$

—where, in both cases, **a**, **b**, **c** are any individual variables, and $\mathbf{b}_0, \mathbf{c}_0, \mathbf{b}_1, \mathbf{c}_1$ are the first four individual variables in alphabetic order after the latest, in alphabetic order, of the variables **a**, **b**, **c**.

It should be noticed that these definitions do not introduce the notations Σ and Π as functional constants, or as names of the propositional functions which in the systems A^0 and A^1 were denoted by Σ and Π. In fact the system A^2 does not contain names of these propositional functions, and the definitions do not assign any formula of A^2 which is abbreviated by the letter Σ or the letter Π standing alone. Only the complete notations $\Sigma(\mathbf{a}, \mathbf{b}, \mathbf{c})$ and $\Pi(\mathbf{a}, \mathbf{b}, \mathbf{c})$ are abbreviations of formulas of A^2.

Nevertheless, for every theorem of A^0 or A^1 there is a corresponding theorem of A^2 in which the notations $\Sigma(\mathbf{a}, \mathbf{b}, \mathbf{c})$ and $\Pi(\mathbf{a}, \mathbf{b}, \mathbf{c})$ of A^0 or A^1 are replaced

---

[526]These two definition schemata illustrate a general method that may be used to find expressions in the system to represent a numerical function which, in the informal treatment, would be introduced by means of recursion equations. The first schema, for example, corresponds to the following recursion equations for addition:

$$a + 0 = a$$
$$a + (b + 1) = (a + b) + 1$$

And the second schema similarly corresponds to the following recursion equations for multiplication:

$$a \times 0 = 0$$
$$a \times (b + 1) = a + (a \times b)$$

This method, illustrated in the two definition schemata in the text, was introduced by Hilbert and Bernays in *Grundlagen der Mathematik*, vol. 2 (1939), Supplement IV G, and by Paul Lorenzen in a paper in *Monatshefte für Mathematik und Physik*, vol. 47 (1938–1939), pp. 356–358. Other methods serving the same purpose are due to Dedekind (1888) and to Kalmár (1930, 1940), and might also be adapted for use in the present connection. For an informal exposition of the matter and a brief account of its history, see Kalmár's paper in *Acta Scientiarum Mathematicarum*, vol. 9 no. 4 (1940), pp. 227–232.

The (informally stated) recursion equations themselves for addition and multiplication are due to C. S. Peirce in the paper cited in the preceding footnote.

by the different notations $\Sigma(\mathbf{a}, \mathbf{b}, \mathbf{c})$ and $\Pi(\mathbf{a}, \mathbf{b}, \mathbf{c})$ of A^2.[527] And it is in this sense that we say that A^2 is adequate for elementary number theory and does not require the functional constants Σ and Π as additional undefined terms.

In such a case, where a complex notation introduced by definition carries the false appearance or suggestion that some part of the notation is to be taken as denoting (or otherwise as having meaning in isolation), it is usual to speak of *contextual definition*. For example, by the two definition schemata just given, the letters Σ and Π are contextually defined: they acquire significance only in the particular contexts $\Sigma(\mathbf{a}, \mathbf{b}, \mathbf{c})$, $\Pi(\mathbf{a}, \mathbf{b}, \mathbf{c})$, and not in isolation or in other contexts. Similarly, by earlier definition schemata in this section—whether those introduced in connection with A^0 and A^1 or those for A^2—the letters Z_0 and Z_1 are contextually defined.

On the other hand, D6 (for example) would not ordinarily be called a contextual definition of the sign $\equiv$, and D24 would not be called a contextual definition of $=$. The difference is that, in the notations which are introduced by D6 and D24, there is nothing which suggests that either of the signs $\equiv$ or $=$ standing alone is significant in any way (e.g., as an abbreviation of a formula of a logistic system).

Thus the contextuality of the definitions of Σ and Π arises from the fact that, in the above definition schemata, we introduced the same parentheses and commas for use after the letters Σ and Π that we also use after functional variables and functional constants. The contextual character of the definitions might be avoided by changing the notations $\Sigma(\mathbf{a}, \mathbf{b}, \mathbf{c})$ and $\Pi(\mathbf{a}, \mathbf{b}, \mathbf{c})$ to, say, $\Sigma_{\mathbf{abc}}$ and $\Pi_{\mathbf{abc}}$. However, the convenience of using ordinary parentheses and commas outweighs the possible deceptiveness,[528] and the present explanation should serve to preclude misunderstandings.

We turn now to consideration of another and different point of view towards the postulates of a mathematical theory, which is possible in certain connections, and which also requires explanation here. In order to distinguish the two we may, from the point of view which we have so far been explaining, speak of *postulates as added axioms of a logistic system* and, from the new point of view, of *postulates as propositional functions*.[529]

---

[527]This will follow from our later detailed treatment of the system A^2, since all the postulates of A^0 and A^1 can be proved as theorems of A^2 when they are modified in the way described (i.e., when the notations $\Sigma(\mathbf{a}, \mathbf{b}, \mathbf{c})$ and $\Pi(\mathbf{a}, \mathbf{b}, \mathbf{c})$ of A^0 and A^1 are replaced by those of A^2).

[528]One aspect of this convenience is in the process of substitution for functional variables. For example, the mere replacement of the ternary functional variable F everywhere by the letter Σ represents what, in the unabbreviated notation, would appear as a considerably more complicated substitution operation (permitted by the rule of substitution for functional variables).

[529]The second point of view, as described below, has long been implicit in the use made of postulates by mathematicians, and in informal expositions of postulate theory, though the logistic method makes possible a more accurate statement of it. This point of view has been emphasized in particular by C. J. Keyser, who speaks in this connection of a "doctrinal function"—see a paper by him in *The Journal of Philosophy*, vol. 15

It is necessary first to introduce the notion of the *representing form* of a postulate belonging to a given system of postulates.[530]

Given a system of postulates, we first select for each of the undefined terms a corresponding variable of the same type (i.e., an individual variable to correspond to an individual constant, and an n-ary functional variable to correspond to an n-ary functional constant), these variables being all different among themselves, and all of them occupying an odd-numbered place (first, third, fifth, etc.) in alphabetic order. To make the procedure definite, we are to select in each case the first available variable in alphabetic order; and where there are several undefined terms of the same type, they are to be taken in their own alphabetic order (the order in which they were originally listed) and the corresponding variables for them are to be introduced in that order. Then we replace each postulate by its universal closure—where the "universal closure" is to be understood in the sense that all the free variables of the postulate are bound by initially placed universal quantifiers, and where therefore in some cases the expression obtained may not be a wff of the underlying logic of the postulates but only of the functional calculus of next higher order. Then in these closures of the postulates we make alphabetic changes of all the variables, replacing a variable that occupies the mth place, in the alphabetic order of variables of its type, by the variable of the same type that occupies the $2m$th place in alphabetic order. Then finally in each postulate we substitute everywhere for the undefined terms (constants) appearing, their corresponding variables. The result of this substitution is the *representing form* of the postulate.

For example, in the system of postulates (A_1), the representing form of the postulate of mathematical induction is the following wff of the pure functional calculus of second order:

$$(G)(x_1)(y_2) \mathbin{\blacksquare} (x_3)F(y_2, x_3, x_3) \supset \mathbin{\blacksquare} (z_1)H(x_1, z_1, z_1) \supset \mathbin{\blacksquare} G(y_2) \supset \mathbin{\blacksquare}$$
$$G(y) \supset_y (\exists z_1)[F(y, x_1, z_1)G(z_1)] \supset (y)G(y)$$

(1918), pp. 262–267. The "abstract" treatment of a system of postulates ("assumptions"), as described by Veblen and Young in the *Introduction* of the first volume of their *Projective Geometry* (first published in 1910), represents substantially the same idea, though the term "propositional function" is not actually used.

Neither Veblen and Young nor Keyser make the distinction which is introduced below between the theorems and the consequences of a system of postulates. Indeed this would hardly have been possible before the work of Tarski and Carnap.

[530]We treat here only the case that the underlying logic is one of the functional calculi of not higher than second order—though extensions to other cases, in particular to one of the functional calculi of higher order, may be made by analogy. The method of extension to functional calculi of higher order will become clear after the explicit formulation of these calculi which is to be given in our next chapter.

After having obtained thus the representing forms of the postulates, we may introduce also in the same way the representing form of any sentence or propositional form,[531] **B**, of the logistic system which consists of the underlying logic together with the postulates. Namely, we apply the same procedure to the wff **B** that we did to each of the postulates.

In order to introduce the notion of a model of a system of postulates, let Γ be the class of representing forms of the postulates, and let F be the pure functional calculus of lowest order in which all the formulas of Γ are wf.[532] Then a *model* of the postulates is a non-empty domain $\mathfrak{I}$ of individuals together with a system of values of the free variables of the representing forms of the postulates which satisfies Γ simultaneously in $\mathfrak{I}$ (or, in other words, which gives the value t simultaneously to the representing forms of the postulates, according to the notion of "value" which is defined in the theoretical syntax of F).

We remark that the various definitions of "value," as introduced for the various pure functional calculi, are coherent in the sense that a wff **A** of a pure functional calculus F has the same value for a given system of values of its free variables, whether **A** is taken as a wff of F or as a wff of one of the pure functional calculi of higher order than F. In regard to the pure functional calculi of first and second orders, this is clear from the definitions already given; and it will continue to hold also for the pure functional calculi of third and higher orders (to be discussed in Chapter VI). Hence in connection with a model of a system of postulates, the representing form of a wff that belongs to the underlying logic may be said to have a *value for* the model, even if this representing form is wf only in a functional calculus of higher order than is required for the representing forms of any of the postulates.

Now given a system of postulates, instead of considering the theorems of the logistic system, we may consider the consequences of the postulates in the following different (and non-effective) sense: A sentence or propositional form, **A**, of the logistic system which consists of the underlying logic together with the postulates is a *consequence* of the postulates if the value of the representing form of **A** is t for every model of the postulates.[533]

From the point of view towards postulate theory which we are now explaining, each postulate is looked upon in effect as a propositional function such that a

[531] In the case of one of the functional calculi, any wff.

[532] For this purpose, the pure functional calculus of first order with equality is to be counted as having an order between the first and the second.

[533] This is the notion of "logical consequence" introduced by Tarski, *Przegląd Filozoficzny*, vol. 39 (1936), pp. 58–68, and *Actes du Congrès International de Philosophie Scientifique* (Paris, 1936), part VII, pp. 1–11.

system of arguments of the propositional function would consist of a non-empty domain $\mathfrak{J}$ of individuals together with a value of each of the free variables of the representing form of the postulate, and the value of the propositional function for these arguments would be the same as the value which they determine of the representing form of the postulate. Similarly the complete system of postulates corresponds to a propositional function, of which a system of arguments would consist of a non-empty domain $\mathfrak{J}$ of individuals together with a value of each of the different free variables that appear in the representing forms of the postulates, the value of the propositional function being t or f according as these arguments do or do not constitute a model of the postulates.[534] The consequences of the postulates—in the above non-effective sense of "consequence"—again correspond to propositional functions in the same way. Syntactically, the mathematical theory to which the postulates lead consists of all the wffs taken together which are consequences.[535] This mathematical theory, however, may be expected to have many interpretations—in fact, since we require a principal interpretation of the underlying logic, each different model

[534]We assume, in making this statement, that the number of undefined terms is finite. Modification to fit the contrary case could be made by considering a binary propositional function of which one argument would be the domain of individuals and the other argument would be the complete system of values of the free variables of the representing forms of the postulates.

[535]Contrast this with the previous point of view, according to which the mathematical theory consists of the theorems.

Objection may indeed be made to this new point of view, on the basis of the sort of *absolutism* which it presupposes—or *Platonism* as Bernays calls it (*L'Enseignement Mathématique*, vol. **34** nos. 1–2 (1935), pp. 52–69; cf. also A. Fraenkel, ibid., pp. 18–32). But it should be pointed out that this Platonism is already inherent in classical mathematics generally, and it is not made more acute or more doubtful, but only more conspicuous, by its application to theoretical syntax. For our definition of the consequences of a system of postulates can be stated for, and treated within, a formalized meta-language which we do not describe in detail here but which can be seen to be not essentially different from formalized languages which are required for the logistic treatment of classical mathematics.

There would certainly be cogent objections (cf. §07) to the proposal to introduce a formalized language by means of the non-effective notion of consequence, and to replace in this way the initial construction of the language within what we have called elementary syntax (§08). But after this formalization of the language (or at least after the formalization of both the object language and the meta-language), the use of the non-effective notion of consequence in the theoretical syntax of the object language is a different matter, and objections to it are on a different level.

It is true that the non-effective notion of consequence, as we have introduced it in theoretical syntax, presupposes a certain absolute notion of ALL propositional functions of individuals. But this is presupposed also in classical mathematics, especially classical analysis, and objections against it lead to such modifications of classical mathematics as mathematical intuitionism (to be discussed in a later chapter) or the partial intuitionism of Hermann Weyl's *Das Kontinuum* (Leipzig, 1918).

(In this latter book, Weyl's objections to the absolute notion of *all* and to the vicious circle which it is held to involve lead him to a position which we may describe roughly as follows, that the simple functional calculi are replaced either by the corresponding predicative functional calculi or by the ramified functional calculi (cf. §58 and footnote 583), Russell's axioms of reducibility (§59) being rejected. As is well known, though he is able to make a partial reconstruction of analysis, Weyl reaches the conclusion that a substantial part of the classical theory is a house built upon sand.)

of the system of postulates yields one interpretation of the mathematical theory.[536] Thus the content of the mathematical theory is not fixed. but is itself to be looked on as the value of a function.[537]

The notions of consistency, independence, and completeness in connection with a system of postulates can be introduced in two different ways, which we may associate with the two different points of view towards postulate theory. We shall distinguish "consistency as to provability" and "consistency as to consequences"—and similarly in the cases of independence and completeness.

The notions of consistency, independence, and completeness as to provability will each depend in an essential way on the choice of the underlying logic as well as on the postulates themselves. But in the case of the corresponding notions as to consequences this dependence can be wholly or partly removed, as we shall see below.

A system of postulates will be said to be *consistent as to provability* if the logistic system which consists of the postulates together with the underlying logic is consistent in one of our earlier senses (§17), say in the sense that there is no wff **A** such that both **A** and $\sim$**A** are theorems.[538]

A system of postulates will be said to be *consistent as to consequences* if there is no wff **A** such that both **A** and $\sim$**A** are consequences of the postulates. Here **A** is a wff of the logistic system which consists of the postulates together with the underlying logic. But the dependence on the underlying logic is removed at once by the following metatheorem (which for the moment we must restrict to the case that the underlying logic is a functional calculus of no more than second order, but which can be generalized later to the case of a functional calculus of higher order):

---

[536]By an interpretation of the mathematical theory we mean, namely, an interpretation of the logistic system which is obtained by adjoining the postulates, as additional axioms, to the underlying logic. This is the same sense in which an interpretation of the postulates is referred to in the last paragraph of §07.

The advantage of economy in the axiomatic method, in that the results obtained hold for all the various interpretations, is a point which has been too often stressed by other writers to need repetition here. (Compare the corresponding remark about the logistic method generally in §07.)

[537]This is Keyser's *doctrinal function*, referred to in footnote 529. An accurate account of this notion from the point of view of the distinction of sense and denotation involves some complexities and will not be attempted here.

[538]Such a notion of consistency, involving the particular symbol $\sim$, is sufficient here because we are considering, as underlying logic, only the ordinary (applied) functional calculi of various orders, in the particular formulations adopted in Chapters III–VI. In order to extend the account to other systems as underlying logics, it will be necessary in each case to specify the sign (primitive or defined) which has to be identified as $\sim$. Or if this cannot be done, then one of the other notions of consistency may be used which is introduced in §17.

**550. A system of postulates is consistent as to consequences if and only if it has a model.

We leave the proof of this to the reader, as well as of the following metatheorem:

**551. If a system of postulates is consistent as to consequences, it is consistent as to provability.

In a system of postulates, the postulate **A** will be said to be *independent as to provability* if it is not a theorem of the logistic system which consists of the postulates other than **A** together with the underlying logic. And **A** will be said to be *independent as to consequences* if it is not a consequence of the other postulates.

Again we leave to the reader the proof of the metatheorems:

552. In a system of postulates, a postulate **A is independent as to consequences if and only if the postulates other than **A** have a model for which the value of the representing form of **A** is f.

553. In a system of postulates, if a postulate **A is independent as to consequences, it is independent as to provability.

The metatheorem **552 provides for the familiar method of establishing the independence of a postulate **A** in a system of postulates by exhibiting a model of the remaining postulates which gives to the representing form of **A** the value f. Such a model is called an *independence example* for **A**.[539]

The similar method of proving consistency of a system of postulates, namely, by exhibiting a model, is also well known.[540] However, it happens in certain important cases that such a proof of consistency, though possible, is of doubtful significance, because in establishing the existence of the model it is necessary to use a meta-language in which equivalents (in some relevant sense) of the postulates and their underlying logic are already present. For example, the consistency of the postulates (A_0), (A_1), or (A_2) may be demonstrated by using the natural numbers in the obvious way to provide a model; but this is a line of argument which evidently would carry no weight at all for one who had real doubts of the consistency of ordinary arithmetic, and which, even if the purpose is only to verify the correct formalization of

---

[539]This method of establishing independence of postulates was used by Peano, *Rivista di Matematica*, vol. 1 (1891), see pp. 93–94, and by Hilbert in his *Grundlagen der Geometrie*, first edition (1899). However, the origin of the method is to be seen still earlier in connection with the non-Euclidean geometry of Bolyai and Lobachevsky—models of the postulates of this geometry, found by Eugenio Beltrami (1868) and Felix Klein (1871), being in effect independence examples for Euclid's parallel postulate.

[540]Cf. the first edition of Hilbert's *Grundlagen der Geometrie*, pp. 19–21.

a theory already admitted informally, seems to accomplish relatively little.

A system of postulates will be said to be *complete as to provability* if the logistic system which consists of the postulates together with the underlying logic is complete with respect to the transformation of **A** into ~**A** (in the sense of §18). In many important cases, however, such completeness is unattainable, as is shown in the incompleteness theorems of Gödel, which were already referred to in the discussion following **546.

A system of postulates will be said to be *complete as to consequences* if, in the case of every wff **A** of the logistic system which consists of the postulates together with the underlying logic, the value of the representing form of **A** either is t for every model of the postulates or is f for every model of the postulates.

The notion of completeness as to consequences, as thus defined, is not wholly free of dependence on the choice of the underlying logic. But such independence of the underlying logic is possessed by still a different completeness notion for postulate systems, namely, that of categoricalness, due to Huntington and Veblen,[541] which we go on to define.

We consider only the case that the undefined terms belong to the notation of a functional calculus of first or second order, or of a functional calculus of first order with equality, i.e., the undefined terms are individual constants or functional constants in the sense of these calculi. However, the extension to higher cases is straightforward (compare footnote 530).

Two models of a system of postulates are said to be *isomorphic* if there is a one-to-one correspondence between the two domains of individuals used in the two models[542] such that the values given in the two models to any

---

[541]E. V. Huntington, *Transactions of the American Mathematical Society*, vol. 3 (1902), see pp. 264, 277–278, 281, 283–284; Oswald Veblen, ibid., vol. 5 (1904), see pp. 346–347. Compare further the remarks of Huntington, ibid., vol. 6 (1905), pp. 209–210. The term *categorical* (now the usual one) appears first in the paper of Veblen, who credits the suggestion of it to John Dewey.

Though the formulation of the idea of categoricalness as a concept applicable to postulate systems generally seems to have been made first by Huntington and Veblen, results are found in the literature much earlier which are tantamount to the categoricalness of particular systems of postulates. Thus paragraph 134 of Dedekind's *Was Sind und was Sollen die Zahlen?* (cf. footnote 525) contains the essentials of the usual proof of categoricalness of Peano's postulates, similar to that which is described in exercise 55.15 below. And the result established by Georg Cantor in the *Mathematische Annalen*, vol. 46 (1895), pp. 510–512, is in effect that a certain system of postulates—his well-known characterization of the continuum—is categorical. (Dedekind speaks of "Bedingungen" and Cantor of "Merkmale," rather than of postulates or axioms.)

[542]Of course it is not excluded as a special case that the two domains of individuals may be the same, in which case the one-to-one correspondence required is some one-to-one correspondence of that domain of individuals on to itself. (We assume the term "one-to-one correspondence" to be familiar to the reader, but an explanation of it may be found, if needed, in footnotes 556, 564.)

particular free variable occurring in the representing forms of the postulates always correspond to each other according to this one-to-one correspondence. I.e., if in the first model the value a is given to an individual variable **a**, and in the second model the value a' is given to **a**, then a must correspond to a' in the one-to-one correspondence between the two domains of individuals; and if in the first model the value Φ is given to an n-ary functional variable **f**, while in the second model the value Φ' is given to **f**, then the propositional functions Φ and Φ' must be so related that, whenever the individuals $a_1, a_2, \ldots, a_n$ of the first domain of individuals correspond in order to the individuals $a_1', a_2', \ldots, a_n'$ of the second domain, the value $\Phi(a_1, a_2, \ldots, a_n)$ is the same as the value $\Phi'(a_1', a_2', \ldots, a_n')$.

Then a system of postulates is said to be *categorical* if all its models are isomorphic.

We leave to the reader the proof of the following metatheorems, which state some obvious connections among the three notions of completeness of a postulate system:

**554. Every system of postulates complete as to provability is complete as to consequences.

**555. Every categorical system of postulates is complete as to consequences.

**556. If a categorical system of postulates has a model $\mathfrak{M}$, then every system of postulates with the same undefined terms and the same underlying logic, if it is complete as to consequences and has the model $\mathfrak{M}$, must be categorical.

Finally, before concluding this section, we have to consider one other way in which postulates are often used. Namely, instead of serving as basis for a special branch of mathematics, a system of postulates may be used in the course of the development of some more general mathematical theory, in the role of a definition of some particular kind of structure which is to be considered in the context of the more general theory.

As an illustration we may take the case of postulates for an *integral domain.*

One system of postulates for an integral domain may be obtained from the postulates (A$_1$) by omitting the postulate of mathematical induction, changing the fifth postulate to $(\exists x)\Sigma(x, y, z)$, changing the twelfth postulate to $(\exists x)(\exists y) \sim . x = y$ and adding the postulate $x_1 = x_2 \supset . F(x_1) \supset F(x_2)$. The fourth postulate then becomes non-independent and may be omitted.

Thus, retaining the same definition schemata that were used in connection with (A_0), we have the following system of twelve postulates, which we shall call (ID):

$$(\exists z)\Sigma(x, y, z)$$

$$\Sigma(x_1, x_2, y_1) \supset . \Sigma(x_2, x_3, y_2) \supset . \Sigma(y_1, x_3, z) \supset \Sigma(x_1, y_2, z)$$

$$\Sigma(x, y, z) \supset \Sigma(y, x, z)$$

$$(\exists x)\Sigma(x, y, z)$$

$$(\exists z)\Pi(x, y, z)$$

$$\Pi(x_1, x_2, y_1) \supset . \Pi(x_2, x_3, y_2) \supset . \Pi(y_1, x_3, z) \supset \Pi(x_1, y_2, z)$$

$$\Pi(x, y, z) \supset \Pi(y, x, z)$$

$$\Pi(x_1, y, z) \supset . \Pi(x_2, y, z) \supset Z_0(y) \vee x_1 = x_2$$

$$(\exists x)\Pi(x, y, y)$$

$$\Pi(x_1, x_2, y_2) \supset . \Pi(x_1, x_3, y_3) \supset . \Sigma(x_2, x_3, y_1) \supset . \Sigma(y_2, y_3, z) \supset \Pi(x_1, y_1, z)$$

$$(\exists x)(\exists y) \sim . x = y \quad .$$

$$x_1 = x_2 \supset . F(x_1) \supset F(x_2)$$

When these postulates are used, not as basis for their own branch of mathematics, but in order to introduce in the context of a more general theory the term "integral domain," or a notation serving the same purpose, they must be rewritten in the form of a definition schema. It would usually be necessary to be able to speak not only of the individuals as forming an integral domain with respect to a pair of operations (in the roles of addition and multiplication) but also of any class of individuals as forming an integral domain. Thus the definition schema must introduce a notation, say id(**f**, **g**, **h**), in which **f** and **g** are ternary functional variables and **h** is a singulary functional variable.

And for values Φ, Ψ, and Θ of **f**, **g**, and **h**, it must be possible to understand id(**f**, **g**, **h**) as expressing that Θ is an integral domain with respect to Φ and Ψ, in the usual sense of those words in informal treatments of algebra.

As derived from the particular system of postulates (ID), this definition schema for id(**f**, **g**, **h**) is the following:[543]

[543]The definiens has an evident relationship to the representing form of the conjunction of the postulates (ID), but differs in several ways, in particular it has one more free functional variable. As here given, it does not parallel the postulates quite perfectly, some obvious simplifications by P and F^1 having been made.

D24. $\mathrm{id}(\mathbf{f}, \mathbf{g}, \mathbf{h}) \rightarrow \mathbf{h}(x) \supset_x \centerdot \mathbf{h}(y) \supset_y \centerdot \mathbf{h}(z) \supset_z \centerdot \mathbf{h}(x_1) \supset_{x_1} \centerdot \mathbf{h}(y_1) \supset_{y_1} \centerdot$
$\mathbf{h}(x_2) \supset_{x_2} \centerdot \mathbf{h}(y_2) \supset_{y_2} \centerdot \mathbf{h}(x_3) \supset_{x_3} \centerdot \mathbf{h}(y_3) \supset_{y_3} \centerdot (\exists z)[\mathbf{h}(z)\mathbf{f}(x, y, z)]$
$[\mathbf{f}(x_1, x_2, y_1) \supset \centerdot \mathbf{f}(x_2, x_3, y_2) \supset \centerdot \mathbf{f}(y_1, x_3, z) \supset \mathbf{f}(x_1, y_2, z)]\ [\mathbf{f}(x, y, z) \supset$
$\mathbf{f}(y, x, z)]\ (\exists x)[\mathbf{h}(x)\mathbf{f}(x, y, z)]\ (\exists z)[\mathbf{h}(z)\mathbf{g}(x, y, z)]\ [\mathbf{g}(x_1, x_2, y_1) \supset \centerdot$
$\mathbf{g}(x_2, x_3, y_2) \supset \centerdot \mathbf{g}(y_1, x_3, z) \supset \mathbf{g}(x_1, y_2, z)]\ [\mathbf{g}(x, y, z) \supset \mathbf{g}(y, x, z)]$
$[\mathbf{g}(x_1, y, z) \supset \centerdot \mathbf{g}(x_2, y, z) \supset \centerdot [\mathbf{h}(z) \supset_z \mathbf{f}(y, z, z)] \vee \centerdot \mathbf{h}(x) \supset_x \centerdot$
$\mathbf{h}(y) \supset_y \centerdot \mathbf{f}(x_1, x, y) \supset \mathbf{f}(x_2, x, y)]\ (\exists x)[\mathbf{h}(x)\mathbf{g}(x, y, y)]$
$[\mathbf{g}(x_1, x_2, y_2) \supset \centerdot \mathbf{g}(x_1, x_3, y_3) \supset \centerdot \mathbf{f}(x_2, x_3, y_1) \supset \centerdot \mathbf{f}(y_2, y_3, z) \supset$
$\mathbf{g}(x_1, y_1, z)]\ (\exists x)(\exists y)(\exists z)(\exists x_1)[\mathbf{h}(x)\mathbf{h}(y)\mathbf{h}(z)\mathbf{h}(x_1)\mathbf{f}(x, z, x_1){\sim}\mathbf{f}(y, z, x_1)]$
$\centerdot [\mathbf{h}(x) \supset_x \centerdot \mathbf{h}(y) \supset_y \centerdot \mathbf{f}(x_1, x, y) \supset \mathbf{f}(x_2, x, y)] \supset \centerdot F(x_1) \supset_F F(x_2)$

In many informal treatments of abstract algebra in the literature, systems of postulates for a group, a ring, an integral domain, a field, etc. enter in this way—in the role not of axioms but of definitions which, in a corresponding formalized treatment, would appear as definition schemata analogous to D24. And the formalization of such a treatment of abstract algebra would then be a development within a pure functional calculus of second order, say F_2^{2p}, with perhaps the axiom of well-ordering of the individuals and an axiom of infinity (see §§56, 57), one or both, as added axioms. Or it may well be necessary for the sake of some parts of the development to use a functional calculus of higher order than the second—this will depend on just what the content of abstract algebra is conceived to be. In any case, abstract algebra is thus formalized within one of the pure functional calculi, and in this sense we may say if we like that it has been reduced to a branch of pure logic.

Many other branches of mathematics are customarily treated in a similar way, so that their formalization brings them entirely within one of the pure functional calculi. And though it is more natural or more usual in some cases than others, it seems clear that every branch of mathematics might be treated in this way if we chose. For example, instead of deriving elementary number theory from the postulates (A_1) in the role of axioms added to an underlying logic, we might transform these postulates into a definition of the term "an arithmetic" (in the formalized treatment, a definition schema), and then re-state and re-prove all the usual theorems of elementary number theory as general theorems about "an arithmetic."[544]

Thus it is possible to say that all of mathematics is reducible to pure logic, and to maintain that logic and mathematics should be characterized, not as different subjects, but as elementary and advanced parts of the same subject.[545]

[544]As long as it is desired only to reproduce (in this sense) the theorems of A^1 within F_2^{2p}, no added axioms are necessary. Also the theorem expressing that *there exist at most one arithmetic* (to within a one-to-one correspondence) requires no added axioms. But for some other theorems, in particular for the theorem expressing that *there exist at least one arithmetic*, an axiom of infinity and perhaps also the axiom of well-ordering of the individuals will be necessary.

[545]There is also another sense (that of Frege and Russell) in which it is often maintained that mathematics is reducible to logic. This is reserved for discussion in a later section. But in the meantime it should be remarked that the issue is at least partly one

EXERCISES 55

55.0. Extend the principles of duality, *372–*374, to a logistic system obtained by adding arbitrary postulates to an applied functional calculus of second order as underlying logic. Carry out the proof in such a way as to include as a special case a proof of the principles of duality for the pure functional calculus of second order.

55.1. Prove the following as theorems of the logistic system A^0 without making use of any of the postulates (A_0) (thus also as theorems of F^{1h}):

(1) $$x = x$$

(2) $$x = y \supset . y = z \supset . x = z$$

(3) $$x = y \supset . Z_0(x) \supset Z_0(y)$$

55.2. Prove the following as theorems of the logistic system A^0, using only the first four of the postulates (A_0):

(1) $$Z_0(x) \supset . Z_0(y) \supset . x = y$$

(2) $$\Sigma(x, y, z_1) \supset . \Sigma(x, y, z_2) \supset . z_1 = z_2$$

(3) $$x = y \supset . y = x$$

55.3. Prove the following (in order) as theorems of the logistic system A^0, using only the first five of the postulates (A_0):

(1) $$(\exists x) Z_0(x)$$

(2) $$z_1 = z_2 \supset . \Sigma(x, y, z_2) \supset \Sigma(x, y, z_1)$$

55.4. With the aid of the results of preceding exercises (if needed), prove the following as a theorem of A^0, using only the first nine of the postulates (A_0):

$$\Pi(x, y, z_1) \supset . \Pi(x, y, z_2) \supset . z_1 = z_2$$

55.5. With the aid of the results of preceding exercises, prove the following as theorems of A^0, using only the first eleven of the postulates (A_0):

(1) $$Z_0(y) \supset \Pi(x, y, y)$$

(2) $$z_1 = z_2 \supset . \Pi(x, y, z_2) \supset \Pi(x, y, z_1)$$

(3) $$y_1 = y_2 \supset . \Pi(x, y_2, z) \supset \Pi(x, y_1, z)$$

(4) $$Z_0(z) \supset . \Pi(x, y, z) \supset Z_0(x) \vee Z_0(y)$$

(5) $$(\exists x) Z_1(x)$$

---

of decision as to terminology. It is also possible to hold, for example, that an axiom of infinity is outside the province of logic, and that logic ends and mathematics begins as soon as such an axiom is added.

55.6. Prove the following theorems of the logistic system which is obtained by adding the first eleven of the postulates (ID) to F^{1h} as underlying logic:

(1) $$\Sigma(x, y, z_1) \supset . \Sigma(x, y, z_2) \supset . z_1 = z_2$$

(2) $$\Sigma(x_1, y, z) \supset . \Sigma(x_2, y, z) \supset . x_1 = x_2$$

(3) $$Z_0(x) \supset . Z_1(y) \supset \sim . x = y$$

55.7. Show that every theorem of A^1 which contains no functional variables is also a theorem of A^0.

55.8. In the system of postulates (A_1), establish the independence of the seventh postulate (the associative law of multiplication) by means of the following independence example. The individuals are the four natural numbers 0, 1, 2, 3, and addition and multiplication are as given in the following tables:

+	0	1	2	3
0	0	1	2	3
1	1	0	3	2
2	2	3	0	1
3	3	2	1	0

×	0	1	2	3
0	0	0	0	0
1	0	1	3	2
2	0	3	2	1
3	0	2	1	3

(I.e., more explicitly, in the representing forms of the postulates the functional variables corresponding to Σ and Π are F^3 and H^3 respectively; and in the model which constitutes the independence example, the value of F^3 is the propositional function Φ such that $\Phi(a, b, c)$ is t if and only if $a + b = c$ according to the first of the above tables, and the value of H^3 is the propositional function Ψ such that $\Psi(a, b, c)$ is t if and only if $a \times b = c$ according to the second of the above tables.)

55.9. In the system of postulates (A_1), establish the independence of the eighth postulate (the commutative law of multiplication) by means of the following independence example. The individuals are 0 and all the complex numbers $\alpha + \beta i$ in which α and β are positive rational numbers. The sum is taken in the usual way. A product is 0 if either factor is 0, and otherwise $(\alpha + \beta i) \times (\gamma + \delta i) = \alpha\gamma + \beta\gamma i$ (the products $\alpha\gamma$ and $\beta\gamma$ being taken in the usual way).

55.10. Establish the independence of the remaining postulates of (A_1) by means of independence examples.

55.11. Establish the independence of the postulates of (ID) by means of independence examples.

55.12. Express by means of a wff of A^1 that the only ternary relation, among natural numbers a, b, c, that satisfies the second pair of recursion equations of footnote 526 and the further condition that c is uniquely determined when a and b are given is the ternary relation Φ such that $\Phi(a, b, c)$ holds when and only when $a \times b = c$ (in the sense that any ternary relation among natural numbers satisfying the two recursion equations and the further condition is formally equivalent to Φ).

55.13. Prove the wff of the preceding exercise as a theorem of A^1. (Make use of the postulate of mathematical induction.)

55.14. For A^1, suppose that the signs 0 and 1 are introduced by contextual definition, according to the following definition schema. Of the signs $a_1, a_2, \ldots, a_n$, let some (possibly) be 0's, let others (possibly) be 1's, and let the remainder be individual variables (not necessarily all different); let $\mathbf{f}$ be any n-ary functional variable or functional constant, let $\mathbf{x}$ and $\mathbf{y}$ be the first two (distinct) individual variables in alphabetic order that do not occur among $a_1, a_2, \ldots, a_n$, and let $\mathbf{b}_1, \mathbf{b}_2, \ldots, \mathbf{b}_n$ be obtained from $a_1, a_2, \ldots, a_n$ by replacing the sign 0 everywhere by $\mathbf{x}$ and the sign 1 everywhere by $\mathbf{y}$; then

$$\mathbf{f}(a_1, a_2, \ldots, a_n) \rightarrow (\exists \mathbf{x})(\exists \mathbf{y})[Z_0(\mathbf{x})Z_1(\mathbf{y})\mathbf{f}(\mathbf{b}_1, \mathbf{b}_2, \ldots, \mathbf{b}_n)].$$

For expressions which abbreviate wffs of A^1 according to this definition schema, establish as a derived rule a rule of substitution for individual variables, allowing to be substituted for an individual variable not only another individual variable but also one of the signs 0, 1.[546]

55.15. Prove that the system of postulates (A_1) is categorical. (*Suggestion*: In one model let Φ and Ψ be the values of the functional variables that correspond to Σ and Π respectively, and in a second model let Φ' and Ψ' be the values of the functional variables that correspond to Σ and Π respectively. In the first domain of individuals there must be two unique individuals 0 and 1, distinct from each other, such that $\Phi(0, b, b)$ and $\Psi(1, b, b)$ hold for all individuals b of the domain. In the second domain of individuals there must be two unique individuals 0′ and 1′, distinct from each other, such that $\Phi'(0', b, b)$ and $\Psi'(1', b, b)$ hold for all individuals b of the domain. The required one-to-one correspondence between the two domains is that in which 0 corresponds to 0′, and 1 corresponds to 1′, and

---

[546]Compare footnote 528.

This definition schema may be thought of as a modified form of a special case of Russell's contextual definition of *descriptions*, i.e., of his schema for contextual definition of the notation $(\iota x)\mathbf{A}$; see the *American Journal of Mathematics*, vol. 30 (1908), p. 253. In this special case, following the remark of Herbrand in *Comptes Rendus des Séances de la Société des Sciences et des Lettres de Varsovie*, Classe III, vol. 24 (1931), p. 33, we are able to simplify the definitions (contained in the schema) by taking advantage of the theorems, $Z_0(x) \supset . Z_0(y) \supset x = y$ and $Z_1(x) \supset . Z_1(y) \supset x = y$, of A^1.

whenever a corresponds to a', and $\Phi(a, 1, c)$ and $\Phi'(a', 1', c')$ hold in the respective domains, then c corresponds to c'. The proof proceeds in the meta-language by the method of mathematical induction.)

55.16. Hence show that the postulates (A_0) with F^{1h} as underlying logic are complete as to consequences.

55.17. Show that the postulates (A_0) are not categorical because, besides the obvious model with the natural numbers as the individuals, there is also the following model. The individuals are the positive and negative integers and 0. The value of the functional variable corresponding to Σ is the ternary relation that holds among a, b, c if and only if $|a| + |b| = |c|$. And the value of the functional variable corresponding to Π is the ternary relation that holds among a, b, c if and only if $|ab| = |c|$.

55.18. The non-categoricalness of the postulates (A_0) as established in the preceding exercise may be thought to be of relatively trivial character, since there does exist a one-many correspondence between the domains of individuals of the two models such that the values of the functional variables corresponding to Σ, and to Π, in the two models correspond to each other according to the one-many correspondence. The second model could moreover be excluded by taking an appropriate simple applied functional calculus of first order with equality as underlying logic, and replacing "$x_1 = x_2$" in the fourth and ninth postulates by "$I(x_1, x_2)$". Let the system of postulates so obtained from (A_0) be called (A_I), and let the logistic system obtained by adding them to the appropriate simple applied functional calculus of first order with equality be called A^I. (1) Show that the postulates (A_I) are complete as to consequences. (2) By means of the metatheorems of exercise 48.22, establish the non-categoricalness of (A_I), and hence also the non-categoricalness of (A_0) in a less trivial sense.[547]

55.19. Let V be a binary functional constant, and consider the system of postulates consisting of the single postulate $(x)(y)V(x, y)$, added to a simple applied functional calculus of first order, having V as its one functional constant, as underlying logic. Show that this system of postulates is complete as to consequences but not categorical.

55.20. Show that in every model of the following system of postulates the domain of individuals is finite, but that there exist models with an arbitrarily large finite domain of individuals. There is one undefined term, a binary functional constant S. The underlying logic is an applied functional

---

[547]This is a special case of the result of Skolem stated in footnote 452. His proof, by a different method from that suggested here, is given in *Fundamenta Mathematicae*, vol. 23 (1934), pp. 150–161.

calculus of first order with equality, having among its primitive symbols all propositional and functional variables, and S as its one functional constant. The postulates are:

$$S(x, y) \supset . S(x, z) \supset . y = z$$
$$F(x) \supset . F(y) \supset_y [S(y, z) \supset_z F(z)] \supset (y)F(y)$$

55.21. To the postulates of the preceding exercise let the following infinite list of postulates be added:

$$(\exists y)S(x, y)$$
$$S(x_1, x_2) \supset . x_1 \neq x_2$$
$$S(x_1, x_2) \supset . S(x_2, x_3) \supset . x_1 \neq x_3$$
$$S(x_1, x_2) \supset . S(x_2, x_3) \supset . S(x_3, x_4) \supset . x_1 \neq x_4$$
$$S(x_1, x_2) \supset . S(x_2, x_3) \supset . S(x_3, x_4) \supset . S(x_4, x_5) \supset . x_1 \neq x_5$$
$$\cdots\cdots\cdots\cdots\cdots\cdots$$

Show that the resulting system of postulates is consistent as to provability but not consistent as to consequences.[548]

55.22. The following are informally stated postulates for *partial* order, with a relation *precedes* as the one undefined term:[549]

No individual precedes itself.
If a precedes b and b precedes c, then a precedes c.

From these a system of postulates for *simple order* is obtained by adding the following third postulate:[550]

---

[548]This is an adaptation of an example due to Tarski—see *Monatshefte für Mathematik und Physik*, vol. 40 (1933), pp. 97–112. By making use of the Gödel incompleteness theorems (to be treated in a later chapter), it is also possible to find a finite system of postulates which is consistent as to provability without being consistent as to consequences.

[549]The name "partially ordered class" is taken from the German "teilweise geordnete Menge" of Felix Hausdorff's *Grundzüge der Mengenlehre* (Leipzig, 1914), p. 139, where the general notion of partial order (as distinguished from the treatment of particular cases of it) seems to have been first introduced.

[550]This definition of simple order should perhaps be credited to C. S. Peirce, who, in the *American Journal of Mathematics*, vol. 4 (1881), p. 86, gives a closely related definition, in terms of a relation analogous to $\leqq$ (rather than to $<$ as in the exercise above).

A definition of simple order in terms of the relation *precedes* (analogous to $<$) is given by Const. Gutberlet in the *Zeitschrift für Philosophie und Philosophische Kritik*, new series, vol. 88 (1886), pp. 183–184. The same definition is used also by Georg Cantor in the *Mathematische Annalen*, vol. 46 (1895), p. 496 (or see his *Gesammelte Abhandlungen*, p. 296); and it is probable that Gutberlet may have taken the definition from a manuscript of Cantor (see *Gesammelte Abhandlungen*, pp. 388, 482–483), though his own statement about the matter is not entirely clear. Both Gutberlet and Cantor state explicitly only the last two of the three postulates given above, but the additional condition that no element precedes itself is tacitly intended, at least by Cantor, as is

If a and b are any two different individuals, either a precedes b or b precedes a.

From these in turn a system of postulates for *well-ordering* is obtained by adding the fourth postulate:[551]

In any non-empty class of individuals there is a first individual, i.e., an individual that precedes all the others in the class.

(1) With a binary functional constant R denoting the relation of preceding, restate these postulates in the notation of an appropriate functional calculus of first order. (2) Hence, by the method which is used in the text to transform the postulates (ID) into the definition schema D24 for id(**f**, **g**, **h**), find expressions for each of the following, in the notation of the pure functional calculus of second order: *the class* Ψ (of individuals) *is partially ordered by the relation* Φ; *the class* Ψ *is simply ordered by the relation* Φ; *the class* Ψ *is well-ordered by the relation* Φ.

55.23. In the case of each of the following systems of postulates found in the literature, restate the postulates in the notation of an appropriate functional calculus (of not higher than second order), using the indicated functional constants as the undefined terms:

(1) Postulates for Euclidean plane geometry. Veblen and Young, *Projective Geometry*, Volume 2, §66, pp. 144–146. O, denoting the ternary relation among A, B, C, that A, B, C are in the order $\{ABC\}$; C, denoting the quaternary relation among A, B, C, D, that AB is congruent to CD. (Omit the continuity postulate, XVII. In stating the postulate XVI, use may be made of the postulates (A_2), as they are stated above, but modified as required, in particular by replacing the functional constant S by a binary functional variable.)

(2) The same postulates with the following continuity postulate added: If K is a non-empty class of points of a line a, if B and C are points of a such that every point X of K is in the order $\{XBC\}$, there is a point A of a such that every point X of K distinct from A is in the order $\{XAC\}$, and no point Z of a in the order $\{ZAC\}$ has the property that every point X of K is in the order $\{XZC\}$.

(3) Postulates for (real) projective plane geometry. H. S. M. Coxeter, *The Real Projective Plane*, 2.21–2.25 (p. 12), 3.11–3.16 (p. 22), and 10.11 (p. 138). P, denoting the class of points; L, denoting the class of lines; I, denoting the

---

clear from *Mathematische Annalen*, vol. 49 (1897), p. 216 (or *Gesammelte Abhandlungen*, p. 321).

The condition that no element precedes itself is of course replaceable by the condition that not both x precedes y and y precedes x. In this form the three postulates are given explicitly by B. I. Gilman (a student of Peirce) in *Mind*, n.s., vol. 1 (1892), pp. 518–526; and by Giovanni Vailati in *Rivista di Matematica*, vol. 2 (1892), p. 73.

[551]The notion of a well-ordered class is due to Georg Cantor in *Grundlagen einer Allgemeinen Mannigfaltigkeitslehre*, Leipzig, 1883, p. 4 (or *Mathematische Annalen*, vol. 21 (1883), p. 548, or *Acta Mathematica*, vol. 2 (1883), p. 393, or *Gesammelte Abhandlungen*, p. 168). Cantor's definition of well-ordering is somewhat different from, but equivalent to, what is now the usual definition by means of the fourth postulate above. Moreover Cantor at first merely presupposed the notion of simple order in giving the definition of well-ordering. But a definition of simple order was supplied in 1895, as explained in the preceding footnote.

relation of incidence; S, denoting the quaternary relation of separation.[552]

(4) Postulates for Euclidean three-dimensional geometry. David Hilbert, *Grundlagen der Geometrie*, seventh edition (1930), §§1–8. O, C, P, L, and I as in parts (1) and (3); π, denoting the class of planes; ι, denoting the relation of incidence between points and planes; K, denoting the senary relation among A, B, C, A', B', C', that the angle ABC is congruent to the angle $A'B'C'$. Special attention must be given to the postulate of linear completeness ("Axiom der linearen Vollständigkeit"), whose expression in the notation of a functional calculus of no higher than second order offers some difficulty, and of which some restatement or modification may be necessary in order to render such expression possible.

(5) Postulates for (real) projective three-dimensional geometry. Mario Pieri, *Memorie della Reale Accademia delle Scienze di Torino*, ser. 2 vol. 48 (1899), pp. 1–56. J, denoting the ternary relation among a, b, c, that c is on the straight line joining a and b.

(6) E. V. Huntington's postulates 1–14 for "the theory of real quantities," *Transactions of the American Mathematical Society*, vol. 4 (1903), pp. 358–370.

(7) Church's postulates for "the second ordinal class," or second number class, of Cantor, *Transactions of the American Mathematical Society*, vol. 29 (1927), p. 179.

(8) A. Lindenbaum's postulates for a metric space,[553] *Fundamenta Mathematicae*, vol. 8 (1926), p. 211; given also by C. Kuratowski, *Topologie I*, first edition (1933), pp. 82–83, or second edition (1948), p. 99. (For the introduction of the notion of real number, make use of the postulates of part (6) of this exercise, or of some other system of postulates serving the same purpose.)

(9) Postulates for a complete space,[553] obtained from the foregoing by adding the postulate that is given by Kuratowski, *Topologie I*, first edition, p. 196, or second edition, p. 312.

55.24. In a *many-sorted functional calculus*[554] (of first or higher order) there are individual variables of more than one sort, the different sorts being distinguished by superscripts, and an infinite list of individual variables of each sort being available. Say in an n-sorted functional calculus the individual variables of the first sort are $x^1, y^1, z^1, x_1^1, \ldots$; those of the second sort are $x^2, y^2, z^2, x_1^2 \ldots$; and so on, up to $x^n, y^n, z^n, x_1^n, \ldots$ as individual variables of the nth sort. There

[552]As in part (1), the notion of an infinite sequence of individuals which enters in these postulates may be provided for by making use of the postulates (A_2) appropriately modified. (Compare the procedure in the text in transforming the postulates (ID) for an integral domain into a propositional form with three free variables, expressing that a class is an integral domain with respect to two ternary relations.)

[553]The notion of a metric space and that of a complete (or complete metric) space are due to Maurice Fréchet, though in a different terminology. See his thesis in the *Rendiconti del Circolo Matematico di Palermo*, vol. 22 (1906), pp. 1–74, and a paper in the *Transactions of the American Mathematical Society*, vol. 19 (1918), pp. 53–65.

[554]See a paper by Arnold Schmidt in the *Mathematische Annalen*, vol. 115 (1938), pp. 485–506. (*Added in proof*. See also improved treatments of the same topic by Arnold Schmidt in the *Mathematische Annalen*, vol. 123 (1951), pp. 187–200, and by Hao Wang in *The Journal of Symbolic Logic*, vol. 17 (1952), pp. 105–116.)

are (or may be) then n sorts of singulary functional variables, again distinguished by superscripts, an infinite list of each sort; thus F^1, G^1, H^1, F^1_1, . . . as singulary functional variables of the first sort, F^2, G^2, H^2, F^2_1, . . . as singulary functional variables of the second sort, and so on. And where **a** is an individual variable and **f** is a singulary functional variable, **f**(**a**) is wf if and only if **f** and **a** are of the same sort. There are (or may be) n^2 sorts of binary functional variables, distinguished by superscripts thus: $F^{1,1}$, $G^{1,1}$, $H^{1,1}$, $F^{1,1}_1$, . . . ; $F^{1,2}$, $G^{1,2}$, $H^{1,2}$, $F^{1,2}_1$; and so on. And, for example, $F^{2,1}(\mathbf{a}_1, \mathbf{a}_2)$ is wf if and only if $\mathbf{a}_1$ is an individual variable of the second sort and $\mathbf{a}_2$ is an individual variable of the first sort. Similarly there are (or may be) n^3 sorts of ternary functional variables, and so on.

For the principal interpretation of an n-sorted functional calculus, there must be a non-empty domain of individuals of the first sort, which is the range of the individual variables of the first sort; a non-empty domain of individuals of the second sort, which is the range of the individual variables of the second sort; and so on (n domains of individuals altogether). The various functional variables then have ranges consisting of propositional functions in a way which will be obvious by analogy with the principal interpretations already given (in Chapters III and V) for the one-sorted functional calculi of first and second order.

For an applied (as distinguished from a pure) n-sorted functional calculus there may also be individual and functional constants, each of which must belong to a particular sort in the same way as the individual and functional variables.

Taking the remaining primitive symbols to be the eight improper symbols listed at the beginning of §30, we may use for an n-sorted functional calculus of first order the formation rules of §30, except that 30ii is modified in the way indicated in the first paragraph of this exercise; the rules of inference and axiom schemata may then be the same as in §30, except that to *306 the requirement is added that **b** must be of the same sort as **a**. An n-sorted functional calculus of second order may be formulated similarly, with appropriate provision added for quantification of functional variables.

(1) State in full a primitive basis for an n-sorted functional calculus of second order, as closely as possible analogous to the primitive basis for F^2_2 given in §50. For a fixed k, show that those theorems of the system which contain individual variables of only the kth sort are (apart from trivial notational differences) the same as the theorems of F^2_2.

(2) With a two-sorted functional calculus of second order as underlying logic, taking the individuals of the first sort to be the points, and the individuals of the second sort to be the lines, and taking as undefined terms a binary functional constant I denoting the relation of incidence and a quaternary functional constant S denoting the relation of separation, state Coxeter's postulates for projective plane geometry (see 55.23(3)) in this notation.

(3) Similarly, state Hilbert's postulates for Euclidean three-dimensional geometry (see 55.23(4)), with a three-sorted functional calculus of second order as underlying logic, the three sorts of individuals being the points, the lines, the planes.

(4) Similarly, state the postulates for a metric space (55.23(8)) with a two-sorted functional calculus of second order as underlying logic, the two sorts of individuals being the points of the space and the real numbers.

(5) The logistic system of part (2) of this exercise (i.e., the logistic system obtained by adding the indicated postulates to the underlying logic) is in an appropriate sense equivalent to the logistic system of 55.23(3). In a like sense, the logistic system of part (3) is equivalent to that of 55.23(4), and the logistic system of part (4) is equivalent to that of 55.23(8). Explain in what sense the equivalence holds. And state and prove a general metatheorem establishing the appropriate equivalence in all such cases. (*Cf.* the papers of footnote 554.)

56. Well-ordering of the individuals. Returning to consideration of the pure functional calculus of second order $\mathrm{F}_2^{2\mathrm{p}}$, we now take up the question of axioms expressible in the notation of the pure functional calculus of second order, alone, which—for some purposes or in some connections—it may be desirable to adjoin to $\mathrm{F}_2^{2\mathrm{p}}$ as additional axioms.

One such axiom, the possible addition of which to $\mathrm{F}_2^{2\mathrm{p}}$ we shall wish to consider, is an axiom to the effect that the individuals can be well-ordered.

In order to express this, we may make use of the definition of well-ordering which was given in 55.22, writing the conjunction of the universal closures of the four postulates of 55.22, replacing the undefined term "precedes" everywhere by the functional variable F^2, and then prefixing the existential quantifier $(\exists F^2)$ to this conjunction. The resulting expression may be simplified, however, by omitting the third postulate, which can be shown to be non-independent. Thus we obtain the axiom (w) which is written below.

The *axioms of choice* are reserved for discussion in connection with the functional calculi of higher order, although certain special cases of an axiom of choice can be stated already in the notation of the functional calculus of second order and summarized in an axiom schema.[555] We anticipate this discussion here so far as to say that it will follow, from the axioms of choice, not only that the individuals can be well-ordered but also various higher domains—in particular that the singulary propositional functions (classes) of individuals can be well-ordered, the binary propositional functions of individuals, and so on—and conversely that the axioms of choice will follow from such assumptions of well-ordering.

However, our present axiom (w) must not be considered as representing a special case or a weak form of an axiom of choice. For the effect when we add it as an axiom to $\mathrm{F}_2^{2\mathrm{p}}$ is just that we restrict the interpretation to such domains of individuals as are capable of being well-ordered, a procedure which should be acceptable even to those who distrust or prefer not to assume any axiom of choice.

---

[555]The axiom schema in question is

$$(\mathbf{x})(\exists \mathbf{f})\mathbf{A} \supset (\exists \mathbf{g})(\mathbf{x})\check{S}^{\mathbf{f}(\mathbf{x}_1, \mathbf{x}_2, \ldots, \mathbf{x}_n)}_{\mathbf{g}(\mathbf{x}, \mathbf{x}_1, \mathbf{x}_2, \ldots, \mathbf{x}_n)}\mathbf{A}|$$

where $\mathbf{x}, \mathbf{x}_1, \mathbf{x}_2, \ldots, \mathbf{x}_n$ are distinct individual variables, $\mathbf{f}$ is an n-ary functional variable, $\mathbf{g}$ is an $(n+1)$-ary functional variable, and $\mathbf{A}$ is a wff containing no bound occurrences of either $\mathbf{g}$ or $\mathbf{x}$. It is given by Hilbert and Ackermann, *Grundzüge der theoretischen Logik*, second edition (1938), p. 104, and third edition (1949), p. 111; also in the paper of Ackermann mentioned in the last paragraph of footnote 507.

Thus the following *axiom of well-ordering of the individuals*—or axiom (w), as we shall also call it—is to be considered as a possible added axiom:

$$(\exists F) \mathbin{.} (x) \sim F(x, x) \mathbin{.} F(x, y) \supset_{xy} [F(y, z) \supset_z F(x, z)] \mathbin{.} G(x) \supset_{Gx} (\exists y) \mathbin{.} G(y) \mathbin{.} G(z) \supset_z F(y, z) \vee y = z$$

Following a method of naming that we adopt as systematic, we call the resulting logistic system $F_2^{2p(w)}$ when the axiom (w) is added to the logistic system F_2^{2p}.

EXERCISES 56

56.0. Restate (w) as an equivalent axiom in prenex normal form, with only four different individual variables, one singulary functional variable, and one binary functional variable.

56.1. Prove the statement made in the text that the third of the four postulates for well-ordering (55.22) is non-independent.

56.2. State and prove as a theorem of $F_2^{2p(w)}$ that the individuals can be simply ordered. (Use the definition of simple order given in 55.22.)

56.3. It follows from axiom (w) that every relation between individuals has a many-one subrelation with the same domain.[556] Expressed in the notation of the functional calculus of second order, this is:[557]

$$(\exists G) \mathbin{.} G(x, z) \supset_{xz} F(x, z) \mathbin{.} F(x, z) \supset_{xz} (\exists z_1) \mathbin{.} G(x, z) \equiv_z z = z_1$$

Prove this as a theorem of $F_2^{2p(w)}$.

56.4. Prove the same theorem in the logistic system that is obtained by adding to F_2^{2p} the axiom schema of footnote 555.

57. Axiom of infinity. A wff of one of the functional calculi may be considered as *an axiom of infinity* if it is valid in at least one infinite domain of individuals but is not valid in any finite domain of individuals.

Of the pure functional calculus of first order with equality, and therefore also of the pure functional calculus of first order, there is in fact no wff

---

[556]One relation is said to be a *subrelation* of a second one if it formally implies the second one, in the sense of formal implication explained in §06. A relation R is said to be *many-one* if for every member a of the domain of R there is a unique corresponding member b of the converse domain of R such that a bears the relation R to b. Moreover a relation is said to be *one-many* if its converse is many-one; and *one-to-one* if both it and its converse are many-one. (See further the explanation of terminology in footnote 517.)

[557]Compare Hilbert and Ackermann, *Grundzüge der Theoretischen Logik*, second edition (1938), formula g on page 104, and third edition (1949), formula g on page 111.

which may thus be considered an axiom of infinity.[558] Therefore the various axioms of infinity which we discuss in this section are wffs only of the pure functional calculus of second order.[559]

An effect of adjoining an axiom of infinity to F^{2p} as additional axiom is of course to restrict the interpretation to domains of individuals which are infinite. We prefer to take an axiom of infinity which imposes no great further restriction on the interpretation, beyond the exclusion of finite domains of individuals.[560]

Consider for example the wff which results when we write the conjunction of the universal closures of the postulates (A_2) of §55, replace the functional constant S everywhere by the functional variable G^2, and then prefix the existential quantifier $(\exists G^2)$ to this conjunction. As an axiom added to F_2^{2p}, this wff would restrict the domain of individuals not merely to be infinite but moreover to be enumerably infinite. This is too severe a restriction for what we regard as the purpose of an axiom of infinity. But a more acceptable axiom of infinity, namely $(\infty 3)$ (or $(\infty 4)$) below, may be obtained by treating similarly four (or three) of the five postulates (A_2).

Of various alternative axioms of infinity which we might consider adjoining to F_2^{2p} as additional axioms, we list here the five following, $(\infty 1)$–$(\infty 5)$:

$(\infty 1)$ $$(\exists F)(x_1)(x_2)(x_3)(\exists y) \mathbin{.} F(x_1, x_2) \supset [F(x_2, x_3) \supset F(x_1, x_3)] \mathbin{.} \sim F(x_1, x_1)\ F(x_1, y)$$

$(\infty 2)$ $$(\exists F)(x)(\exists y)(z) \mathbin{.} F(z, x) \supset F(z, y) \mathbin{.} \sim F(x, x)\ F(x, y)$$

$(\infty 3)$ $$(\exists F) \mathbin{.} (x)(\exists y) F(x, y) \mathbin{.} F(x, y) \supset_{xy} [F(x, z) \supset_z y = z] \mathbin{.} F(y, x) \supset_{xy} [F(z, x) \supset_z y = z] \mathbin{.} (\exists x)(y) \sim F(y, x)$$

---

[558]For an axiom of infinity that would be valid in an enumerably infinite domain of individuals, this is a corollary of exercise 48.24. The same result can be obtained for an axiom of infinity valid only in a non-enumerably infinite domain of individuals by making use of an axiom of choice in the meta-language and following the method of exercise 48.22—as was done by Leon Henkin in his dissertation of 1947 and in the paper cited in footnote 465. (Compare further footnote 451.)

[559]There are also axioms of infinity which are wffs only of functional calculi of still higher order, in particular the "Infin ax" of *Principia Mathematica* as it would be reproduced in our notation, and the three axioms of infinity that correspond to Tarski's definitions of finiteness I, II, III in *Fundamenta Mathematicae*, vol. 6 (1924), pp. 46, 93.

[560]It immediately suggests itself to introduce a more restricted notion of *an axiom of infinity*, defining an axiom of infinity (syntactically) as a wff which is valid in every infinite domain but not in any finite domain. This might indeed be done in a suitable meta-language. But from the point of view of justifying or explaining a preference for one proposed axiom of infinity over another, the effect is less satisfactory than might have been expected. For under the more restricted notion of an axiom of infinity the decision as to which of the wffs $(\infty 1)$–$(\infty 5)$ given below actually is to be classed as an axiom of infinity depends on what are taken as definitions of "infinite" and "finite" for the meta-language; and also on the axioms of the (ultimately formalized) meta-language, in particular on the presence and the form of axioms playing the role of axioms of infinity and of choice. (Compare a similar remark by Mostowski, *Comptes rendus des Séances de la Société des Sciences et des Lettres de Varsovie*, Classe III, vol. 31 (1938), p. 16.)

(∞4) $(\exists F) \centerdot (x)(\exists y)F(x, y) \centerdot F(y, x) \supset_{xy} [F(z, x) \supset_z y = z] \centerdot (\exists x)(y)\sim F(y, x)$

(∞5) $\sim(\exists G)(F) \centerdot G(x, y) \supset_{xy} [G(x, z) \supset_z y = z] \centerdot F(x) \supset_x \centerdot F(y) \supset_y [G(y, z) \supset_z F(z)] \supset (y)F(y)$

Of these, (∞1) has an obvious relationship to the example of Bernays and Schönfinkel of a wff of the pure functional calculus of first order which is satisfiable in an infinite domain of individuals but not in any finite domains;[561] (∞2), to Schütte's example of such a wff of the pure functional calculus of first order;[562] (∞3) and (∞4), to the Peano postulates, (A_2); and (∞5), to the postulates of exercise 55.20.[563]

(∞3) expresses the existence of a one-to-one correspondence between the individuals and a proper subclass of the individuals.[564] Therefore it may also be thought of as derived from the Peirce-Dedekind definition of an infinite class as one having a one-to-one correspondence with a proper subclass.[565]

(∞4) expresses the existence of a one-many correspondence of the individuals to a proper subclass of the individuals, and thus represents a modified form of the Peirce-Dedekind definition of an infinite class.

It is not to be expected that these and other axioms of infinity which we might consider will turn out all to be equivalent to one another in the sense that the (material) equivalence of any two of them is a theorem of F_2^{2p}. In fact, let **B** be called *weaker than* **A** if **A** ⊃ **B** is a theorem of F_2^{2p} but

[561]In the paper cited in footnote 481. Compare also exercise 43.5(2).

[562]In his paper cited in footnote 430. Compare also exercise 43.5(1).

[563]The axiom (∞5) of the writer's monograph of 1944 has here been simplified in accordance with a suggestion made by Paul Bernays in a letter of August 31, 1945. The idea of the axiom, that the individuals cannot be arranged in a closed cyclic order, is taken from Dedekind's second definition of finiteness, which was given in the preface to the second edition (1893) of his *Was Sind und was Sollen die Zahlen?*, and concerning which see further §7 of a paper by Alfred Tarski in *Fundamenta Mathematicae*, vol. 6 (1924), pp. 83–93, and a paper by Jean Cavaillès, ibid, vol. 19 (1932), pp.143–148.

[564]One class is said to be a *subclass* of a second one if all its members are members of the second one (compare footnote 556); and if in addition there is at least one member of the second class which is not a member of the first class, then the first class is said to be a *proper subclass* of the second one. By a *one-to-one correspondence* between two classes is meant a one-to-one relation, in the sense of footnote 556, having one class as its domain and the other class as its converse domain. (These are terms familiar in mathematical writing generally, such as we have often assumed to be known to the reader without the need for special explanation. In particular, the notion of a one-to one correspondence has been used in **439 and its proof and in the definition of categoricalness in §55, the notions of many-one and one-many correspondence in §23 and in 55.18.)

[565]C. S. Peirce, *American Journal of Mathematics*, vol. 7 (1885), p. 202; Richard Dedekind, *Was Sind und was Sollen die Zahlen?* (1888), paragraph 64. As Dedekind points out in the preface to his second edition, the one-to-one correspondence of an infinite class to a proper subclass was first exhibited by Bernard Bolzano in his *Paradoxien des Unendlichen* (1851) and was known also to Cantor in 1878, but neither of these authors has the proposal to make this the definition of an infinite class.

$\mathbf{B} \supset \mathbf{A}$ is not a theorem of F_2^{2p}. Then according to a result due to Andrzej Mostowski[566] and B. A. Trachtenbrot,[567] there is no weakest axiom of infinity, i.e., more exactly, given any axiom of infinity, there exists a weaker axiom of infinity.[568]

As regards the particular axioms of infinity, (∞1)–(∞5), some of the implications and equivalences which hold among them are indicated in the following exercises (together with similar considerations concerning a few additional axioms of infinity introduced in the exercises). These are stated in each case in the form that a particular axiom of infinity is a theorem of the logistic system obtained from F_2^{2p} by adding one of the other axioms of infinity, with or without also the axiom of well-ordering of the individuals. But in view of the deduction theorem, they could also be put (without important difference) in the form that certain implications and equivalences are theorems of F_2^{2p}.

EXERCISES 57

57.0. Restate (∞5) as an equivalent axiom in prenex normal form, with no free variables, and with the shortest prefix that can be obtained by use of propositional calculus and elementary laws of quantifiers in a straightforward process of reduction.

57.1. According to the result just stated (without proof) in the text, if $\mathbf{A}$ is any axiom of infinity there exists an axiom of infinity $\mathbf{B}$ such that $\mathbf{A} \supset \mathbf{B}$ but not $\mathbf{B} \supset \mathbf{A}$ is a theorem of F_2^{2p}. Assuming this, show that, if $\mathbf{A}$ is any axiom of infinity, there exists an axiom of infinity $\mathbf{B}$ such that $\mathbf{A} \supset \mathbf{B}$ but not $\mathbf{B} \supset \mathbf{A}$ is a theorem of $F_2^{2p(w)}$.

57.2. Restate in the notation of F_2^{2p} (as closely as possible) the following informally stated axioms of infinity:

(∞6) There is a subclass of the individuals isomorphic to the natural numbers as given by the Peano postulates.

---

[566] *Comptes Rendus des Séances de la Société des Sciences et des Lettres de Varsovie*, Classe III, vol. 31 (1938), pp. 13–20.

[567] *Doklady Akadémii Nauk SSSR*, vol. 70 (1950), pp. 569–572.

[568] Both Mostowski and Trachtenbrot deal with logistic systems different from F_2^{2p}, and they treat directly the question of a strongest definition of finiteness (of a class) rather than that of a weakest axiom of infinity. The result stated in the text is thus not explicitly contained in their papers but must be inferred from them. Both papers are moreover abstracts in which proofs are not given of the results announced, but what are perhaps sufficient indications to make possible a reconstruction of the proofs are given in Trachtenbrot's paper and in Mostowski's review of it in *The Journal of Symbolic Logic*, vol. 15 (1950), p. 229.

(∞7) If the individuals can be simply ordered, they can be put into a simple order in which there is no last individual.[569]

(∞8) There exists a one-many correspondence of a class of individuals to itself that is not a one-to-one correspondence of that class of individuals to itself.

(∞9) The individuals cannot be simply ordered in such a way that in every non-empty class of individuals there is both a first individual and a last individual.[570]

(∞10) There exist at least two different individuals, and there exists a one-many correspondence of the ordered pairs of individuals to the individuals.[571] (*Suggestion*: A correspondence between the ordered pairs of individuals and the individuals may be thought of as a ternary propositional function and thus represented by a ternary functional variable.)

57.3. Prove each of the following as a theorem of $F_2^{2p(\infty 3)}$: (∞6); (∞1); (∞2); (∞5).

57.4. Prove (∞3) as a theorem of $F_2^{2p(w)(\infty 4)}$.

57.5. (1) Prove (∞9) as theorem of $F_2^{2p(\infty 5)}$. (2) Prove (∞5) as a theorem of $F_2^{2p(\infty 9)}$.

57.6. Prove (∞3) as a theorem of $F_2^{2p(w)(\infty 5)}$.[572]

57.7. Prove (∞5) as a theorem of $F_2^{2p(\infty 4)}$.

57.8. Prove (∞5) as a theorem of $F_2^{2p(\infty 2)}$.

58. The predicative and ramified functional calculi of second order.

Objections against the absolute notion of *all*—as it is involved, e.g., in the notion of *all* classes of individuals, without qualification—have already been discussed briefly in footnote 535. There is much difference of opinion among mathematicians regarding the significance of these objections, some holding them to be pointless and others believing that they cast serious doubts on the methods used and the results obtained in large parts of classical mathematics. Our purpose in this section is not to debate the question of significance but to make a proposed definition of these objections—or of one form of them—by

---

[569]This axiom of infinity is suggested by the definition of finiteness which was given by H. Weber and slightly simplified by J. Kürschak. See *Jahresbericht der Deutschen Mathematiker-Vereinigung*, vol. 15 (1906), p. 177, and vol. 16 (1907), p. 425. The relationship should also be noticed to (∞1), which asserts the existence of a partial order of the individuals in which there is no last individual.

[570]Suggested by Paul Stäckel's definition of finiteness. *Jahresbericht der Deutschen Mathematiker-Vereinigung*, vol. 16 (1907), p. 425.

[571]Suggested by Tarski's definition of finiteness *E*, in *Fundamenta Mathematicae*, vol. 30 (1938), p. 162. As here stated, however, the axiom has been modified by using a one-many correspondence in place of the one-to-one correspondence of Tarski's definition *E*.

[572]That this cannot be done without using axiom (w) follows from a result obtained by Mostowski in his dissertation, *O Niezależności Definicji Skończoności w Systemie Logiki*, published as a supplement to *Annales de la Société Polonaise de Mathématique*, vol. 11 (1938), pp. 1–54.

formulating as a logistic system the weakened functional calculus of second order to which they lead.

In the form in which we wish to take them here, these objections may be said to have originated in Henri Poincaré's condemnation of what he called *impredicative definitions*, i.e., "définitions par . . . une relation entre l'objet à définir et *tous* les individus d'un genre dont l'objet à définir est supposé faire lui-même partie (ou bien dont sont supposés faire partie des êtres qui ne peuvent être eux mêmes définis que par l'objet à définir)."[573] This was afterwards embodied in Russell's *vicious-circle principle*,[574] that "no totality can contain members defined in terms of itself," or "whatever contains an apparent variable must not be a possible value of that variable."[575] Also Weyl objects in a similar way to what he takes to be a vicious circle in classical analysis.[576]

As understood by Russell in particular (and by Whitehead and Russell in *Principia Mathematica*) the vicious-circle principle constitutes a restriction upon the possible range of a propositional or functional variable, and hence a restriction upon substitutions for such a variable. The application of this to the functional calculus of second order affects primarily the axiom schemata *509 and leads first to the *predicative functional calculus of second order* and then to the *ramified functional calculi of second order*, as these are formulated below.[577]

---

[573]The quotation is from a paper by Poincaré in "*Scientia*," vol. 12 (1912), see p. 7. For the earliest statement of Poincaré's objection against impredicative definitions ("définitions non prédicatives") see the *Revue de Métaphysique et de Morale*, vol. 14 (1906), p. 307.

[574]It is not certain, however, that the vicious-circle principle of Russell is the same thing that Poincaré intended, since Poincaré never made a systematic development of his ideas in this direction and the examples which he gives (informally) of impredicative definition are not sufficient to determine what would have been his verdict regarding other examples of what might be considered impredicative definition.

In a paper in the *Revue de Métaphysique et de Morale*, vol. 17 (1909), pp. 461–482 (afterwards reprinted as Chapter IV of *Dernières Pensées* (1913)), there is a discussion by Poincaré of Russell's "hierarchy of types," i.e., of the (higher-order) ramified functional calculus which Russell introduced as based on the vicious-circle principle. From this it is perhaps fair to infer that Poincaré regarded the vicious-circle principle as being in general accord with his own ideas; but that he was unwilling to accept without reservations the ramified functional calculus which Russell proposed as embodying it—even if modified by omission of Russell's axioms of reducibility, discussed in our next section.

Apparently Poincaré (unlike Weyl) believed or hoped that all of classical mathematics could be developed without resort to impredicative definition once the postulates of arithmetic, including the postulate of mathematical induction, are granted. Compare his paper in *Acta Mathematica*, vol. 32 (1909), pp. 195–200, especially §5, pp. 198–200.

[575]See the *American Journal of Mathematics*, vol. 30 (1908), p. 237. The term "apparent variable" is used by Russell in the sense in which we have been using "bound variable" (cf. footnote 28), and the second quoted statement of the vicious-circle principle is therefore to be rendered in our terminology as follows: a wff which contains a bound variable must not denote one of the values in the range of that variable.

[576]See the explanation in his paper, "Der Circulus Vitiosus in der Heutigen Begründung der Analysis" in the *Jahresbericht der Deutschen Mathematiker-Vereinigung*, vol. 28 (1919), pp. 85–92.

[577]We shall not try here to decide upon or state the semantical rules for a principal interpretation of any of the predicative or ramified functional calculi of second order. But we remark that, in order to accord with the motivation as just described, it is

It should be noticed that the functional calculus of first order remains unaffected by the vicious-circle principle.

The predicative functional calculus of second order in the formulation $F_2^{2!}$ has the same primitive symbols and the same wffs as the simple functional calculus of second order F_2^2. A distinction is made of pure and applied, of singulary, binary, etc., predicative functional calculi of second order in the same way as for the simple functional calculus of second order; but where the term "predicative functional calculus of second order" is used without qualification, we shall understand that propositional variables and functional variables of all kinds—singulary, binary, ternary, etc.—are contained, with or without individual and functional constants.

The four rules of inference *500–*503 remain the same for $F_2^{2!}$ as for F_2^2.

The axioms of $F_2^{2!}$ are given by the following seven axiom schemata (the relationship of which to the axioms and axiom schemata of F_2^2 will be evident):

$\mathbf{A} \supset . \mathbf{B} \supset \mathbf{A}$

$\mathbf{A} \supset [\mathbf{B} \supset \mathbf{C}] \supset . \mathbf{A} \supset \mathbf{B} \supset . \mathbf{A} \supset \mathbf{C}$

$\sim\mathbf{A} \supset \sim\mathbf{B} \supset . \mathbf{B} \supset \mathbf{A}$

$\mathbf{A} \supset_{\mathbf{a}} \mathbf{B} \supset . \mathbf{A} \supset (\mathbf{a})\mathbf{B}$, where **a** is a variable of any kind that is not a free variable of **A**.

$(\mathbf{a})\mathbf{A} \supset \mathop{S}^{\mathbf{a}}_{\mathbf{b}}\mathbf{A}|$, where **a** is an individual variable, **b** is an individual variable or an individual constant, and no free occurrence of **a** in **A** is in a wf part of **A** of the form $(\mathbf{b})\mathbf{C}$.

---

necessary to abandon the idea that a sentence denotes one of the two truth-values t and f, and hence to avoid taking these two truth-values as the values of the propositional variables, or classes and relations in extension as values of the functional variables. The standpoint of Russell in 1908, and of Whitehead and Russell in the first edition of *Principia Mathematica*, is apparently best represented by supposing sentences to denote propositions, taking propositions as values of the propositional variables, and properties and relations in intension as values of the functional variables. And the more extensional view advocated by Russell in his *Introduction to the Second Edition* of *Principia Mathematica* (cf. footnote 590) can perhaps be represented by means of an infinite list of truth-values, namely, two truth-values t_m and f_m for each level m, the values of the propositional variables of mth level being t_m and f_m, and the values of the n-ary functional variables of mth level being functions (in extension) having the ordered n-tuples of individuals as their range, and having t_m and f_m (one or both) as values.

It must be added at once that the foregoing is not a reproduction of Russell's own account of semantical matters, especially as found in the introduction and appendices to the second edition of *Principia Mathematica*. But it is rather a first step of a contemplated attempt to fit the ramified functional calculi into our own semantical program and to provide for them "semantical rules" of a kind to which Russell would certainly not consent.

$(\mathbf{p})\mathbf{A} \supset \check{S}^{\mathbf{p}}_{\mathbf{B}}\mathbf{A}|$, where **p** is a propositional variable, and **B** contains no bound propositional or functional variables.

$(\mathbf{f})\mathbf{A} \supset \check{S}^{\mathbf{f}(\mathbf{x}_1, \mathbf{x}_2, \ldots, \mathbf{x}_n)}_{\mathbf{B}}\mathbf{A}|$, where **f** is an n-ary functional variable, and $\mathbf{x}_1, \mathbf{x}_2, \ldots, \mathbf{x}_n$ are distinct individual variables, and **B** contains no bound propositional or functional variables.

The characteristic feature is the restriction upon substitution for propositional and functional variables that is contained in the last two schemata, the restriction, namely, that **B** must not contain bound propositional or functional variables. And indeed if this restriction were removed we would obtain merely another system of axioms and rules for the simple functional calculus of second order.

The predicative functional calculus of second order is the same (apart from trivial notational differences) as the ramified functional calculus of second order and *first level*, and its propositional and functional variables are said to be *predicative*, or *of the first level*.[578] In the ramified functional calculi of second order and higher levels, additional propositional and functional variables are introduced, of successively higher levels, the leading idea being that in substituting for a propositional or functional variable of

[578]The use of the word "level" here is a departure from the terminology of Russell and of *Principia Mathematica*. In the second-order functional calculi, what we call the level of a propositional or functional variable is the same thing that Whitehead and Russell call the order. But in general, and especially in connection with the ramified functional calculi of higher order, we understand by the *level* of a functional variable what would be called in the terminology of Whitehead and Russell the amount by which the order of the functional variable exceeds the order of the variable of highest order which may stand in any of the *argument places* (i.e., in any one of the places between parentheses following the functional variable).

We shall not use the word "order" in this connection except in the sense in which we speak of functional calculi of first *order*, of second *order*, and so on (a use of the word very different from that of Whitehead and Russell). Also we shall use the word "type" in a way which differs from the usage of Whitehead and Russell, and which is suited rather to the simple functional calculi than to the ramified functional calculi. Namely—as will be explained more fully in Chapter VI—all the functional variables which appear in the functional calculi of first (or second) order are said to be of the *first type-class*, the new functional variables which are introduced in the functional calculus of third order are said to be of the *second type-class*, and so on, a new type-class of functional variables being introduced in each successive functional calculus of odd order. There is in our terminology no distinction of type among propositional variables, all of them being of the same type—though in the ramified functional calculi there are propositional variables of different levels. Likewise all individual variables are of the same type (and of a different type from propositional and functional variables). Two functional variables, one m-ary and the other n-ary, are of the same type, if they are both of the first type-class and $m = n$, or if they are of the same higher type-class and $m = n$ and the variable which may stand in each argument place (in order) after one of them is of the same type as the variable which may stand in the corresponding argument place after the other.

given level, the wff **B** which is substituted may contain bound propositional and functional variables of lower levels only. Thus the ramified functional calculus of second order and second level, $F_2^{2/2}$ (in our present formulation of it), contains propositional and functional variables of the first level and of the second level. Similarly $F_2^{2/3}$ contains propositional and functional variables of three different levels, and so on. The ramified functional calculus of second order and level ω, $F_2^{2/\omega}$, contains all the propositional and functional variables of all (finite) levels.

The primitive bases of these (and other) ramified functional calculi may be given simultaneously in the following way.[579]

The primitive symbols are first the eight following:

$$[\quad \supset \quad] \quad \sim \quad (\quad , \quad) \quad \forall$$

Then there is an infinite list of individual variables, the same as for F_2^2:

$$x \quad y \quad z \quad x_1 \quad y_1 \quad z_1 \quad x_2 \quad \ldots$$

Then there are or may be propositional variables of various levels, namely, either no propositional variable, or all propositional variables of not more than a certain maximum level, or all propositional variables of all levels. Explicitly, the symbols admitted as propositional variables are the following, where the superscripts indicate the level and where for any particular level used the list of variables is infinite:

$$\begin{array}{ccccccc} p^1 & q^1 & r^1 & s^1 & p_1^1 & q_1^1 & \ldots \\ p^2 & q^2 & r^2 & s^2 & p_1^2 & q_1^2 & \ldots \\ p^3 & q^3 & r^3 & s^3 & p_1^3 & q_1^3 & \ldots \\ \cdot & \cdot & \cdot & \cdot & \cdot & \cdot & \cdot \end{array}$$

Then for each n there are or may be n-ary functional variables of various levels ($n = 1, 2, 3, \ldots$). Namely, there are either no n-ary functional variables, or all n-ary functional variables of not more than a certain maximum level, or all n-ary functional variables of all levels. The explicit symbols admitted as functional variables are as follows, where the first numeral in the superscript indicates whether the functional variable is singulary, or

[579]This formulation should be compared not only with the original formulation of Russell (in the paper cited in footnote 454) and that in *Principia Mathematica* but also with the formulation of Hilbert and Ackermann in *Grundzüge der Theoretischen Logik*, first edition (1928), and that of Frederic B. Fitch in *The Journal of Symbolic Logic*, vol. 3 (1938), pp. 140–149. All of these differ from our present formulation of (say) $F_2^{2/\omega}$ in not being restricted to what we here call the second order. Fitch's formulation moreover contains notations and axioms which are designed to include in the system in some sense a formulation or partial formulation of arithmetic, so that his system is more nearly comparable to our $A^{2/\omega}$ (see below) than to $F_2^{2/\omega}$.

binary, or ternary, etc., and the second numeral in the superscript indicates the level:

$$\begin{array}{cccccc} F^{1/1} & G^{1/1} & H^{1/1} & F_1^{1/1} & G_1^{1/1} & \ldots \\ F^{1/2} & G^{1/2} & H^{1/2} & F_1^{1/2} & G_1^{1/2} & \ldots \\ F^{1/3} & G^{1/3} & H^{1/3} & F_1^{1/3} & G_1^{1/3} & \ldots \\ . & . & . & . & . & . \\ F^{2/1} & G^{2/1} & H^{2/1} & F_1^{2/1} & G_1^{2/1} & \ldots \\ F^{2/2} & G^{2/2} & H^{2/2} & F_1^{2/2} & G_1^{2/2} & \ldots \\ F^{2/3} & G^{2/3} & H^{2/3} & F_1^{2/3} & G_1^{2/3} & \ldots \\ . & . & . & . & . & . \\ F^{3/1}, & \text{and so on.} \end{array}$$

Then finally there may be individual constants or functional constants or both—where for each functional constant it must be given what its level is and whether it is singulary, binary, ternary, etc.

The formation rules are the same as for the simple functional calculus of second order, F_2^2, with the understanding that the level of a functional variable or functional constant is to be ignored. In particular, e.g., if $\mathbf{f}$ is an n-ary functional variable or functional constant, and $\mathbf{x}_1, \mathbf{x}_2, \ldots, \mathbf{x}_n$ are individual variables or individual constants (or both), then $\mathbf{f}(\mathbf{x}_1, \mathbf{x}_2, \ldots, \mathbf{x}_n)$ is wf, regardless of the level of $\mathbf{f}$.

The same abbreviations of wffs and in particular the same definitions are used as for F_2^2. But in D20 and D21 the propositional variable s is replaced by the propositional variable s^1, of the first level. And in D22 and D23 the functional variable F^1 is replaced by the variable $F^{1/1}$, of the first level.[580]

As a further abbreviation in writing wffs, the superscripts of the propositional and functional variables may be omitted ordinarily. In order to make this possible, the level which is to be given to a particular variable may be specified in words. Or following *Principia Mathematica* we may write an exclamation point after a letter to indicate that it represents a variable which is predicative, or of the first level.

The four rules of inference are the same as for F_2^2, i.e., *500–*503.

The axioms are given by seven axiom schemata, closely analogous to those given above for the predicative functional calculus of second order. Indeed the first five axiom schemata are exactly the same as for $\mathrm{F}_2^{2!}$—except of course that **A**, **B**, **C** are now wffs of the particular ramified functional calculus of second order whose axioms are being given, and in the fourth schema

---

[580]If desired, we might introduce also notations $=_2$ and $\neq_2$, replacing the variable F in D22 and D23 by $F^{1/2}$. Likewise $=_3$ and $\neq_3$, and so on.

a is a variable of any kind belonging to the particular ramified functional calculus of second order (subject to the condition that **a** is not free in **A**). The sixth and seventh axiom schemata are modified as follows:

$(\mathbf{p})\mathbf{A} \supset \check{S}^{\mathbf{p}}_{\mathbf{B}}\mathbf{A}|$, where **p** is a propositional variable, the bound propositonal and functional variables of **B** are all of level lower than that of **p**, and the free propositional and functional variables of **B** are of level not higher than that of **p**.

$(\mathbf{f})\mathbf{A} \supset \check{S}^{\mathbf{f}(\mathbf{x}_1, \mathbf{x}_2, \ldots, \mathbf{x}_n)}_{\mathbf{B}}\mathbf{A}$, where **f** is an n-ary functional variable, and $\mathbf{x}_1, \mathbf{x}_2, \ldots, \mathbf{x}_n$ are distinct individual variables, and the bound propositional and functional variables of **B** are all of level lower than that of **f**, and the functional constants and the free propositional and functional variables of **B** are all of level not higher than that of **f**.

The ramified second-order functional calculi of various levels are specified as follows by means of a maximum level of propositional and functional variables and functional constants. $\mathrm{F}^{2/1}_2$ has all the first-level variables and may have first-level functional constants, but has no variables or constants of higher level (thus it differs only trivially from $\mathrm{F}^{2!}_2$). $\mathrm{F}^{2/2}_2$ has all the propositional and functional variables of first and second levels and may have functional constants of these levels, but has no variables or constants of higher level; and so on. $\mathrm{F}^{2/\omega}_2$ has all the propositional and functional variables of all levels and may have functional constants of any level.

Of particular interest is the pure ramified functional calculus of second order and level ω, $\mathrm{F}^{2/\omega p}_2$, having as primitive symbols all the possible kinds of variables listed above, and no constants. Also logistic systems obtained from $\mathrm{F}^{2/\omega p}_2$ by adding one of the axioms of infinity (∞1)–(∞4), with F taken as a variable of the first level, or the infinite list of axioms obtained from axiom (w) by taking F to be of the first level and G of all possible levels (successively), or both.[581] Also further, logistic systems obtained from $\mathrm{F}^{2/\omega p}_2$ by adding functional constants and postulates containing them.

[581]That part of the system of *Principia Mathematica* which does not go beyond second-order functional calculus is approximately represented by the pure ramified functional calculus of second order and level ω, with the addition of an axiom of infinity and the infinite list of axioms obtained from axiom (w) as described, and the further addition (at least for the first edition of *Principia*) of the axioms of reducibility given in §59 below. For the complete system of *Principia*, functional variables of higher type-class (cf. footnote 578) added, but again ramified, or divided into levels; the axiom of infinity is restated in a different form (the "Infin ax" referred to in footnote 559) requiring functional variables of higher type-class; the axioms obtained, as described in the text, from axiom (w) are superseded by Russell's *multiplicative axioms* (equivalent to axioms of choice); and (at least for the first edition) axioms of reducibility with variables of higher type-class are included, in addition to those given in §59.

Of this last kind is the system of *ramified second-order arithmetic*, $A^{2/\omega}$ which we go on to formulate briefly before concluding this section.[582]

The system $A^{2/\omega}$, in its intended interpretation, probably would not be acceptable to the authors of *Principia Mathematica*, since they require that the natural numbers be defined and their properties proved rather than postulated. On the other hand it can be thought of as in accord with the program of Weyl in *Das Kontinuum*,[583] or with the ideas of Poincaré, both of whom accept the elementary methods of arithmetic, including proof by mathematical induction, as being (to quote Weyl) *ein letztes Fundament des mathematischen Denkens.*

As stated, $A^{2/\omega}$ is obtained from $F_2^{2/\omega p}$ by adding functional constants as undefined terms, and postulates. The functional constants are Σ and Π, ternary functional constants of the first level, the same as the functional constants Σ and Π of the systems A^0 and A^1. The postulates are the first twelve postulates of A^1 unaltered, and an infinite list of postulates obtained from the thirteenth postulate of A^1 (the postulate of mathematical induction) by taking the function variable F^1 to be of all possible levels, successively.[584]

$A^{2/\omega}$ may be called, more fully, a formulation of the *ramified second-order arithmetic of level* ω; and ramified second-order arithmetics of lower levels may be obtained by specifying a maximum level of propositional and functional variables. For example, the wffs of $A^{2/2}$ are the same as the wffs of $A^{2/\omega}$ which contain no propositional or functional variables of level higher than the second; and the postulates of $A^{2/2}$ are fourteen in number, being the same as the postulates of $A^{2/\omega}$ which contain no propositional or functional variables of level higher than the second.

The system of *predicative second-order arithmetic*, $A^{2!}$, is obtained from the *pure predicative functional calculus of second order*, $F_2^{2!p}$, by adding the undefined terms and postulates of A^1. Thus it differs only trivially from $A^{2/1}$.

For $A^{2!}$ and for the various ramified second-order arithmetics, as here formulated, including $A^{2/\omega}$, the same definitions are used as for the corresponding second-order functional calculi $F_2^{2!}$, $F_2^{2/\omega}$, etc., except that the def-

---

[582]The systems A^0 and A^1, formulated in §55, remain unaffected by adoption of the point of view of ramification or division into levels.

[583]Compare footnote 535.

Weyl describes a formulation of ramified arithmetic of second order, and of higher order, and holds such a system to be admissible. However, the actual developments of the book are largely within a predicative second-order arithmetic that would seem to be essentially equivalent to the system $A^{2!}$ (see below), though differing in details of the formulation. At some places this must be extended to a predicative third-order arithmetic, by adding variables of the next higher type-class, but allowing them to appear only as free variables.

[584]I.e., more explicitly, F^1 is replaced by $F^{1/1}$ to obtain the first postulate of the infinite list, by $F^{1/2}$ to obtain the second one, and so on.

inition schemata D22 and D23 are discarded, and the notation $[\mathbf{a} = \mathbf{b}]$ is defined rather in the same way as was done for A^0 and A^1 in §55. The notations $Z_0(\mathbf{a})$ and $Z_1(\mathbf{a})$ are also defined in the same way as for A^0 and A^1.

EXERCISES 58

58.0. Prove as a theorem of $\mathrm{F}^{2!}_2$:

$$x = y \supset . \, y = x$$

58.1. Where F is a singulary functional variable of arbitrary level,[585] prove as a theorem of $\mathrm{A}^{2/\omega}$:

$$x = y \equiv . \, F(x) \supset_F F(y),$$

i.e., written more fully, $\Sigma(x, z, x_1) \supset_{zx_1} \Sigma(y, z, x_1) \equiv . \, F(x) \supset_F F(y)$. (Use the postulates of mathematical induction.)

58.2. Where F and G are singulary functional variables of different levels, show that the following is a theorem of $\mathrm{A}^{2/\omega}$ but not of $\mathrm{F}^{2/\omega}_2$:

$$F(x) \supset_F F(y) \equiv . \, G(x) \supset_G G(y)$$

(*Suggestion:* Consider an interpretation of $\mathrm{F}^{2/\omega}_2$ according to which the individuals are the natural numbers; the values of the propositional variables are t and f, the same for all levels; the first-level propositional functions, i.e., the values of the first-level functional variables, are those which make no distinction among the different individuals, thus only the null class and the universal class, the null relation and the universal relation, and so on; the second-level propositional functions are those additional propositional functions which make no distinction among the different individuals except the distinction of odd and even; the third-level propositional functions are those additional propositional functions which distinguish the individuals as congruent to 0, 1, 2, or 3 modulo 4; and so on.[586] In a manner which will now be familiar to the reader, the semantical argument so obtained can be converted into a syntactical independence proof.)

59. Axioms of reducibility.

Because they were unable otherwise to develop classical mathematics within their system in the manner which they desired, Russell in 1908,[587] and later

---

[585]Thus "F" becomes in effect a syntactical variable, having $F^{1/1}$, $F^{1/2}$, $F^{1/3}$, ... as values; and a theorem schema summarizing an infinite list of theorems is established. This is an instance of what the authors of *Principia Mathematica* call *typical ambiguity*.

The infinite list of postulates of mathematical induction, of $\mathrm{A}^{2/\omega}$, might also be conveniently summarized in a single postulate schema by means of typical ambiguity. (Compare footnote 584.)

[586]Of course not meant to be a principal interpretation but a special interpretation serving the purpose of the particular independence proof.

[587]In the paper cited in footnote 454.

Whitehead and Russell in *Principia Mathematica*, were led to supplement the ramified functional calculi by the addition not only of an axiom of infinity and multiplicative axioms (as explained in footnote 581) but also of the famous axioms of reducibility. The content of the axioms of reducibility is, for a propositional function of arbitrary level, that there exists a formally equivalent propositional function of the first level (the intended interpretation being such that the formal equivalence of propositional functions is not alone sufficient to render them identical). This has been much criticized,[588] in particular on the ground that the effect is largely to restore the possibility of impredicative definition which the distinction of levels was designed to eliminate. Indeed, as many have urged,[589] the true choice would seem to be between the simple functional calculi and the ramified functional calculi without axioms of reducibility. It is hard to think of a point of view from which the intermediate position represented by the ramified functional calculi with axioms of reducibility would appear to be significant. And in the *Introduction to the Second Edition of Principia Mathematica* (1925), Russell in fact recommends abandonment of the axioms of reducibility.[590] Nevertheless, because of their historical importance, it seems desirable to give these axioms here (as far as they fall within the second-order functional calculi).

Syntactically, the *axioms of reducibility*, intended as axioms to be added to $F_2^{2/\omega}$, are the following doubly infinite list, where G is always a functional variable of the first level, and F is of arbitrary higher level:

$$(\exists G) \mathbin{.} F(x) \equiv_x G(x)$$
$$(\exists G) \mathbin{.} F(x, y) \equiv_{xy} G(x, y)$$
$$(\exists G) \mathbin{.} F(x, y, z) \equiv_{xyz} G(x, y, z)$$
$$\cdots\cdots\cdots\cdots$$

[588]In particular by Leon Chwistek in *Przegląd Filozoficzny*, vol. 24 (1921), pp. 164–171; Hilbert and Ackermann, *Grundzüge der Theoretischen Logik*, first edition (1928), pp. 114–115; Adolf Fraenkel, *Einleitung in die Mengenlehre*, third edition (1928), pp. 259–263; W. V. Quine in *Mind*, n.s., vol. 45 (1936), pp. 498–500, and in a paper in *The Philosophy of Alfred North Whitehead* (1941), see pp. 151–152.

[589]Chwistek in the paper cited in the preceding footnote; F. P. Ramsey in a paper in the *Proceedings of the London Mathematical Society*, ser. 2 vol. 25 (1926), pp. 338–384, reprinted in his *The Foundations of Mathematics and other Logical Essays* (1931), pp. 1–61; Rudolf Carnap in *Abriss der Logistik* (1929), §9. See also scattered remarks by Carnap and Hans Hahn in *Erkenntnis*, vol. 2 (1931), pp. 73, 97, 145.

[590]With the suggestion that their place could be partly taken by axioms of extensionality, expressing in the notation of the ramified higher-order functional calculi that formally equivalent propositional functions of the same type and the same level are identical. (Cf. footnote 577.)

Compare also *The theory of constructive types*, a paper in the *Annales de la Société Polonaise de Mathématique*, vol. 2 (1924), pp. 9–48, and vol. 3 (1925), pp. 92–141, in which Chwistek takes what Russell describes as the heroic course of dispensing with the axioms of reducibility without adopting any substitute, thus undertaking to base a logistic treatment of mathematics on the system of *Principia Mathematica* without these axioms.

(There are also axioms of similar form intended to be added to the higher order functional calculi.)

These axioms are not all independent, since those which contain singulary functional variables can be proved by using those which contain binary functional variables, and so on down the list. Also among those which contain n-ary functional variables (with fixed n), it is obvious that one in which F is of lower level can be proved by using one of those in which F is of higher level.

EXERCISES 59

59.0. In the logistic system $F_2^{2/\omega(r)}$, obtained by adding the axioms of reducibility to $F_2^{2/\omega}$, prove

$$F(x) \supset_F F(y) \equiv . G(x) \supset_G G(y),$$

where F and G are singulary functional variables of different levels.

59.1. In $F_2^{2/\omega}$ (thus without using axioms of reducibility), prove

$$(\exists q) . p \equiv q,$$

where p and q are propositional variables of different levels.

59.2. In the logistic system obtained by adding to $F_2^{2/\omega}$ the axioms of reducibility containing binary functional variables, prove as theorems the axioms of reducibility containing singulary functional variables.

59.3. In the logistic system obtained by adding to $F_2^{2/\omega}$ the axiom schema of footnote 555, **f** and **g** being taken as predicative functional variables, show that the axioms of reducibility are theorems. Discuss the question of prescribing the levels of **f** and **g** in this axiom schema so as to avoid obtaining from it any theorem $(\exists G) . F(x) \equiv_x G(x)$ in which G is of lower level than F.

Index of Definitions

This index includes references to all passages in which a new term is introduced or the meaning or usage of a term is explained, whether by definition or otherwise. Both the terms used in the meta-language and the characters used in the various object languages are covered (excepting letters used as variables, and parentheses, brackets, commas); and also references to terminology which is employed by others but not adopted in this book. The numbered definitions D1–D25 pertaining to the object languages are indexed according to their numbers, at the proper place in alphabetic order (under the letter D). And designations used for particular logistic systems, consisting of a capital letter with superscripts or subscripts or both, are included in the index also, each at its proper place in alphabetic order.

In order to fix the alphabetic arrangement the following special rules are adopted. Greek letters are treated as coming in alphabetic order after all the English letters, and are arranged among themselves in their own alphabetic order. Arabic numerals are treated as coming after both English and Greek letters, and are arranged among themselves in order of increasing magnitude (of the corresponding numbers — thus 0 first, then 1, then 2, and so on). Special characters (including inverted letters treated as special characters) are arranged in an arbitrary order after English letters, Greek letters, and Arabic numerals. To the distinction between capital and small letters, roman and italic letters, unaccented and accented letters no attention is ordinarily paid in alphabetizing; but where two entries would otherwise coincide in the alphabetic order, the rule is followed to put capital letters before small letters, and then to put roman letters before italic letters, and then to put unaccented letters before accented letters. Parentheses, the hyphen, the solidus, exclamation point, comma, and other punctation marks are also ordinarily ignored in alphabetizing, but where (in spite of all preceding rules) two entries would otherwise coincide in the alphabetic order, the entry lacking the parentheses or other punctuation mark is placed before the other entry. Finally, account is taken of superscripts and subscripts only in cases in which (in spite of all preceding rules) two or more entries would otherwise coincide in alphabetic order; in such cases, the entries are arranged in the alphabetic order of their superscripts; or if the superscripts are identical, or non-existent, the entries are arranged in the alphabetic order of their subscripts.

The references given in the index are by section number (indicated by "§"); or by the number of an axiom, rule of inference, theorem, or metatheorem (indicated by "†", "*", or "**" forming part of the number); or by numbered footnote (indicated by "n."); or by numbered exercise (indicated by the occurrence of a period as part of the number). Footnote 152 of §10 (see page 70) should be consulted for an explanation of the system of numbering which is used in the text. And it should be noticed in particular how the number of any section indicates in what chapter it will be found, and the number of any axiom, rule, theorem, or metatheorem indicates in what section it will be found. As explained in footnote 152, exercises are placed immediately after the section that is indicated by the number of the exercise — so that, e.g., 46.0 is the first exercise in the collection of exercises (called "Exercises 46") which follow §46, and 46.19 is the twentieth exercise in the same collection.

In searching for a reference from the index, page numbers may be ignored, and there may be used instead the numbers of chapters, sections, and exercise collections which are given at the top of the pages of the text. When the reference is to a footnote, observe that footnotes 1–149 are in the Introduction and that beyond that the number of the footnote indicates directly in what chapter it falls, numbers in the one hundreds being for footnotes in Chapter I, those in the two hundreds for footnotes in Chapter II, and so on.

Index of Authors

Errata

On page 66, in line 12, read: "Else there will be simple elementary true propositions . . ."

On page 142 it should have been pointed out that Wajsberg's paper, cited in footnote 211, contains an error that is not easily set right. However, the metatheorem that is stated in the next-to-last paragraph of the text on page 142, and a similar metatheorem for the formulation F^{1i} of intuitionistic functional calculus of first order, were proved by Curry in the *Bulletin of the American Mathematical Society*, vol. 45 (1939), pp. 288-293, and the proof is reproduced by Kleene in *Introduction to Metamathematics*.—Since Curry's proof depends on Gentzen's *Hauptsatz* for *LJ*, the remark should be made that it is not the use of Gentzen's *Sequenzen* but the *Hauptsatz* itself that is essential, as the *Sequenzen* can of course be eliminated by the definitions on page 165 (with $m = 1$ for the intuitionistic case), and the *Hauptsatz* therefore proved in a form that is directly applicable to formulations of the ordinary kind without *Sequenzen* (compare Curry, loc. cit., and Kurt Schütte in the *Mathematische Annalen*, vol. 122 (1950), pp. 47-65).

On page 150, the parenthetic explanation at the end of the statement of the metatheorem **272 must be changed to read as follows: "(i.e., every application of the rule of substitution is one of a chain of successive substitutions that are applied to one of the axioms of P_2)."

On page 171, add after line 2: "These substitution notations will be used not only when $\mathbf{\Gamma}$, $\mathbf{\Gamma}_1, \mathbf{\Gamma}_2, \ldots, \mathbf{\Gamma}_n$ are well-formed formulas, but when they are formulas consisting of variables or constants standing alone, and even possibly in other cases also. The condition that $\mathbf{A}$ shall be well-formed must be retained, at least in the case of the dotted S."

On page 257, in connection with case X of the decision problem, it should have been pointed out that this case covers only a finite number of wffs that differ otherwise than by alphabetic changes of bound variable or transformations of the matrix by propositional calculus or both. This detracts somewhat from the interest of the case, as the decision problem of a finite class of wffs is always solvable in principle, by the trivial procedure of listing the valid and invalid formulas in the class. But in case X the finite number is large, and there is still an interest in finding a practicable decision procedure.

On page 269, after the word "which" at the end of line 11, insert the words "the premisses are inconsistent, as well as cases in which"; for although exercise 46.22 is deliberately stated in such a way that some additional valid inferences will be found, beyond the traditional categorical syllogisms, there is no point in including the inferences which are valid only because the premisses are inconsistent.

On page 299, replace the words "those of F^2_2," at the beginning of line 20, by "all variants of the axioms of F^2_2."

On page 335, the definition which is given in 55.14 requires the further condition that there is either at least one 0 or at least one 1 among the signs $a_1, a_2, \ldots, a_n$. For otherwise $\mathfrak{f}(a_1, a_2, \ldots, a_n)$ is already a wff, and may not without confusion be used to abbreviate another wff.

On page 336, the writer is indebted to Hugues Leblanc for pointing out that the results asserted in 55.16 and 55.18(1) are erroneous. The exercises may profitably be amended by asking the reader in each case to show the opposite, given the consistency of A^0 and A^I, given that completeness as to provability fails (compare the remark on page 329), and making use of **453 for A^0 and 48.7 for A^I.

On page 352, in the seventh line of footnote 581, insert the word "are" before "added."